Plant Physiology
Fourth Edition

Robert M. Devlin
University of Massachusetts

Francis H. Witham
The Pennsylvania State University

Wadsworth Publishing Company
Belmont, California
A Division of Wadsworth, Inc.

Plant Physiology, Fourth Edition was prepared for publication by the following people:

Production Editor: *Robine Storm van Leeuwen*

Interior and Cover Designer: *Trisha Hanlon*

Cover Photographer: *Richard H. Gross*

Research and drawing of line drawings, labels, and graphs: *Chris Mari van Dyck,* scientific illustrator

Typesetting by *Bi-Comp, Incorporated;* covers printed by *Lehigh Press Lithographers;* text printed and bound by *Halliday Lithograph*

Library of Congress Cataloging in Publication Data

Devlin, Robert M.
    Plant physiology.

    Includes bibliographies and index.
    1. Plant physiology.   I. Witham, Francis H.
II. Title.
QK711.2.D48   1983   581.1   82-24197
ISBN 0-87150-765-X

ISBN 0-87150-765-X

Printed in the United States of America
6 7 8 9 10 — 93 92 91 90

In memory of my father, Patrick C. Devlin, and my mother, Katherine Martin Devlin.
—R.M.D.

For Dr. Carlos O. Miller—scientist, teacher, and friend.
—F.H.W.

Preface

Plant Physiology, Fourth Edition presents a modern introduction to plant physiology today. It blends current research with a sense of historical perspective and does so in a book of manageable size that is designed with the needs of the beginning student in mind.

As in other scientific fields, research in plant physiology proceeds at an accelerated pace. In presenting the basic principles of plant physiology, we have noted the most recent research being conducted, as well as provided the student with the classical pioneering research—an area that is often given short shrift in textbooks.

In order to keep this book to a practical length, we have endeavored to provide only enough detail to stimulate students' curiosity and make them aware of research trends. We encourage students who are interested in a particular aspect of plant physiology to consult the end-of-chapter suggested readings, which provide sources for further study as well as guides to areas needing more research.

Based on responses from users of the previous editions, we have reorganized the chapters so that they provide a more logical approach to the study of plant physiology. The book begins with a discussion of the cell and related basic background material. The next six chapters cover the physical processes at work in plants. The following nine chapters build on this material in treating plant biochemistry and metabolism, and the last seven chapters logically cover plant growth and development.

Other changes in this fourth edition are the movement of the discussion of colloids and of pH and buffers to appendices to improve the flow of material and to offer more flexibility in covering these topics. Also, we have included end-of-chapter questions to help students review the chapter material and organize their understanding of the concepts covered.

We have tried to show that plant physiology is not just an academic discipline but also a science with applications important to our everyday life. We introduce the students to this aspect of the field, showing them how exciting and vital a science it is.

We have kept our writing style as clear and as straightforward as possible so that our readers can more readily share the enjoyment we feel for this field. We have been most fortunate in securing the collaboration of a gifted scientific illustrator, Chris Mari van Dyck, whose drawings give life to our words and a sense of the beauty of plant physiology.

We have also been very fortunate to obtain the assistance of the following reviewers: John Barber, Tulane University; Norman Mitchell, Loma Linda University—La Sierra Campus; Robert Neil, University of Texas at Arlington; Jerry McClure, Miami University at Oxford; Michael Strauss, Northeastern University; Don Miles, University of Missouri at Columbia; Murray E. Duysen, North Dakota State University. In a field as broad as ours and with new research surfacing daily, no one can aspire to be an expert in all areas. Our reviewers kept us informed, honest, and accurate. We willingly give them credit for this, but we also hold them blameless for any shortcomings this book might have. Many more people helped us. If we don't list them all it is not for lack of gratitude but for lack of space. To them all, we give our thanks.

For the preparation of this manuscript, we wish to thank Jean-François Vilain, biology editor, who prodded us when necessary and encouraged us when needed. We especially wish to thank Robine Storm van Leeuwen, production editor, for her monumental patience and editorial skills.

Robert M. Devlin
Francis H. Witham

Contents

Chapter 19 — Gibberellins

Chapter 20 — Cytokinins, Ethylene, and Abscisic Acid

Chapter 1

Plant Cells: Structure and Function

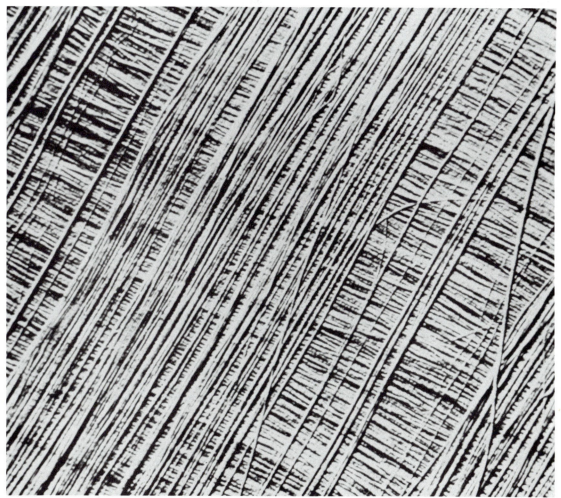

Electron micrograph of cell wall of *Valonia macrophysia* showing orientation of cellulose fibers. *Courtesy of K. Mühlethaler, Institut für Zellbiologie, Zurich.*

The cell is the basic structural and functional unit of life. This concept, which is part of the cell theory, was proposed by the botanist Matthias Schleiden and zoologist Theodor Schwann during the early part of the nineteenth century, some twenty years before the publication of Darwin's theory of evolution. These two theories serve as the unifying basis for the modern-day science of biology.

In unicellular plants and animals, the cell is a complete organism; in the so-called higher life forms, the *multicellular organisms*, there is a collection of different cells that precisely regulate growth and development (*morphogenesis*) through their chemical interactions and specialized functions. It is not surprising that the size and shape of a plant are largely determined by the number, morphology, and arrangement of a plant's cells. Nor is it surprising that an inseparable relationship exists between cellular architecture and cellular function. For example, the conductive tissues of a plant are made up of cells structurally equipped for the rapid transport of water and nutrients.

Although the specialized products and functions of cells are numerous, cells are remarkably alike in that they consist of many of the same kinds of chemicals and structures, which are similarly contained within a membrane, the plasmalemma. Another similarity is the presence of the nucleic acids—deoxyribonucleic acid (DNA) and ribonucleic acid (RNA)—which serve as major components of the informational machinery in all living cells. Thus the *prokaryotes*, organisms having cells without a defined nucleus, as well as the *eukaryotes*, organisms with cells containing a defined nucleus, have much in common. Even those organisms that are apparent exceptions to the cell theory, such as the coenocytic (multi-nucleate) algae and fungi, contain nuclei, mitochondria, plastids, and other membranous structures. The life functions of the coenocytes are not radically different from other organisms and often provide the observing scientist with insights into the cellular functions of all plants.

Since the understanding of the physiology of plants resides in an understanding of the basic structural and functional unit of life, we must examine the structural features of a typical plant cell. The sophisticated techniques of electron microscopy have clearly revealed these features.

"Typical" Plant Cell

There is no such entity as a "typical" plant cell. The similarity of plant cells, however, allows us to produce a composite cell that contains numerous structures found in living cells. Thus our composite cell, as shown in Figure 1–1, is characterized by the presence of a *cell wall* and internal area—the *protoplasm*—which consists of the *cytoplasm* and *nucleus*. We refer to these protoplasmic components collectively as the *protoplast*. Scientists often separate protoplasts from their cell walls and use the former in physiological and biochemical studies.

The cytoplasm is bounded by a membrane, the *plasmalemma*. The nucleus, in turn, is surrounded by a complex membrane system termed the *nuclear envelope*. Throughout the cytoplasm we can see structures termed *cytoplasmic organelles,* which include *mitochondria, plastids, ribosomes, microtubules,* and *microbodies.* (We will define and differentiate these terms later in the chapter.) Within the cytoplasm there are also other conspicuous membranous structures known as the *endoplasmic reticulum* and *Golgi*

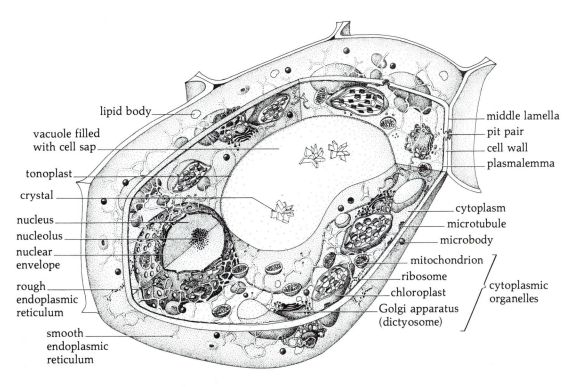

lipid body

vacuole filled
with cell sap

tonoplast

crystal

nucleus

nucleolus

nuclear
envelope

rough
endoplasmic
reticulum

smooth
endoplasmic
reticulum

middle lamella

pit pair

cell wall

plasmalemma

cytoplasm

microtubule

microbody

mitochondrion

ribosome

chloroplast

Golgi apparatus
(dictyosome)

cytoplasmic
organelles

Figure 1–1. Idealized composite plant cell.

apparatus. The latter are often near the nucleus. The cytoplasmic organelles and membranes are located in the *ground substance*, an unstructured colloidal medium consisting of a variety of biochemicals (see Appendix A).

Although there are materials that are dissolved in the protoplasm, much of the particulate phase of the protoplasm is colloidal in nature. Indeed, the protoplasm is usually referred to as a colloidal complex and exhibits many of the properties attributed to colloidal systems. The colloidal character of the protoplasm is mainly due to the presence of proteins. The immense surface area provided by proteins dispersed in the protoplasm produces the necessary condi-

tions for adsorption, chemical movement, and, in turn, reactions required for life. The colloidal system is an essential feature of living matter. For a brief explanation of colloids and their properties, see Appendix A.

Vacuoles are membrane-enclosed areas within plant cells that are filled with a watery fluid, the cell sap, and are scattered throughout the cytoplasm of young plant cells. Older plant cells exhibit a large, central vacuole, the contents of which are enclosed within a single membrane, the *tonoplast.* These water-filled vacuoles contain dissolved chemicals, including sugars, salts, pigments, waste products of metabolism, and even crystals.

Cell Wall

With only a few exceptions, organisms require mechanical support of some kind to maintain a definite form. In plants and animals, the water pressure generated within cells is not always sufficient to maintain the entire organism's structural integrity. In the animal world, support is often provided by an exoskeleton within which other cells are confined or by an endoskeleton to which other cells cling. In plants, however, each individual cell is enclosed in a relatively rigid structure, the cell wall. As we will discuss in detail later, the rigidity of the cell wall as well as the pressure of the water in the vacuoles of plant cells are responsible for the high turgor pressures that develop and assist in mechanical support of the entire organism.

In addition to providing mechanical support, plant cell walls exhibit other important functions that are a part of the dynamic interactions between the external environment and the protoplast. For example, cell walls are involved in absorption and transport of water and minerals, in secretions, and in certain enzymatic activities. Plant pathologists also believe that cell walls and wall components play an important role in disease resistance by inhibiting the penetration of would-be parasites.

The living protoplast produces and maintains the components of the cell wall. Of course there are also cells, in which the protoplast is no longer present, that are spe-

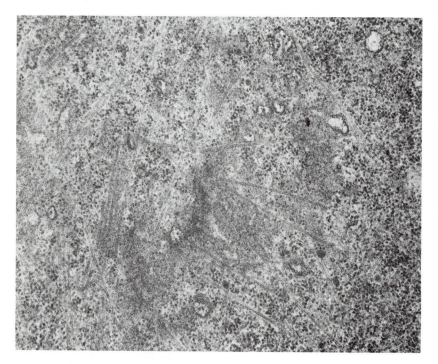

Figure 1-2. Electron micrograph of section cut tangential to transverse cell wall of *Phleum* root cell. Note numerous microtubules in cytoplasm oriented in a plane parallel to transverse wall. Magnification 31,000 ×.

From Biophoto Associates/ Dr. Myron C. Ledbetter/ Brookhaven National Laboratory.

cialized in conduction and support functions. Conducting cells of the xylem, such as the *tracheids*, which have no protoplast and consist of a thick secondary wall, have become highly specialized through the differentiation process.

The components of the cell wall are produced by the protoplast and deposited adjacent to the external surface of the plasmalemma. The chief structural component of the cell wall is *cellulose*, a polysaccharide consisting of thousands of sugar molecules. Pectic substances, hemicelluloses, lignin, suberin, and proteins, including enzymes, are the other major components of cell walls. We will consider the chemical nature of these major cell wall constituents in a later chapter.

Cell Wall Formation

Middle lamella. Cell wall formation is initiated during telophase, the last stage of mitosis. As we can see in Figure 1–2, microtubules in the cytoplasm migrate toward the equatorial region of the cell. These microtubules are part of a system or aggregation of fibrils, called the *phragmoplast*, that forms between the daughter nuclei. At the early stages of cell division, or *cytokinesis*, small droplets or *vesicles* become oriented at the equator of the original mother cell along the phragmoplast and fuse to form the cell plate (Figure 1–3). The vesicles, which are derived from Golgi bodies, probably contain the pectic substances. These vesicles participate in the formation of the first layer, the

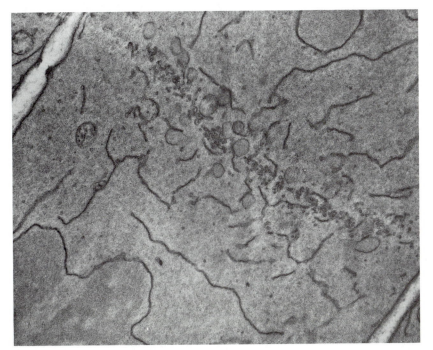

Figure 1–3. Electron micrograph showing early stage of cell plate formation in telophase of dividing onion root tip cell. Developing cell plate extends diagonally from lower right to upper left. Endoplasmic reticulum is present on both sides of cell plate. Magnification 9,400 ×.

From K. Porter and R. Machado. 1960. Biophys. Biochem. Cytol. 7:167.

middle lamella, which is composed of pectic substances that ultimately cement adjoining cells together. During early formation the middle lamella is primarily jellylike due to the high percentage of *pectic acid*, a molecule that is composed of approximately a hundred or less α-D-*galacturonic acid* molecules (see Chapter 11). Other compounds found in the middle lamella are the insoluble salts of pectic acid—*calcium* and *magnesium pectate*—the *pectins*, and low amounts of *protopectins*. The pectins, found primarily in the middle lamella and primary walls, consist of 200 or more galacturonic acid derivatives per molecule in which the C_6 carboxyl groups are esterified with methyl groups (see Chapter 11). The protopectins, found mostly in the primary wall, are similar to but of higher molecular weight than the pectins. "Hardening" of the middle lamella, which takes place during later stages of wall formation, is due to the presence of calcium and magnesium salts of pectic acid and to the infiltration of polysaccharides, including cellulose and sometimes lignin. The characteristic softening of ripening fruit is accompanied by an increase in the solubility of pectic substances of the middle lamella. These substances probably lose their binding properties due to reactions mediated by *pectolytic enzymes*, which increase in activity as a fruit matures.

Primary wall. Soon after formation of the middle lamella, the cell increases in volume and enlarges. Enlargement is accompanied and followed by the impregnation of the middle lamella with three types of substances: (1) *cellulose*; (2) *hemicelluloses*, which include a variety of polysaccharides, such as xylans, arabans, and galactans; and (3) glycoproteins, consisting of carbohydrates, proteins, and other substances. This deposition results in the formation of a thin layer, 1 to 3 μm in thickness. This layer, which lies next to the inside surface of the middle lamella and the outside surface of the plasmalemma, is termed the primary wall. To gain further perspective, remember that the middle lamella is located between the primary walls of adjacent cells. Also, many cells in plants contain only primary walls, and these walls do not necessarily undergo further development. The meristematic cells, epidermal cells, and cells involved in metabolism are of this type.

Enlargement of the cell is characterized mainly by the stretching of the primary wall. One type of stretching, which takes place during or soon after wall formation, is reversible (as in a rubber band) and is characterized by the maintenance of cross-linkages between cell wall components. Reversible stretching is said to be dependent on the elastic properties of the wall. A second type of stretching is irreversible and is characterized by cell wall deformation, or the breaking of linkages between wall components, and leads to irreversible lengthening of the wall. This second type of stretching is irreversible because of the displacement of existing wall components and the impregnation of additional cellulose and other substances into the wall space that is created as a result of cell wall deformation and of the reestablishment of cross-links after stretching ceases. Irreversible stretching is said to be dependent on the plastic properties of the cell wall incurred by wall deformation. New wall material is added during or after wall stretching by means of two processes: *intussusception*, which is the incorporation, directly into the wall space, of wall chemicals produced in the cytoplasm; or *apposition*, which is the formation of layers on existing layers.

An analysis of the primary walls of *Avena coleoptile* cells by Bishop, Bayley, and Setterfield (7) showed that hemicelluloses were present in much greater concentration than pectic substances. Similarly, Ray (29) and Albersheim (1, 2) have demonstrated that primary walls have a lower concentration of pectic substances. This information suggests that hemicelluloses and other components of the primary wall play a more important role in the initial stages of cell growth than was previously thought. Indeed, the hemicellulose *xyloglucan* appears to function as an important cross-link in the cell wall structure. Xyloglucan is hydrogen bonded to cellulose and covalently linked to the pectic polymers (4, 20). This latter association, however, is now believed to be doubtful.

In a study of the wall composition of onion root tip cells, Jensen (19) found that, although cell walls of provascular cells had a high concentration of pectic substances and hemicelluloses, the cell walls of the cortex and protoderm had a low concentration of these compounds. Although all of the common constituents of the cell wall are present in every primary wall, their relative concentrations appear to vary with the type of cell. Also, cell walls contain a significant amount of structural protein that appears to be particularly rich in the two amino acids proline and hydroxyproline.

Plasmodesmata and pit fields. Plasmodesmata (singular: plasmodesma) are cytoplasmic strands between cells and are formed as the vesicles at the equator of the cell harden around endoplasmic reticulum strands during cell plate formation. These strands, which traverse cells, are believed to function as very effective transport pathways of water and other substances. Plasmodes-mata may be found as aggregates in parts of the walls referred to as *primary pit fields*, which are "thin" areas in the cell wall. The pits opposite each other in the primary cell walls of adjacent cells are known as *pit pairs* (see Figure 1–5) and, with the middle lamella between them, constitute collectively the *pit membrane*. In those cells that also have secondary walls, the pits may be either simple or bordered. The difference between the two is that when the secondary wall is formed somewhat over the pit cavity, it gives the appearance, when viewed head-on, of the bottom surface of a saucer. Simple pits are not obscured by an overgrowth of the secondary wall.

Secondary wall. In parenchymatous cells once the primary wall is formed, the cessation of elongation and deposition of wall materials are evident. However, in other cells, such as developing tracheids and fibers, the wall continues to thicken after elongation stops, as layers of cellulose and lignin are deposited to form the *secondary wall*. The secondary wall, which may range from 5 to 10 μm in thickness, causes the entire wall to become much less flexible and, finally, almost inelastic. We can, thus, understand why cell elongation ceases with secondary wall formation. Further, the thickening secondary wall in some instances fills in much of the cellular volume and causes death and disintegration of the protoplast. The disappearance of the protoplast results in the formation of a *lumen* (cavity)—a structure that is characteristic of supporting fibers and the tracheids of the xylem.

Many secondary walls contain lignin, a noncarbohydrate polymer derived from such phenylpropane compounds as coniferyl, *p*-coumaryl, and sinapyl alcohols.

These alcohols are present in the wall, with the matrix of hemicelluloses and other substances cross-linked to cellulose. Lignin, second in abundance only to cellulose in the entire plant, is the most abundant chemical present in secondary cell walls and is important because it adds rigidity to their structure. However, in some plants almost pure cellulose is laid down in cell wall layers. A familiar example is the cotton fiber in which more than 90 percent of the wall's dry weight is pure cellulose. Some plant cell walls are covered with cutin or impregnated with suberin or waxes—materials that protect the cell from excessive water loss. Certainly, the cuticle of leaf and stem surfaces is of prime importance in this respect.

Cellulose and Other Cell Wall Components

Cellulose is a polysaccharide that consists of repeating molecules of the six-carbon sugar, β-D-glucose (see Chapter 11). The hierarchy of the cellulose arrangement in the cell wall, based upon increasing observable organization, begins with the simple cellulose chain and continues with the micelle, microfibril, and macrofibril (see Figure 1–4). A *cellulose chain* is believed to number one to three thousand cellulose molecules, with each molecule connected to the next by a β-1,4-linkage (see Chapter 11). Cellulose chains form crystalline structures termed *micelles*. One micelle is composed of

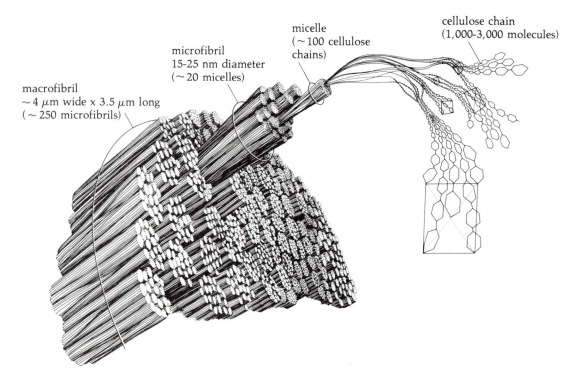

micelle
(~100 cellulose
chains)

cellulose chain
(1,000-3,000 molecules)

microfibril
15-25 nm diameter
(~20 micelles)

macrofibril
~4 μm wide x 3.5 μm long
(~250 microfibrils)

Figure 1–4. Arrangement of molecular cellulose chains into macrofibrils, microfibrils, and micelles.

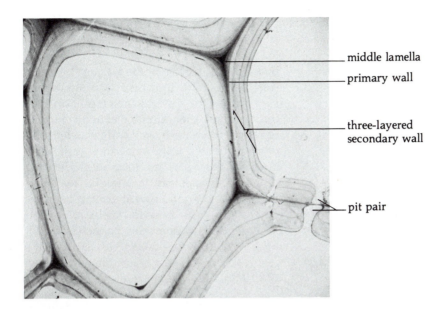

middle lamella

primary wall

three-layered
secondary wall

pit pair

Figure 1–5. Electron micrograph of banana root stele cells.

Courtesy of W.C. Mueller, University of Rhode Island.

approximately 100 chains of cellulose, arranged in a latticelike formation. The micelle is considered to be the smallest structural unit of the cell wall. The next level of organization, the *microfibril*, is composed of approximately twenty micelles. The cross section of a microfibril measures approximately 15 to 25 nanometers in diameter. One estimate is that approximately 2,000 cellulose chains make up a microfibril. A collection of approximately 250 microfibrils are organized into a *macrofibril*. Macrofibrils resemble a woven rope, are about 4 μm wide and 3.5 μm long, and provide the cell wall with considerable strength.

Removal of all noncellulosic material from the cell wall causes very little change in cell shape or in most of the mechanical properties of the wall, thereby indicating that the noncellulosic components are interspersed within the cellulose framework. A single cotton fiber, which is visible to the naked eye, may have as many as 1,500 microfibrils and as many as 7.5×10^8 individual molecular cellulose chains.

As a result of studies performed in Albersheim's laboratory (3, 4, 20) on the primary cell wall constituents of suspension-cultured sycamore cells, plant scientists were provided with an early model of the arrangement of cell wall components and the mechanism of cellular extension. Xyloglucans are thought to be noncovalently cross-linked through hydrogen bonding to the cellulose micelles—a feature of many cell walls that seems to be extremely important for cellular enlargement by cell wall deformation. As we have already noted, cell wall deformation results from the breakage of cross-links between the xyloglucans and cellulose micelles. Other important implications derived from the model may help to understand cell wall structure.

We can distinguish three distinct layers in the secondary wall, each one having a different microfibril arrangement. For example, in the banana tracheid wall (see Figure 1–5), we can discern five layers: the middle lamella, a thin primary wall, and a three-layered secondary wall, as well as pit pairs.

We can account, then, for nine layers of wall separating the cell cavities of two adjacent tracheids.

Wall Synthesis

Construction of the cell wall requires that substances produced in the cytoplasm migrate through the plasmalemma to the wall areas. Electron microscopic evidence provided by Ramsey and Berlin (28) indicates that the matrix wall components penetrate the plasmalemma by a process similar to reverse pinocytosis. The cell wall proteins, which are synthesized on the rough endoplasmic reticulum and cell wall polysaccharides, hemicelluloses, and pectic substances produced in the Golgi apparatus, seem to be incorporated into secretory vesicles that ''bud'' off the Golgi apparatus or the endoplasmic reticulum and then fuse with the plasmalemma in such a manner that the products are released to the outside wall area (Figure 1–6).

Many investigators have noted that the cellulose microfibrils are laid down in parallel patterns, particularly during later stages of wall formation. Ledbetter and Porter (22), in their initial finding of microtubules in the cytoplasm of *Phleum* root cells (see Figure 1–2), indicated that the organelles were very close to the cytoplasm-wall interface and oriented parallel to the microfibrils laid down in the cell wall. The location of microtubules at the cytoplasm-wall interface and their parallel orientation with the cellulose microfibrils in the cell wall certainly support the current conclusion that microtubules are directly involved in cellulose microfibril orientation in the cell wall.

Membranes

We must emphasize the fact that most of the cellular activities are dependent on the organization of various chemical components within bounding membranes or membranes of cellular organelles and endoplasmic reticulum. As they have done with other perplexing problems in biology, scientists have focused considerable attention on the structure of membranes, and they have done so

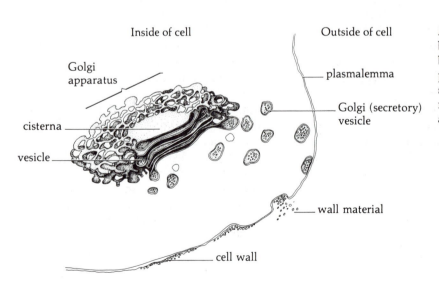

Inside of cell   Outside of cell

Golgi apparatus

plasmalemma

cisterna

Golgi (secretory) vesicle

vesicle

wall material

cell wall

Figure 1–6. Vesicles budding off Golgi bodies, fusing with plasmalemma, and subsequent release of materials into wall area.

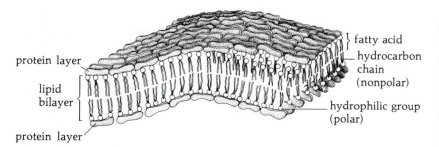

protein layer

lipid bilayer {

protein layer

} fatty acid

hydrocarbon chain (nonpolar)

hydrophilic group (polar)

Figure 1–7. Unit bilayer membrane model of Danielli and Davson showing bimolecular lipid layer between monomolecular layer of protein at each surface. Note association of polar ends of lipid with protein at each surface. Nonpolar hydrocarbon ends of lipid constituents interact toward center.

especially through the widespread use of the electron microscope.

To explain the structure of membranes, Davson and Danielli (11, 12) provided the first definitive model—that of a bimolecular lipid layer associated with protein (Figure 1–7). They reasoned that the surface tension and permeability properties of membranes indicated that their structure consisted of a major lipid component that accounted for the cellular penetration of nonpolar, or nonsurface-charged, substances. Their observations also indicated that a protein layer existed on the two surfaces of the membrane for the purpose of transport of polar, or surface-charged, substances and other compounds that contained both polar and nonpolar parts. Their

membrane model depicted a double layer of lipid, sandwiched between two layers of protein. There is unequivocal evidence (based on freeze-etching, permeability, and membrane structural changes) that Davson and Danielli's "sandwich" membrane model is not truly representative of all membrane structure. Even though this "unit membrane" model did not account for the dynamic changes in membrane permeability, it provided the basis for many of the experiments leading to our current understanding of membrane structure and dynamics.

Figure 1–8 illustrates the most widely accepted model membrane today—the fluid mosaic model. The protein components may be structural and enzymatic and they

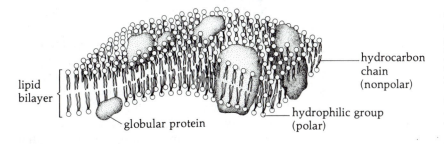

lipid bilayer {

globular protein

hydrocarbon chain (nonpolar)

hydrophilic group (polar)

Figure 1–8. Fluid mosaic model. Small spheres and vertical lines represent phospholipids. Large protein bodies are interspersed at surface and through entire membrane. Carbohydrates and other components are also interspersed in phospholipid medium.

may differ considerably from one cellular organelle to another. As a matter of corroboration of the phospholipid and protein natures of membranes, we can construct synthetic membranes with known proteins and lipids. The fluid mosaic model accounts for the dynamic properties of membranes with respect to the transport of hydrophobic (water-fearing) and hydrophilic (water-loving) materials, the presence of enzymatic components, and the changes in permeability that scientists have observed. In other words, this model is consistent with the fluid and dynamic properties of natural membranes. The membrane consists of a double layer, or bilayer, of phospholipids, with their hydrophobic hydrocarbon tails oriented inward, and globular proteins interspersed among the phospholipids, much like variously weighted Ping-Pong balls in a puddle of viscous fluid. The protein components may be structural or enzymatic and may differ considerably, in quality and quantity, from cellular organelle or membrane system to another or from one surface of a given membrane section to another.

The model also accounts for the dynamic nature of membranes in that both the components and surface area may change as reflected by changes in permeability and enzymatic activities at the cell surfaces of the organism. Consequently, the proteins and the components are not considered to be fixed but may float in and on the phospholipids, thereby creating a mosaic of substances. The proteins are of all sizes, from those that are embedded in the lipid medium only slightly to those that span the complete biomolecular phospholipid medium. The proteins may also be partially hydrophilic and partially hydrophobic. Toward the surfaces of the membrane we would expect to find primarily hydrophilic exposed ends of protein. When a protein is

clearly associated with the lipid layer, we would expect hydrophobic interactions, especially within the middle of the membrane.

The fluid mosaic model makes provision for the presence of other membrane components, such as carbohydrates and protein derivatives. The carbohydrates in plant cell membranes may be extremely important in various cell-surface transformations necessary for the penetration or exclusion of certain substances. As we shall see later, membranes may contain enzymes, carriers, proton pumps, structural proteins, and high-energy compounds that facilitate the exclusion from as well as the movement of minerals and chemicals into and out of plant cells.

There is no question that the amounts of lipid, protein, and other components in membranes may change radically from one moment to the next just by changes in the relative amounts of hydrophobic and hydrophilic groups. Thus membranes are *differentially permeable*—that is, they regulate the passage of diverse substances specifically. Although the term semipermeable is often used, it is not descriptive of biological membranes. Semipermeable implies that the membrane is permeable to many substances but with no specificity; and the term also fails to describe the true, dynamic nature of membranes. Instead, scientists use the term *differentially permeable* widely to describe that property of a membrane whereby the passage of diverse materials is regulated into and out of the cell, organelles, and vacuole at different rates, depending on the relative affinity (solubility) of those materials for the lipid and protein contituents. In fact, some materials may pass through membranes so slowly that they are said to be excluded. There are other factors that also contribute to the permeability of membranes, but their involvement is far too extensive to consider here.

Transport is termed *passive* when a substance passes through the membrane without the cell's expending metabolic energy. Diffusion, ion exchange, Gibbs-Donnan effect, and mass flow are all believed to be forms of passive transport.

Some materials may be accumulated in the cell or secreted to the external environment by means of active transport. This movement of materials across membranes requires an expenditure of cellular energy and the presence of receptors or carriers that transport molecules usually but not always against concentration gradients. Energy-requiring carrier systems are called *pumps* and have been the subject of scientific exploration for some time. With the fluid mosaic model in mind, we will now explore the membrane systems and organelles of plant cells.

Plasmalemma

Although the cell wall seems to separate the cell from its environment, many materials pass through via pores, plasmodesmata, or simply by the wetting action of water. Bordered by the cell wall, there is a thin, delicate, flexible structure, the cytoplasmic membrane, or plasmalemma, that encloses the cytoplasm and an array of cellular components.

Because of the similarity of the plasmalemma and the cytoplasm, we have difficulty differentiating between the two under the light microscope. However, when using proper staining techniques, we can clearly see the plasmalemma with the electron microscope. The plasmalemma encloses the cellular contents and regulates the passage of materials into and out of cells.

Endoplasmic Reticulum

The cytoplasm of the cell is interlaced by an elaborate membrane-bound vesicular system called the endoplasmic reticulum (ER). The vesicles appear as a system of one-unit membranes, surrounding cavities that differ in size and shape and often giving the appearance of a network of tubules (see Figure 1–9). In some parts of the cytoplasm, the vesicles appear as flattened sacs called *cisternae* (singular: cisterna), sometimes filled with fluid. Although maintaining its general appearance, the endoplasmic reticulum may become modified during development and during certain activities of the cell. For example, during high cellular synthetic activities, numerous ribosomes may become associated with the ER. When ribosomes are attached to the endoplasmic reticulum, they constitute part of the *rough endoplasmic reticulum*. In this association the ribosomes are involved directly in the synthesis of *polypep-*

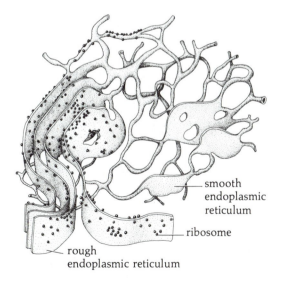

smooth endoplasmic reticulum

ribosome

rough endoplasmic reticulum

Figure 1–9. Structure of endoplasmic reticulum.

tides (proteins), which are secreted into the lumen, also referred to as the *intercisternal space* of the ER. In cell wall synthesis, specific polypeptides appear to be released from the ribosome surfaces and move into a lumen of the ER and then to an associated Golgi apparatus. In some instances the ER is not associated with ribosomes and is termed *smooth endoplasmic reticulum*. The smooth ER plays an especially important role in the synthesis and assembly of glycolipids.

According to several observations, the lumena of the endoplasmic reticulum are continuous with the nuclear envelope and extend to the cell surface (35, 38). In fact, observers have found membranes of this system in the primary walls of some cells and even extended to neighboring cells (36, 37, 38). Whaley and his colleagues (36, 37) initially pointed out that the inclusion of the nuclear membrane with the endoplasmic reticulum provides for extensive surface contact between nuclear material and the cell cytoplasm, and serves as a communicative system within the cell. Where some strands of the endoplasmic reticulum extend from one cell to the next, the nuclei of both cells may be said to be in direct contact with one another because the area between the double nuclear membranes are continuous with those of the ER.

A three-dimensional view of the cell shows that the endoplasmic reticulum divides the cytoplasm into numerous, small cavities. This compartmentalization of the cytoplasm has drawn a great deal of attention in recent years. Within these compartments certain enzymes and metabolites may be accumulated or excluded—a circumstance, perhaps, of vital importance to the cell. We will see in a later chapter, for example, that overloading the system with a certain metabolite and excluding another can force a reaction to proceed in a certain direction. Although their knowledge is incom-

plete, plant scientists fully appreciate the importance of the endoplasmic reticulum to the general functioning of the cell.

Golgi Apparatus

The Golgi bodies, or dictyosomes, as seen in electron micrographs (see Figure 1–10), often appear in cross section as two distinct structures: a stack of five to fifteen flattened, membrane-bound *cisternae* and several small, spherical vesicles that seem to group around the edges of the cisternae (36, 37). Collectively, the Golgi cisternae and vesicles (Golgi bodies) are termed the *Golgi apparatus*. The vesicles "pinch off" from the surface of the cisternae membranes.

The membranes of the Golgi bodies resemble somewhat those of the endoplasmic reticulum. Indeed, some fusion between the Golgi cisternae and those of the endoplasmic reticulum takes place (17). Investigators also suggest that the small vesicles associated with the Golgi cisternae fuse with these

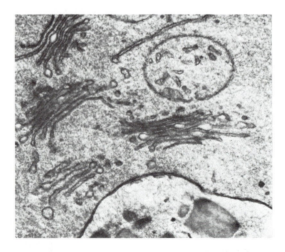

Figure 1–10. Electron micrograph of Golgi apparatus and vesicles in cortical cell of radish root.
Courtesy of M.A. Hayat, Kean College of New Jersey.

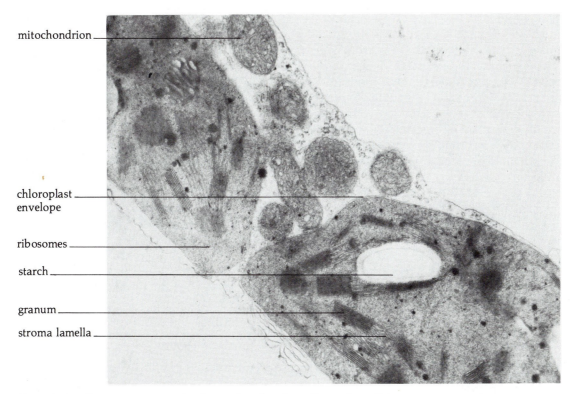

mitochondrion —

chloroplast —
envelope

ribosomes —

starch —

granum —

stroma lamella —

Figure 1–11. Electron micrograph of mitochondria from Kentucky tall fescue (*Festuca arundinacea*). Note six mitochondria (round to oval) between two chloroplasts. Cristae appear as clear areas throughout mitochondrion.

Courtesy of R. Zimmerer, Juniata College.

combined cisternae or fuse with each other to form cisternae of the ER.

The Golgi apparatus has not as yet been isolated in the pure state; however, electron micrograph studies indicate that this system of membranes is involved in secretory processes. Specifically, the vesicles, containing cell wall precursors (for example, polysaccharides, proteins, and other cellular chemicals) synthesized or accumulated in the cisternae, migrate, at the completion of mitosis, to the cell plate or toward the cell surface, fuse with the plasmalemma, and deposit cell wall materials at the plasmalemma–cell wall interface. Thus both the Golgi bodies and the endoplasmic reticulum play an important role in cell wall formation.

Mitochondria

With the possible exception of the nucleus and the chloroplast, the mitochondrion has been the most extensively studied organelle of the cell. We will concern ourselves at this time more with the structure of the mitochondrion than with its function. The latter is covered in detail in Chapter 16 on respiration.

Mitochondria, which are pleomorphic (many-formed) bodies (see Figure 1–11), are bounded by two, unit membranes. These membranes enclose an inner matrix and average approximately 0.5 to 1.0 μm in width and 3.0 to 8.0 μm in length. Numerous folds in the inner membrane project deep into the matrix. Some of these folds completely bridge the interior of the mitochondrion and, as scientists have observed in the laboratory when they have sliced mitochondria thinly, the folds appear to connect with the inner membrane on the opposite side. The projecting folds of the inner membrane are collectively called the *cristae*.

Analyses of the contents of mitochondria reveal the presence of phospholipids, the nucleic acids DNA and RNA, the Krebs cycle enzymes, various substrates, the cytochromes, and other components of the electron transport system (ETS). In fact it has been estimated that there are at least 2,000 complete sets of the Krebs cycle enzymes in one mitochondrion.

The mitochondrion provides a great deal of the cell's usable energy. As might be expected where cellular activity is high, mitochondria are dense, as exemplified by meristematic cells in which mitochondria are found in abundance. What do we mean when we say that mitochondria provide the cell with usable energy? When proteins, fats, and carbohydrates are broken down in the cytoplasm, the resulting products are oxidized, with the liberation of carbon dioxide, water, and energy. In mitochondria, much of the energy released is conserved in the form of high-energy phosphate bonds. The most important compound in this respect is *adenosine triphosphate* (ATP) (see Chapter 16). The advantage of storing energy in this compound is that it can be released and utilized quite readily to drive the energy-consuming reactions of the cell. We will discuss details relating to the synthesis of ATP in mitochondria chloroplastids and the cytoplasm in later chapters.

Because of the complex structural organization found in mitochondria and because of the similarity of organization in mitochondria from a variety of species, we can only assume a close relationship between form and function. For example, oxidative phosphorylation ceases on loss of integrity of the double-membrane structure. The reactions of the Krebs cycle, which occur in mitochondria, are dependent on the double-membrane structure (41), although the enzymes involved in these reactions can easily be extracted from the soluble matrix. It is interesting to note that fragments of mitochondria are capable of carrying out some but not all of the oxidations of the Krebs cycle (15, 16). As we shall see in Chapter 16, the inner membrane may be further structured in a fashion that is especially important to the production of ATP by means of *oxidative phosphorylation*.

Both mitochondria and chloroplasts are delimited by a double membrane, both produce ATP, and both contain uncomplexed DNA and RNA that is usually 70S ribosomes dissimilar to the nucleic acids of the nucleus of the cells within which they reside. For example, the mitochondrial DNA isolated from Mung beans, turnips, sweet potatoes, and onions differs from nuclear DNA isolated from the same plants (32). Of further interest is the fact that both mitochondria and chloroplasts are able to divide and grow somewhat independently of the nucleus. Obviously, the nucleic acids are of fundamental importance in the storage and transmission of information for some protein synthesis, functions that necessitate the presence of nucleic acids in a self-replicating body. Nevertheless, both chloroplasts and mitochondria cannot develop and survive independently of the cell nucleus.

Plastids

Plastids are membranous, cellular organelles characteristic of plants. They are generally round, oval, or disc-shaped bodies about 4 to 6 μm in diameter and observable under the light microscope. Two unit membranes, called an *envelope,* are at the surface. Internally, plastids consist of a membrane system and matrix. Plastids are classified as proplastids, leucoplasts, amyloplasts, chloroplasts, or chromoplasts. *Proplastids* give rise to plastids. *Leucoplasts* are nonpigmented plastids, devoid of chlorophyll and carotenoids, and are prevalent in cells of certain plant organs, including leaves, roots, and storage organs. When plastids play an extensive role in starch biosynthesis, as in the cells of potato tubers and the endosperm of corn kernels, they are termed *amyloplasts.* Leucoplasts, which also produce proteins, oils, and other substances, can develop chlorophyll and become *chloroplasts* upon exposure to light.

Another kind of plastid, the *chromoplast,* is pigmented. Chromoplasts are plastids that contain carotenoid pigments only. The function of chromoplasts is obscure, but they are responsible for the coloring of autumn leaves, flowers, and fruit. In ripening fruit or fruit peel, for example, the internal membrane structure and chlorophyll of the chloroplasts is lost while carotenoids accumulate to form the chromoplasts. A familiar example of the conversion of chloroplasts to chromoplasts is in ripening tomato berries.

Chloroplasts are perhaps the most important organelles for sustaining all life because of their function of collecting light energy and converting it into chemical energy (*photosynthesis*). Although chloroplasts will be considered more extensively in a later chapter on photosynthesis, let us briefly consider some of the terminology that applies to their structure (Figure 1–12). Chloroplasts are bounded by a double-membrane envelope, with the internal structure consisting of membranes and the nonmembrane area, or *stroma.* In the chloroplast

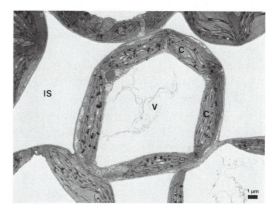

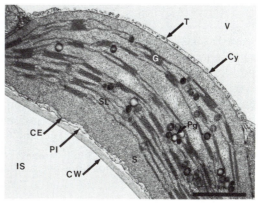

Figure 1–12. Electron micrograph of mesophyll cells (*top,* magnification 2,900 ×) and mesophyll cell chloroplast (*bottom,* magnification 14,500 ×) from alfalfa (*Medicago sativa*) leaf. Note chloroplast (C), intercellular space (IS), vacuole (V), cell wall (CW), chloroplast envelope (CE), cytoplasm (Cy), granum (G), mitochondrion (M), plasmalemma (Pl), plastoglobulin (Pg), stroma (S), stroma lamellae (SL), tonoplast (T).

Courtesy of R. Rufner, Massachusetts Agricultural Experiment Station, University of Massachusetts.

there is an elaborate structure of membranes that resemble simple, flattened sacs called *stroma lamellae*. Other membranes are more concentrated in areas of the chloroplast and form stacks of disklike, flattened sacs called *thylakoids*. *Grana* (singular: granum) are collections of 5 to 50 thylakoids and appear as stacks of miniature pancakes. The *grana thylakoids* are often connected to the stroma lamellae.

As with the mitochondria, chloroplasts (and plastids in general) contain DNA and RNA, the latter often seen as 70S ribosomal particles. Therefore, as we would expect, plastids may arise from the division of existing plastids or in some cases from small organelles known as *proplastids*.

Ribosomes

Characteristics of Ribosomes

Associated with the endoplasmic reticulum and found "free" in the cytoplasm or in mitochondria and plastids are submicroscopic, spheroidal particles called *ribosomes* (see Figure 1–11). Ribosomes isolated from pea seedlings are about 0.1 to 0.3 μm in diameter and consist of 50 to 60 percent of ribonucleic acid (RNA) and 40 to 50 percent of protein. In essence, ribosomes are multimolecular aggregations of RNA and protein.

Biochemists often characterized ribosomes on the basis of their subunit sedimentation (S) constants. In fact, under certain experimental conditions such as the use of low magnesium preparations, scientists can disassociate a suspension of ribosomes into subunits (examples are 40S and 60S). The RNA comprising the ribosomal structure is referred to as *ribosomal RNA* (*rRNA*) while the "coded" RNA at the surface of the ribosome, involved in peptide synthesis, or *translation*, is termed *messenger RNA* (*mRNA*).

When ribosomes are associated with the endoplasmic reticulum, the latter is termed rough ER. When the endoplasmic reticulum is devoid of ribosomes, it is called smooth ER. Ribosomes are usually found clustered or attached like beads on a string to mRNA, and the clusters, or *polyribosomes*, are active primary sites of peptide synthesis. Rarely, if at all, are single ribosomes the site of protein synthesis in living cells.

Ribosomes versus Microsomes

Before widespread use of the electron microscope, biochemists were able to isolate a cellular fraction that supported peptide synthesis in cell-free preparations (in vitro). This fraction, produced by high speed centrifugation, was said to consist predominantly of particles called *microsomes*, which were active in peptide synthesis.

In actuality, the so-called microsome is a combination of fragmented or sheared membranes and associated ribosomes. It is a good example of the structural distortions that can result from routine biochemical and biophysical techniques, and for this reason researchers are constantly required to consider such effects on their experimental plants. Even though we can consider the microsome an operational term in biochemical studies on protein synthesis, it is not a cellular organelle per se. Historically, it was useful for in vitro experiments and implicated the ribosome with protein synthesis.

Vacuoles

In young, immature cells, such as found in meristematic regions, the cell is generally filled with a dense cytoplasm. Scattered throughout the cytoplasm are numerous, small vacuoles, which appear under microscopic examination as clear droplets. As the cell matures and enlarges, the small vacuoles fuse together to form one large, central vacuole that usually fills almost 90 percent of the entire volume of the cell (see Figures 1–12 and 1–13). When there is a central vacuole, the cytoplasm is pressed up against the cell wall and forms just a thin layer around the vacuole.

The vacuole is bounded by a single membrane, the *tonoplast*, which is differentially permeable but encloses a solution in which numerous materials are suspended. The contents of the vacuoles are collectively referred to as the *cell sap*. The major, but not exclusive, functions of the vacuole are: (1) maintenance of the turgor pressure, which is important for structural support and control of water movement, (2) storage of materials essential to the cell's metabolic activities, and (3) accumulation of cellular metabolic by-products, defensive substances, and toxic materials. Thus the cell sap contains such substances as sugars, organic acids, mineral salts, gases, pigments (anthocyanins), alkaloids, fats and oils, tannins, and (on occasion) crystals (calcium oxalate, for example).

The cell sap is often acidic but may range from a pH of 1.0 to as high as 11.0 depending on the components present. For this reason, the cell sap often complicates cytological and biochemical studies because the dissolved materials in the vacuoles and the low pH interfere with staining enzyme analyses and product extraction.

The tonoplast undoubtedly plays an important role in the biochemical activities of plant cells. For example, the accumulation of hydrogen ions and storage of toxic

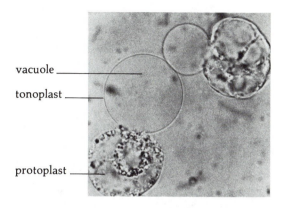

Figure 1–13. Light micrograph showing release of vacuole by osmotic lysis of tobacco protoplasts.
From I.J. Mettler and R.T. Leonard. 1979. Plant Physiol. *64:1114.*

components suggest the presence of "pumps" and membrane carrier systems that are involved in the penetration of a wide variety of materials into the vacuole but may not allow transport from within the vacuole back into the cytoplasm. The permeability dynamics of the tonoplast are vital to the functioning of the entire plant cell. The tonoplast is also known to engulf particles and even organelles such as mitochondria by reverse pinocytosis, with their subsequent digestion by the vacuole enzymes. The anthocyanins present in vacuoles are water soluble, are responsible for the color of numerous flowers, fruit, and vegetables, and are also responsible for some of the predominant colors in plant leaves during the autumn. Because of their changes in color at various pH's, anthocyanins were used as the first indicator of pH in plant and animal extracts. Thus the vacuole is much more than a dumping site in the cell and is even involved in the breakdown and recycling of cellular compounds.

Microtubules

Microtubules are elongated, hollow, non-membranous structures, are approximately 10 to 20 nm in diameter, and are macromolecules made up of the protein α,β-*tubulin*. Microtubules, attached to the kinetochore (centromere) of chromosomes and present in the spindle fibers during mitosis, are involved in the separation and migration of homologous chromosomes to opposite poles during telophase. They help in cell wall formation by directing the alignment of the cellulose microfibrils as these microfibrils are deposited (see Figure 1–2). Microtubules are substructures of flagella and cilia in plant cells that are motile, such as gametes of lower land plants and algae.

Microbodies: Glyoxysomes, Peroxisomes, and Spherosomes

Glyoxysomes, peroxisomes, and spherosomes are small (about 1 to 2 nm in diameter), dense structures called microbodies. They are bounded by a single membrane and, unlike chloroplasts or mitochondria, do not have any observable internal membrane structure. These organelles, however, often have a very dense, proteinaceous interior. *Glyoxysomes* are primarily found in tissues of oil-bearing seeds, where fat is converted to carbohydrate, a process that is catalyzed by enzymes of the glyoxylate cycle. The enzymes that are characteristic of

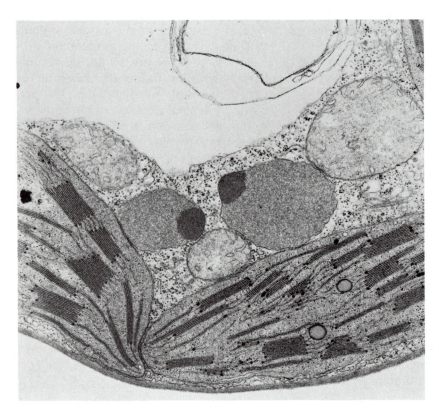

Figure 1–14. Electron micrograph of peroxisomes (microbodies) in leaf cells of tobacco (*Nicotiana tabacum*). Note nucleoids (of unknown function) in microbodies. Magnification 26,000 ×.

Micrograph by S.E. Frederick. Courtesy of E.H. Newcomb, University of Wisconsin.

the glyoxylate cycle—isocitrate lyase, malate synthetase, aconitase, citrate synthetase, glycolate oxidase, malate dehydrogenase and catalase—have all been found in glyoxysomes (14).

Peroxisomes are very similar in appearance to glyoxysomes, and both structures contain a number of the same enzymes. Peroxisomes function in the metabolism of the glycolate produced by chloroplasts during photosynthesis. Evidence shows that peroxisomes are associated with photorespiration, a process characteristic of C_3 but not C_4 plants. Also, localization of photorespiration in the plant cells corresponds to areas of dense numbers of peroxisomes (Figure 1–14).

Spherosomes are small, enzyme-containing particles that are found in the cytoplasm of plant cells. In addition to the enzyme hydrolase, spherosomes contain other hydrolytic enzymes classified as proteases, ribonucleases, phosphatases, and esterases. The primary function of spherosomes in the cell appears to be the storage and transport of lipids. The spherosomes of plant cells are somewhat similar to the lysosomes of animal cells. However, although they do contain a number of similar enzymes, their overall enzyme complements are quite different, and the two particles should be thought of as distinct from one another. Figure 1–15 shows spherosomes isolated from peanuts.

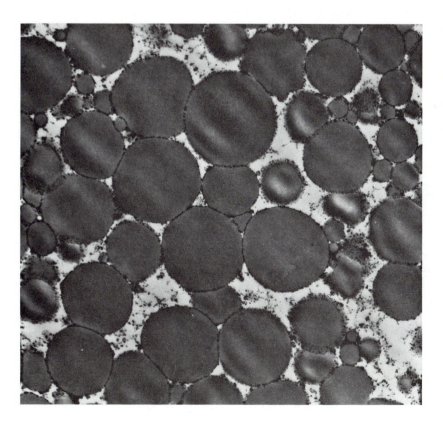

Figure 1–15. Electron micrograph of isolated spherosomes from peanuts.
From L.Y. Yatsu and T.J. Jacks. 1972. Plant Physiol. 49:937–943. Print courtesy of the Southern Regional Research Center, USDA.

Nucleus

Ever since its discovery by Robert Brown in 1835, the nucleus of the cell has attracted the interest and curiosity of thousands of investigators. Its main interest as a subject for study resides in the fact that it has the controlling influence on the heredity and activity of the cell. The nucleus controls or directs the synthesis of all proteins, including the enzymes that catalyze most if not all of the metabolic reactions of the cell.

In the immature cell, the nucleus is a spherical body embedded in the cytoplasm of the cell. In the mature, living plant cell, the nucleus is generally located to one side of the cell since it is pressed against the cell wall by the vacuole. Generally, the nucleus, approximately 5 to 10 μm in diameter, appears to be slightly flattened under these conditions.

The nucleus is bounded by a double membrane, the *nuclear envelope*. Electron microscope studies have revealed two very interesting features of the nuclear envelope: it is continuous with the endoplasmic reticulum, and the nuclear envelope contains relatively large pores in its structure (Figure

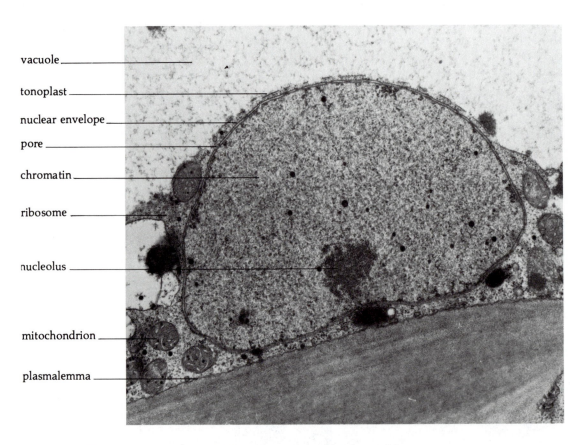

vacuole
tonoplast
nuclear envelope
pore
chromatin
ribosome
nucleolus
mitochondrion
plasmalemma

Figure 1–16. Electron micrograph of nucleus from calamondin fruit peel.
Courtesy of K.B. Evensen, The Pennsylvania State University.

1–16). Direct communication clearly exists between the cytoplasm and the nucleoplasm.

The nucleoplasm is composed of a structured and a structureless phase. The structured phase is a network of threads called *chromatin* and is made up of DNA and proteins. This phase of the nucleoplasm appears either as a network or as distinct chromosomes, depending on whether or not the nucleus is in mitosis. The structureless phase of the nucleoplasm appears as a granular substance and is generally referred to as the *nuclear sap*. Substantial amounts of DNA, RNA, lipids, phospholipids, and particularly proteins consisting of the basic histones have been found in the nuclei, along with several hydrolytic enzymes, such as ribonuclease, dipeptidase, and phosphatase.

During interphase the nucleus contains one or more *nucleoli*; the number present depends on the species being observed. For example, the onion cell nucleus generally contains four nucleoli. The nucleoli become evident during the telophase stage of mitosis as a result of the activity of the *nucleolar organizers*, which are particular areas on specific chromosomes referred to as *nucleolar chromosomes*. Nucleoli are present in nondividing nuclei but seemingly disappear during mitosis.

Chemical analysis of the nucleolus shows that it is primarily composed of ribosomal RNA (rRNA) subunits and proteins. Nucleolar RNA is of chromatin origin (8). In fact, the region of DNA containing the gene informational sequence for rRNA synthesis is the nucleolar organizer. Also, it is interesting to note that nucleolar RNA synthesis is primarily directed toward the assembly of ribosomes. Observers have not as yet detected a membrane for the nucleolus, although they have observed dense and fibrous areas.

Questions

1–1. What is the major cellular difference between prokaryotic and eukaryotic organisms?

1–2. Beginning from the outside surface, list the parts of a "typical" plant cell. Name some of the plant cell types that do not possess one or more of the "typical" plant cell features.

1–3. List some of the known and possible functions of the cellular components indicated in your answer to question 1–2.

1–4. List the names of the chemicals that may be components of the cell wall.

1–5. What is the significance of calcium and magnesium pectate in multicellular plants?

1–6. Describe how the middle lamella and primary cell wall are formed (general events) and what the important functional properties of each structure are.

1–7. Explain the structural changes of the primary cell wall that are necessary for cellular enlargement.

1–8. Name the structures in plants that connect and provide a translocation network between living cells. From what material are those structures constructed and where specifically might we look in the cell to observe them?

1–9. Explain the events involved in secondary wall formation. What are the major functions of the secondary wall?

1–10. What is the most widely accepted model to explain membrane structure today? What percentage of organelles of a "typical" plant cell do you think are comprised of membranes?

1–11. Membranes are said to be important with respect to compartmentalization in cells. What do you think this term means and of what significance is it to the functioning of plant cells?

1–12. Explain the following terms: rough and smooth endoplasmic reticulum, cisterna, vesicle, crista, stroma lamella, granum thylakoid, proplastid, polyribosome, tonoplast, nuclear envelope, chromatin, and chromosome?

1–13. Name the different types of plastids that may be found in plant cells. List their functions.

1–14. Explain why the biochemical research on plant cells is complicated by the presence of vacuoles.

1–15. What structural and chemical features do the microbodies in plant cells have in common? How do they differ?

1–16. Cytological features of plant cells suggest that events that take place in the cytoplasm and at the cell surface influence and possibly regulate the activities of the nucleus. What are these features and what are some of the activities of the cytoplasm that might serve to regulate nuclear activity?

1–17. After reading Appendix A, explain the terms *lyophilic* and *lyophobic*. How do colloids differ from solutions and suspensions?

1–18. What are some of the properties of colloids? What role do they play in living cells?

1–19. What are the major components of plant cells that contribute to the protoplasmic colloid? How might they be precipitated out?

Suggested Readings

Albersheim, P. 1975. The wall of growing plant cells. *Sci. Amer.* 232(4):80–95.

Beevers, H. 1979. Microbodies in higher plants. *Ann. Rev. Plant Physiol.* 30:159–193.

Esau, K. 1977. *The Anatomy of Seed Plants*, 2nd ed. New York: Wiley.

Galun, E. 1981. Plant protoplasts as physiological tools. *Ann. Rev. Plant Physiol.* 32:237–266.

Gunning, B.E.S., and A.R. Hardham. 1982. Microtubules. *Ann. Rev. Plant Physiol.* 33:651–698.

Haupt, W. 1982. Light-mediated movement of chloroplasts. *Ann. Rev. Plant Physiol.* 33:205–233.

Kirk, I., and B.E. Juniper. 1965. The ultrastructure of the chromoplasts of different color varieties of *Capsicum*. In T.W. Goodwin, ed., *Biochemistry of Chloroplasts*. New York: Academic Press.

Ledbetter, M.C., and K.R. Porter. 1970. *Introduction to the Fine Structure of Plant Cells*. New York: Springer-Verlag.

Lott, J.N.A., with J.T. Darley. 1976. *A Scanning Electron Microscope Study of Green Plants*. St. Louis, Mo.: Mosby.

McGilvery, R.W. 1979. *Biochemistry: A Functional Approach*. Philadelphia: Saunders.

Metzler, D.E. 1977. *Biochemistry*. New York: Academic Press.

Possingham, J.V. 1980. Plastid replication and development in the life cycle of higher plants. *Ann. Rev. Plant Physiol.* 31:113–129.

Preston, R.D. 1979. Polysaccharide conformation and cell wall function. *Ann. Rev. Plant Physiol.* 30:55–78.

Swanson, C.P., and P.L. Webster. 1977. *The Cell*, 4th ed. Englewood Cliffs, N.J.: Prentice-Hall.

Thompson, W.W., and J.M. Whatley. 1980. Development of Nongreen Plastids. *Ann. Rev. Plant Physiol.* 31:375–394.

Chapter 2

Diffusion, Osmosis, and Imbibition

Permanent overhead sprinkler irrigation system for Florida citrus grove. *Courtesy of USDA—Soil Conservation Service.*

The processes we are about to consider are remarkably important to the maintenance of plant life—that is, the reproduction and survival of plants. In fact, the diffusion of gases, water, and nutrients between the plant and its external atmosphere and among the cells affects essentially all biochemical processes in plants. The extent of diffusion is in turn controlled by definite energy changes that are consistent with the laws of *thermodynamics*, the science that deals with energy changes in chemical and physical processes. In this chapter, we will consider the energy requirements and changes that are fundamental to the process of diffusion in living organisms.

Three Laws of Thermodynamics

The first law tells us that energy can be converted from one form to another and indicates in relative terms that work can be accomplished from a given amount of energy, but it does not provide more information about the work process. The second law of thermodynamics states that heat cannot be converted into work without leaving a change (disorder) in some part of the system. This second law allows us to predict whether a process can occur spontaneously or without the addition of energy from an external source. At this point, we can only direct attention to the importance of thermodynamics with respect to the kinds of processes necessary for plant life. An indepth discussion of the mathematics necessary to explain the laws fully is not warranted here.

We are able to determine by simple inspection that certain kinds of processes are spontaneous. For example, the uncoiling of a tightly wound watch spring, the flow of water downhill, the expansion of gas in volume, and the dissolving of sugar in water are obviously spontaneous reactions. But less obvious are other spontaneous processes in plants, such as water moving from cell to cell and being translocated to the top of a plant against gravity. Understanding the second law of thermodynamics helps us to analyze such a process and, particularly, to predict the potential for the occurrence of a process.

We know from the second law that spontaneous processes require an initial state of relatively higher energy than is required in the final state. A simple way to state it is that spontaneous reactions occur along an energy gradient. Further, these reactions can be harnessed to do work while they are taking place and can be reversed only if externally supplied energy is introduced into the system. As spontaneous reactions take place, however, a net loss in the capacity to do work occurs, and a change toward a more random or disorganized state takes place—that is, as molecules become less highly ordered, potential energy decreases and the process slows down or stops. We refer to this increase in random arrangement as *entropy*. We may also think of it as a loss in potential capacity to perform work. When maximum entropy for a process is reached, the process is said to be at equilibrium.

The third law of thermodynamics states that the absolute entropy of most substances is zero at absolute zero ($-273.18°C$). In plants, however, entropy and energy are not usually measured as absolute values in the analysis of many physiological processes. To evaluate and understand most processes, plant scientists rely heavily on knowledge concerning the relative states of energy and entropy with respect to initial and final states of a given chemical reaction or physical reaction. In this regard, *Gibbs free energy* (*G*) represents the amount of energy that is available to accomplish work and, as we would surmise, is derived from the relationship of various factors—*entropy* (*S*), *absolute temperature* (*T*), *total internal energy* (*E*), *pressure* (*P*), *and volume* (*V*)—as set forth in the following equation:

$$G = E + PV - TS$$

Where:

E = internal energy (the sum of electronic, nuclear, rotational, vibrational, and translational kinetic)

P = pressure in atmospheres or bars

V = volume in liters

T = absolute temperature (273.18° + °C)

S = entropy

Although we cannot usually calculate Gibbs free energy and entropy as absolute values, we can often study a process on the basis of relative differences in free energy, where ΔG equals the difference in free energy between final and initial states ($\Delta G = G_2 - G_1$). For further clarification, let us consider water flowing down- and uphill through a siphon (Figure 2–1).

As water flows downhill through the siphon, we know that the process is taking place from a higher energy level (G_1) to a lower energy level (G_2) because the force of water falling can be used to accomplish work. Also, if we were to return the water to container A, it would require energy. Therefore, as water flows downhill, $G_2 - G_1 = -\Delta G$, with the minus sign indicating a release of energy (*exergonic reaction*) during this spontaneous process.

If we now try to drive water back up the siphon from container B (now G_1) to

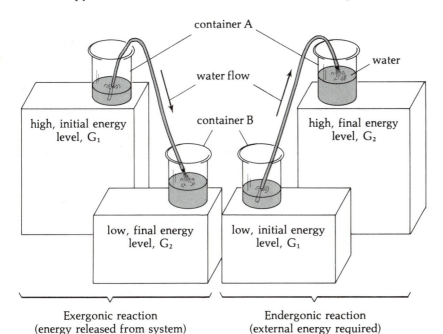

container A

water

water flow

container B

high, initial energy level, G_1

low, final energy level, G_2

low, initial energy level, G_1

high, final energy level, G_2

Exergonic reaction
(energy released from system)
$G_2 - G_1 = -\Delta G$

Endergonic reaction
(external energy required)
$G_2 - G_1 = +\Delta G$

Figure 2–1. Energy changes as water flows from high and low energy levels through a siphon.

container A (G_2), or uphill, $G_2 - G_1 = +\Delta G$, with the plus sign indicating an input of energy, or an *endergonic reaction*, an energy-requiring nonspontaneous process. In biological systems endergonic and exergonic reactions are often coupled and account for important syntheses and assimilation reactions (see chapters on photosynthesis and respiration).

Types of Energy

We may classify energy as electronic, nuclear, rotational, vibrational, and translational kinetic. We can determine the extent of *electronic energy* by the movement of electrons in specified orbitals relative to the nucleus of a given atom. Electron excitation results from the atomic absorption of given amounts of energy, which causes an electron to jump to a higher orbital with a possible spin change or simply causes an electron to change spin without displacement to another orbital. Often, the electron's jump to a higher orbital (energy level) is accompanied by changes in rotational and vibrational energy.

Although electron excitation cannot normally be accomplished within the temperature ranges encountered in living organisms, excitation of plant pigments by light energy is a notable exception. The excitation of chlorophyll represents the primary event of a series in which light is converted to chemical energy. Photosynthesis is the best-known but not the only process that relies on pigment excitation. It is interesting to note that pigment excitation is rapidly followed by a return to *ground state*—that is, the condition attained when an electron returns to its "normal" or most stable spin and orbit, with a discharge of a packet of energy, which may be conserved or released as light.

Nuclear energy, based on the state of atomic nuclei, is not encountered in the study of the physical and chemical reactions of plants (with the exception of radioactive tracer chemicals), and we will, therefore, not consider it further. In dealing with organic molecules, however, *rotational energy* (atoms moving around each other) and *vibrational energy* (atoms moving toward or away from each other) are critical. These two kinds of energy are characteristic of molecules with two or more atoms and involve the movement of atoms in molecules with respect to each other.

Translational Kinetic Energy

Of particular significance to the diffusion process and to plants is translational kinetic energy because it is the driving force behind the straight-line movement of molecules of gases, liquids, and solutions.

At temperatures above absolute zero (0°K or −273.18°C), all components of matter are in motion—that is, they possess a certain amount of *translational kinetic energy*. This motion is random; the molecules or atoms move in all directions, and in many cases collide with one another. Consider, for example, the air we breathe, which is primarily a mixture of nitrogen, oxygen, and carbon dioxide molecules. These molecules are constantly moving in a random manner and they occasionally collide with one another. Nitrogen molecules are much more abundant than oxygen molecules; and carbon dioxide molecules are extremely rare, and account for only 0.03 percent of the mixture. All three types of molecules, however, are mixed uniformly in the atmosphere.

If we were to open a bottle of perfume, the perfume molecules evaporating from the surface of the liquid would diffuse among the air molecules and would eventually become uniformly mixed with them. The perfume molecules are able to do this because they, too, are in constant motion. After complete evaporation of the perfume and complete dispersal of the perfume molecules among the air molecules, a new dynamic system develops, consisting of randomly moving nitrogen, oxygen, carbon dioxide, and perfume molecules. The process that allows for the initial distribution of the perfume molecules is of course diffusion. This spontaneous process is fundamental to the movement of a multitude of diverse substances in plants, and we will consider diffusion from several different viewpoints.

Diffusion

Beginning students of plant science are often provided the following definition of diffusion: the net movement of a substance from an area of its own high concentration into another area of lesser concentration as a result of the random, translational kinetic motion of molecules, ions, or atoms. At a very fundamental level of understanding, this definition is adequate, but it does not give a true indication of the basic energy requirements of the process. Let us consider the pressure of gases.

Gas Pressure

One of the most common demonstrations of gas pressure is found in the barometer, an instrument for measuring atmospheric pressure. If we fill a glass tube with mercury and then invert it, with its open end under the surface of mercury contained in a shallow dish, the mercury in the tube

will fall to a certain height (Figure 2–2). At sea level, the height at which a column of mercury will stand in a glass tube is 760 mm. In other words, the weight of the gas (air) above the surface of the mercury in the dish shown in Figure 2–2 is sufficient to push the column of mercury up the glass tube to a height of 760 mm. The average air pressure at sea level is known as the *standard atmospheric pressure* and is expressed as 760 mm of mercury or as 1 atm.

A good example of pressure being exerted visibly by confined gas is in an inflated balloon. The rubber membrane of the balloon is only slightly permeable to nitrogen and oxygen, the gases most abundant in air. When we inflate a balloon, the air molecules become more concentrated, and an increase in pressure results. The pressure exerted by a gas in a closed container is the sum of the pressures exerted by a great number of molecules as they hit the walls of the container simultaneously. If we increase the concen-

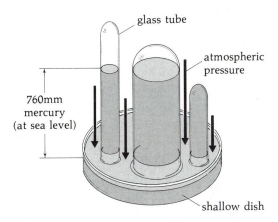

Figure 2–2. Average height of column of mercury in a barometer is 760 mm at sea level. Height of column does not depend on diameter of glass tube. Tube on right is too short to allow any mercury to run out.

tration of the gas in the container, many more of the gas molecules will collide with the walls at any one time. Obviously, this results in an increase in pressure. The walls of the balloon will stretch in order to compensate for the increase in pressure, thereby giving visual evidence of pressure exerted by a gas.

At normal pressures and temperatures, gas molecules are widely separated; the number of collisions that would interfere with the diffusion of one gas into another is therefore limited. This fact is easily appreciated when we consider the extent to which a gas can be compressed. The air that fills a classroom, for example, could easily be compressed into a test tube and still remain in the gaseous state. The compression of the gas, however, would increase the free energy of the gas molecules, hence the number of collisions with each other and with the sides of the container. Essentially, the pressure would be greatly increased.

If the gas were released suddenly, it would diffuse rapidly outward. If you don't believe energy is involved, blow up a balloon and release it suddenly! The force with which the gas molecules escape might be described as the *diffusion pressure*. Although this term was coined to describe the activity (pressure) of a diffusing gas, liquid, or solute, plant physiologists no longer use it.

Chemical Potential

From the standpoint of Gibbs free energy considerations, we can state that the diffusion of gas released from the balloon depends on the difference in free energy of the gas inside the balloon (G_1) and free energy on the outside (G_2). In other words, this spontaneous reaction ($G_2 - G_1 = -\Delta G$) occurs along an energy gradient (high to low). However, instead of using the free energy term, we can more meaningfully equate the amount of free energy to a given amount of a substance by using the term *chemical potential*. The chemical potential is the amount of free energy per gram molecular weight of a substance, which, in this example, would be the gas.

Hence, from our discussions concerning energy considerations, we should use the following, more-precise definition of diffusion. Diffusion is the net movement of a substance from an area of its own high chemical potential into another area of its lower chemical potential, which is due to the random, translational kinetic motion of molecules, ions, and atoms. The direction of diffusion of a substance is determined entirely by the differences in chemical potential of that substance and is independent of the diffusion of other substances.

Let us again use the rubber balloon to illustrate this principle. Suppose we inflate a balloon with nitrogen gas. Since the rubber walls of the balloon are relatively impermeable to nitrogen, the confined nitrogen in the balloon will have a relatively high chemical potential. Carbon dioxide, unlike nitrogen, can readily pass through a rubber membrane. If we allow the nitrogen-filled balloon to stand in air, the carbon dioxide in the air will diffuse into the balloon until an equilibrium is established. Carbon dioxide diffuses into the balloon because its chemical potential in the air is higher than its chemical potential in the balloon (which was zero). The inward diffusion of carbon dioxide occurs even though the chemical potential of the nitrogen gas confined in the balloon is considerably higher than that of the carbon dioxide in the air.

The importance of independent diffusion to the plant will become more apparent in the succeeding chapters. Let us remember that diffusion is dependent on a chemical potential gradient. We can envisage the energy gradient itself as a steep, downhill energy slope that becomes less steep (increased entropy) as the process proceeds (see Figure 2–1). Net movement (diffusion) will cease when the gradient no longer exists—that is, when equilibrium is attained ($\Delta G = G_2 - G_1 = 0$). Also, any factor that influences the chemical potential gradient will similarly affect diffusion.

Factors Affecting Rate of Diffusion of Gases

Temperature. The rate at which a gas will diffuse increases as temperature increases. An increase in temperature causes an increase in the kinetic energy (chemical potential) of the gas molecules—that is, a rise in temperature is accompanied by an increase in the velocity at which the gas molecules move.

The effect of temperature on a physical or chemical reaction is often determined by its *temperature coefficient* or Q_{10}. The Q_{10} is expressed as the ratio of the velocity of a process at a given temperature to that at 10°C lower ($V_T/V_T - 10$°C). The following equation may be used to calculate the Q_{10} of biological processes.

$$\log Q_{10} = \left(\frac{10}{T_2 - T_1}\right) \log \frac{K_2}{K_1}$$

Where:

T_2 = the higher temperature

T_1 = the lower temperature

K_2 = rate of reaction at higher temperature

K_1 = rate of reaction at lower temperature

One major advantage to be gained from calculating Q_{10} values is that knowing these values sometimes makes it possible to determine whether a process is purely physical or chemical. For example, Q_{10} values slightly greater than one indicate a physical process such as diffusion and photochemical reactions. Photochemical reactions depend on energy supplied by light and moderate temperatures. An increase in temperature does not provide sufficient energy to cause electronic displacement necessary for the reaction. The Q_{10} for chemical reactions that take place at physiological temperatures often approach 2 and higher. Thus Q_{10} determinations may be used to distinguish chemical reactions from purely physical ones.

Density of diffusing molecules. The rate at which gases diffuse under constant conditions varies widely with different gases and is related to the density of the gas. *Graham's law of diffusion* summarizes this principle: The rates of diffusion of gases are inversely proportional to the square roots of their densities. On the basis of this law, the following relationship can be written:

$$\frac{r_1}{r_2} = \frac{\sqrt{d_2}}{\sqrt{d_1}}$$

where r_1 and r_2 are the diffusion rates of gases that have the densities d_1 and d_2, respectively. If we apply this equation to the gases hydrogen and oxygen, we find:

$$\frac{r_h}{r_o} = \frac{\sqrt{d_o}}{\sqrt{d_h}} = \frac{\sqrt{16}}{\sqrt{1}} = \frac{4}{1}$$

Since the density of oxygen is sixteen times that of hydrogen, the rate of diffusion of hydrogen is four times that of oxygen.

We can illustrate Graham's law easily in the laboratory (see Figure 2–3). If we insert a cotton plug at both ends of a glass tube and simultaneously soak one cotton plug with ammonium hydroxide (NH_4OH) and the other plug with hydrochloric acid (HCl), we will have a system in which two gases (NH_3 and HCl) are diffusing toward each other at rates dependent on the mass of their molecules. The spot where the two gases meet is indicated by a white ring of solid ammonium chloride (NH_4Cl). As Figure 2–3 shows, the ammonium chloride ring is closer to the HCl end of the tube. We would anticipate this result since the density of HCl is almost twice that of NH_3.

Solubility in diffusion medium. The more soluble a substance is in the diffusion medium, the faster it will diffuse. However, if the diffusion medium is concentrated, there is increased resistance to diffusing substances. Also the distance (expanse) of the medium or area through which substances move will be directly proportional to the rate of diffusion. It is interesting to note that the solubility of gases in liquids decreases as temperature increases. For example, at 760 mm pressure, 0.04889 liter of oxygen will dissolve in 1 liter of water at 0°C, 0.03891 liter at 10°C, 0.03102 liter at 20°C, 0.02608 liter at

30°C, and 0.01761 liter at 80°C. Indeed, boiling is one common laboratory practice for removing gases from a liquid. With the exception of very soluble gases, however, the solubility of gases in liquids increases with an increase in pressure. This property of gases is the basis of *Henry's law:* The mass of a slightly soluble gas that dissolves in a definite mass of a liquid at a given temperature is very nearly directly proportional to the partial pressure of that gas.

The carbonated beverage business has made a practical application of Henry's law. The dissolution of carbon dioxide in the beverage is accomplished under a pressure of about 5 atm and the beverage is then capped. When the cap is removed, the pressure above the solution drops to 1 atm, and the gas escapes in the form of bubbles from the now supersaturated solution. The release of bubbles of gas from a liquid is known as *effervescence.*

The extreme solubility of some gases in water makes them exceptions to Henry's law. The reason for their extreme solubility is that the gas and water interact, as is easily evidenced in some cases by the release of energy in the form of heat. Ammonia (NH_3) and sulfur dioxide (SO_2) are examples of very soluble gases. Let us examine the reactions of these gases with water:

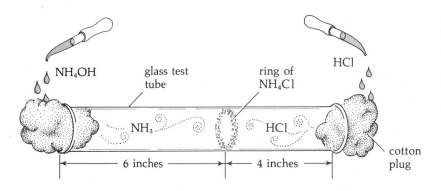

Figure 2–3. Graham's law. Ring of ammonium chloride represents point where two gases, HCl and NH_3, meet after diffusing from their respective cotton plugs.

$NH_3 + H_2O = NH_4OH$

$SO_2 + H_2O = H_2SO_3$

When ammonia and water interact, the product formed is ammonium hydroxide (NH_4OH); when sulfur dioxide and water interact; the product formed is sulfurous acid (H_2SO_3). Since a great deal of water is removed by the chemical reaction in each case, we can easily see why Henry's law does not apply to the very soluble gases. In the case of ammonia and water, almost half of the water enters into reaction, and what is left is actually a concentrated solution of ammonium hydroxide.

Chemical potential gradient. In general, the steeper the chemical potential gradient, the faster the rate of diffusion. The steepness of the gradient is controlled by the difference in concentrations of substance available for diffusion between one area and another and the connecting distance between these two areas through which diffusion will occur. In fact, any factor that will increase or decrease the chemical potential gradient (e.g., concentration, pressure, temperature) will influence the rate of diffusion.

The factors that control the rates of diffusion of gases largely control the rates of diffusion of liquids and solids as well. However, in addition to temperature, molecular density, diffusion medium, and chemical potential gradient, other factors (in particular the size and solubility of the diffusing molecule) affect the diffusion of solutes in solvents, liquids in liquids, and gases in liquids.

Water: Structure, Properties, and Interactions

To understand the different physiological processes related to the diffusion of water,
we must review the fundamental chemical and physical properties of water and its interaction with other substances. Water is a substance that may justifiably be called the fluid of life. It comprises over 90 percent of the chemical content of many organisms, and it participates either directly or indirectly in all metabolic reactions. Water is a remarkable compound with unique properties that result from its molecular configuration and hydrogen bonding.

Molecular Structure and Hydrogen Bonding

The water molecule is composed of two hydrogen atoms covalently bonded to one side of an oxygen atom. Since the mean angle between the hydrogens (105°) is not rigid, water can absorb large quantities of heat and be subject to other physical stresses without breakage of the bonds. Water is a *polar* molecule and as with other polar molecules it has a *surface charge*. Obviously, water is a *dipolar* substance in that the hydrogen pole is positively charged and the other pole is negatively charged due to the electrophilic (electron-attracting) properties of oxygen. Because of the asymmetrical distribution, water molecules associate with each other (*cohesion*) and "wet" other substances (*adhesion*). It is precisely these interactions that are critical to the movement of water in soils and the translocation of water in plants.

The attraction of the positive hydrogen atoms of one water molecule for the negative oxygen atom of another water molecule results in a *hydrogen bond*. Although a hydrogen bond is stronger than an association of molecules through van der Waals forces, it is considerably weaker than a covalent or electrovalent bond. There is virtually no limit, however, to the number of water molecules that can associate through hydrogen bonding. To use a fanciful example, a lake may be thought of as a loosely associated, gigantic molecule rather than as an accumulation of discrete water molecules.

Properties of Water Important to Plants

Hydrogen bonding of water produces a dipolar molecule and favors the formation of a latticelike structure that enables the packing of many atoms into a small area and stabilizes the molecular structure of water. Also, the presence of hydrogen bonds is directly responsible for the high heat of fusion, the high specific heat, and the high heat of vaporization of water. The energy required to break hydrogen bonds for the melting of ice, for the heating of water, and for the evaporation of water is considerably greater than the energy needed to overcome the van der Waals forces that are normally found in the weak association of molecules of ethane, ether, and benzene. Hydrogen bonding is responsible for the adherence of water molecules to such substances as glass, cellulose (cell walls), and clay micelles. These materials are readily wet because the water molecules have easy access to exposed oxygen atoms on the surfaces and can readily form hydrogen bonds. On the other hand, water-repellent fabrics and hydrocarbons, such as waxes, do not wet easily because very little hydrogen bonding can occur. Water is also a liquid at room temperature (25°C) and is lighter as a solid than as a liquid because of the hydrogen bonding. Have you ever thought of how ice forms on a lake (from top to bottom) and how that manner of ice formation is related to the survival of the organisms present in the water?

A property of water that is very important to the living cell is its solvent action. Because it forms a solution with a vast array of compounds, water is sometimes referred to as the *universal solvent*. The solvent action of water is an effect of its ability to form hydrogen bonds due to the asymmetrical distribution of its charges. In solution with water, such compounds as sugars, alcohols, and amino acids—which contain oxygen atoms, hydroxyl (—OH) groups, or amino (—NH_2) groups—form hydrogen bonds with the molecules of water. The polar nature of the water molecule causes salts to be held in solution through charge interaction (*ionization*); salts dissolved in water exist in the form of positive and negative ions.

The solvent action of water is of tremendous importance for the living plant. The essential elements necessary for normal plant growth, the compounds necessary for energy transfer and storage, and the components of structural compounds all require water as a translocation and reaction medium. These materials are dissolved in water and, in this form, distributed throughout the plant. The processes of diffusion, osmosis, and imbibition are intimately associated with the essential function of the translocation of water and solutes from site of origin to site of activity. Indeed, physiological processes operate in dilute aqueous solutions and suspensions; and the reactions are, therefore, under control of the physical and chemical laws that govern the activities of dilute solutions and suspensions.

Solutions

When we stir sucrose (common table sugar) into a glass of water, a clear solution of sugar in water results. We can distinguish two components of this system, a *sol-

ute (sucrose) and a *solvent* (water). The solute is dissolved in the solvent, thus an association exists between the two. In this and any other solution, the molecules of solute are evenly dispersed throughout the solvent, and the resulting solution is a homogeneous mixture of solute and solvent molecules. Although the molecules of the solute and solvent are constantly in motion, their movement is random. But keep in mind that the total translational kinetic energy (energy of motion) of the solvent molecules within the entire solution will be less than that of the same amount of pure solvent because of the solvent-solute association at any given time. The solute does not settle out but remains evenly dispersed at the expense of kinetic energy of the solvent molecules. Anyone who has ever mixed some solutions in a test tube will recognize that energy changes are involved in the solution process as evidenced by the spontaneous heating or cooling of the test tube.

When we add only a small amount of solute to a solvent, a dilute solution is formed. To make the solution more concentrated, we add more solute. At a given temperature and pressure, only a certain amount of solute can form a solution with a given amount of solvent. When this amount of solute is present, the solution is said to be saturated.

Now suppose we stir a small amount of an ionic substance, such as sodium chloride (common table salt), into water. Although a solution is formed, its formation is slightly different from that of the solution of sucrose and water. Sucrose is a nonionic substance and therefore remains intact in water. Sodium chloride (NaCl), however, is an ionic substance and undergoes *ionization* when placed in water—that is, the sodium chloride molecule dissociates to form sodium and chloride ions.

To illustrate a point about the energetics of solutions, let us assume that a salt dissociates 100 percent in water. Then twice as many water molecules will be involved in the solution process as are required when an equivalent amount of sugar (which is nondissociating) is placed in solution. In addition, because solvent molecules now interact with two particles for every one introduced, more energy, derived from the kinetic energy of the solvent molecules, will be required proportionately for the solution process. Understanding this simplistic concept of solutions is essential for grasping, the mechanics of the diffusion (osmosis and imbibition) of water in biological systems.

Diffusion of Water: Osmosis and Imbibition

Although the two forms of diffusion are similar, osmosis and imbibition are unique phenomena and play their own individual roles in plant development. *Osmosis* may be thought of as a special type of diffusion, the movement of water through a differentially permeable membrane. Although we can include solvents other than water within the general definition of osmosis, our primary concern is the osmosis of water in plants. *Imbibition* is a special type of diffusion in which an adsorbent is involved.

Osmotic Potential

The osmotic process can be demonstrated and measured by a very simple apparatus termed an *osmometer* in which two compartments are separated by a differentially permeable membrane. Let us assume that water, but not dissolved solutes such as sucrose, can pass through the membrane.

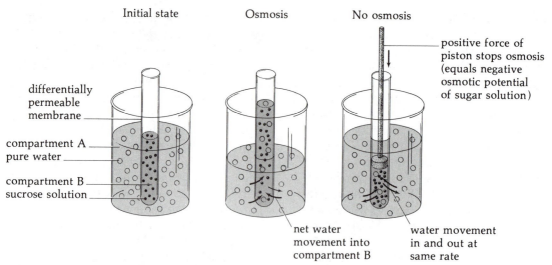

Figure 2–4. Osmometer filled with pure water in compartment A and sugar solution in compartment B.

We place pure water in compartment A and a solution of sucrose into compartment B (Figure 2–4). The pure water is said to be *hypotonic* (has lower tonicity, or less solute) to the sucrose solution. Conversely, the sugar solution is *hypertonic* (has higher tonicity, or more solute) to the pure water in compartment A. Since the membrane is permeable to water, water will move in and out of each compartment. However, initially the rate of movement of water moving into compartment B will be greater than the amount moving out because the chemical potential of the pure water is higher with higher translational kinetic energy (more molecules moving) than that of the sucrose solution. In the sucrose solution some of the water is interacting with solute particles, thereby effectively reducing the number of free water molecules and hence decreasing the total translational kinetic energy. Under these circumstances, water will build up in compartment B. As water accumulates, the

sucrose solution in compartment B becomes more and more diluted, and the rate of water moving into compartment B slows down accordingly. There is less and less difference between the chemical potential of the pure water and that of the solution as the process proceeds.

Let us assume we are able to introduce a piston into compartment B and, by applying force, stop water from entering compartment B. The applied force required would be equal to the maximum pressure that might build up if water moved into this particular sugar solution in a confined system. The pressure that a solution would have to build up to increase its chemical potential to that of pure water is termed *osmotic pressure*. Osmotic pressure for a solution is the pressure (energy that is depleted by the solution process) that would have to be applied to stop the diffusion of pure water into the solution under ideal osmotic conditions. Thus osmotic pressure is actually a potential and is not usually reached or measured in plant cells. It is a measure of the absence of energy to do work or capacity to flow in an ideal osmotic situation.

For example, a 1 molal solution of an undissociated substance in a beaker at 0°C may be designated as having an osmotic pressure of 22.4 atm or 22.7 bars. The solution is not exerting pressure but has less energy than pure water, the amount of which depends upon the amount of solute in a given volume of water. The energy lost during the solution process may be restored by the application of external energy via the piston in the osmometer or the influx of water into a confined system such as a plant cell. Thus plant scientists use the term *osmotic potential*, usually designated by ψ_s, to describe the absence of energy in a solution, due to the amount of solvent-solute interactions, as compared with pure water under ideal osmotic conditions. In returning to Gibbs free energy relationships, we can justify the use of a negative sign for osmotic potential value because the solvation process is characterized by:

$$G_2 - G_1 = -\Delta G$$

Where:

G_2 = situation after solvation

G_1 = situation before solvation

Thus a molal sucrose solution at 0°C has an osmotic potential of -22.4 atm or -22.7 bars. These actual values were obtained originally by van't Hoff, who applied the gas laws equation to solutions and calculated osmotic pressures of solutions accordingly:

$$\Pi = \frac{N}{V} \times RT \quad \text{or} \quad \Pi = CRT$$

Where:

Π = osmotic potential

N = number of moles

V = volume in liters

R = gas constant

T = absolute temperature

$C = \dfrac{N}{V}$ = concentration

The negative sign is inserted to denote osmotic potential and to be consistent with the laws of thermodynamics.

The importance of the osmotic potential is that it characterizes a solution in several ways. It indicates the maximum pressure (osmotic pressure) that might develop if the solution were allowed to come to equilibrium with pure water in an ideal osmotic system, and it is proportionately related to the amount of solute in a solution and to the decrease in chemical potential (total free energy) due to solvent-solute interactions.

Turgor Pressure

A rigid, relatively inelastic structure—the cell wall—encloses the plant cell and its differentially permeable plasmalemma. This unique feature of the plant cell allows it to survive in a relatively wide range of osmotic concentrations. In contrast, the animal cell can live only in solutions in which the osmotic concentrations are identical (*isotonic*) or nearly identical to that of the cell contents.

The plant cell, when placed in pure water, swells but does not burst. Because of the negative osmotic potential of the vacuolar solution (cell sap), water will move into the cell and will cause the plasmalemma to be pressed against the cell wall. The actual pressure that develops—that is, the pressure responsible for pushing the membrane against the cell wall—is termed *turgor pressure*. The cell wall, being rigid, exerts an equal and opposite pressure, which we will call *wall pressure*. As a result of this interplay of forces, the plant cell under these conditions is said to be turgid. The first, easily observed sign of a water deficit in a plant is a decrease in the turgor of its leaf cells, which gives the leaves a wilted appearance.

Water Potential

The *chemical potential* is the free energy per mole of any substance in a chemical system. Therefore, the chemical potential of a substance under conditions of constant pressure and temperature depends on the number of moles of substance that is present. In discussing plant-water relations, we generally refer to the chemical potential of water as *water potential* (ψ_w). When we use the term water potential, we are expressing the difference between the chemical potential of water at any point in a system (μ_w) and that of pure water under standard conditions (μ_w°). With the formula:

$$\psi_w = \mu_w - \mu_w^\circ = RT \ln \frac{e}{e^\circ}$$

we can readily determine water potential. In the formula, R is the gas constant (erg/mole/degree), T is the absolute temperature (°K), e the vapor pressure of the solution in the system at temperature T, and e° the vapor pressure of pure water at the same temperature. The expression $RT \ln (e/e^\circ)$ is zero. Knowing this, we can say that pure water has a potential of zero. In biological systems, however, (e/e°) is generally less than zero, making $\ln (e/e^\circ)$ negative. Conse-

quently, the water potential of biological systems is usually expressed as a negative quantity. Since pure, unconfined water is defined as having a potential of zero, any dilution of water with a solute establishes a potential that is less than that of pure water and is expressed as a negative number. Further, the negative number is consistent with the Gibbs free energy difference between pure water and solutions.

We can express both water potentials and chemical potentials in energy units. However, it is more convenient, when dealing with biological systems, to express water potentials in pressure units (atmospheres or bars). We can make conversion from energy units to pressure units by dividing the water potential by the partial molal volume of water (V_w):

$$\frac{\mu_w - \mu_w^\circ}{V_w} = \frac{RT \ln \frac{e}{e^\circ}}{V_w}$$

The units of the above equation are:

$$\frac{erg/mole}{cm^3/mole} = \frac{erg}{cm^3} = dyne/cm^2$$

and

one bar = 0.987 atm = 10^6 dynes/cm^2

If we dissolve some substance such as sugar in pure water contained in a beaker, the resulting solution has an osmotic potential that is lower (more negative) than that of pure water. Since this is an unconfined solution (not under the pressure of a piston or cell wall, the turgor pressure is zero. Consequently, $\psi_s = \psi_w$ and the presence of solute decreases the free energy. What is important here is the proportion of solute particles to water molecules. An increase in solute will produce a more negative osmotic and hence water potential. If we were to construct the system to allow for the establishment of turgor pressure, then the amount of positive pressure generated would offset the effect of solute and make the water potential less negative than that of the osmotic potential.

If we subject both the solution and the pure water to the same pressure, the effect of the imposed pressure is quantitatively the same for both systems. For example, if we subject both systems to a pressure (as in an osmometer) of 6 bars, then the water potential for both systems will become less negative by 6 bars. In fact, pure water would develop a positive water potential.

Relationship of Osmotic Quantities

The following hypothetical situation should help to clarify the relationship of water potential, osmotic potential, and turgor pressure. Solution B, with a water potential (ψ_w) equal to -30 bars, an osmotic potential (ψ_s) of -30 bars, and zero turgor pressure is enclosed by an inelastic membrane that is permeable only to water. Because there is no turgor pressure in this system, the water potential is equal to the osmotic potential ($\psi_w - \psi_s$). This system is submerged in solution A, with an osmotic potential of -10 bars (Figure 2–5). The turgor pressure of solution A is zero because the solution is unconfined. Therefore, the water potential and osmotic potential are equal. As can be seen in Figure 2–5, the water potential of solution A is less negative than the water potential of solution B. Consequently, an energy gradient exists from solution A to B, and a net flow of water will take place from solution A to B or from the solution of less

negative to more negative water potential. Another way we can express the net movement of water in this example is to say that water will move along a free energy gradient or energetically downhill.

Since solution B is enclosed in an inelastic membrane, the turgor pressure will build up, and equilibrium will be reached between the two systems with the entrance of only a small amount of water into the internal solution. The actual turgor pressure (ψ_p) developed in the internal solution will amount to 20 bars and will counteract the -20 bars of the osmotic potential. The wall pressure at this point will also be 20 bars. Since the water potential of the solution in B will be made less negative by the amount of pressure it is subjected to, the water potential of the internal solution should become less negative by 20 bars, thus equaling the water potential of the external solution. Here we can make the general statement that when two aqueous solutions of different osmotic potentials are separated by a membrane permeable to water only, the water potentials will tend to equilibrate. With the above discussion in mind, we can thus state:

$$\psi_w = \psi_s + \psi_p$$

One assumption that we are making here, however, is that the osmotic potential does not change, an assumption based on the observation that in cells or solutions confined by a relatively inelastic membrane or wall, the water gain or loss is not sufficient to dilute or concentrate the solution and lower or raise the osmotic potential. The reverse is true of the turgor pressure, it is affected by slight changes in the solution concentration.

In actuality as water passes through the membrane by osmosis and into a cell, it often encounters some resistance from

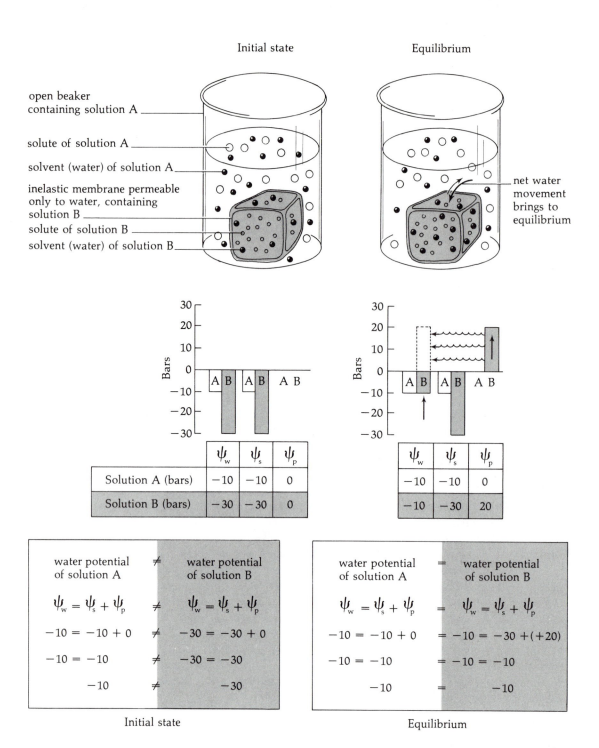

Figure 2–5. Relationship of water potential, osmotic potential, and pressure potential (turgor pressure). Water potentials of solutions A and B are the same at equilibrium.

other substances, a factor that contributes to the *matric potential* (ψ_m). The matric potential may be defined as the energy lost (with respect to pure water) as water diffuses and interacts with other substances in the diffusion medium.

The relationship between all osmotic quantities, then, is:

$$\psi_w = \psi_s + \psi_m + \psi_p$$

Since the ψ_m is not appreciable and difficult to measure in osmotic systems, we consider it to be negligible when solving problems of osmosis in plant cells. As we shall see later, the matric potential is important to the imbibition process.

From the above equations, we can see that when the turgor pressure (ψ_p) equals (in number but not in sign) the osmotic potential (ψ_s) of a solution, the water potential of that solution equals zero.

If an aqueous solution with an osmotic potential of -10 bars is enclosed in an inelastic membrane and submerged in pure water ($\psi_w = 0$), a turgor pressure of 10 bars will be reached in the internal solution when the two systems reach equilibrium— that is, at equilibrium the water potential of the internal solution will be zero.

Initial state: $-10 = -10 + 0$

$$\psi_w = \psi_s + \psi_p$$

Equilibrium: $0 = -10 + 10$

In the examples given so far, we have been using hypothetical situations in which the solution is enclosed by an inelastic membrane. However, the cell wall of the plant cell is elastic to some degree, and a certain increase in volume is realized when a flaccid cell becomes fully turgid. With an increase in volume, there is a dilution of the cell sap, thus a raising of the osmotic po-

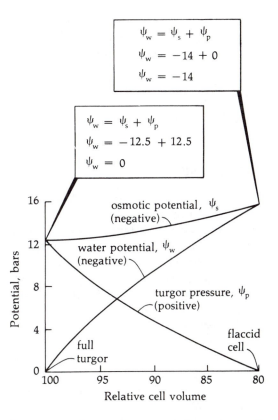

Figure 2–6. Changes that occur when plant cell takes up water. When osmotic potential and turgor pressure are equal in magnitude but different in sign, water potential of cell sap is zero.

tential of the sap. However, the equation $\psi_w = \psi_s + \psi_p$ is still accurate because there is an equating of water potentials. Figure 2–6 shows the changes that occur when a plant cell takes up water. In a flaccid cell ($\psi_p = 0$), the osmotic potential of the cell sap is equal to its water potential. If this cell is placed in pure water, water moves into the cell, causes an increase in turgor pressure, and thus causes a certain amount of elastic stretching of the cell wall. With an increase in cell volume (caused by stretching of the cell wall), there is a dilution and consequent rise in the osmotic potential of the cell sap. At the point where the osmotic potential equals but is opposite in sign to the turgor pressure, and the water potential equals zero, the cell is said to be fully turgid. There is no further increase in the volume of the cell at this point.

Plasmolysis

When we place a living plant cell in a solution with an osmotic potential identical to that of its own cell sap (an *isotonic solution*), the appearance of the cell remains normal in every respect. However, if the water potential (ψ_w) of the surrounding solution is less negative than that of the cell sap (*hypotonic*) or more negative than that of the cell sap (*hypertonic*), we can easily observe several changes in cell structure. For example, if we immerse epidermal tissue from the leaves of *Rhoeo* or *Zebrina* in a hypertonic solution of sucrose, we can observe the plasmalemma pulling away from the cell wall. We can see this easily because of the pigmentation of the vacuolar contents of the leaf cells of these plants.

Let us examine in a little more detail what happens in this situation. First, the water inside the cell has a greater free energy and thus a greater tendency to flow outward. Second, the cell and vacuolar membranes are practically impermeable to sucrose but readily permeable to water. Third, the cell wall will allow the free passage of both sucrose and water. Thus there will be a net movement of water out of the cell vacuole and into the external solution; water will move from a region of less negative (high) to a region of more negative (low) water potential. This movement of water results in a loss of turgor, a shrinking of the vacuole, and a pulling away of the cell

membrane from the cell wall. *Incipient plasmolysis* is the initial pulling away of the membrane from the cell wall. At this point the turgor pressure is zero. If the process continues, there will be a tendency for the cell wall to be pulled toward the cytoplasm because of the cohesive and adhesive properties of the water between the cell wall and plasmalemma. This cell is then said to be under tension, and the turgor pressure becomes negative. Eventually, the forces exerted by the retracting plasmalemma will become greater than those between the water molecules of the wall. Complete plasmolysis follows, with the plasmalemma being pulled entirely away from the wall. If the plasmolysis is not extensive, however, plasmolyzed cells usually can be *deplasmolyzed*—that is, if a cell that has been plasmolyzed is placed in a hypotonic solution, it will regain its turgidity.

A different condition develops if a living plant cell is placed in a solution that is hypotonic to the cell sap. In this situation, water moving from a region of less negative water potential (the external solution) to one of more negative water potential (the cell sap) will enter the cell and cause it to become more turgid. Since the cell wall is elastic to some degree, the cell volume will increase slightly. The turgor pressure of the cell will also, of course, increase. Because the increase in volume of the cell in a hypotonic solution is generally very small, it is difficult to observe any differences in appearance between a plant cell in an isotonic solution and a plant cell in a hypotonic solution.

Osmosis between Cells

Let us assume two cells adjacent to one another are protected from any negligible evaporation. The cell sap of cell A has an osmotic potential of −14 bars and turgor pressure of 4 bars. Cell B has a water potential of −16 bars and osmotic potential of −24 bars. The final situation for both cells is determined on the basis of $\psi_w = \psi_s + \psi_p$ and is summarized in Figure 2–7.

The important point is that since both cell solutions are confined, the water potentials of each will tend to come to equilibrium, with changes in turgor pressure. Water will then flow from cell A to cell B, or from a cell solution of a ψ_w of −10 bars to that of a −16 bars. In problems of this kind, we assume that the volume change is not sufficient to change the osmotic potential. Even though not entirely accurate, we can use an approximate calculation of the osmotic condition between cells to predict the direction of osmosis.

Osmotic Potential Measurements

The *boiling point* of an aqueous solution is higher than that of pure water, the *vapor pressure* of the water in a solution is lower than that of pure water, and a solution freezes at a lower temperature (*freezing point depression*) than pure water. These factors, called the *colligative properties* of solutions, are interrelated, and the extent to which each factor is affected is directly proportional to the number of dissolved particles (molecules or ions) present. Therefore, a measure of any one of these factors is an indirect measure of the osmotic potential because it is also one of the colligative properties of solutions. We do not generally use boiling point elevation to measure the osmotic potential of the cell sap. However, we can measure the vapor pressure depression and freezing point depression of expressed

plant juices with a considerable degree of accuracy. For example, the theoretical freezing point depression of a 1 molal solution composed of nonionized solute has a freezing point depression of −1.86°C and a theoretical osmotic potential of −22.7 bars (−22.4 atm). We can easily arrive at an equation relating these two factors—freezing point depression and osmotic potential—and we can use this equation to determine the osmotic potential of a solution of unknown concentrations. Thus,

$$\psi_s = \frac{-22.7 \times \Delta_{fp}}{-1.86}$$

In this equation, Δ stands for the observed freezing point depression of the unknown solution. If, for example, some plant juice is expressed and found to have a freezing point depression of 1.395, the osmotic potential of this solution would be:

$$\psi_s = \frac{-22.7 \times -1.395}{1.86} = -17.025 \text{ bars}$$

The determination of a solution's osmotic potential by determination of its freezing point is termed *cryoscopy*, and the technique is referred to as the *cryoscopic* method.

A less strenuous method of determining the osmotic potential of cell contents makes use of the *plasmolytic phenomenon*. A graded series of solutions, covering a certain range of osmotic potentials (water potentials) is prepared; the solutions, which are usually of sucroses, are prepared so as to provide a graded series in which some of the solutions are hypotonic and others are hypertonic to the cells to be treated. Strips of plant tissue, preferably tissue containing anthocyanins, are placed in the different solutions and after a time (approximately 30 minutes) are placed under the microscope.

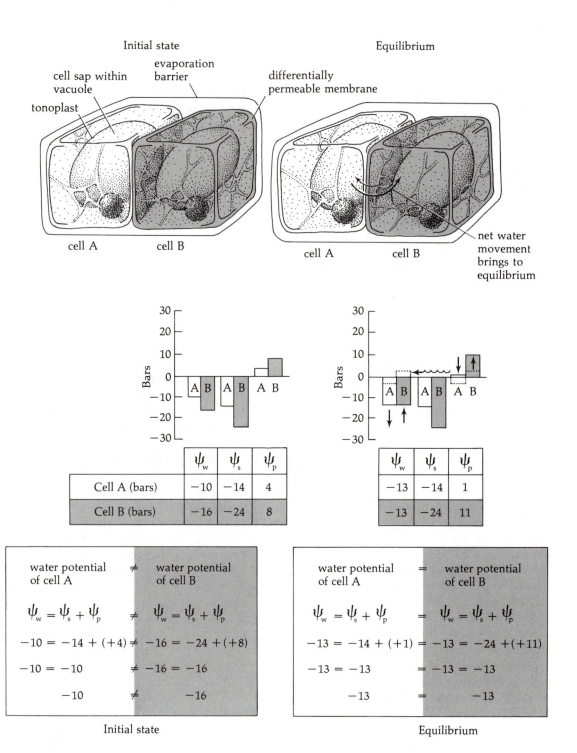

Figure 2–7. Hypothetical relationship of water potentials, osmotic potentials, and pressure potentials between adjacent cells before and after equilibrium. Assumptions are that cells do not dry out and that there is no appreciable effect of volume change directly on osmotic potentials.

Examination of the strips of tissue from the different solutions will show some in which all of the cells are turgid, some in which nearly all of the cells are plasmolyzed, and some in which about 50 percent of the cells are just beginning to show signs of plasmolysis (incipient plasmolysis). At incipient plasmolysis, the turgor pressure of the cell is zero, and the osmotic potential of the cell contents is equal to the water potential of the cell and to the water and osmotic potentials of the external solution.

Water Potential Measurements

Water potential, the sum of all osmotic quantities, is the major determinant of osmosis in plants and the easiest of the osmotic quantities to measure. We will consider now the most common methods used to determine the water potential of cells and organs of plants.

Volume Method

The volume method of measuring water potential is based on changes in linear dimensions (length) of a tissue when it is placed in solutions of different osmotic potentials. When solutions are placed in a beaker, there is no turgor pressure; consequently, the solutions are unconfined, and $\psi_w = \psi_s$. This situation is not true for plant cells. Strips of root, fruit, or leaf tissue, 3 to 4 cm long and of the same width, are measured and placed in a series of different concentrations of sucrose solutions for about 1 hour. The tissues are removed and remeasured. The change in length is then plotted against the known osmotic potentials of the solution. The water potential of the solution $\psi_w = \psi_s + 0$ in which the tissue

does not change in length is the same as the water potential ($\psi_w = \psi_s + ?$) of the tissue.

Gravimetric Method

This method, which is essentially the same as the volume method, involves the placement of preweighed plant tissue (potato tuber cylinders, for example) into a graded series of solutions of sucrose or other *osmoticum* (osmotically active solute) at known osmotic potential ($\psi_w = \psi_s + \psi_p$; $\psi_p = 0$) (see Figure 2–8).

A representative sampling of tissues is incubated for a predetermined time in the solutions, removed, and reweighed. The

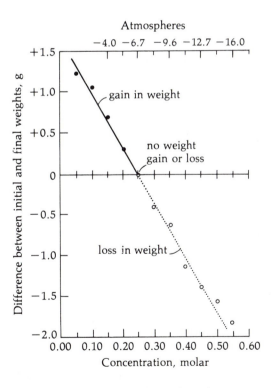

Figure 2–8. Hypothetical results of measurements of water potential of potato cylinders.

weight gain or loss is plotted against the water potential ($\psi_w = \psi_s$) of each solution. When the points are connected, the intercept at the abscissa (through zero) represents the water potential of the tissue, with zero weight gain or loss. The water potential of the solution corresponding to the intercept point is equal to that of the tissue.

Chardakov's, or Falling Drop, Method

Two graded series of sucrose solutions (ranging from 0.15 to 0.50 molal in increments of 0.5 molality) are placed in test tubes set up in duplicate (see Figure 2–9). Homogeneous plant tissue is placed into each test tube of one of the series (test series). Only a drop of methylene blue is mixed into each solution of the second series (control series). Plant tissue is not added to the control series and the dye does not appreciably change the osmotic potentials.

After the tissue has incubated 15 to 30 minutes, it is removed from each tube. The actual time of incubation can be just long enough for osmosis to proceed and change the concentration of each solution in the test series; the attainment of equilibrium is not necessary. After the tissue is removed, a small drop of the respective control series solution is introduced below the surface of its corresponding test solution. If the drop rises in the test solution, it means that the drop is lighter and that the tissue incubation solution is more concentrated—an indication that water from the solution entered the tissue. Conversely, if the drop falls, it means that the test solution is lighter—an indication that water has left the tissue and diluted the solution. In this latter instance, the water potential of the solution initially is more negative than that of the tissue. Ac-

cordingly, if the density of the drop from the methylene blue solution is the same as that of the test solution, the drop will diffuse into the solution uniformly. At this point (called the null point), the water potential of the tissue and solution is equal.

It is also possible to determine the solution changes with a refractometer (*refractometer method*) instead of the falling drop. The refractometer is used to measure directly the concentration changes that take place in the tissue incubation solutions. No change in concentration indicates, of course, that the solutions have the same water potential as that of the tissue cells. This method does not require the methylene blue series of solutions. Also, experimental error due to technique is minimized. A refractometer, however, is not always available.

Vapor Pressure (Thermocouple Psychrometer) Method

The vapor pressure method is based on the fact that tissue will not gain or lose water to the atmosphere when the vapor pressure of air corresponds to the water potential of the tissue. The most extensively used apparatus is constructed for measurements of humidity inside a closed chamber containing two thermocouple junctions. One remains at the temperature of the air in the chamber, the other cools rapidly when a weak current is passed through the two junctions. Moisture from the air in the chamber will eventually condense on the cooling thermocouple. The drop of moisture then acts as a "wet bulb." The water potential of the air in the chamber is equal to the difference between the temperature of the "wet bulb" and that of the dry thermocouple.

Step 1. Set up test and control series.

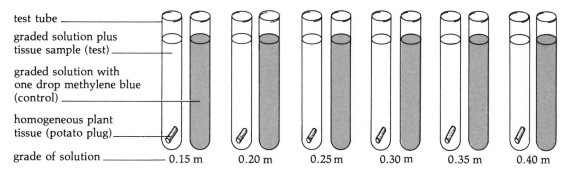

test tube

graded solution plus
tissue sample (test)

graded solution with
one drop methylene blue
(control)

homogeneous plant
tissue (potato plug)

grade of solution ———— 0.15 m   0.20 m   0.25 m   0.30 m   0.35 m   0.40 m

Step 2. Incubate series for 15 to 30 minutes.

Step 3. Remove tissue and introduce drop of control solution into test solution.

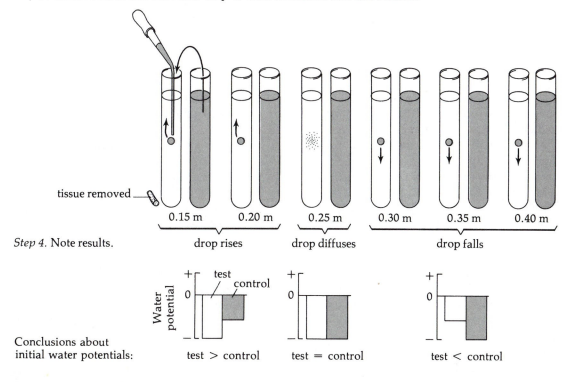

tissue removed

0.15 m   0.20 m   0.25 m   0.30 m   0.35 m   0.40 m

Step 4. Note results.

drop rises   drop diffuses   drop falls

Water potential

test / control

test > control   test = control   test < control

Conclusions about
initial water potentials:

Figure 2–9. Chardakov's, or falling drop, method. Drop rising in test solution indicates test solution became more dense than its counterpart. Water diffused out of test solution into tissue, which initially had a more negative ψ_ω.

Reverse is true when drop descends in test solution. Diffusion of drop in test solution indicates no change, hence water potentials were equal initially.

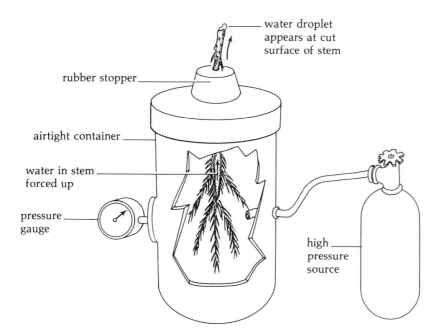

water droplet
appears at cut
surface of stem

rubber stopper

airtight container

water in stem
forced up

pressure
gauge

high
pressure
source

Figure 2–10. Pressure bomb is used to measure water potentials or plant moisture stress (PMS). When stem of shoot is cut, water (under tension) recedes from cut surface. Pressure required for drop to appear at cut surface is equal to average water potentials of leaves.

A leaf disk is removed from the leaf and equilibrated in the chamber for approximately 20 minutes. The relative humidity of the equilibrated chamber is determined, and the chamber temperature and dry bulb minus wet bulb temperature are recorded. The reading is corrected to 25°C, and the water potential determined from a calibration chart. A calibration chart is needed for each psychrometer chamber and is usually compiled by the taking of measurements of distilled water and 1 molal NaCl. Each reading is corrected to 25°C, and the results are plotted as the reading in microvolts on the abscissa and water potential on the ordinate. For example, distilled water equals 0 water potential and 1 molal NcCl equals −46.4 bars at 25°C.

Pressure Bomb

The pressure bomb is a device that is used to determine the plant moisture stress

and the water potential of a leafy shoot and is based on the assumption that the water column in a plant is almost always under tension because of the pull exerted by the osmotic influences (water potential) of the living cells of the leaves (see Figure 2–10). If the tension is high, the water potential of the leaf cells is very negative. When a stem is cut, the water column is disrupted, and because the water column is under tension, it will recede back into the stem toward the leaves. The shoot is placed in a chamber, with the cut end protruding through an airtight hole. Pressure is increased within the chamber, and the water column within the twig is forced back to the cut surface. The pressure in the chamber is then carefully recorded.

The pressure required to force the water to appear at the cut surface is equal to the tension (but with the opposite sign) of the water column at the time the shoot was cut. If low pressure is sufficient to force water to the cut surface of the shoot, the living cells, primarily of the leaves, have slightly negative water potentials, with the shoot being under relatively low moisture stress. But if high pressure is required to force water to the cut surface the moisture stress (tension) is relatively high due to very negative water potentials of the leaf cells.

Imbibition

Another diffusion process of water in plants is termed *imbibition*. As with osmosis, imbibition may be considered a special type of diffusion since the net movement of water is along a diffusion gradient. In the case of imbibition, however, an adsorbent is involved. If dry plant material is placed in water, a noticeable swelling takes place and sometimes amounts to a considerable increase in volume. Anyone who has experienced a sticking wooden door or window frame during a prolonged period of humid weather has encountered the phenomenon of imbibition; dry wood is a particularly good adsorbent.

Tremendous pressures can develop if an adsorbent is confined and then allowed to imbibe water. For example, dry wooden stakes, driven into a small crack in a rock and then soaked, can develop enough pressure to split the rock. Indeed, this was a form of quarrying used to great advantage in the past.

Conditions Necessary for Imbibition

Two conditions appear to be requisite for imbibition to occur: (1) A water potential gradient must exist between the surface of the adsorbent and the liquid imbibed, and (2) a certain affinity must exist between components of the adsorbent and the imbibed substance.

Very negative water potentials exist in dry plant materials. For example, water potentials of −900 bars have been recorded in some dry seeds. Therefore, when this material is placed in pure water, a steep water potential gradient is established, and water moved rapidly to the surface of the adsorbent. As water continues to be adsorbed, the water potential becomes less negative until, theoretically, it finally equals that of the external water. At this point an equilibrium is established, imbibition ceases, and water moves to and from the adsorbent in equal quantity.

An adsorbent does not necessarily imbibe all kinds of liquids. For example, dry plant materials immersed in ether do not swell appreciably. Rubber, however, does imbibe ether and will swell appreciably if submerged in it. But rubber will not imbibe water. The obvious implication is that certain attractive forces must exist between components of the imbibant and the imbibed substance.

A considerable amount of colloidal material is present in both living and dead plant cells (see Appendix A). Proteins and polypeptides are hydrophilic colloids—that is, they have a strong attraction for water. In addition, plant cells possess a considerable amount of carbohydrate, in the form of cellulose and starch, to which water is strongly attracted. The adsorption of water to the surfaces of these hydrophilic colloids is of major importance to the imbibition process. Seeds, which are particularly high in colloidal material, are very good adsorbents. Indeed, water is brought in to the germinating seed largely through the process of imbibition. The water potential of a biological system is made more negative by the presence of these adsorptive, or water-binding, materials. To these materials or to the forces they generate, the term matric potential (ψ_m) is applied. In discussion of plant-water relations, the term *matric potential* has replaced the old term *imbibition pressure* and is somewhat analogous to osmotic potential. As might be expected, the water potential of dry plant materials such as seeds is quite negative.

Matric Potential

Matric potential is analogous to osmotic potential in that it represents the potential maximum pressure that an adsorbent will develop if submerged in pure water (4). The actual pressure that develops when water is imbibed may be thought of as turgor pressure (*pressure potential*). With these facts in mind, we can present the following equation:

$$\psi_w = \psi_m + \psi_p$$

This equation is, of course, similar to the one used for osmotic systems, where water potential is equal to osmotic potential plus turgor pressure. Remember that matric potential is always negative. No turgor pressure develops in an unconfined adsorbent, and the above equation under these conditions simplifies to:

$$\psi_w = \psi_m$$

The matric potential of air-dried seeds, such as cocklebur, may approach −1000 bars (5, 6). If we immerse seeds such as these in pure water, the water potential of the very small amount of water in the dry seeds would be nearly −1000 bars. After imbibition ceases, the water potential of the

external and internal water is zero. On the other hand, if we submerge seeds containing water with a water potential of −500 bars in a solution of NaCl with an osmotic potential of −50 bars (water potential equals −50 bars), the water potential of the seed water at equilibrium will be −50 bars. As in osmotic systems, the water potentials tend to equilibrate.

Factors Affecting Rate and Extent of Imbibition

The rate and extent of imbibition is affected primarily by temperature and by the osmotic potential of the substance to be imbibed. Temperature does not affect the amount of water taken up by the adsorbent, but it does have a definite effect on the rate of imbibition; an increase in temperature causes an increase in the rate of imbibition (see Figure 2–11).

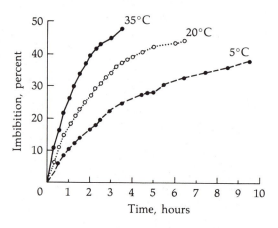

Figure 2–11. Rate of imbibition by *Xanthium* seeds at different temperatures.
Reprinted from Botanical Gazette 69, figure 3–7 by C.A. Shull, by permission of The University of Chicago Press. Copyright 1920. The University of Chicago Press.

Both the amount of water imbibed and the rate of imbibition are affected by the osmotic potential of the substance to be imbibed. The addition of a solute to pure water causes a more negative water potential. This addition has the effect of altering the water potential gradient between the solution water and the adsorbent. The water potential gradient is less steep than it would be if the same adsorbent were submerged in pure water. Similarly, a decrease in the water potential gradient will bring about a decrease in the rate at which water is imbibed and thus the amount of water taken up. Some data by Schull (5) on the effects of osmotic potential in imbibition by air-dried cocklebur seeds are shown in Table 2–1.

Table 2–1. Imbibition by air-dried cocklebur seeds as affected by different osmotic potentials. Source: *From C.A. Schull. 1916. Measurement of the surface forces in soils.* Bot. Gaz. 62:1.

Molar Concentration	Water imbibed after 48 hrs (% of dry weight)	Osmotic Pressure (atm)
H_2O	51.58	0.0
0.1*M* NaCl	46.33	3.8
0.2*M* NaCl	45.52	7.6
0.3*M* NaCl	42.05	11.4
0.4*M* NaCl	40.27	15.2
0.5*M* NaCl	38.98	19.0
0.6*M* NaCl	35.18	22.8
0.7*M* NaCl	32.85	26.6
0.8*M* NaCl	31.12	30.4
0.9*M* NaCl	29.79	34.2
1.0*M* NaCl	26.73	38.0
2.0*M* NaCl	18.55	72.0
4.0*M* NaCl	11.76	130.0
Sat. NaCl	6.35	375.0
Sat. LiCl	−0.29	965.0

Volume and Energy Changes

The volume of an adsorbent increases as a result of imbibition. However, the total volume of the system (the volume of the water in which the adsorbent is submerged plus the volume of the adsorbent) is always less after imbibition than before imbibition starts. We can easily demonstrate this fact by placing air-dried seeds in a graduated cylinder containing water, reading the initial volume, and comparing it with the volume of the system after imbibition ceases. The reason for this difference in volume is that water molecules adsorbed to the surfaces of colloidal material present in the adsorbent are held relatively tightly. As a consequence, they are packed closer together, and the result is a decrease in the volume of the system.

As a result of the tight adsorption of the water molecules, some of the kinetic energy possessed by these molecules is lost. This energy loss shows up in the system as heat. Therefore, there is always an increase in temperature as a result of imbibition.

Questions

2–1. What primary chemical feature of the water molecules is responsible for many if not all of the physical and chemical properties of water? Explain.

2–2. List the properties of water and why they are important to plants.

2–3. Define the following terms: solute, solvent, ionization, and dissociation.

2–4. A closed, saclike membrane, completely filled by a solution with an osmotic potential of −27 bars, is immersed in a solution with an osmotic potential of −21 bars. Assume that the membrane is permeable to water only and that the osmotic potentials will not change with osmosis. What will be the water potential of the internal solution at equilibrium? The osmotic potential? The turgor pressure? Answer the same questions for an external solution with an osmotic potential of −16 bars and for one with an osmotic potential of −20 bars.

2–5. Which osmotic quantity is most important in determining the directions of osmosis? Explain.

2–6. Cell A has an osmotic potential of −15 bars and is immersed in a solution with an osmotic potential of −10 bars. Cell B has an osmotic potential of −8 bars and is immersed in a solution with an osmotic potential of −6 bars. Both cells are first allowed to come to equilibrium with the solution in which each is immersed (the volume of which is assumed to be large) and they are then removed and brought into intimate contact. Assume there will be very little evaporation. In which direction will water diffuse? Why?

2–7. If an NH_4Cl solution is injected directly into the cell sap of a plant cell, the sap becomes more acidic; but if the cell is immersed in the NH_4Cl solution, the sap becomes more basic. Explain.

2–8. A chain of cells, each of which has an osmotic potential of −10 bars, is arranged so that one terminal cell dips in a solution with an osmotic potential of −5 bars and the other in a solution with an osmotic potential of −8 bars. The volume of these solutions is very large in comparison to the size of the cells. Evaporation is prevented. Will any movement of water occur? Explain.

2–9. A cell has an osmotic potential of −12 bars. Water evaporates from it until the cell walls are pulled inward so that the cell is subjected to a tension of −4 bars. What then are the water potential, osmotic potential, and turgor pressure of the cell?

2–10. Cells A, B, and C, having osmotic potentials of −7, −11, and −5 bars, respectively, constitute a chain of three cells in the order named. A part of the lowest cell, C, dips into a solution with an osmotic potential of −3 bars. Neither of the other

two cells is in contact with the solution, which is large in volume in comparision with the cells. Evaporation from the cells is prevented. What will be the osmotic potential, water potential, and turgor pressure of each cell at equilibrium?

2–11. If all three of the cells (question 2–10) were completely immersed in the solution, what would be the water potential, osmotic potential, and turgor pressure of each cell at equilibrium?

2–12. What effect will a change of starch to sugar in a cell have on its water potential? What effect will an increase in the permeability of the cell membranes to solutes have on its water potential? What effect will an increase in the permeability of the cell membranes to water have on its water potential?

Suggested Readings

Bewley, J.D. 1979. Physiological aspects of desiccation tolerance. *Ann. Rev. Plant Physiol.* 30:195–238.

Brown, R.W., and B.P. Van Haversen, eds. 1971. Psychrometry in water relations research. *Proceedings of the Symposium on Thermocouple Psychrometers.* Agr. Exp. Sta., Utah State University.

Fischer, R.A., and N.C. Turner. 1978. Plant productivity in the arid and semiarid zones. *Ann. Rev. Plant Physiol.* 29:277–317.

Kozlowski, T.T., ed. 1968–1978. *Water Deficits and Plant Growth.* Vols. 1–5. New York: Academic Press.

Meidner, H., and D.W. Sheriff. 1976. *Water and Plants.* New York: Wiley.

Slatyer, R.O. 1967. *Plant-Water Relationship.* New York: Academic Press.

Sutcliffe, J. 1968. *Plants and Water.* London: Edward Arnold.

Weatherley, P.E. 1970. Some aspects of water relations. In R.D. Preston, ed., *Advances in Botanical Research.* New York: Academic Press.

Wiebe, H.H. 1971. *Measurement of Plant and Soil Water Status.* Bull. 484. Agr. Exp. Sta., Utah State University.

Chapter 3

Absorption and Translocation of Water

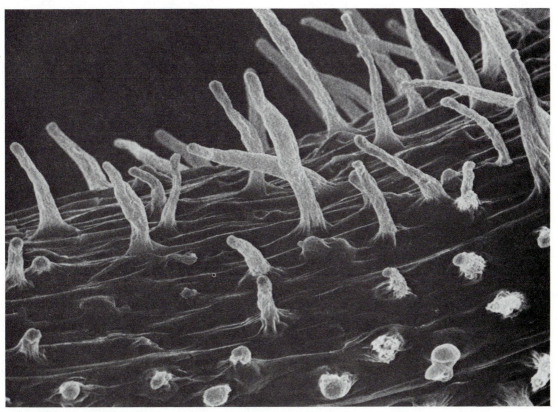

Scanning electron micrograph of root hairs of radish (*Raphanus sativus*). *From J.N.A. Lott. 1976.* A Scanning Electron Microscope Study of Green Plants. *St. Louis, Mo.: Mosby. Courtesy of J.N.A. Lott, McMaster University.*

During the plant's life cycle, large amounts of water are continuously absorbed from the soil and moved (translocated) through the plant. It is ironic that so much water passes through the plant but is lost to the atmosphere by transpiration. Nevertheless, a small portion is used for physiological processes.

In this chapter we shall study the absorption and translocation of water within the plant system. Although it is relatively simple to account for the rise of water in the shorter herbaceous and shrubbery plants, the problem becomes much more complex when we try to explain how water reaches the tips of the tallest trees. As we shall see, the process is reasonably understood in that water translocation depends on the adhesive and cohesive properties of water and the difference in water potentials from the soil all the way through the leaves and into the atmosphere. Although, physically, water runs uphill to the tops of trees (sometimes approaching a height of 400 feet), energetically, it follows a downhill gradient. The establishment of such a water potential gradient is influenced by several factors, particularly those of the soil and the atmosphere.

Soil Factors Affecting Absorption of Water

The most important soil factors that affect water absorption by roots are temperature, osmotic potential of the solution, aeration, concentration of CO_2, and availability of water. Although atmospheric conditions may also affect absorption, conditions in the soil are generally the limiting factors in the absorption of water by roots.

Temperature

Soil temperature has a profound influence on rates of water absorption. For over two hundred years, observers have known that soils with low temperatures absorb less water, but only relatively recently have scientists explained this phenomenon. Apparently, the inhibitory influence of low soil temperature on water absorption is manifested in several ways. At lower temperatures, water is more viscous, a factor that reduces its mobility, protoplasm is less permeable to water (39), and root growth is inhibited. The combined effect of these factors causes a reduction in water absorption at low temperatures.

We can demonstrate this inhibitory effect quite readily in the greenhouse. If we place a layer of crushed ice on the surface of soil in which a vigorous coleus plant is growing and if conditions are good for transpiration, the plant will wilt in two hours. If we then remove the ice, the plant will regain its turgidity within an hour.

Osmotic Potential of Soil Solution

Knowing that water is absorbed because a water potential gradient exists between the soil solution and the cell sap of the interior root cells, we can understand why the osmotic potential (salt concentration) of the soil solution is an important factor in water absorption. Indeed, if the water potential of the soil solution is more negative than that of the cell sap of the root cells, water will move out of the plant instead of being absorbed.

Some plants (halophytes) have a greater tolerance than others for high salt concentrations in the soil solution. It is significant to note that the osmotic potential of the cell sap of halophytes may be a good deal more negative than that found in other plants.

Aeration

If a tobacco field is saturated by a heavy rainfall and then exposed to bright sunlight, the leaves of the tobacco plants, in many instances, will severely wilt in a short period of time (26). Tobacco growers term this phenomenon *flopping*. The flopping of tobacco leaves is most severe under conditions of poor drainage and is caused by a retardation of water absorption as a result of the displacement of soil gases by the water, thereby leaving the roots poorly aerated. When transpiration occurs at a rapid rate in the bright sunlight, the combined effect of accelerated water loss and retarded water absorption results in the development of a *water deficit* in the plant.

Root growth and metabolism are definitely retarded under conditions of low oxygen tension. Although inhibited root growth would have a significant effect on water absorption under conditions of prolonged periods of poor aeration, immediate effects would be insignificant. However, a reduced metabolic rate of the root and, thus, the capability of the root to take up and accumulate salt would seriously affect its water absorption. An adverse effect of lowered oxygen tensions on water absorption is also probable.

Concentration of CO_2

The accumulation of CO_2 in the soil appears to have a greater inhibitory effect on water absorption than do lowered oxygen tensions. Apparently, an increase in

CO_2 causes an increase in the viscosity of protoplasm and a decrease in the root's permeability to water (15, 34), thereby bringing about a retardation in water absorption. Kramer and Jackson (26) found that sunflower and tomato plants wilted more rapidly when the soil air was replaced by CO_2 than when the soil was replaced by nitrogen.

Although increased CO_2 concentration in the soil atmosphere may have a detrimental effect on water absorption, we should not place too much emphasis on this factor. Toxic amounts of CO_2 are not likely to accumulate in the soil atmosphere under field conditions.

Availability of Water

Not all of the water in the soil is available to the plant. As the soil water in the immediate area of the root system is depleted, the absorption of water by the plant becomes more and more difficult due to the decreased diffusion gradient. Eventually, the physical factors that hold water to the soil become stronger than the physical factors that are involved in the uptake of water by the plant.

Before discussing soil water–plant relationships, we should become familiar with the terms *field capacity*, *permanent wilting percentage* (*PWP*), and *total soil moisture stress* (*TSMS*). Kramer (24) defines *field capacity* as the water content of a soil after it has been thoroughly wetted and then allowed to drain until capillary movement of the water has essentially ceased. *Permanent wilting percentage* is the percentage of soil water left when the leaves of a plant growing in the soil first exhibit the symptoms of permanent wilting—that is, the leaves do not regain turgor when placed in a saturated atmosphere.

Wadleigh and Ayers (46) introduced the term *total soil moisture stress*. They defined TSMS as the sum of osmotic potential of the soil solution and soil moisture tension. By *soil moisture tension*, we mean those gravitational, adsorptive, and hydrostatic forces that hold water to the soil. Figure 3–1 shows the field capacity and PWP of a clay and a loam soil.

Work performed in the early part of the twentieth century established that field capacity and PWP differ with the type of soil tested. Contrasting two widely different types of soil, we find, for example, that clay has a much higher field capacity and PWP than does sand. However, scientists also thought that field capacity and PWP were soil-moisture constants for any particular type of soil. Although this statement is undoubtedly true for field capacity, it is questionable for PWP. The PWP of a soil appears to differ, depending upon the test plant used. Slatyer (36) has indicated that plant osmotic factors rather than soil factors determine the PWP of a soil. A mesophytic leaf may have an osmotic potential of about −20 bars, while the osmotic potential of leaves of some halophytes may exceed −200 bars (21). This large difference in osmotic potentials is indicative of the differing capacity of various plants to "draw" water from the soil. In other words, the PWP of a soil is dependent on the plant's internal water relations with respect to its capacity to absorb water from the soil and is not, as earlier thought, a soil-moisture constant.

During the day, as the soil water in the near vicinity of the root surface is depleted, the TSMS increases (see Figure 3–2). The TSMS decreases during the night (night recovery) as water moves from the remaining soil mass to the surface of the root. The water potential of the plant follows the same pattern; it is more negative during the day and less negative at night. However, the plant water potential always remains more negative than the TSMS—an essential condition if water is to be drawn into rather than out of a plant. As the soil dries a little more each day, the TSMS and the plant water potential become progressively more negative.

As we can surmise, the daily, progressively developing negative water potentials, coupled with progressively slower night recovery, lead to an increasingly apparent loss of turgor by the leaves. Finally, a point is reached at which the TSMS attains a level equal in magnitude to the osmotic potential of the plant leaves (we will assume this to be −14 bars). Recovery of turgor at this level is impossible because the water potential–TSMS equilibrium established at night is at a water potential that allows for only

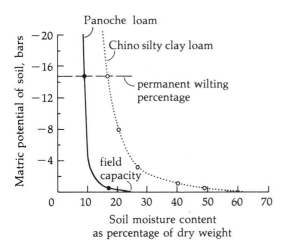

Figure 3–1. Matric potentials of sandy loam and clay loam soils plotted over water content.

Data for Panoche loam from C.H. Wadleigh et al. 1946. U.S. Dept. Agr. Tech. Bull. 925; data for Chino loam from L.A. Richards and L.R. Weaver. 1944. J. Agr. Res. 69:215.

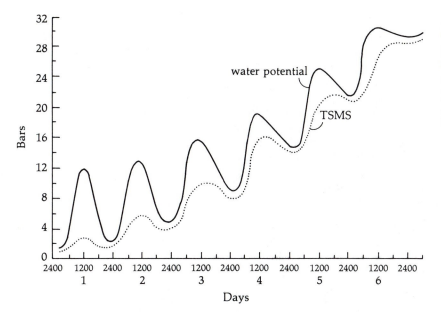

Figure 3–2. Diurnal changes in plant water potential and TSMS as soil is dried progressively from field capacity.

From R.O. Slatyer. 1957. Bot. Rev. 23:585.

zero turgor pressure. The PWP is obtained at this point. We can redefine PWP, then, as the soil water content present when the plant water potential and the soil TSMS are at equilibrium, and the turgor pressure of the plant leaves is zero (37, 38).

Although water is not readily available to the plant at levels above field capacity and at levels below the PWP, the plant may take up some water under these circumstances (20, 36, 37, 38). However, plant growth will essentially cease at the level of the PWP, and death through desiccation will occur unless water is added to the soil (thereby lowering the TSMS).

Absorption of Water

Under natural conditions, practically all water absorption by rooted plants takes place through the root system. The area of young roots where the most absorption takes place is the root hair zone (see Figure 3–3). Water diffused into the root hair and, to a lesser degree, into other root epidermal cells as a result of a water potential gradient. As long as the water potential of the root cell sap is more negative than that of the soil solution, water continues to enter the cell. An increase in the solute concentration of a cell or a decrease in its turgor pressure will cause a more negative water potential to develop in its cell sap. As a result, the uptake of water will increase. Most water absorption, therefore, appears to occur through the mediation of osmotic mechanisms.

Since the root systems of different plants may vary a great deal in appearance and extent of soil penetration, there is no doubt that their water-absorbing capacity also differs. Some plants have root systems that penetrate deep into the soil, other plants form a dense network of branch roots that do not penetrate deep but cover a large area of soil at a shallow depth.

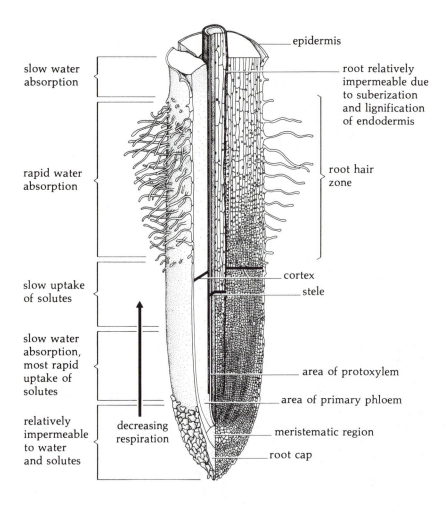

slow water absorption

rapid water absorption

slow uptake of solutes

slow water absorption, most rapid uptake of solutes

relatively impermeable to water and solutes

decreasing respiration

epidermis

root relatively impermeable due to suberization and lignification of endodermis

root hair zone

cortex

stele

area of protoxylem

area of primary phloem

meristematic region

root cap

Figure 3–3. Areas of root involved in absorption and translocation of water.

The root hair zone is the area of the root where most water absorption takes place—in other words, this is the area of greatest permeability. Root hairs are very delicate structures and commonly last for only a short period of time (less than two days). Root hairs that persist, although relatively rare, have been observed on some plant species (8). These root hairs, however, become thick-walled and to a certain degree lignified and suberized with age, factors that extensively limit their ability to absorb water.

In a growing root system, there are a large number of root tips through which absorption takes place. The root tips represent the growing area of the root. In the older tissues of the root, several mm back from the tip, secondary thickening takes place, and a periderm layer of highly suberized cells develops. The permeability of the root is greatly impeded by this layer. Obviously, most of a plant's root system does not absorb water very efficiently.

Although by far the most efficient water absorption takes place at the unsuberized root tip, under certain circumstances a significant amount of water may also be taken up by suberized areas of the root (23). Many investigators (23) have noted that only a very small percentage of the root system of certain trees is unsuberized, making it necessary for suberized tissues to take up water so the tree is adequately supplied with water. Addoms (1) observed that the suberized roots of the yellow poplar (*Liriodendron tulipifera* L.), the sweet gum (*Liquidambar styraciflua* L.), and the shortleaf pine (*Pinus echinata* Mill.) absorbed a dye solution. She pointed out that there are three ports of entry for water through suberized roots: lenticels, breaks around branch roots, and wounds. The capacity for water absorption by suberized root areas of different plants is, therefore, quite possibly related to the extent to which the root's anatomy allows for the development of these avenues of entrance.

Path of Water Movement through Root

Water taken up into the root hairs and other epidermal cells in or near the root hair zone moves from these cells through the cortex tissue, the endodermis, the pericycle, and finally into the xylem (see Figure 3–4). A great deal of the water taken up by the root hairs moves along the walls of the cortical cells. The water's movement along the endodermal cell walls is impeded, however, by the presence of the *Casparian strip*, a band of suberin in the inner surface of the transverse and radial primary walls of the endodermal cells (Figure 3–5). Water is diverted into the endodermal cell and follows a water potential gradient to the pericycle and into the conducting cells of the xylem. The xylem tissue of the roots connects directly with the xylem tissue of the stem. The water thus moves out of the root and into the stem.

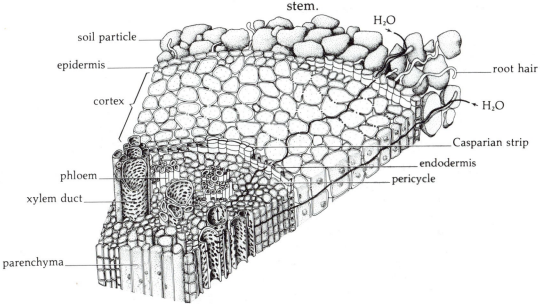

Figure 3–4. Cross section of root through root hair zone illustrating path of water movement from soil to xylem.

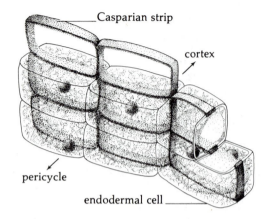

Figure 3–5. Endodermis and orientation of Casparian strip.

Anatomy of Xylem Tissue

For well over a hundred years, scientists have recognized the xylem as the tissue most involved in water translocation. Several different types of cells, living and nonliving, comprise xylem tissue. Of these, the tracheary elements are the most prevalent, and it is through these cells that practically all water translocation takes place. Also found in the xylem tissue are xylem fibers (sclerenchyma) and living parenchyma cells.

Tracheids and vessels. The vessel elements and tracheids are the cells most involved with water translocation (see Figure 3–6). Both are more or less elongated, have lignified secondary wall, and are dead when they are mature and functional. Since vessel elements and tracheids are dead at maturity, there is no interfering protoplast in the lumina of the cells—a situation that allows for the efficient translocation of relatively large amounts of water. Perforated end walls are characteristic of vessel elements

but do not occur in tracheids. However, tracheids are well supplied with bordered pits. In the more-developed vessel elements, the end walls may be entirely missing, thus leaving nothing to obstruct the passage of water through the cell.

If we took a large number of vessel elements and stacked them end to end, we would have a long tubelike structure. The long, tubelike structures, resulting from a

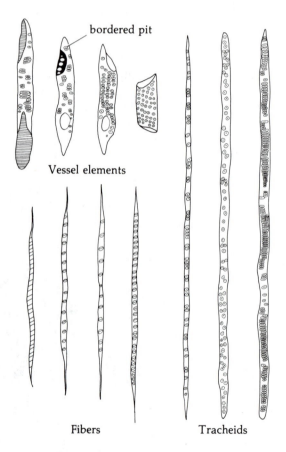

Figure 3–6. Vessel elements, tracheids, and xylem fibers found in xylem tissue.

From K. Esau. 1958. Plant Anatomy. *New York: John Wiley & Sons, Inc. Reproduced by permission.*

series of vessel elements being attached to one another by their end walls, is called a *vessel* or *xylem duct* (see Figure 3–7). The vessels of the xylem tissue form a network of ducts that extends to all areas of the plant and gives all living cells an easily accessible supply of water. This network is of primary importance to the plant, not only for the maintenance of turgor but also for the translocation of other substances that may be carried from cell to cell by the moving water (e.g., essential mineral elements).

The vessel system is the principal pathway by which water is translocated in the angiosperms. However, vessels are not present in the conifers, and in this group the tracheids form the principal pathway for water translocation (see Figure 3–8). Tracheids are long spindle-shaped cells, with sharply inclined end walls. The end walls of the tracheids overlap each other and through the mediation of bordered pits provide a continuous pathway for the movement of water. As we might expect, the movement of water in a group of tracheids as compared to a vessel system is probably much less direct and meet with more resistance. Nevertheless, water flow does not seem to be impeded in the taller trees, many of which are conifers and which, therefore,

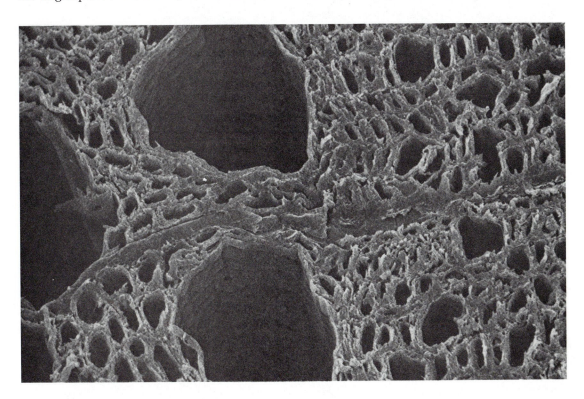

Figure 3–7. Scanning electron micrograph of xylem of oak showing vessels in summer wood and spring wood. Spring wood has vessels with large diameters.

From J.N.A. Lott. 1976. A Scanning Electron Microscope Study of Green Plants. St. Louis, Mo.: Mosby. Courtesy of J.N.A. Lott, McMaster University.

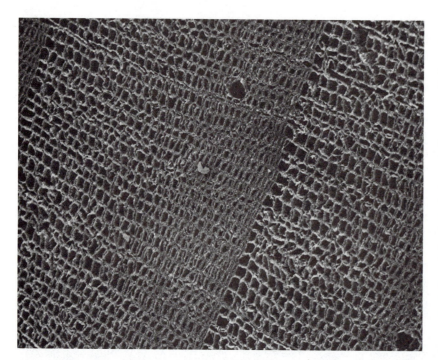

Figure 3–8. Scanning electron micrograph of wood of Douglas fir. (*Pseudotsuga menziesii*). Secondary xylem is evident in production of annual ring. Spring wood consists of large lumina and thin secondary walls. Summer wood consists of tracheids with small lumina and thick secondary cell walls. Note resin duct and ray. *From J.N.A. Lott. 1976.* A Scanning Electron Microscope Study of Green Plants. *St. Louis, Mo.: Mosby. Courtesy of J.N.A. Lott, McMaster University.*

are completely devoid of vessels. In Figure 3–8 growth of the secondary xylem is evident in the production of annual rings. The spring wood consists of tracheids with large lumina and thin secondary cell walls. In contrast, the summer wood is characterized by cells with small lumina and very thickened secondary cell walls. The growth of the tracheids is directly related to the seasonal growing conditions, particularly the availability of water.

Although vessels and tracheids are oriented in the plant in a vertical direction with respect to their long axis, and water movement is predominantly in this direction, lateral water movement does take place. Numerous pits through which water may pass perforate the side walls of vessel elements and tracheids. Generally, where cells lie alongside each other, pits occur in pairs and are called, appropriately, *pit pairs*. Thus where pits lie adjacent to each other, water may move from cell to cell laterally. Since

pit pairs may occur between two vessel elements, two tracheids, a tracheid and a vessel element, a tracheid or vessel element and living parenchyma cells, and so on, water can be easily distributed throughout all the tissues of a plant.

Xylem fibers. The xylem fiber is a long, thin, tapering cell with a very thick, lignified cell wall and is dead at maturity. The primary function of the xylem fiber is support, and it is doubtful if any significant amount of water is moved through this cell. Nevertheless, it is possible for some water to pass through xylem fibers since they are in association with each other and with the tracheids and vessel elements via pit pairs.

Xylem parenchyma. Living parenchyma cells may be found interspersed in the conducting cells or as components of the xylem rays and are generally referred to as wood and ray parenchyma, respectively. One obvious function of xylem parenchyma is the storage of food. Starch is accumulated toward the end of the growing season and is then depleted during the cambial activity of the following growing season (14). The xylem ray parenchyma greatly facilitates the lateral transport of water and nutrients. The living parenchyma cells of the xylem may have a vital role in the translocation of water. Later in this chapter we will discuss this interesting suggestion in more detail.

Path of Water Movement in Leaf

The xylem of the stem is divided and subdivided many times to form a complex network of water-conducting tissues, which finally ends in the fine veins, or *vascular bundles*, of the leaf (see Figure 3–9). As we will discuss later, bundle sheath cells surrounding the vascular bundles are characteristic of C_4 plants and are important in carbon diox-

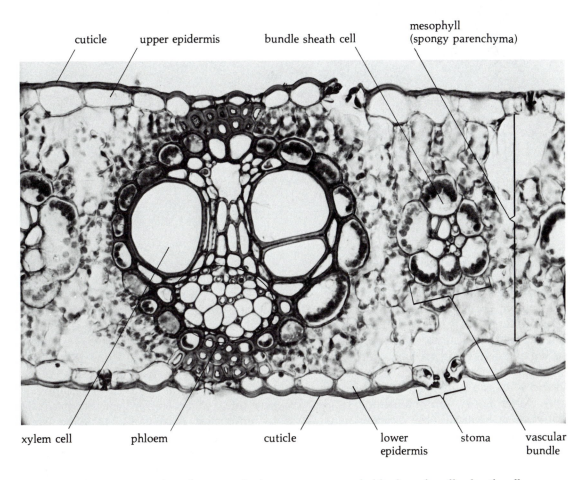

Figure 3–9. Transverse section of monocot leaf—corn (*Zea mays*). Note closed vascular bundles surrounded by large bundle sheath cells.

Courtesy of C.J. Hillson, The Pennsylvania State University.

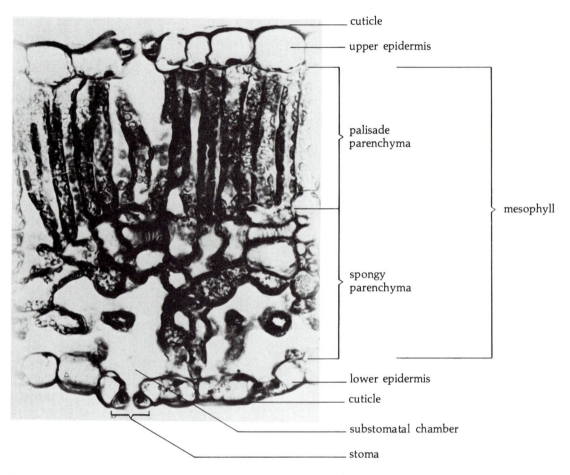

cuticle

upper epidermis

palisade parenchyma

mesophyll

spongy parenchyma

lower epidermis

cuticle

substomatal chamber

stoma

Figure 3–10. Transverse section of dicot leaf.   *Courtesy of C.J. Hillson, The Pennsylvania State University.*

ide fixation. Dicot and monocot leaves also show differences in mesophyll structure. The typical dicot leaf has a mesophyll consisting of palisade and spongy parenchyma. Monocots do not exhibit such extensive differences in size and shape of mesophyll cells. Water moves from the tracheid of the leaf veins into the mesophyll cells, where a small portion is used by the cells and a large amount evaporates, in the form of water vapor, from the surfaces of the mesophyll cells into intercellular spaces. Much of this water

vapor escapes from the substomatal chamber through the stomata into the surrounding atmosphere as a result of a very steep water vapor gradient (see Figure 3–10).

In a rapidly transpiring plant, the xylem vessels and tracheids are generally in a state of *negative pressure*, or tension. Although the rate of transpiration is often similar to the rate of absorption, as Figure 3–11 indicates, under a variety of circumstances, transpiration can and does exceed absorption. The suction force, created by the rap-

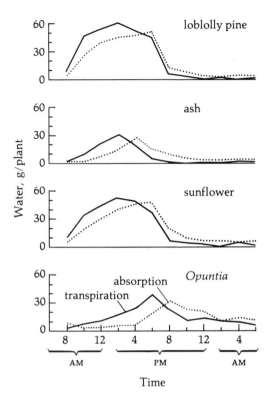

Figure 3–11. Rates of transpiration and absorption (grams/plant) in loblolly pine, ash, sunflower, and *Opuntia* on clear hot summer day.
From P.J. Kramer. 1937. Am. J. Bot. 24:10.

idly moving columns of water, is transmitted to the root, and water is literally pulled into the root from the soil. The water potential of the cell sap becomes more negative as it is subjected to increasing negative pressure (tension); this relationship may be expressed in the following equation:

$$\psi_w = \psi_s + (-\psi_p)$$

The natural consequence of a more negative water potential is an increased water uptake.

Water uptake in the manner described occurs as a result of activity (transpiration) in the shoot. The root merely acts as an adsorbing and absorbing surface. This phenomenon is clearly supported by the fact that the shoot can absorb water through dead roots and may, in fact, take it up at a faster rate. Also, the resistance to water uptake by living roots may be due to the living cells of the root (24).

Absorption of Water by Aerial Parts of Plant

The absorption of water both in liquid and vapor forms occurs, to a small extent, through the aerial parts of most plants. The extent to which this occurs may be dependent on the water potential of the leaf cells and the permeability of the cutin layer (17). For example, Roberts, Southwick, and Palmiter (33) found that the cutin layer on the leaves of the McIntosh apple was not continuous but occurred in lamellae parallel to the outer epidermal walls. Interspersed with the parallel layers of cutin, they found parallel layers of pectinaceous material of good water-absorbing capacity. Not only was this material present with the cutin layer at the surface of the leaf, but it extended vertically to the vein extensions within the interior of the leaf. It thus formed a continuous path for water from the surface to the vascular tissue. Obviously, the permeability of the cutin layer of the McIntosh leaf to water is rather good.

Some investigators believe that water absorbed by the leaves can travel in a "negative" direction through the plant and can actually diffuse through the roots into the soil. Studies by Breazeale, McGeorge, and Breazeale (4, 5, 6, 7) demonstrated that both tomato and corn plants are capable of moving water absorbed by the leaves back into the soil. This activity could, of course, only occur along water potential gradients favoring movement in this direction.

Apoplast and Symplast

The terms *apoplast* and *symplast* were originally introduced by Münch (30) in his studies on the flow of water and solution in plants. To modern plant scientists the terms are convenient for describing the path of absorbed and translocated water and solution. In our discussion, we have mentioned that water may be translocated across the root cortex through a system of interconnecting cell walls and intercellular spaces before reaching the Casparian strip of the endodermal walls. Once through the endodermal and pericycle cells, the water wets the xylem cell walls as well. Münch referred to the nonliving continuum, which included all the nonliving cells and cell walls of the xylem, as the *apoplast*. Currently, plant scientists consider the apoplast a system that includes all nonliving cells and all walls and intercellular spaces in roots and shoots (stems and leaves) where water and dissolved solutes might translocate. Obviously, since living cells are not included, water translocation within this system is not directly due to osmosis per se but to capillary action or, as with solutes, to free diffusion.

Movement of water and solutes into the living cells of the plant, however, is due to osmosis (water), free diffusion (passive uptake of solutes), or active uptake (solutes). This living continuum in the plant, including the plasmodesmata and elements within the cytoplasmic membrane, Münch termed the *symplast*. We will consider these two systems in later chapters dealing with mineral uptake.

Water that is absorbed moves from the soil to the interior of the root along an increasingly negative water potential gradient—that is, water moves through the root epidermis and into the cortex because of increasing energy gradients established by proportional solute concentration. We may wonder why the salt content of the interior cells is higher than that of the exterior root cells. The absorption and accumulation of salt by root cells requires metabolic energy (see Chapter 6). A theory by Crafts and Broyer (9) suggests that there is a decreasing O_2 and energy gradient and an increasing CO_2 gradient from the cortex to the stele (conducting cells). Metabolic activity would be at a minimum, then, in the interior cells in the immediate area of the xylem ducts. Since energy is required to accumulate and hold salt against a concentration gradient, the stele cells, in contrast to the cortex cells, favor the loss of solute upward. Because diffusion back through the impervious Casparian strip is impossible, there is a unidirectional loss of salt into the lumina of the xylem cells. Water would also follow this unidirectional path and would diffuse from the less negative water potential of the soil solution to the more negative water potential of the sap in the interior of the root.

Translocation of Water
Root Pressure

The stump of a recently felled tree or a detopped herbaceous plant will often give visual evidence of root pressure. We may observe xylem sap under pressure exuding from the cut end of the stump. If a well-watered tomato plant is detopped and the stump is attached with a rubber sleeve to a glass tube containing some water, we can

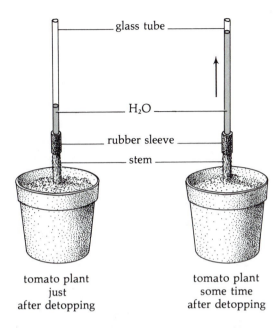

Figure 3–12. Demonstration of root pressure. Note rise of liquid (exudate) in glass tube.

see this pressure at work. Figure 3–12 shows that in such an experiment, water is actually "pushed" up the glass tube.

Stocking (40) defined *root pressure* as a pressure developing in the tracheary elements of the xylem resulting from the meta-

bolic activities of roots. Root pressure is, therefore, referred to as an active process. The movement of water up the stem as a result of root pressure is due to osmotic mechanisms that are created as a result of the active absorption of salt by the roots.

Root pressure, which is developed due to the accumulation of solute in the xylem ducts, appears to be affected by factors that also affect respiration (e.g., oxygen tension, narcotics, auxin, and respiration inhibitors). Several investigators (18, 19, 35, 45) have observed an autonomic, diurnal fluctuation in exudations caused by root pressure. Figure 3–13 provides an example of the rhythmic nature of root pressure exudations. Note the close agreement between the periodicity of root pressure and the exudation rate, which can be defined as the relative speed of a liquid released at the cut surface of a stem.

Detopped tomato plants, with their root systems immersed in solutions of different salt concentrations, exhibit different exudation rates (2); lower exudation rates result when roots are immersed in solutions of lower salt concentration. Vaadia (45) has suggested that the diurnal fluctuation of exudation rates is caused by a periodicity of salt transfer into the xylem. Obviously, this

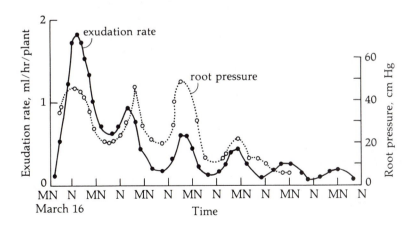

Figure 3–13. Diurnal fluctuation in exudation rate and root pressure of topped sunflower plants. *From Y. Vaadia. 1961. Physiol. Plant. 13:701.*

would cause a periodicity in the magnitude of the osmotic potential of the xylem ducts, which would have the effect of a change in the rate of water absorption in accordance with a change in water potential gradients.

We should stress that water absorbed in this manner does not require a *direct* expenditure of energy. The energy is expended in the absorption and accumulation of salts. However, the water potential is the driving force in water absorption.

Some earlier workers attempted to explain the rise of water in plants as being primarily a result of root pressure. Scientists currently believe that the magnitude of pressure developed is not sufficient to "push" water to the heights reached by most trees. Although plant scientists have observed values higher than 6 atm (48), root pressures in excess of 2 atm are seldom obtained. Indeed, root pressures of any magnitude are conspicuously absent from conifers, which are among the tallest of trees. In addition, estimates of the ability of root pressure to raise water to appreciable heights do not take into account the friction encountered in the passage of water through the xylem ducts.

Another reason root pressure probably is not a major factor in the rise of water in plants is that exudation rates are generally much slower than normal transpiration rates. Also, xylem sap under normal conditions is generally under tension instead of pressure, an observation that supports the argument that root pressure is not an important factor in water translocation. We should mention here, however, that when conditions for transpiration are poor, root pressure may be a significant factor in the movement of water. In some plants, liquid water loss, as in guttation, a phenomenon that is caused by root pressure, is most no-

ticeable under conditions unfavorable for transpiration.

Vital Theory

Early investigators believed that the ascent of water in plants was under the control of "vital activities" in the stem. This belief was most likely stimulated by the fact that living cells are present in the xylem tissue (xylem parenchyma and xylem ray cells). Experiments by Strasburger (41, 42) and others, however, have caused modern botanists to discount the vital theory on water translocation. Strasburger, for example, demonstrated that stems in which the living cells have been killed by the uptake of poisons are still capable of water translocation.

Cohesion-Tension Theory

Imagine, if you will, a long, hollow glass tube, one end of which is submerged in a beaker of water. The tube is filled with water so that there is an unbroken connection between the water in the beaker and in the tube. If we place a thoroughly soaked sponge at the other end of the tube so that a connection is made between the water held by the sponge and the water in the tube, an unbroken column of water can be "pulled" up from the beaker. We may accelerate this process by using a fan to move dry air over the sponge and by using a heat lamp to increase the temperature of the area immediately surrounding the sponge. The rate at which the water moves up the tube is directly related to its rate of evaporation from the sponge. As water evaporates from the sponge, it is replaced by water from the tube, which, in turn, is replaced by water from the beaker (see Figure 3–14).

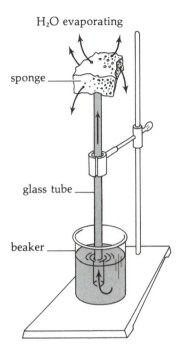

H₂O evaporating

sponge

glass tube

beaker

Figure 3–14. Demonstration of cohesion-tension theory. Water evaporating from surface of sponge will be replaced by water from glass tube, which in turn will be replaced by water from beaker.

How is it possible to pull a column of water up a tube without having the column break? Why doesn't the column of water, when under tension (being pulled), pull away from the glass wall of the tube? To answer these questions we must understand the cohesive and adhesive properties of water. The water molecules cohere to each other and, at the same time, adhere to the glass wall of the tube. Therefore, the water column will not break until its cohesive strength and adhesive strength are overcome by gravitational pull on the column or by an interruption of the column by air.

Let us now compare this physical example with a plant growing in soil in a natural habitat. The water in the beaker could be compared with the soil water. The glass tube is somewhat analogous to the tracheary tissue of the plant (the vessels more closely fitting this analogy). The evaporating surface of the sponge is similar to the evaporating surface of the leaf mesophyll. If we presume that an unbroken column of water exists between the soil water and the water of the leaf tissue, we can see how water could be translocated from the soil. As water evaporates from the leaf's mesophyll cells, it causes a decrease in the water potentials (makes them more negative) of the mesophyll cells in direct contact with the air spaces of the leaf. The water lost by the cell surfaces is replaced by water moving from cell to cell within the mesophyll along water potential gradients in an energetically favorable direction. Finally, the movement of water within the leaf will be transmitted to the water in the xylem elements in the veins, hence exerting a pull, or putting water in the xylem tissue in a state of tension. This state of tension continues through unbroken columns of water (due to the cohesive and adhesive properties of water) from the leaves to the root system. The water potential of water in living cells of root systems and along the cell walls become more negative in relation to the water potential of the soil, thus promoting absorption.

Do we have any evidence that the contents of xylem vessels are, in fact, under tension in a normal transpiring plant? Direct evidence is lacking, since direct measurement of tension with known methods would disrupt the continuity of the water columns and thus eliminate the tension that might be present. However, indirect evidence that the xylem contents of a transpiring plant are in a state of tension is plentiful.

Thut (44) demonstrated that a leafy shoot cut under water and sealed to a mercury manometer could support a column of mercury above barometric level. The column of water attached to the mercury in the manometer would have to be in a state of tension under the above circumstances. If a woody twig of a rapidly transpiring plant is cut, the water in the xylem elements "snaps" away from the area of the cut, thereby indicating that the water in the stem is under tension (44). This phenomenon is the basis for the pressure bomb procedure used to determine water potential. Perhaps the most striking demonstration that water is under tension in transpiring plants is in dendrographic measurements of diameter variations in tree trunks (see Figure 3–15).

When water in the xylem elements is under tension, it will, because of its adhesive properties, cause a shrinkage in the diameters of the xylem cells. Although this decrease in diameter is not measurable for the individual xylem element, the total effect can be recorded by means of a dendrograph. This instrument gives a continuous record of changes in the diameter of a trunk over a period of time. As might be expected, there is a decrease in diameter during periods of high transpiration and an increase in periods of low transpiration. Note in Figure 3–15 that when transpiration was relatively low in late May and early June, only slight variation in trunk diameter was recorded. However, in July, when temperature and transpiration increased, variations in trunk diameter were obvious.

Assuming that we are convinced that water—due to its cohesive and adhesive properties and to the anatomy of the xylem tissue—can be pulled up in an unbroken chain through the plant, we should now ask: Can the tensile strength of water support the column of water that would be nec-

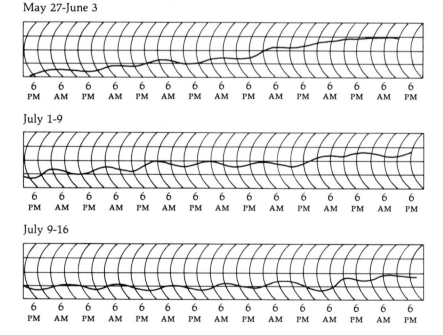

Figure 3–15. Changes in relative diameter of American beech (*Fagus grandifolia* Ehrh.) with time, as measured by dendrograph. Data were collected for 3-week durations. Note decrease in diameter during periods of high transpiration and increase in diameter during periods of low transpiration.
From H.C. Fritts. 1958. Ecology. *39:705.*

essary to reach the tops of the highest trees? As we might expect, the answer is Yes. Measurements of the tensile strength of water have exceeded 300 bars. To raise water to the top of a 400-foot tree would require a difference in pressure of about 13 bars between the top and bottom. In addition, water encounters friction as it moves through the xylem tissue. Although this friction is considerable, the tensile strength of water is sufficient to overcome the frictional and gravitational forces encountered in its vertical rise in a plant.

The cohesion-tension theory, first introduced by Dixon (11, 12), is the most plausible explanation today of the translocation of water in plants. Root pressure is capable of moving water upward in a plant, but not in the quantity and to the heights necessary for most plants. Probably the strongest argument for the cohesion-tension theory is that it is the only theory that can account for the quantity and the rate of water movement in a vigorously transpiring plant. We must remember, however, that any physiological event (water loss, solute buildup or translocation, mineral absorption) that indirectly or directly causes water potentials to become more negative with increased water potential gradients established from one area to the next will influence water movement. The most important thermodynamic property of water with respect to water flow in biological systems is its water potential. Water will tend to move according to the law of thermodynamics in a direction from higher (less negative) to lower (more negative) water potentials.

In one area of the plant (roots and leaves), the process of water translocation is characterized by an energetically favorable process (osmosis), which is linked through the adhesive and cohesive properties of water to the endergonic process of lifting and pulling water to the tops of the highest trees against the forces of gravity and anatomical resistance. Water translocation is not unlike the linkage of exergonic and endergonic chemical reactions necessary for organismal energy capture and utilization. In the latter situation, the link is provided by the peculiar and complementary physical and chemical properties of the participating chemicals.

Questions

3–1. Name the soil factors that influence the water absorption of roots. Explain the action of each factor on the water absorption process.

3–2. Define the following terms: field capacity, permanent wilting percentage, total soil moisture stress. Are differences in these measurements a function of different soils or different plants? Explain.

3–3. What is the most significant condition between soil and plant that favors the absorption of water by roots? Explain.

3–4. Should field crops be fertilized during times of drought? Explain.

3–5. Describe the path of water from an epidermal cell in a root through the various root, stem, and leaf tissues until it is finally released as vapor into the atmosphere.

3–6. As a review of general botany, name the cell types found in the xylem, phloem, cortex, cambium, mesophyll, and vascular bundle.

3–7. What are the apoplast and symplast systems of a plant? Why were these terms introduced?

3–8. Explain the principal mechanisms that account for the translocation of water in plants. Consider the importance of osmosis, water potential gradients, and the cohesive and adhesive properties of water.

3–9. Is transpiration absolutely necessary for the translocation of water in plants? How do plant cells receive water when transpiration is minimal?

3–10. When water is being translocated at relatively high velocity and is subjected to high tension in the conducting elements, what prevents the breaking of the water columns?

3–11. In providing cut flowers for decoration purposes, what is the reason for cutting the plants under water?

3–12. Certain plant pathogens are known to cause wilting in plants. What possible influence might such organisms have on the translocation of water in plants?

Suggested Readings

Dainty, J. 1976. Water relations of plant cells. In U. Lüttge and M.G. Pitman, eds., *Encyclopedia of Plant Physiology*, 2A. Berlin: Springer.

Dixon, H.H. 1914. *Transpiration and Ascent of Sap in Plants*. London: Macmillan.

Kozlowski, T.T., ed. 1968–1978. *Water Deficits and Plant Growth*. vols. 1–5. New York: Academic Press.

Lüttge, U., and N. Higinbotham. 1979. *Transport in Plants*. New York: Springer-Verlag.

Meidner, H., and D.W. Sherriff. 1976. *Water and Plants*. New York: Wiley.

Slatyer, R.O. 1967. *Plant Water Relations*. New York: Academic Press.

Sutcliffe, J. 1968. *Plants and Water*. New York: St. Martin's Press.

Chapter 4

Water Loss: Transpiration

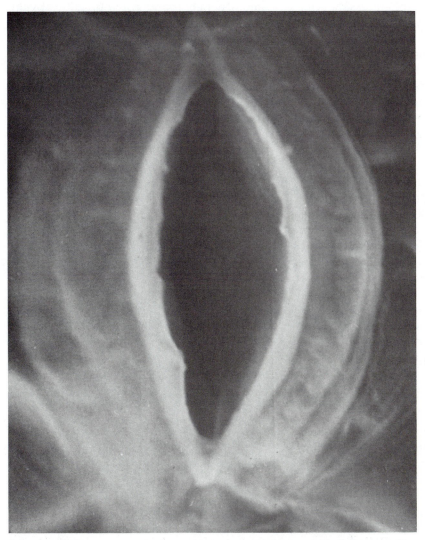

Scanning electron micrograph of open stoma from *Zebrina purpusii* leaf. *Courtesy of Jorg Schönherr, Technical University, Munich.*

Although water is the most abundant constituent of plant tissues (except in dry seeds), only a small portion of the water that is absorbed is retained for metabolic processes. Large amounts of absorbed water are translocated to the leaves and are lost to the surrounding atmosphere. Aside from a given amount of water that provides a temporary contribution to the maintenance of turgidity and the possible translocation of dissolved minerals, water use in plants is inefficient and can endanger their survival. Indeed, for this reason a great percentage of the earth's surface is unsuitable for plant growth and agriculture. Large amounts of water are continuously lost from the plant due to the anatomical features of plants, particularly those of the leaves.

Although transpiration is the phenomenon most responsible for the excessive water loss from plants, other processes involved are guttation, secretion, and bleeding. We will not consider secretion and bleeding further except to mention that *secretion* is the loss of liquid water (solutions) from glands and nectaries, and *bleeding* describes water loss from wounds. The amount of water loss associated with these two processes is negligible.

Guttation

Plants growing in a moist, warm soil and under humid conditions will often exhibit droplets of water along the margin or tip of their leaf blades. The loss of liquid water in this manner is called *guttation*. The factors that favor guttation are high water absorption, high root pressure, and reduced or no transpiration. In other words, under these conditions water absorption greatly exceeds transpiration, and water is literally "pushed" up the xylem ducts and out through specialized structures called *hyda-*

thodes, which are specialized pores found at the extreme tip of the leaf and usually adjacent to a vein. When water uptake exceeds water loss, a hydrostatic pressure is built up in the xylem ducts, and water must escape by whatever path is available. Hydathodes are generally found at the tips of leaf veins and therefore represent excellent ports of exit for water being pushed up from the roots (see Figure 4–1).

Water that exudes from a hydathode does so as a result of hydrostatic pressure developed in the sap of the xylem ducts and not as a result of any local activity on the part of the hydathode or surrounding tissues. However, there are openings in various organs of plants through which water is

Figure 4–1. Guttation, or water loss as a liquid, from edges of strawberry leaf. Special conditions of water availability, osmotic factors, and relative humidity promote guttation.
Courtesy of J. Arthur Herrick.

actively secreted—that is, the cells surrounding the pore actively participate in pushing water through the pore. These structures are sometimes called water *glands*.

We speak of the liquid escaping through hydathodes as water. However, guttation liquid is not pure water but a solution containing a great number of dissolved substances (Table 4–1). When guttation wa-

Substance	Rye (mg/l)	Wheat (mg/l)	Barley (mg/l)
P	1.1	0.7	2.3
K	18.0	27.0	30.0
Na	0.5	0.8	1.1
Ca	1.5	3.0	4.8
Mg	1.5	1.5	2.4
Mn	0.02	0.02	0.05
Fe	0.4	0.15	0.07
Cu	0.04	0.03	0.03
B	0.04	0.05	0.08
Zn	0.02	0.05	0.05
Mo	0.001	0.002	0.003
Al	0.06	0.08	0.09
NO_3^-	1.0	1.0	1.0
phosphate	2.0	0.9	1.0
NH_4^+	5.6	5.0	8.9
arabinose	2.5	5.6	4.1
fructose	10.3	4.4	1.8
galactose	10.3	7.6	4.0
glucose	18.7	2.6	38.7
ribose	1.0	tr	1.0
sucrose	3.8	4.9	0.0
xylose	1.8	2.0	0.2
succinic acid	ca. 10	ca. 10	ca. 10
aspartic acid	2.2	0.5	3.6
asparagine	2.5	1.9	9.5
glutamic acid	0.7	0.0	0.0
glutamine	0.8	0.3	1.2
biotin	0.002	0.001	0.018
choline	0.30	0.06	1.9
inositol	9.0	0.25	4.5
p-aminobenzoic acid	0.00006	0.00005	0.002
pantothenic acid	0.040	0.085	0.08
pyridoxine	0.01	0.0005	0.0001
riboflavin	0.00025	0.0002	0.0002
thiamine	0.00006	0.00005	0.0025
uracil	0.0	0.0	1.6
pH	5.0	5.5	6.7

Table 4–1. Analysis of composition of guttation liquid from seedlings of Rosen rye, Genesee wheat, and Traill barley.

Source: *From J.L. Goatley and R.W. Lewis. 1966. Composition of guttation fluid from rye, wheat, and barley seedlings. Plant Physiol. 41:373.*

ter evaporates rapidly, the dissolved materials may sometimes be seen as precipitates on the surface of the leaf. Sometimes the salts precipitated on the leaf surface are redissolved and taken into the interior of the leaf. Generally, the salt concentration is very high under these circumstances and may cause injury to the leaf (10, 23). Although it is not shown in Table 4–1, plant scientists have detected certain phytohormones, such as the cytokinins, in root exudates.

Transpiration

Water is lost from a plant, primarily in the form of vapor, by *transpiration*. The loose arrangement of the living, thin-walled mesophyll cells, which results in an abundance of intercellular space, provides an ideal condition for the evaporation of water from internal leaf surfaces. Part of the epidermal surface of the leaf is made up of a great number of microscopic pores called *stomata*. The stomatal pores open into the intercellular spaces of the leaf and provide an uninterrupted path from the interior of the leaf to the external environment. We can visualize the transpiration stream as an unbroken column of water being pulled from the soil, through the roots, up the xylem ducts, out of the mesophyll cells to their surface, into the intercellular spaces as vapor, and through the stomatal pores into the atmosphere.

In addition to stomatal transpiration, water is lost, also as a vapor, directly from the surfaces of leaves and herbaceous stems and through *lenticels*, small openings in the corky tissue covering stems and twigs. The former is called *cuticular transpiration* and the latter *lenticular transpiration*. In cuticular transpiration water vapor diffuses directly through the *cuticle,* a waxlike layer of cutin covering the surface of leaves. This layer greatly retards water loss, and without it any water retention by the plant would be almost impossible. The cuticle, although retarding water loss, is somewhat permeable to water vapor. The extent of cuticular transpiration varies greatly among different species of plants. In plants with leaves possessing a thick cutin layer, this form of transpiration is insignificant. However, plants with thin cutin layers may suffer severe water deficits when conditions favoring high transpiration are prevalent. Generally, the cutin layer is thicker in sun leaves and plants of dry habitats as compared to shade leaves and plants of moist habitats.

The amount of water lost through cuticular and lenticular transpiration is insignificant when compared to the amount of water lost through stomatal transpiration. Only under very dry conditions, when the stomata are closed, can water loss through the cuticle and lenticels be considered important. However, lenticular transpiration may cause some desiccation in those trees that shed their leaves at the onset of winter. During a cold winter, water absorption by roots is at a minimum, thus the importance of lenticular transpiration is increased.

Magnitude of Transpiration

The amount of water a plant actually uses is small compared to the large quantities it transpires. Indeed, transpiration rates of some herbaceous plants are so great that, under favorable conditions, the entire volume of water in a plant may be replaced in the course of a single day (39). For example, it has been estimated that a single corn plant may transpire up to 54 gallons of water in one growing season. At this rate, a single acre of corn could transpire the equivalent of 15 inches of water during one growing season. The amount of water lost varies to some extent from one species to the next, as the data in Table 4–2 illustrates.

Table 4–2. Water loss by transpiration per single plant for five kinds of plants during growing season.
Source: *Reprinted with permission of The Macmillan Company from* Fundamentals of Plant Physiology, *by J.F. Ferry and H.S. Ward. Copyright 1957, The Macmillan Company.*

Kind of Plant	Transpiration during Growing Season (gal)
cowpea	13
Irish potato	25
winter wheat	25
tomato	34
corn	54

In a discussion of water loss by plants, Kozlowski (28) has cited data that dramatically emphasize the tremendous amounts of water that are lost by trees. For example, an average forest in the southern United States may lose as much as 8,000 gallons of water per acre per day (41). Cummings (9) estimated that a single, 48-foot-high, open-grown silver maple tree may transpire as much as 58 gallons per hour.

These figures indicate the importance of good water management in agricultural practices. Economic losses caused by crop failures are great during prolonged periods of drought. In a hungry world, this matter is of considerable importance.

Measurement of Transpiration

We can use several methods to measure transpiration, as Figure 4–2 illustrates. These methods measure either the water absorbed or the water vapor transpired by a plant. The first approach takes advantage of the accordance between absorption and transpiration rates under most conditions. However, there are several exceptions to this rule.

Weighing method. Perhaps the simplest way to measure transpiration is merely to weigh a potted plant at the beginning and at the end of a prescribed period of time (see Figure 4–2a). The soil surface should be covered and the pot wrapped with some water-repellent material, such as aluminum foil, to retard evaporation from surfaces other than the plant. The loss of weight by the plant over a short period of time will be almost completely due to transpiration. Gain or loss of weight due to photosynthesis or respiration is insignificant. When using this method, we are restricted to small plants that can be conveniently grown in a pot. For field work, however, plant scientists often use a very large balance known as a *lysimeter*. A big plant may be grown in a large container filled with soil, which is placed on a weighing platform. The amount of water lost from both the plant and the soil, termed *evapotranspiration*, is measured by weighing the container. The lysimeter method is the most precise method for measuring transpiration and evapotranspiration in the field.

The amount of transpiration of excised parts of plants, such as leaves, fruits, and branches can also be measured. A plant part is excised, immediately weighed, and then, after a short period of time, weighed again. Although relative rates of transpiration may be compared in this manner, transpiration of an excised organ frequently deviates from the normal transpiration of the intact plant. In the initial stages, the rate of transpiration of an excised organ may exceed normal rates, probably because of the release of tension in the xylem ducts. After a short period of time, however, transpiration rates will fall off because of a decrease in the water content of the tissue, stomatal closure, permeability changes, and so on.

Potometer. This method works on the premise that, generally, the rate of water absorption is very nearly equal to the rate of transpiration. A portion of a shoot of coleus, geranium, or some other suitable plant is sealed in a water-filled glass vessel. The glass vessel has two other outlets, a graduated capillary tube and a water reservoir (see Figure 4–2b). Before the rate of transpiration (more exactly, the rate of absorption) can be measured, the entire apparatus is filled with water so that no air spaces are

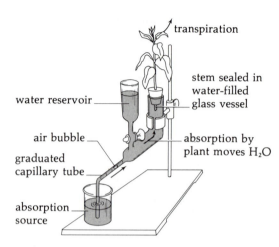

b. Potometer method

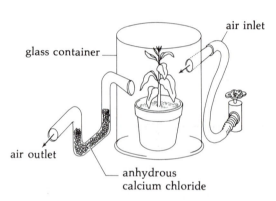

c. Collection and weighing of water vapor

Figure 4–2. Methods of measuring transpiration.

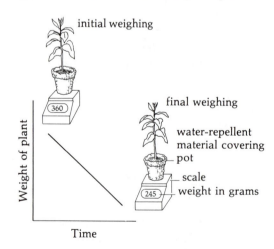

a. Weighing method

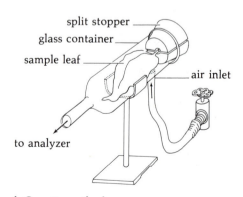

d. Cuvette method

present. This may be accomplished by manipulating the stopcock, which controls the flow of water into the vessel from the reservoir. An air bubble is then introduced into the capillary tube. As transpiration proceeds, the air bubble will move along the capillary tube and give a measure of the rate of transpiration. The potometer method is ideal for observing the effects of different environmental factors (temperature, light, air movement) on transpiration rates. However, its reliability is limited because it actually measures water absorption rather than transpiration; under certain circumstances the two can vary considerably.

Collection and weighing of water vapor. Transpiration may be measured by enclosing the plant in a glass container, so that water vapor can be trapped and weighed (see Figure 4–2c). Air of known moisture content is passed over the plant through an opening in the glass container and passed out over some preweighed water-absorbing material, such as anhydrous calcium chloride. The continuous stream of air passing over the plant keeps the moisture content of the enclosed air approximately equal to that of the surrounding atmosphere. The moisture content of the air passed over the plant is measured by passing it through the same apparatus minus the plant. The difference in weight between the calcium chloride before and after air is passed through it is a measure of the moisture content of the air. The difference in weight between the calcium chloride receiving air passed over the plant and calcium chloride receiving air passed through the apparatus without the plant is a measure of transpiration.

Cuvette. This method is similar in principle to the previous method except that it is designed to measure transpiration of a single leaf (see Figure 4–2d). The cuvette method is ideal for laboratory work when the experimenter is interested in following the effect of different factors (light, temperature, and humidity) on the transpiration process. Air of a known humidity is introduced into the cuvette, passed over the leaf, and collected after it exits. The relative humidity is determined and provides a measure of the rate of transpiration. Obviously this method, as is the case with the potometer method, is useful in the laboratory but not in the field.

In the field, scientists often use *tent chambers*, equipped with suitable built-in air inlets and outlets and temperature-sensing devices to measure transpiration of large plants. Air of known water content is passed into the tent and over the plant. As the water exits, its relative humidity is measured. The increase in water content in the air is a good measure of the transpiration process.

Cobalt chloride. In this method, transpiration is indicated by a change in color rather than a change in weight. Filter paper disks are impregnated with a slightly acidic 3 percent solution of cobalt chloride and thoroughly dried. When dry, paper impregnated in this manner will be blue in color; when the paper is exposed to humid air, it will gradually change to pink. In a similar manner, the cobalt chloride–treated paper will gradually change from blue to pink when exposed to a transpiring leaf surface. The rate of color change is indicative of the rate of transpiration.

The cobalt chloride method can be used for measuring only the relative rates of transpiration of different plants. Due to modifications of different environmental conditions, transpiration rates indicated by this method may deviate considerably from the actual transpiration rates. The surface of the leaf covered by the paper is subjected to practically no air movement, a reduction in light, and a steeper vapor pressure gradient.

The cobalt chloride method, although rarely used today in studies on relative transpiration rates, does have historical significance and contributed to our knowledge of the transpiration process.

Stomatal Mechanisms of Opening and Closing

The epidermal surface of a leaf bears a great number of pores called *stomata*. The stomata are microscopic and are bordered by two specialized epidermal cells called *guard cells*, which control the opening and closing of the stomata. When fully opened, the stomatal pore may measure from 3 to 12 μ in width and from 10 to 40 μ in length (30). The surface of a leaf, depending on the species, may contain from 1,000 to 60,000 stomata per square centimeter. As large as these numbers seem, the stomatal pores are so small that they occupy, when fully opened, only 1 to 2 percent of the total leaf surface. Stomata are more frequently found on the under surface of leaves, but in many species they are found on both surfaces (Table 4–3).

With the exception of a few aquatic types, all angiosperms and gymnosperms have stomata (11). Functional stomata have been found among the cycads (50), ferns (60), horsetails (18), liverworts, and mosses (15). Apparently, stomata are widespread in the plant kingdom, the algae and fungi being the only groups lacking them.

Stomatal Movement

The mechanism by which the opening and closing of the stomatal pore is accomplished has been the subject of numerous investigations. The stomatal movement is generally understood to be a direct response to increases or decreases in the osmotic potential of the guard cells. The changes in water potentials that result from these osmotic changes cause water to move in or out of the guard cells. If water moves in, the cells expand (become *turgid*); if water moves out, they go *flaccid*. When the guard cells are turgid, the stoma is open; when the guard cells are flaccid, the stoma is closed. To effect this movement of water, an exchange must take place between the guard cells and the surrounding mesophyll and epidermal cells. The development of a more negative osmotic potential in the guard cells would cause a water potential gradient to develop

Plant	Upper Epidermis	Lower Epidermis
apple *(Pyrus malus)*	none	38,760
bean *(Phaseolus vulgaris)*	4,031	24,806
corn *(Zea mays)*	6,047	9,922
oak *(Quercus relutina)*	none	58,140
orange *(Citrus sinensis)*	none	44,961
pumpkin *(Cucurbita pepo)*	2,791	27,132
sunflower *(Helianthus annuus)*	8,527	15,504

Table 4–3. Number of stomata per square centimeter of leaf surface.

Source: *From C.L. Wilson and W.E. Loomis. 1962. Botany. Holt, Rinehart and Winston, New York.*

between the guard cells and their neighboring cells. Water would diffuse into the guard cells, causing them to become more turgid. The development of a less negative osmotic potential in the guard cells would, of course, cause a water potential gradient to develop in the opposite direction, and water would flow out of the guard cells into the neighboring cells. The factors causing the fluctuation of the osmotic potential of the cell will be discussed later in this chapter.

Anatomy and Cytology of Stoma

Although changes in turgor provide the motive force for the opening and closing of stomata, an unusual feature of the guard cell wall causes the stomata to open in the manner that they do. The cellulose microfibrils making up the wall of a guard cell are arranged radially (from the pore outward and relatively perpendicular to the pore) rather than longitudinally, or along its length. This orientation of the cellulose microfibrils is termed *radial micellation*. In addition, the guard cell wall adjacent to the stomatal pore is thicker than the outer wall, possibly due to the termination and overlapping of radiating microfibrils in this area. Thus an increase in turgor pressure causes those relatively nonthickened and more elastic areas of the wall to stretch. Conversely, the resistance of the thicker cell wall at the surface adjacent to the other guard cell and the radial arrangement of the cellulose microfibrils predispose the guard cells to increase in length and stretch away from each other, thereby forming an aperture, the *stomatal pore*.

The appearance of the guard cells differs characteristically from the surrounding epidermal cells (see Figure 4–3). The guard cells of some plant species, particularly the grasses, are dumbbell shaped and are associated with epidermal cells that also differ in appearance from the rest of the epidermal cells. These epidermal cells are called *subsidiary cells* or *accessory cells*.

One other distinguishing feature of the guard cell is the presence of chloroplasts. Epidermal cells do not possess chloroplasts. Microspectrophotometrically obtained absorption spectra taken from individual guard cell chloroplasts are similar to those obtained from mesophyll chloroplasts, thereby indicating the presence of both chlorophyll a and chlorophyll b (61). The photosynthetic process, or noncyclic photophosphorylation (to be discussed later), takes place in the guard cell, but at a reduced rate compared to a mesophyll cell.

Factors Affecting Stomatal Movement

The environmental factors having the greatest influence on the opening and closing of stomata are light; presence of potassium, chloride, and hydrogen malate; CO_2 concentration; water deficits and abscisic acid; and temperature.

Light. Generally, the stomata of a leaf are open when exposed to light and remain opened under continuous light unless some other factor becomes limiting. When darkness returns, the stomata close. The amount of light necessary to achieve maximal stomatal openings varies with the species but is usually considerably less than is needed for maximal photosynthetic activity. For example, a light intensity of 250 footcandles is all that is necessary to achieve maximal stomatal opening in tobacco leaf tissue (63). A

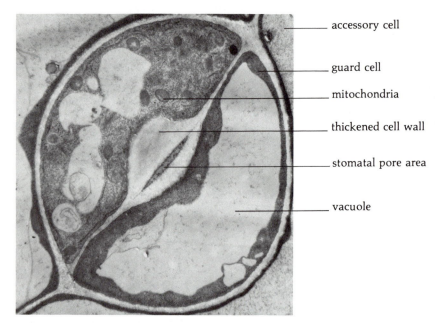

_____ accessory cell

_____ guard cell

_____ mitochondria

_____ thickened cell wall

_____ stomatal pore area

_____ vacuole

Figure 4–3. Electron micrograph and drawing (left) of stoma from garden bean (*Phaseolus vulgaris*). Note large and small vacuoles in guard cells. Stoma on right is characteristic of some grasses. Note barbell-shaped guard cells.

Electron micrograph courtesy of R.P. Zimmerer, Juniata College.

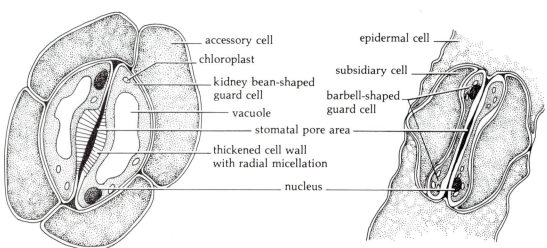

accessory cell

chloroplast

kidney bean-shaped guard cell

vacuole

stomatal pore area

thickened cell wall with radial micellation

nucleus

epidermal cell

subsidiary cell

barbell-shaped guard cell

much higher light intensity is needed to obtain even an average rate of photosynthesis in this species. Indeed, the stomata of some plant species may be induced to open by bright moonlight (33). Plants with crassulacean acid metabolism (CAM plants) exhibit different stomatal behavior in that the sto-

mata are open at night and closed during the day. We will consider these plants later.

There are numerous exceptions to the general rule that stomata exposed to light remain open and those in darkness remain closed. The stomata of some plants are closed only for about three hours after sunset. Potato, pumpkin, onion, and cabbage are examples of plants with this type of stomata. However, their stomata can be induced to close even at midday if the plants are wilted. The stomata of *Equisetum* remain open continuously, even under severe wilting conditions (30). The stomata of most cereals open for only one to two hours during the day. In fact, Brown and Pratt (6) report that many grasses native to dry habitats possess stomata that appear never to open perceptibly.

A study of the effect of different wavelengths of light on the opening of tobacco leaf stomata suggests that some wavelengths are much more effective than others. No opening occurred when stomata were exposed to either far-red or ultraviolet irradiation (64). Good stomatal opening was obtained in the red and blue regions of the spectrum but no opening in the green region. Essentially the same results were obtained with *Senecio*. The response of stomata to different wavelengths bears a resemblance to the action spectrum for adenosine triphosphate (ATP) synthesis in isolated chloroplasts (4). We will return to an in-depth study of ATP production in chloroplasts when we consider photosynthesis.

How does light bring about the opening of the stomata? Early workers assumed that guard cells, when exposed to light and warmth, increased their output of osmotically active substances through the process of photosynthesis. An extension of this idea was that a more negative osmotic potential and a subsequent increase in turgor cause the stomata to open. However, photosynthesis occurs in the guard cell at a reduced rate and certainly does not account for the amount of osmotically active substances required to produce the response of stomata to light. Several workers observed the

strange phenomenon that the starch content of guard cells is high in the dark and low in light (32, 33, 48). In the other epidermal cells and the mesophyll cells exactly the opposite situation exists (20).

Sayre, in his work with *Rumex patientia* (48), also noted that the stomata are sensitive to changes in pH. Generally, a high pH favors opening and a low pH favors closing of the stomata. Subsequently, observers noted that illumination of the guard cells in many species resulted in an increase in pH, and a return to darkness induced a lowering of pH in the guard cells (49, 51). A high pH is accompanied by a decrease in starch and an increase in osmotically active reducing sugars and result in an increase in turgor. The converse responses occur when the pH is lowered. This pH effect is explained by Yin and Tung (62), who obtained evidence of the presence of a phosphorylase in chloroplasts that favors the degradation of starch at pH 7.0 and favors starch synthesis at pH 5.0. Superficially, it seems that a pH-dependent phosphorylase regulates the starch-sugar content and hence the water movement into and out of the guard cells. Very small amounts of osmotically active sugars have been extracted from guard cells. Additional work suggests that phosphorylase is involved primarily with starch degradation rather than with its synthesis (34). To date no explanation based on sugar buildup in the guard cells adequately accounts for guard cell turgidity changes.

Potassium, chloride, hydrogen, and organic acids. Evidence indicates that the turgidity of guard cells of many species of plants is regulated by K^+, Cl^-, H^+, and organic acids.

When plants are exposed to light, the guard cells accumulate large amounts of K^+ (12, 22). Although the process is not entirely understood, light appears to stimulate the active movement of K^+ ions from surrounding cells into the vacuoles of the guard cells. Potassium accumulation is also accompanied by starch degradation, organic acid (mostly malic acid) buildup (45), and an increase in pH. The K^+ accumulation seems to be due to the operation of an *active exchange process* in which protons (H^+) are "pumped" out of the guard cell into the accessory cells. Further, in some but certainly not all species, as the K^+ ions migrate, they are accompanied by Cl^-; the anion apparently moves in response to the electrical differential created by the K^+ uptake into the guard cells.

Organic acids, primarily malic acid, build up in the guard cells of illuminated leaves as protons (H^+) move from the guard cells into the accessory cells. The organic anions (mostly malate) within the guard cells are neutralized by the influx of potassium. Thus in guard cells, osmotic and hence water potentials become very negative due to the presence of K^+, Cl^-, potassium malate or dipotassium malate or both.

A water potential gradient is established from the accessory cells to the guard cell sap that results in water movement, increased turgidity, and stomatal opening. What causes the organic acids to build up is a matter of conjecture at this time. One suggestion is that relative carbon dioxide levels in the guard cells regulate the synthesis of organic acids (31).

CO_2 concentration. Stomata are quite sensitive to variations in CO_2 concentration. For example, under experimental conditions stomatal opening can be induced even in darkness by significantly reducing CO_2 concentration below that of normal air (28, 44, 45, 53). An increase in CO_2 concentration above that found in air will cause stomata to close even in the light. Indeed, stomatal closure can be induced by merely breathing on leaves (35).

It appears that CO_2 concentration in the leaf intercellular spaces rather than in the external air primarily controls stomatal movement. Stomata that have been closed because of exposure to high concentrations of CO_2 do not rapidly open when transferred to a CO_2-free atmosphere in the dark. The logical assumption is that CO_2 concentration in the leaf intercellular spaces remains high and inhibits stomatal opening. However, exposure to light causes the stomata to open because the intercellular CO_2 is consumed in photosynthesis. Stomata located in the nonchlorophyllous regions of variegated leaves respond to light and dark more slowly than do those located in the chlorophyllous regions. Presumably, this is due to a slower change of CO_2 concentration in the intercellular spaces associated with the nonchlorophyllous regions.

Figure 4–4 presents a flowchart of the possible events involved in stomatal opening.

Water deficits and abscisic acid. Whenever the rate of transpiration exceeds the rate of absorption for any period of time, a water deficit is created in the plant. This deficit may take place even under conditions favoring good water absorption and usually results in *incipient wilting*—a condition in which wilting of the leaves has set in, although the wilting is not visible to the eye. Under conditions of water deficit, the stomatal pores of many mesophytes will close, thereby reducing transpiration significantly. In addition, observers have noted that abscisic acid (ABA) will accumulate in the leaves of water-stressed plants (16, 17, 59). And when

Figure 4–4. Proposed mechanisms for opening of stomata. Increased concentration of K^+, Cl^-, and malate in guard cells of some species are supported by experimental evidence. Formation of soluble sugars from starch hydrolysis may not be direct mechanism of stomatal opening (dotted lines).

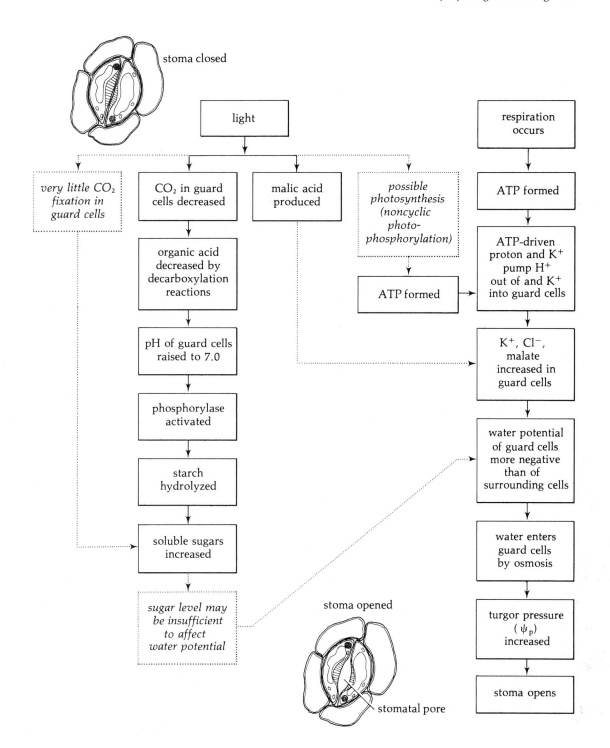

stoma closed

light

respiration occurs

very little CO₂ fixation in guard cells

CO₂ in guard cells decreased

malic acid produced

possible photosynthesis (noncyclic photo-phosphorylation)

ATP formed

organic acid decreased by decarboxylation reactions

ATP formed

ATP-driven proton and K⁺ pump H⁺ out of and K⁺ into guard cells

pH of guard cells raised to 7.0

K⁺, Cl⁻, malate increased in guard cells

phosphorylase activated

starch hydrolyzed

water potential of guard cells more negative than of surrounding cells

soluble sugars increased

water enters guard cells by osmosis

sugar level may be insufficient to affect water potential

stoma opened

turgor pressure (ψ_p) increased

stomatal pore

stoma opens

ABA is applied to intact leaves or epidermal strips, it will stimulate stomatal closure. Thus plant scientists now believe that the phytohormone ABA is a primary regulator of the stomatal apparatus in water-stressed plants. The most striking illustration of ABA activity comes from studies with an X-irradiated induced wilty tomato mutant. This mutant wilts under slight water stress, but its stomata close only after ABA applications. Although evidence is building rapidly to support the role of ABA in water-stressed plants, questions concerning its action remain unanswered. We do not as yet know the stimulus that causes increased levels of ABA, the site of synthesis in the cell, and the mechanism by which ABA regulates stomatal closing.

Temperature. When all other factors are not limiting, an increase in temperature causes an increase in stomatal opening. Wilson (57) demonstrated that the stomata of *Camellia*, privet, and cotton remain closed under continuous light when temperatures are lower than 0°C. As temperatures are increased, stomatal opening in all three plants increases. However, in cotton and onion there is a decline in stomatal opening at temperatures exceeding 30°C (see Figure 4–5). Stomatal closure at these temperatures may be due to higher intercellular CO_2 concentrations caused by increased rates of respiration (19).

Factors Affecting Rate of Transpiration

Other features of the plant, besides stomatal movement, are known to affect the rate of transpiration. The root-shoot ratio and the area and structure of the leaf have considerable influence on the loss of water from

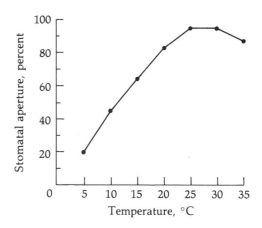

Figure 4–5. Effect of temperature on opening of cotton leaf stomata under constant light.
From C.C. Wilson. 1948. Plant Physiol. *23:5.*

plants. Environmental factors—light, humidity of the air, temperature, wind, and availability of soil water—will affect the steepness of the water potential gradient between the internal and external atmosphere of the leaf and will, therefore, affect the rate of transpiration.

Plant Factors

Root-shoot ratio. In a situation where all of the conditions for good transpiration are present, efficiency of the absorbing surface (root surface) and evaporating surface (leaf surface) controls the rate of transpiration. If water absorption lags behind transpiration, a water deficit will occur in the plant, which, in turn, will reduce transpiration. Parker (42) found that transpiration increases with increase in root-shoot ratio (see Figure 4–6). Sorghum typically transpires at a higher rate per unit of leaf surface than does corn. Miller (40) has pointed out that secondary root development is much

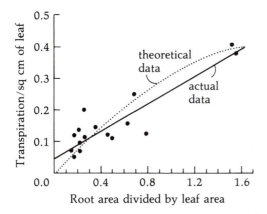

Figure 4–6. Grams of water lost per square centimeter of leaf area per day (loblolly pine) is plotted over root area divided by leaf area in square centimeters.

From J. Parker. 1949. Plant Physiol. *24:739.*

more advanced in sorghum than it is in corn, and that this development may be the main reason for the different transpiration rates in these two plants. In other words, the sorghum root system provides more water to the shoot than does the corn root system.

Leaf area. We can perfectly logically assume that the greater the leaf area, the greater will be the magnitude of water loss. This assumption is correct, although proportional agreement between leaf area and water loss is absent (30). On a per unit area basis, smaller plants often transpire at a greater rate than do larger plants. Note in Table 4–4 (40) that although the larger plant lost more water, the amount of water loss per unit area was greater in the smaller plant.

The removal of leaves from a plant (thereby decreasing the leaf area) may increase the rate of transpiration per unit leaf area of that plant. Thus investigations by Cullinan (8) and Kelley (24) showed that the pruning of fruit trees increases their transpiration rates per unit of leaf area, but that total water loss is greater in the unpruned trees. Presumably, this situation arises from the fact that the root system of the pruned trees is providing a greater amount of water to a smaller number of leaves, thus increasing transpiration efficiency.

Leaf structure. Plants native to dry habitats generally exhibit a number of structural modifications, particularly in their leaves. Thus leaves of xerophytic plants may possess a thick cuticle, thick cell walls, well-developed palisade parenchyma, sunken stomata, a covering of dead epidermal hairs, and so on. That these features affect water loss may be easily shown by allowing detached xerophytic and mesophytic leaves

Plant	Leaf Surface (cm²)	Total H₂O Lost over Six-hr Period (g)	Transpiration Rate (per m² per hr, g)
Pride of Saline corn	14,568	918	629
Sherrod White Dent corn	12,989	784	723

Table 4–4. Total water loss during six-hour period and water loss per square meter per hour by two different strains of corn.

Source: *From* Plant Physiology *by E.C. Miller. Copyright © 1938 McGraw-Hill Book Company. Used with the permission of McGraw-Hill Book Company.*

to dry together under the same conditions. Visible wilting will be observed in the mesophytic leaves long before it is observed in the xerophytic leaves. The resistance of xerophytic leaves to water loss and wilting is primarily a function of cutin layer thickness and efficiency. Under dry conditions, the stomata are closed and cuticular transpiration becomes the main avenue for water loss.

Many investigators have observed that with adequate water supply, the rates of transpiration of xerophytic species may be higher than those of mesophytic species. This situation may be due, in part, to the higher number of stomata per unit area and the more extensive venation of xerophytic leaves as compared to mesophytic leaves. We may even observe the difference in stomatal frequency and venation in the same plant species grown under moist and dry conditions. Another factor that may result in higher rates of transpiration by xerophytic leaves is their greater internal evaporating surface—that is, more cell wall surface is exposed to the internal atmosphere than is generally the case in mesophytic leaves (61, 62).

We must remember, however, that the higher rates of transpiration in xerophytic plants as compared to mesophytic plants are found only under those conditions where the stomata are opened. Perhaps it would be more correct to say that the rate of stomatal transpiration of xerophytes exceeds that of mesophytes.

Environmental Factors

The most important environmental factors that influence the rate of transpiration are light, humidity of the air, temperature, wind, and availability of soil water. Al-though we will discuss the individual effects of each one of these factors, transpiration by the plant in its natural habitat is generally under the influence of several of these factors at any one time. One factor may augment or negate the effect of another factor.

Light. Light occupies a prominent position among those factors influencing transpiration since it has a dominating effect on stomatal movement. The stomata of a plant exposed to light are opened and allow transpiration to proceed. In the dark, stomata are closed and transpiration essentially ceases. The effects of other environmental factors are, therefore, dependent on the presence of light. In certain exceptional cases, particularly CAM plants, however, the stomata are closed in light.

Humidity of air. Before covering the effect of humidity on transpiration, we should discuss some of the terms used to express the moisture content of the air. Most of us are familiar with the term *relative humidity*. Since a direct proportionality exists between vapor pressure and the concentration of water vapor in the atmosphere, relative humidity is an expression of the ratio of the actual vapor pressure to the vapor pressure of the atmosphere when saturated at the same temperature. For example, the atmosphere at 20°C is saturated at a vapor pressure of 17.54 mm Hg and has a relative humidity of 100 percent. If the relative humidity at 20°C is 50 percent, then the vapor pressure would be 8.77 mm Hg, and at 10 percent relative humidity it would be 1.754 mm Hg.

For our purposes, the term *vapor pressure* rather than the term *relative humidity* more precisely defines the situation. For example, at 50 percent relative humidity, the

Temperature, °C	Vapor Pressure, mm Hg, at Different Values of Relative Humidity					
	0	20%	40%	60%	80%	100%
0	0	0.92	1.83	2.75	3.66	4.58
10	0	1.84	3.68	5.53	7.37	9.21
20	0	3.51	7.02	10.52	14.03	17.54
30	0	6.36	12.73	19.09	25.46	31.82
40	0	11.06	22.13	33.19	44.25	55.32

Table 4–5. Relationship between vapor pressure and relative humidity at different temperatures.

vapor pressure of the atmosphere may have any number of values, depending on the temperature (Table 4–5). The vapor pressure at 50 percent relative humidity and at 40°C is 27.66 mm Hg; at 0°C, it is 2.29 mm Hg. The difference in pressure is 23.37 mm Hg. If we relate these figures to the rate of evaporation from a moist surface into surrounding air at 50 percent relative humidity, we can see that evaporation would take place much more rapidly at 40°C than at 0°C. The *vapor pressure gradient* would be much steeper at 40°C (55.32–27.66 mm Hg) than at 0°C (4.58–2.29 mm Hg).

A change in temperature or vapor pressure can change the relative humidity. A rise or fall in temperature with no change in vapor pressure will cause a drop or rise, respectively, in relative humidity. A rise or fall in vapor pressure with no change in temperature will be accompanied by a rise or fall, respectively, in relative humidity.

The internal atmosphere of a leaf is generally considered to be saturated or nearly saturated. The external atmosphere is usually in an unsaturated condition. A vapor pressure gradient (water potential gradient) exists, therefore, between the internal and external atmospheres, and water vapor will diffuse through the stomata from the area of high vapor pressure (less nega-

tive water potential) to the area of low vapor pressure (more negative water potential). The steeper the gradient, the more rapidly transpiration will proceed. If the vapor pressure of the external atmosphere is kept constant, the steepness of the vapor pressure gradient may be increased or lowered by increasing or lowering, respectively, the temperature of the atmosphere. In other words, the external atmosphere can hold more water vapor at a higher temperature and less at a lower temperature. However, the water content of the internal atmosphere of the leaf will remain at or near the point of saturation at both the high and low temperatures. Therefore when the water content or vapor pressure of the external atmosphere remains constant, an increase in temperature brings about an increase in the vapor pressure gradient.

If the external atmosphere has a vapor pressure of 8.77 mm Hg, it would be equivalent to 50 percent relative humidity at 20°C. At 20°C, the internal atmosphere has a vapor pressure of 17.54 mm Hg. If we increase the temperature to 30°C and keep the vapor pressure of the external environment constant at 8.77 mm Hg, the difference in vapor pressure between the internal and external atmospheres increases. By lowering the temperature to 10°C and again keeping the

Internal and External Temperatures (°C)	Internal Atmosphere, mm Hg (100% relative humidity)	External Atmosphere, mm Hg (vapor pressure = 8.77 mm Hg)	Difference, mm Hg (vapor pressure gradient measure)
10	9.21	8.77	0.44
20	17.54	8.77	8.77
30	31.82	8.77	23.05

Table 4–6. Hypothetical situation showing vapor pressure difference between internal and external leaf atmospheres.

vapor pressure of the external atmosphere constant, we can decrease the difference between the two, as shown in Table 4–6.

If the temperature remains constant, the vapor pressure gradient (water potential gradient) between the internal and external atmospheres may be increased or lowered by lowering or increasing, respectively, the vapor pressure of the external atmosphere. Of course, if the vapor pressures of the internal and external atmospheres are the same, no transpiration will take place.

Temperature. If all other factors are constant, an increase in temperature within a certain physiological range almost always brings about an increase in the rate of transpiration. This phenomenon is due to the effect of temperature on stomatal movements and water potential gradients. Stomata generally close at temperatures approaching 0°C and open increasingly as temperatures increase, up to about 30°C (see Figure 4–5).

In addition to its effect on the opening of stomata, an increase in temperature steepens the gradient between the internal and external atmospheres of the leaf.

In the discussion so far, we have assumed that the temperature of the leaf and the air is the same. However, this assumption is not always correct. Fleshy or relatively thick plant structures, such as fruits, stems, and thick leaves, often reach temperatures above that of the surrounding air

when they are exposed to sunlight (30). This factor has the effect of steepening the vapor pressure gradient between the internal and the external atmospheres.

Because temperatures are higher during the day and because light induces sto-

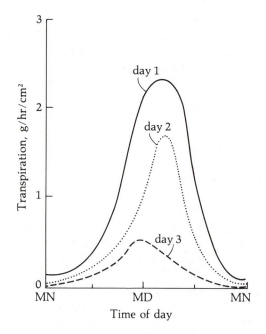

Figure 4–7. Changes in diurnal rhythm of transpiration in bean (*Phaseolus vulgaris*) as soil becomes increasingly dry over 3-day period. Time is midnight (MN) and midday (MD).

Figure 4–10 from W.M.M. Baron. 1967. Physiological Aspects of Water and Plant Life. London: Heinemann. Reprinted by permission.

mata to open, transpiration rates almost always exhibit a clear diurnal rhythm, high rates occurring at midday and low rates at night (see Figure 4–7).

Wind. Air in the immediate area of a transpiring leaf becomes more and more concentrated with water vapor. We know that under these circumstances the vapor pressure gradient is lowered and the rate of transpiration decreases. However, if wind disperses the water vapor that is concentrated in the area of the leaf, the rate of transpiration will again increase.

Increase in the rate of transpiration as a result of wind is not proportional to the wind velocity, as Figure 4–8 shows (47). Several investigators (14, 36, 46) have shown that when plants are suddenly exposed to wind, there is a sharp increase in the rate of transpiration, followed by a gradual falling off of this increase. This phenomenon indicates that the effect of wind on transpiration may be rather complex.

Wind blowing over an evaporating surface has a significant cooling effect, a circumstance that lowers the vapor pressure gradient and, thus, the rate of transpiration. In addition, winds of high velocity may possibly cause stomatal closure. The wind can cause an increase in the rate of transpiration and can also cause stomatal closure. It, therefore, exerts both a negative and a positive influence on transpiration.

Availability of soil water. For a short period of time, absorption of water by the plant may lag behind the release of water via transpiration without noticeably affecting the plant. If this condition is prolonged, however, a water deficit will develop, and the plant will wilt. The availability of soil water to the roots of a plant and the efficiency of its absorption have a profound influence on the rate of transpiration.

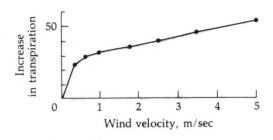

Figure 4–8. Effect of wind velocity on transpiration rate of *Chamaecyparis obtusa*.
From T. Satoo. 1962. *Wind, transpiration and tree growth.* In T. Kozlowski, ed., **Tree Growth.** New York: Ronald Press. Reprinted by permission of John Wiley & Sons, Inc.

Significance of Transpiration
Cooling Effect

The value or lack of value of transpiration to the plant has been a subject of debate among plant physiologists for quite some time. Some have argued that the cooling effect of transpiration keeps the plant from being overheated. However, plants growing under conditions where transpiration is negligible do not overheat, suggesting that the cooling effect of transpiration is of no real significance in the dissipation of heat as far as the plant is concerned. There are certainly other means by which heat may be dissipated—that is, conduction, convection, and radiation.

Effect on Growth and Development

Winneberger (58) has obtained some indirect evidence indicating that transpiration may have an influence on the growth of some plants. He observed that buds of the Hardy pear cease to grow under conditions of high humidity and that, under the same conditions, growth of sunflower plants is reduced to about half of the normal. Since transpiration is negligible under conditions of high humidity, Winneberger concluded that transpiration is a necessary factor in the normal growth of these two plants.

When the rate of transpiration exceeds that of absorption, a water deficit may occur and wilting may take place. This development is, of course, detrimental to the plant and, if carried to extreme, may result in the death of the plant. In temperate climates injury due to desiccation may occur to plants that retain their leaves through the winter months. On some winter and early spring days, air temperatures may be high enough to support a considerable amount of transpiration. However, the soil is usually frozen or nearly so, under these conditions, and water is not absorbed (26, 27). The results may be particularly damaging to conifers.

The amino acid and protein metabolism of a plant can be affected by water stress conditions (2, 7, 25). Not only is protein synthesis inhibited, but there is an acceleration of protein breakdown. For example, Barnett and Naylor (2) found that soluble protein levels in Bermuda grass decreased with increasing water stress. Under the same conditions, there was a significant increase in the levels of free arginine and especially free proline. Possibly, the amino acid proline acts as a storage compound while protein synthesis is inhibited as a result of water stress. On return to normal conditions, excess proline would be utilized to build new protein. Table 4–7 shows changes in the levels of free amino acids in Bermuda grass shoots as water stress is increased.

Amino Acid	Control (μm/g dry weight)	Moderate stress (μm/g dry weight)	Severe stress (μm/g dry weight)
aspartic acid	11.8 ± 8.9	4.5 ± 1.8	8.4 ± 3.5
asparagine; threonine	24.6 ± 4.1	29.8 ± 13.7	64.2 ± 17.1
serine	9.9 ± 2.3	8.3 ± 2.7	11.0 ± 4.5
glutamic acid	28.7 ± 9.2	10.5 ± 4.8	4.7 ± 1.9
N			0.8 ± 0.3
proline	<2.7	30.5 ± 23.9	69.3 ± 33.0
glycine	1.8 ± 1.3	1.7 ± 1.1	1.2 ± 0.7
alanine	31.9 ± 12.3	15.2 ± 3.8	11.6 ± 4.2
cystine			0.6 ± 0.1
valine	2.1 ± 0.7	3.5 ± 1.6	7.0 ± 2.1
isoleucine		0.9 ± 0.4	1.2 ± 0.4
γ-aminobutyric acid	3.2 ± 2.1	7.0 ± 4.1	4.3 ± 1.4
U		0.8 ± 0.2	1.5 ± 0.9
ammonia	94.3 ± 36.6	78.0 ± 54.0	55.4 ± 26.4
lysine	0.5 ± 0.4	0.7 ± 0.1	1.0 ± 0.4
histidine		0.5 ± 0.1	1.4 ± 0.3
arginine		0.8 ± 0.3	2.5 ± 0.1
Totals	211.5	192.9	246.5

Table 4–7. Changes in amounts of free amino acids in common Bermuda grass shoots with increasing water stress.

Source: *From N.M. Barnett and A.W. Naylor. 1966. Amino acid and protein metabolism in Bermuda grass during water stress. Plant Physiol. 41:1222.*

Effect on Mineral Salt Absorption

Because of the presence of mineral salts and water together in the soil, and because both are absorbed by roots, early plant physiologists naturally assumed that salt absorption and transport took place as a consequence of transpiration. Numerous studies in the 1930s, however, clearly established that salt absorption is predominantly an active process (requiring metabolic energy) and that only a small amount of salt is absorbed passively as a result of water uptake (see Chapter 6). Once the absorbed salts have been "dumped" into the xylem ducts of the root, transpiration definitely influences their translocation and distribution in the plant. In this respect, the transpiration stream provides an efficient means of transport and distribution once salts have been absorbed by the roots.

Some authors have maintained that a significant amount of salt absorption does take place passively and is under the influence of transpirational pull. Hylmö (21), Kramer (29), and Pettersson (43) have demonstrated some correlation between the absorption of ions and the rate of transpiration. Currently, plant scientists believe that, although active absorption of salts predominates, some passive absorption also takes place and that this is under the influence of the transpirational pull. However, transpiration is not a necessary driving force for salt distribution in plants since salts continue to be distributed when plants are not transpiring.

Questions

4–1. Name the processes by which water may be lost from plants. Explain each process.

4–2. Indicate the techniques used to measure transpiration. Explain the theory behind each technique.

4–3. Are the guard cells from monocots and dicots similar in appearance? Do they open in much the same manner?

4–4. Of what importance is potassium in the opening and closing of stomata?

4–5. What are some of the past explanations to account for the opening and closing of stomata? Do they have any experimental basis? Is there a general mechanism to explain the opening and closing of stomata? Explain.

4–6. Provide any ideas you might have as to why the stomata of certain crassulacean plants are open at night and closed during the day?

4–7. Indicate some of the factors that influence stomatal opening and closing. How are these factors involved in regulating stomatal behavior?

4–8. What experimental approaches might be studied to develop crop plants that are able to grow in arid environments?

4–9. What are antitranspirants and how might they work?

4–10. Is transpiration necessary for the cooling of plants in temperate zones, particularly at midday? Explain.

Suggested Readings

Aylor, D.E., J.-Y. Parlange, and A.D. Krikorian. 1973. Stomatal mechanics. *Am. J. Bot.* 60:163–171.

Clark, C. 1970. *The Economics of Irrigation.* London: Pergamon Press.

Dixon, H.H. 1914. *Transpiration and the Ascent of Sap in Plants.* London: Macmillan.

Fisher, R.A., and T.C. Hsiao. 1968. Stomatal opening in isolated epidermal strips of *Vicia faba*. II. Responses to KCl concentration and role of potassium absorption. *Plant Physiol.* 43:1958–1968.

Jensen, M.E. 1972. Programming irrigation for greater efficiency. In D. Hillel, ed., *Optimizing the Crop Yield.* New York: Academic Press.

Lüttge, U., and N. Higinbotham. 1979. *Transport in Plants.* New York: Springer-Verlag.

Raschke, K. 1975. Stomatal action. *Ann. Rev. Plant Physiol.* 26:309–340.

Stocking, C.R. 1956. Guttation and bleeding. In W. Ruhland, ed., *Encyclopedia of Plant Physiology* 3:489. Berlin: Springer.

Thut, H.F. 1928. Demonstration of the lifting power of evaporation. *Ohio J. Sci.* 28:292.

Whyte, W.T. 1981. The land and water squeeze on our food. In J. Hayes, ed., *Yearbook of Agriculture.* Washington, D.C.: USDA.

Willmer, C.M., and J.E. Pallas, Jr. 1973. A survey of stomatal movements and associated potassium fluxes in the plant kingdom. *Can. J. Bot.* 51:37–42.

Chapter 5

Detection, Occurrence, and Availability of Essential Elements

Cucumber plants grown commercially in liquid culture (hydroponically). *Courtesy of E.L. Bergman, The Pennsylvania State University.*

In the following three chapters, we will discuss the fundamental principles of mineral nutrition, a subject recognized but not clearly understood early in the history of agriculture. People from primitive agricultural societies undoubtedly observed that adding plant and animal debris to the soil increased crop yield. Woodward's observation (57) in 1699 that plants can survive and grow better in muddy water than in clear rainwater was very puzzling at the time. The ease with which we explain this phenomenon today is based on the cumulative work of many pioneer scientists.

For the actual recognition of the dependence of plants on elements contained in the soil, we must give credit to de Saussure (15). In 1804 de Saussure clearly demonstrated that the inorganic mineral elements contained in the ash of plants are obtained from the soil via the root system. He maintained that nitrogen and mineral elements supplied by the soil were essential to the growth and development of the plant. Despite the strong experimental evidence presented by de Saussure, the contribution of the inorganic components of plant ash to the general welfare of the plant was not fully recognized until the brilliant scientist Liebig gave it his support. Liebig's address to the British Association for the Advancement of Science in 1840 provided the springboard for today's vast knowledge concerning mineral nutrition.

Elements Found in Plants

Essential Elements

In 1830 Sachs and Knop made serious attempts to determine the mineral content of plants experimentally. Using liquid cultures—aqueous nutrient solutions in which they immersed the plant roots—they were able to show that the following ten elements are essential to the plant: carbon (C), hydrogen (H), oxygen (O), nitrogen (N), phosphorus (P), potassium (K), calcium (Ca), sulfur (S), magnesium (Mg), and iron (Fe). These ten elements were generally accepted as all that a plant needed for normal growth and development. However, we know today that minute amounts of at least five other elements are essential to the growth of most plants, and several additional elements are specifically required by certain plants. The method of growing plants in aqueous nutrient solutions as employed by Sachs and Knop is used experimentally and commercially today and is known as *hydroponic culture*.

Trace Elements

Because the analytical methods in the time of Sachs and Knop were crude by today's standards, minute amounts of certain elements in plants, called *trace elements*, could not be detected. In addition, contamination of the water cultures could not be avoided to the extent needed to eliminate trace amounts of minerals. Therefore, elements required in trace amounts remained undetected, and their necessary function in the plant remained unknown. In the early part of the twentieth century, with the use of better measuring devices and greater attention to sterile techniques, the requirements of some of these trace elements began to be observed. Bertrand (8) was the first to observe that the element manganese (Mn) is needed for normal plant growth. By 1939 the necessity of the trace elements Mn, zinc (Zn), boron (B), copper (Cu), and molybdenum (Mo) had been detected in various plants. Thus our list of the elements necessary for the normal growth and development of the majority of plants is C, H, O,

N, P, K, Ca, S, Mg, Fe, and the trace elements Mn, Zn, B, Cu, and Mo.

In addition to these, other elements that are essential for the normal growth of certain (although not the majority of) plants are sodium (Na), aluminum (Al), silicon (Si), chlorine (Cl), gallium (Ga), and cobalt (Co).

Methods of Detection and Physiological Effects

Ash Analysis

To detect some mineral elements of a plant, we can subject the plant to high temperatures (about 600°C) and then analyze its ash content. In the ash, only the mineral elements are present, all of the organic compounds have been decomposed and passed off in the form of gases. The primary elements (carbon, hydrogen, and oxygen) are therefore given off as CO_2, water vapor, and oxygen. In addition to carbon, hydrogen, and oxygen, we cannot accurately detect the element nitrogen with this method since some of it is given off in the form of ammonium or nitrogen gas. All of the other mineral elements that were absorbed from the soil will be present in the plant ash. Table 5–1 provides an example of the ash analysis of the mineral content of *Zea mays*.

Although we may think of the analysis of plant ash as a method of determining the relative quantities of mineral elements in a plant, too many variables are present to give accurate, reliable results. For example, the high temperatures may cause vaporization or sublimation of some of the elements. Generally, the elements in the ash are not present in their pure state but rather in the form of oxides. The qualitative and quantitative analyses of the ash for the dif-

Table 5–1. Ash analysis of Pride of Saline corn plants grown at Manhattan, Kansas.
Source: *From* Plant Physiology *by E.C. Miller. Copyright 1938, McGraw-Hill Book Company. Used with permission of McGraw-Hill Book Company.*

Element	Weight (g)	Total Dry Weight (%)
nitrogen	12.2	1.459
phosphorus	1.7	0.203
potassium	7.7	0.921
calcium	1.9	0.227
magnesium	1.5	0.179
sulfur	1.4	0.167
iron	0.7	0.083
silicon	9.8	1.172
aluminum	0.9	0.107
chlorine	1.2	0.143
manganese	0.3	0.035
undetermined elements	7.8	0.933

ferent elements present is dependent on various chemical treatments. The chance of cumulative erroneous results gathered from these treatments is too great to allow heavy reliance on quantitative data for the majority of minerals obtained from the ash analysis of plant tissue. Finally, we must emphasize that, although ash analysis provides information concerning the relative amounts of minerals present in or taken up (e.g., aluminum and silicon) by the plant, they are not reliable methods for determining the extent of the utilization of these minerals by the plant.

Solution Culture

It did not take scientists long to realize the impracticality of using soil as a medium for growth in any serious study of plant mineral requirements. To render a soil free of the mineral elements used by plants and then control the amounts of nutrients made available to the roots embedded in the soil is impossible. In contrast, solution cultures provide an excellent means for controlling the quantity and relative proportions of mineral salts given to a plant in an experiment. Two other good reasons for using solution cultures in mineral nutrition studies are the excellent solvent characteristics of water and the relative ease with which water can be freed of most contaminating influences.

By using water as a medium, we may set up good quantitative studies of the nutritional needs of plants. However, good results depend on careful attention to small details. Because satisfactory growth may be achieved with extremely small amounts of trace elements, contamination problems are always present. Some of the sources of contamination are the rooting medium, the reagents, the containers, the water, the cutting implements, the seeds, and the dust in the surrounding atmosphere. Obviously, we cannot totally eliminate these contaminating influences, but we can keep them to a minimum.

Several studies have shown that the best containers for solution cultures are made of borosilicate glass or natural polyethylene (20). However, even if we use these materials, we may expect some contamination, such as the presence of boron in borosilicate glass and, perhaps, molybdenum and cobalt in polyethylene. Water distilled in metal stills usually is contaminated with trace amounts of copper, zinc, and molybdenum. Redistillation of water in stills made entirely of borosilicate glass is necessary to remove these elements (40, 52). Another satisfactory method of ridding water

of contaminating trace elements is to pass it over cation and anion exchange resins (21).

In early studies of plant nutrition, the nutrient reagents used presented a major source of contamination. These reagents had to be purified by various means before trace element deficiencies could be demonstrated. Reagents may be purchased today that are pure enough for most studies, but even these contain trace amounts of contaminants.

From this discussion we can see that most of the difficulties encountered in mineral nutrition studies are associated with trace element contamination. A study of deficiencies caused by major nutrients can be easily made because of the relatively large amounts needed for normal growth. In such studies a small amount of contamination is not a serious problem.

The next step is to prepare stock solutions from inorganic salts containing the necessary elements for normal plant growth. Once stock solutions are prepared and the proper containers obtained and filled with deionized water, nutrient solutions may be prepared by simply adding, in the correct proportion, the necessary inorganic salts from the stock solutions. Table 5–2 shows satisfactory formulas for nutrient solutions.

A simple manipulation of the complete formula so that one of the necessary elements is left out provides a solution in which the deficiency symptoms caused by the lack of that element may be studied. As Figure 5–1 illustrates, the roots of the plants to be studied are submerged in the nutrient solution, and the stem projects through an opening cut in the container cover. To give a more rigid system, the stem is generally held stationary in this opening by some inert padding material such as cotton. The container needs to be covered in order to

(1) Salt	gram/liter	Salt	mg/liter
KNO_3	1.02	H_3BO_3	2.86
$Ca(NO_3)_2$	0.492	$MnCl_2 \cdot 4H_2O$	1.81
$NH_4H_2PO_4$	0.23	$CuSO_4 \cdot 5H_2O$	0.08
$MgSO_4 \cdot 7H_2O$	0.49	$ZnSO_4 \cdot 7H_2O$	0.22
		$H_2MoO_4 \cdot H_2O$	0.09
		$FeSO_4 \cdot 7H_2O$ 0.5% ⎱	0.6 ml/liter
		Tartaric acid 0.4% ⎰	(3 × weekly)

(2) Salt	gram/liter	ppm	mM/liter
KNO_3	0.505	K, 195; N, 70	5.0
$Ca(NO_3)_3$	0.820	Ca, 200; N, 140	5.0
$NaH_2PO_4 \cdot 2H_2O$	0.208	P, 41	1.33
$MgSO_4 \cdot 7H_2O$	0.369	Mg, 24	3.0
Ferric citrate	0.0245	Fe, 5.6	0.1
$MnSO_4$	0.002230	Mn, 0.550	0.01
$CuSO_4 \cdot 5H_2O$	0.000240	Cu, 0.064	0.001
$ZnSO_4 \cdot 7H_2O$	0.000296	Zn, 0.065	0.001
H_3BO_3	0.001860	B, 0.370	0.033
$(NH_4)_6Mo_7O_{24} \cdot 4H_2O$	0.000035	Mo, 0.019	0.0002
$CoSO_4 \cdot 7H_2O$	0.000028	Co, 0.006	0.0001
$NaCl$	0.005850	Cl, 3.550	0.1

Table 5–2. Two nutrient solution formulas. Source: *Part (1) from D.I. Arnon and D.R. Hoagland. 1940. Soil Sci. 50:4.* © *1940 The Williams & Wilkins Co., Baltimore. Part (2) from E.J. Hewitt. 1963. In F.C. Steward, ed.,* Plant Physiology. *Academic Press, New York.*

eliminate as much contamination due to atmospheric dust as possible. For good root growth and mineral salt absorption, some means of aeration should be provided.

Solid Medium Cultures

Although a solid medium, such as sand, crushed quartz, or gravel is generally easier to work with than a liquid medium, purification problems are more difficult to cope with (Figure 5–2). Today, however, we can obtain highly purified silica sand and crushed quartz that are very low in available trace elements. The added attraction of a solid culture is that the roots grow in a natural medium and no means of support needs

to be provided (see Figure 5–2c, for example). Nutrient solutions are added to the solid culture in three different ways: by pouring over the surface (*slop culture*), by dripping on the surface (*drip culture*), and by forcing solution up from the bottom of the container (*subirrigation*). In all three systems, the nutrient solutions that are added drain out through an opening in the bottom of the container. In subirrigation, the solution is collected in a reservoir and used repeatedly. The pumping apparatus used in this latter system may be attached to a timing mechanism that can be set to give periodic irrigation to the sand, quartz, or gravel.

Of the three systems, the slop culture is the easiest to manipulate but offers the least control. The drip culture may be set up

Figure 5–1. Tomato (*Lycopersicon esculentum*) plants grown hydroponically in jars. Diagram illustrates aeration procedure.

Courtesy of E.L. Bergman, The Pennsylvania State University.

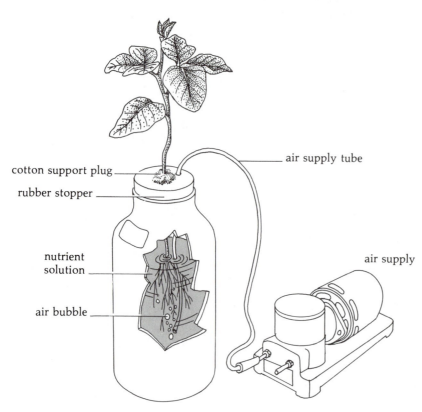

a.

b.

c.

d.

Figure 5–2. Techniques using nutrient cultures for greenhouse plant production. (a) "Pipe dream" method: tomato plants grow up through vertical cylinders, roots extend into horizontal channels and are bathed in flowing stream of nutrient solution. (b) Lettuce plants grow hydroponically in large containers of nutrient medium with suitable aeration. (c) Geraniums grow in gravel culture and periodically receive nutrient medium. (d) Bean plants are supported in styrofoam, roots are bathed in nutrient medium.

Photo (a) courtesy of W. Troxell, Master's thesis, The Pennsylvania State University; photos (b), (c), and (d) courtesy of E.L. Bergman, The Pennsylvania State University.

so that the amount of solution being added is equal to the amount of solution draining off. This system allows for a continuous nutrient supply and partial control of the amount of nutrients reaching the root system. The subirrigation system may be set up to work automatically and also gives partial control of the amount of nutrients reaching the plant roots. The subirrigation system is the most desirable of the three systems, but it is the most expensive and the most difficult to set up initially.

Occurrence of Elements

Because of their relative importance and abundance in the plant, we will not cover carbon, hydrogen, oxygen, and nitrogen in this chapter but will give them more extensive attention in separate chapters.

Phosphorus

Phosphorus is present in the soil in two general forms, organic and inorganic. In the organic form, phosphorus may be found in nucleic acid, phospholipids, and inositol phosphates, compounds that are common to the organic part of the soil. According to present-day knowledge, plants do not absorb organic phosphorus, either from the solid or solution phase of the soil. Therefore, organic phosphorus represents an unusable form of the element with respect to the plant. However, organic compounds are eventually decomposed, and phosphorus is released in an inorganic form that is readily taken up by the plant.

As reported by Wiklander (53), much of the phosphorus of the soil solution is present in the inorganic form, mainly as the phosphate ions $H_2PO_4^-$ and HPO_4^{2-}. The quantity of either ion present is dependent on the pH of the soil solution, the lower pH favors the $H_2PO_4^-$ ion, and the higher pH, HPO_4^{2-}.

Phosphate ions are adsorbed very strongly to the solid phase of the soil; the result is a very low concentration of phosphate in the soil solution. Work with radioactive phosphorus has revealed that continuous exchange reactions occur between the free inorganic phosphate ions of the soil solution and phosphate ions adsorbed to the solid phase (32, 36, 37).

Figure 5–3 presents data from a study by McAuliffe and colleagues (32) of phosphorus exchange reactions between the solid and liquid phases of the soil. A soil sample was first allowed to stand in water for four days to provide sufficient time for any phosphate originally present in the solid and liquid phase to come to equilibrium. A small amount of radioactive ^{32}P inorganic phosphate in solution was then

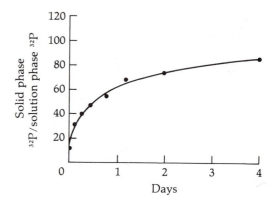

Figure 5–3. Ratio of solid phase ^{32}P to solution phase ^{32}P plotted against time following addition of trace amount of ^{32}P inorganic orthophosphate to suspension of Caribou soil.

Adapted from Soil Science Society of American Journal, *volume 12, 1948, pages 119–123 by permission of the Soil Science Society of America.*

added to the system. Since, as Figure 5–3 shows, most of the ^{32}P was soon adsorbed to the solid phase, we can infer that an equilibrium exists between the phosphorus of the solid and liquid phases of the soil and that most of the phosphorus is adsorbed to the solid phase.

The most important factors controlling the availability of phosphorus are pH of the soil solution, dissolved aluminum and iron, available calcium, anion exchange, and presence of microorganisms.

pH of soil solution. Three different forms of the phosphate ion may be encountered over the pH range found in soil solutions. Under very acid conditions, the monovalent form ($H_2PO_4^-$) is prevalent; the divalent form (HPO_4^{2-}) is present in the intermediate pH range, and the trivalent form (PO_4^{3-}) exists under alkaline conditions. At a pH reading intermediate between two of the ionic levels, two ionic forms of the phosphate ion may be found. Thus it is possible at a pH of 6 to have both the monovalent and divalent phosphate ion present in the soil solution. Although phosphorus is readily absorbed by the plant in the ionic form, phosphate is adsorbed very firmly, thus limiting the supply of phosphate ion to the plant.

Dissolved aluminum and iron. Under very acid conditions, enough soluble aluminum and iron is present to precipitate phosphate as iron and aluminum phosphates, a form of phosphorus that is unavailable to the plant. Strong evidence for precipitation reactions involving aluminum and iron has been presented (14, 23).

Available calcium. Calcium may react with all three forms of the phosphate ion to give the salts monocalcium phosphate ($Ca(H_2PO_4)_2$), dicalcium phosphate (Ca_2HPO_4), and tricalcium phosphate ($Ca_3(PO_4)_2$). Because of its solubility in water, monocalcium phosphate represents a form of phosphorus that is available to the plant. The dicalcium phosphate is only slightly soluble in water but will release phosphorus to the plant. However, tricalcium phosphate, which is formed under alkaline conditions, precipitates phosphate in an almost insoluble form, thereby making it unavailable to the plant. Magnesium acts in much the same manner as calcium, forming mono-, di-, and tri-magnesium phosphates.

The presence of excessive amounts of calcite ($CaCO_3$) in the dry alkaline soils of some of our western states creates a serious problem in phosphorus nutrition. Phosphorus is usually applied to phosphorus-deficient soils as superphosphate. Superphosphate contains available phosphates, such as $Ca(H_2PO_4)_2$, which react with $CaCO_3$ to form the insoluble $Ca_3(PO_4)_2$. Thus phosphorus added in this manner to

an alkaline soil containing $CaCO_3$ is never made available to the plant.

The importance of pH to the availability of phosphorus is brought out strikingly in the above discussion. In acid soils, phosphorus availability is limited by the presence of soluble aluminum and iron, and in alkaline soils its availability is hampered by the formation of insoluble calcium phosphate salts. Thus, for the best results in phosphorus nutrition, soil pH conditions between 6 and 7 are necessary.

Anion exchange. Anion exchange may take place between the minerals contained in the clay micelles of soil and the phosphate ion, a reaction somewhat similar to that involving aluminum and iron hydroxides. Presumably, the anion $H_2PO_4^-$ replaces a hydroxyl anion on the surface of the clay micelle under mild acid conditions.

The addition of hydroxyl ions to the soil, such as occurs in liming operations, will shift the reaction to the left, releasing the phosphate anion and raising the pH, thus also releasing phosphate from aluminum and iron complexes. However, overliming, which may cause a pH rise to over 7, could again tie up phosphate in the form of insoluble calcium phosphates.

Microorganisms. In soils high in organic matter, there is usually a high population of microorganisms. A significant proportion of inorganic phosphate may be "biologically fixed" under these circumstances. The phosphorus temporarily fixed in the organic structures of these organisms is eventually returned to the soil in a bound form. After *mineralization*, or return to free elemental form, it may be utilized again by the plant.

Calcium

Calcium is the major exchange cation of fertile soils (31). However, the major portion of calcium in the soil is found in a non-exchangeable form, chemically bound in primary minerals such as *anorthite* ($CaAl_2Si_2O_8$). Through weathering, this calcium can be made available. *Calcite* ($CaCO_3$ is present in soils of semiarid and arid regions, and insoluble calcium phosphate salts occur in alkaline soils. Some of this calcium is available to the plant, depending on the solubility of the salt and the degree of alkalinity.

Much of the exchangeable calcium of the soil is adsorbed onto the surface of clay micelles. These micelles are commonly thought of as being disk-shaped bodies with a surface-enveloping layer of negative charges. The micelle, as a whole, may be said to be negatively charged. The negative charges of the micelle strongly attract cations such as H^+ and Ca^{2+}, these cations being readily adsorbed to the surface of the micelle (see Figure 5–4).

The reaction shown in Figure 5–4 is reversible—that is, if the hydrogen ion concentration is raised, Ca ions will be released and the H ions will take their place. This phenomenon is known as *cation exchange*.

Other cations, such as Mg^{2+}, Na^+, and K^+, may also become adsorbed to the surface of clay micelles. However, Ca^{2+} appears to be the most active in this respect.

We have already discussed some of the undesirable characteristics of an acid soil. Specifically, we mentioned the activity of soluble aluminum and iron compounds, which tend to tie up free phosphate ions. What may be done to remedy undesirable acid conditions in a soil?

One of the main reasons for an acid condition is the lack of exchangeable metallic cations and a predominance of exchangeable hydrogen ions. Addition of cations, such as calcium or magnesium, can alleviate acid conditions and, at the same time, supply essential elements to the soil. The most effective and economical method of control-

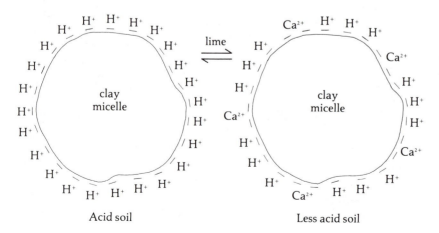

Acid soil   Less acid soil

Figure 5–4. Effect of liming on clay micelles in acid soil. Cation exchange takes place, with some Ca^{2+} being adsorbed to surface of micelle.

ling soil pH is the application of lime. Lime to the chemist is *calcium oxide* (CaO), but to the farmer it is any compound containing calcium or magnesium capable of counteracting the harmful effects of an acid soil (35).

In an acid soil, we have clay micelles with a predominance of exchangeable hydrogen ions adsorbed to their surfaces. With the addition of lime compounds, such as calcium carbonate ($CaCO_3$) or calcium oxide (CaO), many of the hydrogen ions are replaced by calcium ions. In addition, the released hydrogen ions are tied up in the form of water. The final result is a rise in pH and an increase in the supply of exchangeable calcium ions (see Figure 5–5).

Liming has harmful as well as beneficial effects. Overliming a soil may cause the pH of the soil to rise above 7. In sandy soils, for example, where the protective buffering effect of organic matter is absent, harmful

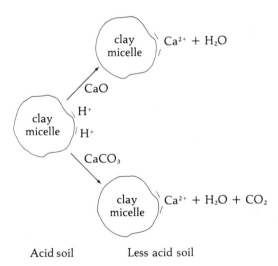

Acid soil   Less acid soil

Figure 5–5. Occurrence of cation exchange between calcium and hydrogen ions resulting from application of lime compounds to acid soil.

conditions may result from overliming. Calcium and phosphate have a tendency to form insoluble calcium phosphate salts under alkaline conditions, thereby making both calcium and phosphate unavailable to the plant. In addition, with a rise in pH to above 7, manganese, iron, zinc, and copper are definitely less available to the plant (29, 30). Boron availability may also be hindered by overliming.

Magnesium

Magnesium is present in the soil in water-soluble, exchangeable, and fixed form and is present in primary minerals (10). Like calcium, it is an exchangeable cation. However, magnesium is much less abundant than calcium. Much less of this cation is adsorbed to clay micelles and hence less is available to the plant through cation exchange. By far the greatest amount of soil magnesium is found in magnesium silicates, a form unavailable to plants until weathering processes cause release of the magnesium in a soluble or available form (7). Longstaff and Graham (27) have studied the availability of fixed magnesium from some minerals; Table 5–3 presents the data. Fixed magnesium in minerals such as magnesite ($MgCO_3$), olivine (($MgFe)_2SiO_4$), and dolomite ($MgCo_3 \cdot CaCO_3$) is available to plants in satisfactory amounts for growth. In fact, *dolomite* and its products are the most popular and economical sources of magnesium fertilizer (17).

Magnesium-deficient areas in the United States are chiefly confined to the sandy soils of the eastern seaboard where agricultural soils have to be furnished periodically with a magnesium supplement such as dolomite. Soils developed on sandstones, granites, and coastal sands are rela-

Mineral	Mg in Plant Tissue (%)	Uptake of Mg (mg/pot)	Condition of Plant
control	0.16	16.0	Mg deficiency
hornblende	0.15	17.5	Mg deficiency
talc	0.19	21.2	Mg deficiency
magnesite	0.20	41.8	normal
olivine	0.24	47.1	normal
dolomite	0.29	51.8	normal

Table 5–3. Uptake of magnesium by soybean plants from some mineral of soil.

Source: *From W.H. Long-staff and E.R. Graham. 1951. Release of mineral magnesium and its effect on growth and composition of soybeans. Soil Sci. 71:167.* © 1951, The Williams & Wilkins Co., Baltimore.

tively low in magnesium; soils developed on basic rock and dolomitic limestones have an abundance of magnesium (7).

Potassium

Potassium is present in the soil in a nonexchangeable (fixed) form, an exchangeable form, and a soluble form. Although there is a relatively high content of this element in the soil, most of it is nonexchangeable and, therefore, unavailable to the plant. When we speak of an element being unavailable, especially with respect to potassium, we mean that utilization of the element in its present form by the plant is not possible. However, availability of potassium in potassium bearing minerals, such as biotite, muscovite, and illite, is made possible through normal weathering processes. Some reports have concluded that the major portion of potassium removed by crops from the soil comes from nonexchangeable sources. Figure 5–6 illustrates the forms of potassium in the soil in the presence of the clay mineral illite.

Wiklander (54) has reviewed the nature and mechanism of potassium fixation and its release in an available form. Through leaching and weathering processes, some of the lattice potassium ions are released. The empty "holes" left by the

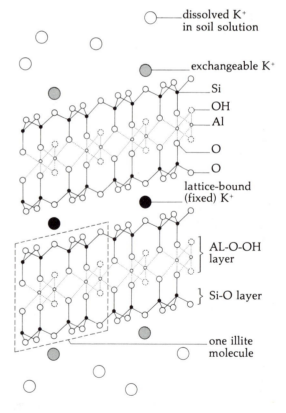

Figure 5–6. Dissolved, exchangeable, fixed and lattice-bound potassium on illite.

migrating potassium ions may be filled by calcium, magnesium, or hydronium ions (H_3O), thereby resulting in partial expansion of the mineral and a depletion in the soil's supply of potassium. When potassium salts are applied to the soil, the "alien ions" are released from the lattice and are replaced by the newly added potassium ions. However, the newly fixed K^+ ions are not held as securely as the original K ions and are, therefore, more available to the plant.

An equilibrium exists between the soluble, exchangeable, and fixed forms of potassium.

soluble K $\rightleftharpoons$ exchangeable K $\rightleftharpoons$ fixed K

As in all equilibria, a change in the concentration of any one of the constituents will cause a shift toward stabilization. For example, depletion of the soluble K in the soil by the plant and soil microorganisms will cause a release of exchangeable K, which, in turn, will cause the slow release of fixed K. This situation is desirable because adsorbed and fixed K, which are not readily leached from the soil, can be made available to the plant.

Sulfur

Soil sulfur is found primarily in the organic fraction (41), but it may also be found in minerals such as pyrite, cobaltite, gypsum, and epsomite and in the soil solution as the sulfate ion (SO_4^{2-}). Sulfur is taken up by the plant as the sulfate ion. Like the phosphate ion, the sulfate ion is weakly adsorbed, the adsorption increasing with a decrease in soil pH. Adsorption is favored by the presence of hydrated oxides of iron and aluminum (54). The sulfate ion is generally thought of as replacing hydroxyl ions in clay minerals, a process known as *anion exchange*. Operations such as liming, which tend to bring the soil pH up by adding hydroxyl ions, cause the release of sulfate ions from the soil particles and their replacement by the hydroxyl ions.

Organic sulfur is made available to the plant through *biological oxidation*. Through the activity of certain microorganisms, sulfur is transformed from the organic form to the sulfate ion, the form of sulfur that higher plants absorb. Soil microorganisms oxidize not only organic sulfur but also sulfide minerals such as ferrous sulfide (FeS). Where there is good aeration, sufficient moisture, and a suitable temperature, FeS can be chemically oxidized to elemental sulfur. The elemental sulfur is then oxidized to sulfate by sulfur bacteria. The two-step oxidation of ferrous sulfide in soil was first demonstrated by Wiklander, Hallgren, and Jonsson (55) and may be written as follows:

$$FeS + H_2O + \tfrac{1}{2}O_2 \xrightarrow[\text{oxidation}]{\text{chemical}} Fe(OH)_2 + S$$

$$2S + 2H_2O + 3O_2 \xrightarrow[\text{oxidation}]{\text{biological}} 2H_2SO_4$$

Biological oxidation in the soil of pyrite (FeS_2) has also been demonstrated, sulfuric acid being the final product (54).

Another source of soil sulfur is sulfur dioxide from the atmosphere, the sulfur being brought to the soil by rain and snow (56). Near industrial centers, this source may reach significant proportions. The direct absorption of sulfur dioxide by the soil (and perhaps by plants) can also be considered a source of soil sulfur (3).

Iron

Soils are usually not deficient in iron, but they may be deficient in exchangeable and soluble forms of iron. Appreciable quantities of iron are present in minerals, in hydrated oxides such as limonite ($Fe_2O_3 \cdot 3H_2O$), and in the sulfide form (10). Iron is most available to the plant in the ferrous form, but significant quantities of the ferric ion may also be absorbed.

The availability of iron to the plant is controlled rather sharply by the soil pH. In acid soils, appreciable amounts of iron are dissolved in the soil solution and are available to the plant. However, in neutral or alkaline soils, iron is much more insoluble. In fact, one of the dangers of overliming is that the resulting increase in pH will cause symptoms of iron deficiency to appear in plants. However, even in soils poor in soluble iron, this element may be available by the direct contact of plant roots with iron-containing soil particles (13).

Manganese

According to Leeper (24), the manganese of the soil may exist in the bivalent, trivalent, and/or tetravalent forms. The bivalent ion may be found dissolved in the soil solution or as an exchangeable ion adsorbed to the soil colloids, both of which are available to the plant. The exchangeable bivalent ion is significant in manganese nutrition since very little of the soil manganese is likely to be found dissolved in the soil water (54). Much of the manganese of the soil is tied up in insoluble compounds in the tri- and tetravalent forms and to a lesser extent in the bivalent form and thus is nonchangeable or unavailable to the plant. Also, manganese combined in the organic form is unavailable. The major portion of the insoluble compounds are tetravalent and trivalent oxides of manganese.

Since it is the reduced form of manganese (bivalent ion) that is absorbed by the plant, poorly aerated, acid soils should favor the availability of manganese. Under these conditions, the tri- and tetravalent forms may be reduced to the bivalent form. Conversely, well-aerated alkaline soils will favor the oxidation of manganese, thus making it unavailable to the plant. Manganese oxides, such as Mn_2O_3 and MnO_2, are formed under these conditions. Obviously, this is another situation where liming the soil, thus raising the pH, may cause the unavailability of an essential element.

The conversion of bivalent manganese to the tri- and tetravalent forms may also occur through biological oxidation (24). The activity of microorganisms in this respect is most prevalent in neutral or slightly alkaline soils, according to Quastel (41); he also reports that the higher valency forms of manganese may also be biologically reduced to the bivalent form and thus made available to the plant. Figure 5–7 illustrates manganese conversion in the soil.

The amount of phosphate in the soil may indirectly affect the availability of manganese. Thus the addition of calcium hydrogen phosphate to soil has been shown to increase the uptake of manganese (9). An increase in soluble manganese because of the formation of soluble manganese phosphates may be the reason for this increased absorption.

Copper

The major portion of the copper of primary rock is present as chalcopyrite ($CuFeS_2$), which is the probable source of natural deposits of copper sulfide in the soil (10). Very little copper is found dissolved in the soil solution. Wiklander (54) estimates that the soil solution of ordinary soils contains 0.01 ppm copper, and the actual water-soluble amount does not exceed 1 percent of the soil.

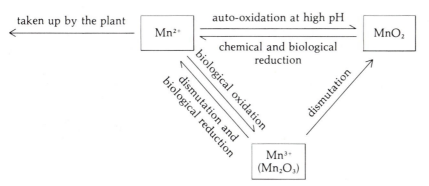

Figure 5–7. Manganese conversion in soil under aerobic conditions. *Data from P.J.G. Mann and J.H. Quastel. 1946. Nature. 158:154.*

The divalent copper cation is adsorbed very strongly to the soil colloids and organic materials of the soil (19), a form in which it is relatively exchangeable. Adsorption of copper as a complex monovalent ion ($CuOH^+$, $CuCl^+$) has been demonstrated in organic soils (28) and on clay minerals (34).

Soil copper also may form very stable complexes with the organic matter of the soil and in this form is nonexchangeable. In addition, copper may be found in nonexchangeable form as a constituent of organic debris or as a component of primary and secondary minerals (54). The unavailability of copper tied up in organic matter has been emphasized by Steenbjerg (45), who pointed out that this may be one of the major causes of copper deficiency in organic soils.

The addition of calcium hydrogen phosphate to the soil appears to cause a decrease in the uptake of copper by sour orange (9). The formation of insoluble copper phosphate may be the reason for this phenomenon.

Zinc

According to Bould (10), zinc occurs in the ferromagnesium minerals, magnetite, biotite, and hornblende. Weathering of these minerals releases zinc in the divalent form, which is readily adsorbed onto soil and organic matter in exchangeable form. Although little is known about the concentration of zinc in the soil solution, it is thought to be quite low.

As with many other essential elements, one of the factors controlling the availability of zinc is the soil pH. The availability of zinc decreases as pH increases; therefore, plants growing in alkaline soils will very likely exhibit symptoms of zinc deficiency. Camp (12) has noted that zinc deficiency may occur in *Citrus* growing in a soil with a pH above 6. An increase in the availability of zinc with a decrease in pH is thought to be a consequence of the action of acids on the solubility of ZnS and $ZnCO_3$ and on the rate of weathering of zinc-bearing minerals (54).

As with copper, the addition of calcium hydrogen phosphate to the soil results in a decrease in zinc uptake by plants (9, 43). One reason given for the decreased uptake is the formation of relatively insoluble zinc phosphate in the soil.

Boron

Boron appears in exchangeable, soluble, and nonexchangeable forms in the soil—that is, as boric acid (H_3BO_3), calcium or manganese borates, and as a constituent of silicates (10, 54). Like zinc, the dissolved boron content in the soil solution is very low. Analyses of different soils have indicated that the amounts of boron in organic soils may be higher than those found in acid soils of humid regions where boron deficiency is likely to occur.

As with manganese and zinc, raising the pH of the soil causes boron to become less available to the plant. The probable reason for this is the formation of insoluble boron compounds. However, this view has been challenged by Drake, Sieling, and Scarseth (16), who claim that over a wide range of pH the solubility of boron is unaffected. Perhaps the solution to this controversy may be found in the well-known observation that liming a soil may cause the unavailability of boron. In a liming operation, the pH of a soil is usually raised, a condition that appears to support the suggestion that raising the pH of a soil makes boron less available. However, liming a soil also increases its calcium content, and Reeve and Shive (42) have found that increasing amounts of calcium in sand cultures causes decreasing uptake of boron in tomato plants. Since liming is a common method of raising soil pH, perhaps the explanation for the observation that increasing the pH decreases the availability of boron is found not in the influence of pH but in the influence of calcium.

The addition of calcium hydrogen phosphate to the soil decreases the uptake of boron just as it decreases the uptake of zinc and copper. Whether this outcome is the result of the addition of calcium or of the addition of phosphate, as was the case with copper and zinc, remains unclear.

Molybdenum

According to Wiklander (54), molybdenum is present in soils in three forms:

dissolved in the soil solution as molybdate ions (MoO_4^{2-} or $HMoO_4^{-}$), adsorbed to soil particles in an exchangeable form, and in a nonexchangeable form as a constituent of soil minerals and organic matter. The amount of molybdenum dissolved in the soil solution is thought to be extremely low. Barshad (6), in an analysis of California soils, found that the water-soluble molybdenum content ranged from 0.3 to 3.9 ppm of dry soil. Even this very small amount is considered to be unusually high (51). In contrast to all of the other trace elements, molybdenum becomes more available with an increase in soil pH (42).

Part of the molybdenum content of the soil exists in the form of three oxides: molybdenum trioxide (MoO_3), molybdenum dioxide (MoO_2), and molybdenum pentoxide (Mo_2O_5), as determined by Amin and Joham (4). Molybdenum in these forms is unavailable to the plant. This is especially true of the more reduced oxides (MoO_2 and MO_2O_5)—but the trioxide may readily be made available by reaction with the cations of the soil. Here we have oxidation making an element more available. This situation can be contrasted with that of manganese in which the reduced state was more available.

The adsorption of molybdenum ions to clay minerals and hydrated oxides resembles that of the sulfate and phosphate anions (54). Thus molybdenum anions will exchange with hydroxyl ions (OH^-) on these substances.

Other Elements

Several studies, first initiated by Osterhout (38, 39), have demonstrated that *sodium* may be essential to the growth of some marine algae. It has been definitely shown that sodium is required for the growth and development of several blue-green algae (2) and of higher plants. Sodium may partially substitute for potassium in both higher (18) and lower plants (1).

Silicon may also be required by some plants. For example, Sommer (44) has demonstrated that the growth of rice and millet improves when silicon is added to the culture medium. Lipman (26) has concluded that silicon improves the growth of barley and sunflower plants. Several algae classes contain silicified structures, silicon is, therefore, considered essential for these plants. It is also believed to be essential for *Equisetum*.

In several early studies *aluminum* was found to improve growth in a variety of plants, as Stiles (50) pointed out. However, aluminum is better known for its toxic rather than its beneficial effects when present in excessive amounts. Thus McLean and Gilbert (33) reported that lettuce, beetroot, timothy, and barley are all sensitive to aluminum toxicity.

Early studies of mineral nutrition concluded that *chlorine* is an essential element to some plants. Lipman (26) showed that chlorine could improve the growth of buckwheat and garden peas. More recently, Broyer and his colleagues (11) demonstrated the necessity of chlorine for the normal growth of tomato plants. They suggested that *bromine* may substitute for chlorine. This suggestion was later confirmed by Ulrich and Ohki (53), who demonstrated that chlorine or bromine is essential for the growth of sugar beet, probably because chlorine is required in photosynthetic oxidation of H_2O.

It is doubtful that any plant requires *gallium*. However, Steinberg (46, 47) demonstrated the need for this element in the fungus *Aspergillus niger* and in a higher plant, *Lemna minor* (Duckweed). However, in later studies (48, 49) he achieved only limited success in demonstrating a gallium requirement in these organisms.

Although *cobalt*, as a component of vitamin B_{12}, is required by some animals, only a few blue-green algae have been shown to require it among plants (22). Stiles (50), however, has described many instances of cobalt toxicity in plants.

Questions

5–1. What is hydroponic culture? How is it useful in mineral nutrition studies? What are some of the problems that might be encountered by scientists using the technique?

5–2. Name some of the current commercial applications (different crops grown) of hydroponics and those that might be made in the future.

5–3. In addition to mineral nutrition, indicate and explain other experimental uses of hydroponics. For example, can chemicals other then elemental salts be introduced into plants through hydroponics?

5–4. Why are the elements Mn, Zn, B, Cu, and Mo referred to as trace elements?

5–5. Explain the terms slop culture, drip culture, subirrigation, and sand culture.

5–6. Name the sources in the soil from which plants derive the following elements: phosphorus, calcium, magnesium, potassium, sulfur, iron, manganese, copper, zinc, boron, and molybdenum.

5–7. Other elements, in addition to the fifteen essential elements indicated, may be necessary for plant growth and development. What might some of these elements be and what are their likely general physiological roles?

5–8. How might you determine whether an element is essential for plant growth and development?

5–9. Overliming will often cause plant deficiencies of a specific element. What is this element and why does liming lead to a deficiency?

5–10. Why is the pH of the soil important to the availability to and uptake of certain elements by the plant?

Suggested Readings

Arnon, I. 1974. *Mineral Nutrition of Maize.* Int. Potash Inst., eds. Bern: Der Bund.

Clarkson, D.T., and J.B. Hanson. 1980. The mineral nutrition of higher plants. *Ann. Rev. Plant Physiol.* 31:239–298.

Epstein, E. 1972. *Mineral Nutrition of Plants: Principles and Perspectives.* New York: Wiley.

Hewitt, E.J., and T.A. Smith. 1975. *Plant Mineral Nutrition.* London: English University Press.

Mengel, K., and E.A. Krikby. 1978. *Principles of Plant Nutrition.* Int. Potash Inst., eds. Bern: Der Bund.

Rains, D.W. 1976. Mineral metabolism. In J. Bonner and J.E. Varner, eds., *Plant Biochemistry,* 3rd ed. New York: Academic Press.

Sutcliffe, J.F., and D. Baker. 1974. *Plant and Mineral Salts.* London: Edward Arnold.

Chapter 6

Mineral Salt Absorption and Translocation

Tomato and pepper plants growing in soil mix of formulated natural peat and blended essential elements. *Courtesy of Fision Horticulture Division, Bramford, Ipswich, Suffolk, England.*

In Chapter 5 we discussed the occurrence and availability of the essential elements in the soil. Our next step is to determine how these elements penetrate the root tissue and how they are transported throughout the plant.

Early workers assumed that inorganic salts were passively carried into the plant with the absorption of water. They also assumed that the translocation of absorbed salts to different areas of the plant was dependent on the transpiration stream. These assumptions, however, could not adequately account for the obvious differences in the salt composition of the plant tissues and the medium in which the plant grew. The suggestion was made that absorption could be explained as an osmotic phenomenon. Osmotically active substances were thought to diffuse along concentration gradients from the soil solution into the plant. The osmotic concentration inside the cell was continuously kept at a low point through the utilization of the absorbed substances in metabolism. The osmotic theory sufficiently explained the absorption but did not account for the rapid translocation of the salts once they were absorbed. Again, the transpiration stream was implicated, this time as only aiding in the dispersal of the salts, not their absorption. Thus early attempts to explain salt absorption and translocation only emphasized physical mechanisms and almost entirely neglected the participation of metabolic energy.

However, during this time the brilliant physiologist Pfeffer (47) made a statement that contrasted sharply with prevalent theories on salt absorption and remarkably foreshadowed a current popular theory. Pfeffer claimed that "the nature of the plasma is such as to render it possible that a substance may combine chemically with the plasmatic elements, thus being transmitted internally,

and then set free again." This statement agrees very nicely with the carrier theory on salt absorption that is generally accepted today.

As is usually the case when one tries to buck the tide of popular thought, this provocative theory on absorption was not taken too seriously, and physical mechanisms and models were continuously produced to explain salt absorption. Work accomplished in the 1930s finally showed that salt absorption is largely dependent on metabolic energy— that is, that the uptake of salt is predominately active. However, passive absorption is still sufficiently important for ion accumulation for us to consider it in detail, along with active uptake.

Passive Absorption

Outer and Apparent Free Space

Salt absorption takes place through the intimate contact of the root system with the soil colloids or soil solution. What are the mechanisms involved in the passage of dissolved inorganic salts from the soil solution into the plant? Numerous investigators have demonstrated passive, or nonmetabolic, absorption of ions. They found that when a plant cell or tissue is transferred from a medium of low salt concentration to a medium of relatively high salt concentration, there is an intial rapid uptake of ions, followed by a slow steady uptake that is under metabolic control (see Figure 6–1). The initial rapid uptake is not affected by temperature or metabolic inhibitors—that is, metabolic energy is not involved. If the above tissue is returned to the low salt medium, some of the ions taken up will diffuse out into the external medium. In other words, a part of the cell or tissue immersed

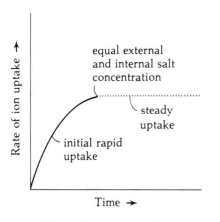

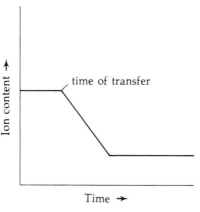

Figure 6–1. (a) Ion uptake by cells in relatively high salt solution. Initial rapid uptake not affected by metabolic inhibitors. (b) Results following cells' return to low salt solution. Only portion of cells is open to free diffusion.

a. Plant cells after transfer from low to high salt concentration

b. Plant cells after transfer from high to low salt concentration

in the salt solution is opened to *free diffusion* of ions. Since free diffusion implies that ions can move freely in or out of the tissue, the part of the tissue opened to free diffusion will reach an equilibrium with the external medium and the ion concentration of this part will be the same as that found in the external medium. The part of a plant cell or tissue that allows free diffusion to take place is referred to as *outer space*.

With the establishment of the concept of outer space, workers turned to the task of calculating the volume of plant cell or tissue involved. They immersed a tissue in a solution of known concentration, allowed it to come to equilibrium, and then determined the amount of salt taken up.

Hope and Stevens (28) found that bean root tips, when immersed in a KCl solution, reached equilibrium in 20 minutes. This reversible diffusion of KCl took place in the absence of metabolic energy, and the volume of tissue involved was considered to include a part of the cytoplasm. Subsequent work by Hope (27) demonstrated that the measured volume of the tissue allowing free diffusion increased when the concentration

of KCl in the external solution increased, and since active transport was inhibited, we can only assume that a passive accumulation of ions against a concentration gradient must have occurred. The term *apparent free space* was introduced to describe the apparent volume accommodating the free diffusion of ions.

How may ions be accumulated against a concentration (chemical potential) gradient without the participation of metabolic energy? Several forms of passive absorption termed *ion exchange, Donnan effect and equilibrium,* and *mass flow of ions* may be responsible for the movement of ions against a chemical potential gradient.

Ion Exchange

Ions adsorbed to the surfaces of the cell walls or membranes of a tissue may exchange with ions from the external solution in which the tissue is immersed. We have already encountered analogous ion-exchange mechanisms between the soil solution and the soil colloids in a previous chapter. Let us suppose, for example, the cation K^+ of the external solution exchanged with a hydrogen ion H^+ adsorbed to the surface of the membrane. Then anions could possibly exchange with free hydroxyl ions in the same manner. Thus ion-exchange mechanisms would allow for a greater absorption of ions from the external medium than could normally be accounted for by free diffusion.

Donnan Effect and Equilibrium

The *Donnan effect and equilibrium* theory takes into account the effect of fixed or non-diffusible ions. Let us take, for example, a membrane that is permeable to some ions and not to others and that separates the cell from the external medium. Let us suppose that on the inner side of this membrane there is a concentration of anions to which the membrane is impermeable (negatively charged proteins are examples of fixed anions). Now, if the above membrane is freely permeable to the cations and anions in the external solution, equal numbers of cations and anions from the external solution will diffuse across the membrane until an equilibrium is established. Normally, this equilibrium would also be electrically balanced. However, additional cations are needed to balance the negative charges of the fixed anions on the inner side of the membrane (see Figure 6–2). Therefore, the cation concentration would become greater in the internal solution than it is in the external solution. Also, because of the excess of negative charges due to fixed anions, the concentration of anions in the internal solution will be less than the concentration of those ions in the external solution.

When the product of the anions and cations in the internal solution is equal to

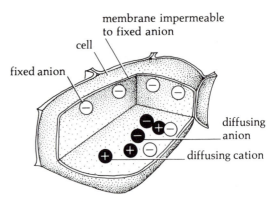

a. Accumulation of cations

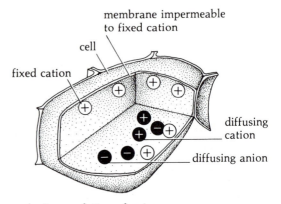

b. Accumulation of anions

Figure 6–2. Ion diffusion across membranes.

that of the anions and cations in the external solution, the Donnan equilibrium is attained according to the following equation:

$$[C_i^+] [A_i^-] = [C_o^+] [A_o^+]$$

where

C_i^+ = cations inside
A_i^- = anions inside
C_o^+ = cations outside
A_o^- = anions outside

Thus, the accumulation of ions against a concentration gradient can occur without the participation of metabolic energy until a Donnan equilibrium is reached. We should recognize, however, that although this mechanism may not necessarily occur in plant tissue as described here, it serves as one explanation to account for the passive accumulation of ions against a concentration gradient in response to an *electrochemical potential gradient*.

Mass Flow of Ions

Some investigators believe that ions can move through roots along with the mass flow of water (30, 31, 35, 36). According to this theory, an increase in transpiration should cause an increase in the absorption of ions. That this is so has been generally accepted (53), but whether the effect of transpiration is direct or indirect is not clear. Some investigators claim that transpiration indirectly affects ion absorption by removing ions after they have been released into the xylem ducts, causing by this dilution an increase in ion absorption activity (9, 10, 26). Opposing this is the suggestion that ions move in mass flow with water from the soil solution through the root and eventually to the shoot. One or both of these mechanisms may be a part of the general picture of salt absorption by plants. To prove or disprove either theory would be very difficult.

Work by Lopushinsky (39) with detopped tomato plants indirectly supports the concept that an increase in transpiration can bring about an increase in salt absorption. By applying different degrees of hydrostatic pressure to detopped tomato root systems enclosed in pressure chambers that contain nutrient solutions of radioactive ^{32}P

and ^{45}Ca, he was able to determine that increases in the hydrostatic pressure brought about an increase in the amount of phosphate and calcium moved into the root xylem. He determined this by analyzing the root exudate for ^{32}P or ^{45}Ca under normal root pressure conditions and under conditions of increased hydrostatic pressure (see Figure 6–3). Although in the above experiments, water is being pushed up the xylem ducts, the system is somewhat similar to one in which water is being pulled up the xylem ducts, as in transpiration. In both cases, an increase in water flow either because of an increase in hydrostatic pressure or in transpirational pull, results in an increase in the total uptake of ions.

From this discussion, we have learned that at least part of the total salt taken up by a plant may result from passive absorption through free diffusion of ions into the apparent free space of a tissue. Accumulation of ions against a concentration gradient is possible under the above circumstances due to ion-exchange mechanisms, or the Donnan effect and equilibrium. The mass flow of ions through root tissue may also be possible with the aid of transpirational pull. All of these mechanisms occur in the absence of metabolic energy.

Active Transport

Direct analyses of the vacuolar sap of plants immersed in solutions of known salt concentration have demonstrated unequivocally that both anions and cations are accumulated by plants against concentration gradients. Furthermore, the extent of accumulation is such that known electrochemical mechanisms, such as ion exchange and the Donnan effect and equilibrium, cannot

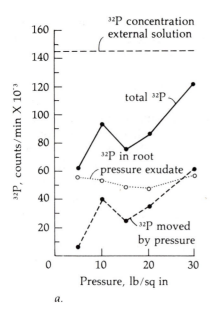

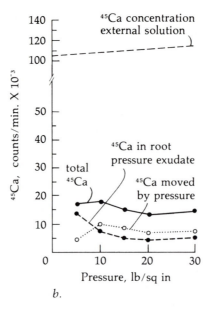

Figure 6–3. Effect of pressure on rate of (a) ^{32}P and (b) ^{45}Ca movement into xylem of tomato roots. ^{32}P or ^{45}Ca in root pressure exudate represents amount of radioactive ions moved into root xylem in absence of applied pressure. ^{32}P or ^{45}Ca moved by pressure represents fraction of total ^{32}P or ^{45}Ca associated with water movement under applied pressure.

From W. Lopushinsky. 1964. Plant Physiol. *39:494.*

account for the extent of accumulation that occurs. Analyses of the ion accumulation in the sap of *Nitella clavata* and *Valonia macrophysa* by Hoagland (24) give an excellent picture both of the accumulation and selective properties of salt absorption mechanisms in plants (see Figure 6–4).

Since ion accumulation is inhibited when the metabolic activity of the plant is inhibited by low temperatures, low oxygen tension, metabolic inhibitors, and so on, we can only assume that ion accumulation as it occurs in plants requires metabolic energy. *Active transport* is the transport of ions with

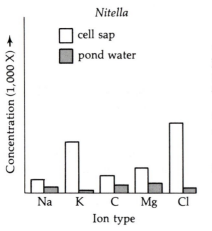

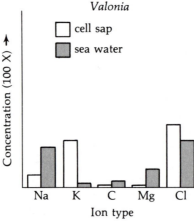

Figure 6–4. Relative concentrations of ions in cell sap of *Nitella clavata* and *Valonia macrophysa*. For comparison and to demonstrate that ions can be accumulated against a concentration gradient, relative concentrations of these ions in growth media are also shown.

the aid of metabolic energy. Various mechanisms have been devised to explain active transport, none of which has been universally accepted. All the suggested mechanisms, however, accept the concept that the active transport of an ion across an impermeable membrane is accomplished through the mediation of a *carrier* compound present in the membrane.

Carrier Concept

Inner space is the space in a tissue or cell to which ions penetrate through the mediation of metabolic energy. Where outer space ends and inner space begins has not been clearly established. However, apparent free space volume measures have suggested that in some instances part of the cytoplasm allows for free diffusion of ions.

The area or barrier between outer and inner space is impermeable to free ions. Passage across this area is thought to require the intercession of specific carriers, which combine with ions in outer space and release these ions in inner space. The barrier may often be the plasmalemma.

The most important feature of the carrier theory is the assumption of an intermediate *carrier-ion complex*, or combination of the carrier and the ion in one form that facilitates ion movement across the impermeable barrier. Ions released into inner space cannot move out and thus are accumulated. Figure 6–5 illustrates the carrier concept in a simplified form.

In this illustration the ion carrier is initially activated. The activation requires ATP and a suitable enzyme. According to some investigators, the enzyme is a kinase (phosphokinase) that affects the phosphorylation

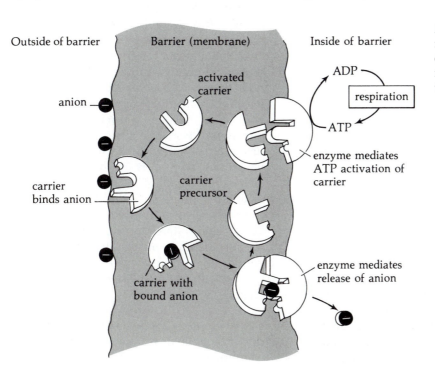

Outside of barrier   Barrier (membrane)   Inside of barrier

anion

carrier binds anion

activated carrier

carrier precursor

carrier with bound anion

ADP

respiration

ATP

enzyme mediates ATP activation of carrier

enzyme mediates release of anion

Figure 6–5. Simplified model of carrier concept for anions. Same mechanism also illustrates cation transport.

(activation) of the carrier. Others may visualize the activation as a conformation (structural) change in the carrier, which favors its complementation with the ion. The activated carrier may complex with an ion at the outer barrier surface to form a carrier-ion complex that is subsequently cleaved at the inner barrier surface. In some ATP models the enzyme (possibly phosphatase) may mediate the splitting off of the phosphorus, with a loss of the ion because of its low affinity for the now nonactivated carrier. Thus the ion is released into inner space (usually the cytoplasm or possibly the vacuole). We may indicate the process taking place in the barrier according to the following steps:

1. carrier $\xrightarrow[\text{ATP} \quad \text{ADP}]{\text{kinase}}$ carrier*

    (carrier activation)

2. carrier* $\xrightarrow{\text{ion (+ or −)}}$ carrier*-ion

    (carrier-ion complex)

3. carrier*-ion $\xrightarrow{\text{phosphatase?}}$ carrier + ion

    (ion release)

The carrier concept has received impressive support by numerous investigators since its formulation by van den Honert in 1937. Three characteristics of salt absorption and active transport strongly support the validity of the carrier concept.

Isotopic exchange. The portion of ion absorption associated with active transport is largely nonexchangeable with ions of the same species in outer space or in an external medium. Radioactive ions have been especially helpful in this observation. As Epstein (18) has pointed out, the fact that not only back diffusion but also isotopic exchange of the actively absorbed ions is prevented suggests a membrane highly impermeable to free ions. Since ions are absorbed, we must attribute their movement across the impermeable membrane to the intervention of carriers. Experiments by Leggett and Epstein (37) clearly demonstrate this analysis.

Leggett and Epstein studied the absorption of sulfate, labeled with ^{35}S, by excised barley roots. They found, after a period of S^*O_4 absorption, that the total S^*O_4 absorbed could be separated into two fractions: diffusible S^*O_4 and actively absorbed S^*O_4. The roots were allowed to absorb labeled sulfate from a solution of $K_2S^*O_4$ for a period of 60 minutes. The total amount of labeled sulfate taken up was determined for some of the root samples. Other samples were allowed to stand in water or solutions of nonradioactive $CaSO_4$ for varying periods of time up to 120 minutes. This period was termed a *desorption period*, during which freely diffusible sulfate moved out of the root tissue. Immersing the roots in solution of $CaSO_4$ allowed for any isotope exchange that might occur. During the desorption period, there was a rapid loss of labeled sulfate, followed by a period in which there was no further loss (see Figure 6–6). The initial rapid loss, of course, was due to diffusion of S^*O_4 from those areas in the root that allow free or reversible diffusion of ions—outer space. That fraction of labeled sulfate remaining indicated those ions that were actively transported to inner space. The labeled sulfate ions in inner space were not able to diffuse out during the desorption period, nor were they able to exchange for the stable isotope SO_4 ion in the solution of $CaSO_4$.

Saturation effects. Some support for the carrier concept comes from observations demonstrating that at increasingly higher con-

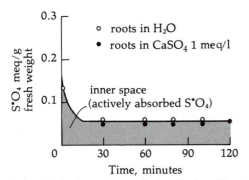

Figure 6–6. Separation of absorbed sulfate into diffusible and actively absorbed sulfate. Before zero time, excised barley roots were exposed to $K_2S^*O_4$, 0.5 meq/liter, for 60 minutes.

From J.E. Leggett and E. Epstein. 1956. Plant Physiol. 31:222.

centrations of salt in the ambient medium, absorption rates appear to approach a limit—in other words, a saturation point is asymptotically approached, at which time all of the active sites on the carriers are occupied. We can immediately see the analogy between this situation and the well-known saturation effect found in enzyme reactions. The fact that a level maximum rate of absorption may be maintained over a relatively long period of time suggests the participation of a *finite* number of carriers working, so to speak, at maximum efficiency—that is, the active sites on the carriers in the above situation are occupied all of the time. As soon as a carrier releases an ion to inner space, it is immediately occupied by an ion from the outer space areas in the tissue. Thus at the saturation point the cycle is kept in continuous motion and cannot be made to proceed faster by an increase in the salt concentration. Figure 6–7 gives an example of the effect of concentration levels on phosphate absorption by *Chlorella* cells.

Specificity. The carrier concept offers a reasonable explanation of the fact that roots absorb ions selectively. That is, ions are absorbed at different rates and have different levels of accumulation in the root tissue, thereby suggesting the presence of specific carriers. This specificity is rather rigid with ions of dissimilar chemical behavior but weak or nonexistent with ions of similar behavior. Thus Epstein and Hagen (19) have shown that the monovalent cations potassium, cesium, and rubidium compete with each other for the same binding site—that is, the rate of absorption of rubidium can be lowered by the addition of potassium or cesium to the nutrient solution. Increasing the concentration of rubidium can overcome the inhibiting effects of the other two cations. Neither sodium nor lithium inhibit rubidium absorption, thereby suggesting different binding sites for these ions. Selenate has been shown to inhibit sulfate absorption, but not phosphate or nitrate absorption (37).

Again, we can find a situation analogous to enzyme-substrate activity. In en-

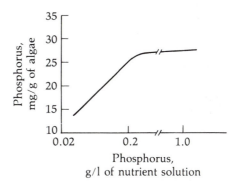

Figure 6–7. Phosphorus contents of *Chlorella* grown in nutrient solutions containing different concentrations of phosphorus.

From H.J. Krauss and J.W. Porter. 1954. Plant Physiol. 29:229.

zyme studies competitive inhibition is well known and is usually explained on the basis of a mutual attraction of a substrate and an inhibitor for active sites on the enzyme. The carrier, like the enzyme, may have a binding site that attracts two or more ions, and it may also differentiate among ions just as an enzyme will among different substrates. The similarities found in the activities of carriers and enzymes is strong support for the carrier concept in active salt absorption.

Ion Pumps

Early workers observed that although salt accumulation is dependent on metabolic energy, there appeared to be no quantitative relationship between salt absorption and respiration. However, Lundegårdh and Burström (42) claimed that such a relationship exists between anion absorption and what they called anion, or salt, respiration. They observed that the rate of respiration increases when a plant is transferred from water to a salt solution. The amount by which respiration is increased over normal or ground respiration by the transfer of a plant or tissue from water to a salt solution is known as *salt respiration*.

The original observations by Lundegårdh and Burström have since been expanded and developed into a workable theory on active salt absorption by Lundegårdh (40, 41). Lundegårdh's theory assumes the following:

1. Anion absorption is independent of cation absorption and occurs by a different mechanism.

2. An oxygen concentration gradient exists from the outer surface to the inner surface of a membrane, thus favoring oxidation at the outer surface and reduction at the inner surface.

3. The actual transport of the anion occurs through a cytochrome system.

Since there is a quantitative correlation between anion absorption and salt respiration and since this correlation does not exist with cation absorption, it was assumed that only anions are actively transported. The inhibition of salt respiration and consequent inhibition of anion absorption by cyanide or carbon monoxide led Lundegårdh to propose that transport of anions is mediated through *cytochrome oxidase* and that cytochromes may be anion carriers.

The cytochrome pump models have not been worked out in detail and may not accurately reflect the way the cytochromes exist in vivo. For example, they are not known to be present in outer membranes. We do know, however, that the cytochromes are intimately associated with the inner membrane structures of organelles (i.e., chloroplasts and mitochondria, as will be pointed out in later chapters).

We will consider two models of salt absorption pumps to show the attempts of early scientists to provide a basis for understanding the carrier concept. We must remember that models are working tools and justifiably are not accepted as accurate mechanisms. Figure 6–8 presents Lundegårdh's cytochrome theory of salt absorption.

According to Lundegårdh's theory, dehydrogenase reactions on the inner surface of the barrier produce protons (H^+) and electrons (e^-). The electrons that are produced move outward via a cytochrome chain, and anions move inward (see Figure 6–8). At the outer surface of the barrier, the reduced iron of the cytochrome is oxidized, losing an electron and picking up an anion. The released electron unites with a proton and oxygen to form water. At the inner barrier surface, the oxidized iron of the cyto-

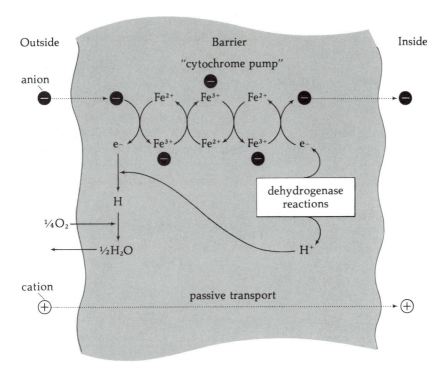

Figure 6–8. Lundegårdh's cytochrome theory on salt absorption. Anions (A⁻) are actively absorbed via a "cytochrome pump." Cations (M⁺) are absorbed passively.

chrome becomes reduced by the addition of an electron released in a dehydrogenase reaction. The anion is released on the inside of the barrier in this last reaction. Cations are absorbed passively to balance the potential difference caused by the accumulation of anions on the inner barrier surface.

Although the cytochrome transport theory does help to visualize metabolic energy participation in ion absorption, a number of investigators do not consider the details scientifically sound. For example, Robertson, Wilkins, and Weeks (52) found that 2,4-dinitrophenol (DNP), an inhibitor of oxidative phosphorylation, increases respiration but decreases salt absorption. This finding implies that ATP production should be included in any theory of ion accumulation. The original proposal that only anions are capable of stimulating respiration has come under considerable attack. For example, Handley and Overstreet (21) found that

both potassium and sodium ions stimulated respiration. Finally, if there is only one carrier for all anions, then competition for binding sites among anions should be apparent. The anions sulfate, nitrate, and phosphate do not, however, compete with one another.

ATP Carrier Mechanism

The finding by Robertson, Wilkins, and Weeks (52) that 2,4-dinitrophenol inhibited salt absorption presents a strong case for the participation of ATP in active salt absorption. Low concentrations of 2,4-dinitrophenol will completely retard ATP formation without affecting or decreasing respiration. Thus ATP may very possibly be essential for ion pumps.

Bennet-Clark (2) has proposed a mechanism for active salt absorption that utilizes ATP. This investigator has suggested that phospholipids may be important in the transport of ions across otherwise impermeable membranes. In this transport, lecithin, a phospholipid, is synthesized and hydrolyzed in a cyclic manner—it picks up ions on the outer surface and releases them on hydrolysis into inner space. The synthesis of at least one of the components of this phosphatide cycle requires ATP (see Figure 6–9). Again, such a model has serious limitations when we apply it to living plants.

For instance, plants do not contain lecithin, choline, or choline esterase. Since plants do have similar substances, however, models of this nature do provide important working concepts.

It seems quite reasonable to assume that ATP provides the energy for transport. A general model by Hodges (25) shows ATP degradation important in the transport of H^+ ions across membranes (see Figure 6–10). As part of the model, the enzyme

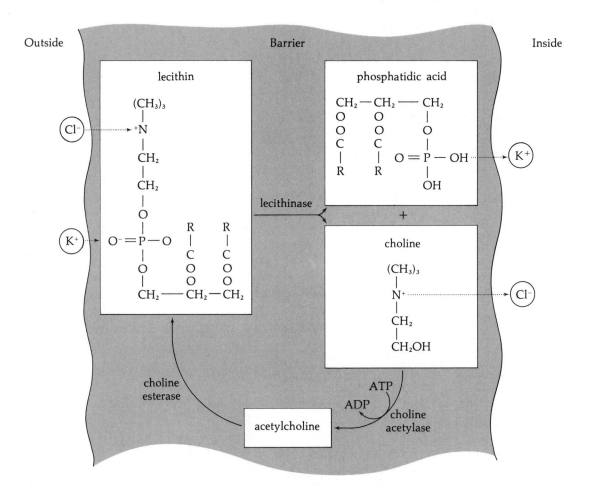

Figure 6–9. Phosphatide cycle. Ions from outer space are picked up by lecithin. Hydrolysis of lecithin-ion complex releases ions to inner space. Lecithin is then resynthesized.

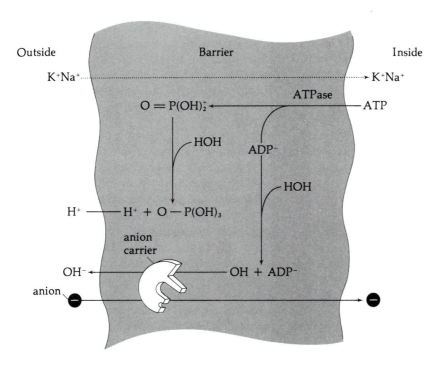

Figure 6–10. Role of ATP in hydrogen ion transport. Note abbreviations: adenosine triphosphate (ATP), adenosine diphosphate (ADP), phosphoryl cation $[O=P(OH)_2]^+$, hydroxyl ion (OH^-), hydrogen (H^+).

From H. Lundegårdh. 1950. Physiol. Plant. 3:103.

ATPase catalyzes the conversion of ATP to a phosphoryl cation and anion ADP^-. The phosphoryl cation $[O=P(OH_2)]^+$ reacts with water, generating H^+. The H^+ ions produced in this manner accumulate outside the cell with a pH gradient generated across the membrane (transmembrane potential), the anions (ADP^-) stay in the cell. In actuality many cells are negatively charged and attract cations that diffuse through the pores (channels) in the membrane (plasmalemma) in exchange for H^+. Higinbotham (22, 23) termed this process *electroosmosis.* The specificity of ion uptake may reside in the nature of the chemicals lining the pores. These chemicals (antibiotics) may interact selectively with some ions; other ions pass or are transported through the membrane. Hodges also proposed that the anion ADP^- in the cytoplasm reacts with water to release OH^-. Accordingly, OH^- might exchange for other anions.

As indicated in Figure 6–9, ATP may be involved by direct energy coupling in the formation of the carrier. However, unequivocal evidence to support this idea is lacking.

Transmembrane Potential and Nernst Equation

A transmembrane potential, or voltage, develops across a membrane due to different concentrations of ions on each side. Since it is a voltage, it can be measured with a set of electrodes and a voltmeter. One needlelike electrode (microelectrode) is used to measure the cell sap through a puncture into the vacuole. The other serves as the reference electrode at the cell exterior. The potential difference between the exterior and the interior of the cell is then measured. We must remember that the potential difference that is measured is due to the properties of the membrane as they influence membrane permeability and most importantly to the transport of ions. The pumping of H^+ ions to the outside of cells and the cells' retention of anions will establish an electrochemical gradient and hence transmembrane potential. As we may deduce from Hodges model (Figure 6–10), the extent of the transmembrane potential directly influences the relative membrane transport of cations, assuming the membrane is permeable to them.

With respect to the transport of ions in response to an electrical gradient across the membrane, their net movement will cease when equilibrium between their electrical and chemical potentials is reached. We can use the *Nernst equation* to analyze this equilibrium and to evaluate whether ion accumulation is due to passive or active uptake. The Nernst equation appears as follows:

$$\Psi_i - \Psi_o = E$$

$$E = \frac{RT}{n \cdot F} \ln \frac{[C_o^+]}{C_i^+} = \frac{RT}{n \cdot F} \ln \frac{[A_i^-]}{[A_o^-]}$$

Ψ_i = electrical charge of cell interior

Ψ_o = electrical charge of cell exterior (medium)

E = transmembrane potential (millivolts)

R = gas constant

T = absolute temperature

F = Faraday constant

n = valence of the ions

C_i = cation concentration (molarity) inside cell

C_o = cation concentration (molarity) outside cell

A_i = anion concentration (molarity) inside cell

A_o = anion concentration (molarity) outside cell

From this equation we can see that when E is less than 1, the cell is negatively charged and $[C_o^+]/[C_i^+]$ must be less than 1. At equilibrium we would expect the cations to have accumulated within the cell sap. Also, under these conditions of E being less than 1, $[A_i^+]/[A_o^-]$ must be less than 1 and the anion concentration is higher on the outside than on the inside of the cell. The significance of these observations, based on the Nernst equation, is that the cell sap can have a higher cation accumulation than that of the outside nutrient medium, without active cation transport. In other words, the cations are not transported against an electrochemical gradient. If, however, the concentration of cations inside is higher than that predicted by the Nernst equation and equilibrium, then a part or all of the process of accumulation was against an electrochemical gradient and is characterized as active transport.

In practice, to determine active or passive transport of ions, the concentration of the ions inside and outside are measured. Also, the transmembrane potential (E) is measured. The ion concentrations are substituted into the Nernst equation to arrive at a calculated transmembrane potential (E-calculated). The following relationship is used to calculate whether the process is active or passive:

$$D = (E\text{-measured}) - (E\text{-calculated})$$

If D (the difference or driving force) is a negative value and the ion is a cation, the cation was accumulated by passive transport. A positive value indicates that the cation was taken up by an active process. If D is negative for an anion, this indicates active uptake; and if D is positive for an anion, this indicates passive uptake. In an intact plant the situation is not always as clear-cut as described here. To determine the process

Ion	E-*calculated*	E-*measured*	D	*Uptake*
cation	−120	−80	−40	passive
cation	−120	−175	+55	active
anion	−120	+100	−220	active
anion	−120	−100	+20	passive

Table 6–1. Theoretical values illustrating calculations for determining if an ion has been accumulated according to passive or active transport.

involved in the ion accumulation according to the Nernst equation, the cells (tissues, organs) must be in equilibrium when measurements are made. This condition is extremely difficult to insure. Nevertheless, Table 6–1 provides some hypothetical numbers to reinforce the method described.

The negatively charged plant cells would indicate that anions are taken up by active means (against an electrical gradient), and cations are transported more often in a passive manner. This observation of cation transport supports the previously stated point.

Factors Affecting Salt Absorption

The physical and biochemical activities of living organisms are subject to the influences of their external and internal environments. Salt absorption is no exception; it is speeded up, slowed down, or kept in dynamic equilibrium by a complex of ever-changing factors. The scientist has learned to consider the influence of individual factors by controlling the environment and studying the effect of the one factor in question. Scientists have done this with the process of salt absorption, and we now have an extensive, if incomplete, picture of how this process might proceed in nature's ever-changing environment. We will discuss the effects of temperature, pH, light, oxygen tension, interaction, and growth on salt absorption.

Temperature

In general, an increase in temperature results in an acceleration of salt absorption. However, the influence of temperature on salt absorption is confined to a relatively narrow range. In addition to accelerating salt absorption, increase in temperature past a maximum point will inhibit and eventually terminate the process (see Figure 6–11). Most likely, the inhibitory effects of high temperatures occur because of the denaturation of enzymes involved either directly in salt absorption or in the synthesis of some necessary component of salt absorption.

Temperature changes affect both passive and active absorption processes. The rate of free diffusion, for example, depends on the kinetic energy of the diffusing molecules or ions, which is, in turn, dependent on temperature. Therefore, lowering of temperature will slow down any process dependent on free diffusion. Low temperatures will, of course, slow down the biochemical reactions found in active transport.

Hydrogen Ion Concentration

The availability of ions in the soil solution is profoundly affected by the hydrogen ion concentration. Ionization of electrolytes

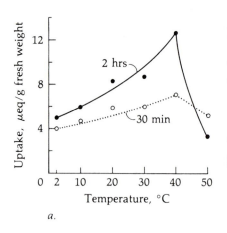

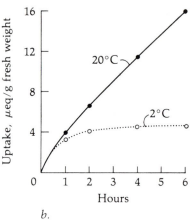

a.   b.

Figure 6–11. (a) Effect of temperature on uptake of potassium ions by washed carrot tissue slices. (b) Uptake of potassium ions by washed carrot tissue slices over prolonged period of time.

Reprinted with permission from J.F. Sutcliffe. Mineral Salts Absorption in Plants. *1962. Pergamon Press Ltd.*

or the valence numbers of different ion species are influenced by changes in pH. For example, the monovalent phosphate ion, $H_2PO_4^-$, is the form of phosphorus most readily taken up by plants. However, as a medium approaches a more alkaline pH, production of first the bivalent phosphate (HPO_4^{2-}) and then the trivalent phosphate (PO_4^{3-}) is favored. The bivalent ion is only sparingly available to the plant; the trivalent ion is not available at all. Since the monovalent ion is absorbed more readily than the bivalent ion, absorption of phosphate is accelerated at an acid pH. Robertson (51) has pointed out that since boron is taken up as the undissociated acid, H_3BO_3, or as the $H_2BO_3^-$ ion, it too must be absorbed more readily at a lower pH. In contrast to the above observations with anions, increase in pH will favor the absorption of cations.

Numerous experiments have shown little pH effect, as judged by growth (51). Marked pH effects most likely occur when ion availability is inhibited. However, if the concentration of ion is high enough, it will be difficult to show a deficiency for that ion in the plant over a physiological range of pH values. Of course, at pH values outside the physiological range, damage to plant tissues and carriers will inhibit salt absorption.

Light

The effects of light on the opening and closing of stomata and on photosynthesis indirectly affect salt uptake. Opened stomata increase the mass flow of water in the transpiration stream and thus may indirectly influence salt absorption. The energy derived from the photosynthetic process provides energy for salt uptake, and the oxygen given off improves conditions for the active absorption of ions.

Oxygen Tension

The active phase of salt absorption is inhibited by the absence of oxygen. Indeed, it was this observation that most strongly supported early theories on active transport. Figure 6–12 shows the strong influence of oxygen on the uptake of phosphate.

Interaction

The absorption of one ion may be influenced by the presence of another ion. In a study of the uptake of KBr by excised barley roots, Viets (58) found that the absorption of potassium is affected by the presence

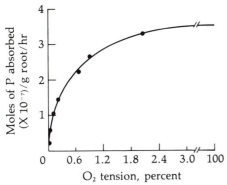

Figure 6–12. Effect of oxygen on uptake of phosphate by excised barley roots in phosphate solutions $1 \times 10^{-4}M$ (pH 4).

From H.T. Hopkins. 1956. Plant Physiol. 13:155.

of calcium, magnesium, and other polyvalent cations in the external medium. Viets noted a dual effect by calcium on the uptake of both potassium and bromine. He found that uptake of potassium and bromine is less in the absence of calcium, but it decreases after the calcium concentration is increased past a maximum point (see Figure 6–13). Overstreet, Jacobson, and Handley (46) also noted this effect of calcium. The absorption of magnesium is also adversely affected by the presence of calcium (45).

Epstein and Hagen (19) have described the interaction of several ions (K, Cs, Li, Rb, and Na) as competition for binding sites on carriers. For example, they found that potassium, rubidium, and cesium compete with one another for a mutual binding site. Lithium and sodium, in contrast, are not competitive since they have different binding sites. It was later found that barium, calcium, and strontium compete with one another for a mutual binding site, which is not involved in the active uptake of magnesium (20).

Interaction among ions appears primarily to be associated with the availability and specificity of binding sites on carriers. If enough binding sites are present, interac-

tion will not be apparent, and ions with mutual binding sites will be taken up with maximum efficiency. Also, if the binding site of an ion is highly specific for that ion, its absorption should not be affected by the presence of other ions.

Growth

For a short period of time, we can study the absorption of salt by plant tissues without interference from growth. However, over a prolonged period of time, salt absorption may be profoundly influenced by growth. The growth of a tissue or plant may increase surface area, number of cells, and synthesis of new binding sites or carriers, factors that will stimulate salt absorption. The increased volume of water taken up by a cell as it matures may dilute the internal concentration of salt and thus increase absorption activity.

When dealing with the growth of a complete plant instead of a tissue, we must

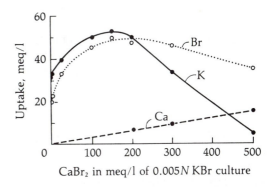

Figure 6–13. Effect of Ca on uptake of K and Br. At low concentrations of Ca, uptake of both K and Br is increased. As concentration of Ca is increased, uptake of K and Br is inhibited.

From F.G. Viets. 1944. Plant Physiol. 19:466.

consider the different phases of development and their influence on salt absorption. For example, as a root ages, much of the surface area formerly involved in salt absorption becomes heavily suberized and unable to take up salt. Vegetative development and the metabolic activity associated with vegetative development make heavy demands on many of the elements. Also, increased vegetative growth is usually accompanied by increased water movement, which may affect the passive absorption and translocation of salts.

Absorption and Translocation

How are the salts translocated in the plant? Nutrient availability in the solid and liquid phases of the soil has been explained in two theories: the *contact exchange theory* and the *carbonic acid exchange theory*. Both theories have been defended and criticized but still remain the best explanations for mineral salt availability to the plant from the soil (see Figure 6–14).

According to the authors of the contact exchange theory, Jenny and Overstreet (32, 33), ions may be exchanged from one adsorbent to another (clay colloid and root) without the participation of free electrolytes—that is, an ion may be adsorbed by the plant root without first being dissolved in the soil solution. These authors explain this as an overlapping of oscillation spaces of adsorbed ions. An ion adsorbed electrostatically to a solid particle, such as a plant root or clay micelle, is not held too tightly but will oscillate within a certain small volume of space. If two adsorbents are close enough, the oscillation volume of an ion adsorbed to one particle may overlap the oscillation volume of an ion adsorbed to the

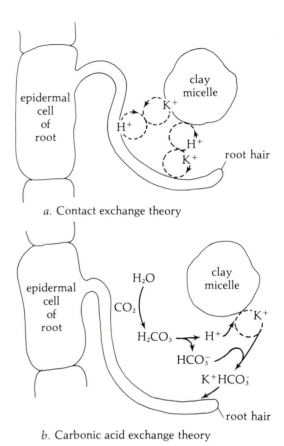

a. Contact exchange theory

b. Carbonic acid exchange theory

Figure 6–14. Contact and carbonic acid exchange theories. Hydrogen from root exchanges with K^+ on surface of clay micelle. Carbonic acid exchange involves formation of carbonic acid from CO_2. The former dissociates and H^+ ions adsorb to clay micelle surface and displace K^+, which may be adsorbed to root cells.

other particle, and an exchange of ions may take place (see Figure 6–14a).

The soil solution plays an important part in the carbonic acid exchange theory in that it provides the medium for exchange of ions between the root and clay micelles. According to this theory, respiratory CO_2, released by the root, forms carbonic acid (H_2CO_3) on contact with the soil solution. In the soil solution, carbonic acid dissociates to form a cation (H^+) and an anion (HCO_3^-). The hydrogen ions diffuse to clay micelles where they may exchange with cations adsorbed to the clay surface. The cations originally adsorbed to the clay surface are released to the soil solution; here they are free to diffuse to the root surface where they may be absorbed in exchange for H^+ or as ion pairs with bicarbonate (see Figure 6–14b).

The actual absorption of salts by roots is both passive and active. The movement of salts into apparent free space is passive, allowing for the free diffusion of ions. There is some confusion as to what area of the cell is occupied by apparent free space. Some investigators, such as Levitt (38), contend that it is confined to the cell walls. Others suggest that part of the cytoplasm may also be included in apparent free space. Inner space, where salts are accumulated to higher concentrations than in the external medium, is thought to comprise part of the cytoplasm and the vacuole. With the above picture in mind, we must now determine how the absorbed salt moves from the outer surface of the root, across the cortex, and into the lumina of the dead, conducting cells of the stele.

Absorbed ions are thought to move rather freely in the root as far as the *endodermis*, where further penetration may be retarded by the Casparian strip. Calculations of the volume of apparent free space by Butler (11) and Epstein (17) have supported the contention that metabolic energy is not required for mineral salts to reach the endodermis. Diffusing ions probably move relatively unhindered through the wet cell walls (apoplast) and plasmodesmata symplast of the cortex cells to the endodermis. In this respect, all of the cytoplasm of the cortex cell may be connected via plasmodesmata; these structures offer excellent pathways for the movement of salt.

An explanation of how salts are transported across the endodermis and passed into the lumina of the xylem vessels, where they are accumulated against a concentration gradient, has been a perplexing problem for many years. Just as the accumulation of salts in the vacuoles of cells is an active process, so also is metabolic energy that is utilized in the accumulation of salts in the xylem vessel. The endodermal cells present a barrier to the passive diffusion of ions, and the Casparian strip is thought to be the controlling feature in this respect (see Figure 6–15). Because of this strip, materials in solution cannot pass either between or through the walls of the endodermal cells, and they cannot pass between the protoplast and the wall because of the tight attachment of the protoplast to the Casparian strip. Therefore, the only route available is through the protoplast.

Scientists have proposed various theories to explain the passage of salts across the endodermis and into the xylem. The most generally accepted theory is based on the supposition that there is a gradient of decreasing O_2 and increasing CO_2 from the cortex to the stele (15). The living cells in the immediate area of the xylem vessels would, therefore, possess a low level of metabolic activity. Since energy is required to accumulate salt against a concentration gradient and to hold this salt, these innermost cells, in contrast to the cortex cells, favor the loss of salts. Thus the implication is that the carrier systems operate from the cortex toward the stele (14). Since diffusion back through the impervious Casparian strip is impossible, there is a unidirectional loss of salt into the lumina of the xylem vessels.

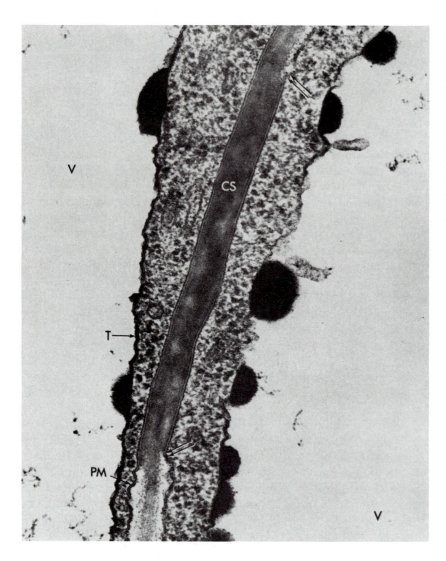

Figure 6–15. Endodermal cells with Casparian strip, a band of lignified and subarized material. Abbreviations are: Casparian strip (CS), plasmalemma (PM), tonoplast (T), and vacuole (V). Magnification 95,200×.

From Biophoto Associates. Dr. Myron C. Ledbetter. Brookhaven National Laboratory.

Circulation of Salts

Salts accumulated in the xylem ducts of the root are translocated to the shoot and, once there, distributed and redistributed throughout the plant. For example, mineral salts that have been deposited in the leaves of a shoot may be withdrawn prior to abscission and may be translocated to other parts of the plant (e.g., reproductive areas or younger leaves). Also, there may be a general redistribution of elements that are highly mobile in the plant.

Generally, the circulation of elements takes place in the vascular tissues. Determining what vascular tissues provide passage for salts from one area of the plant to another presented a difficult task to plant physiologists before the advent of the radioactive tracer. Since the initial use of radioactive tracers, scientists have discovered several different pathways for the translocation of salts. We will discuss the movement of salts in the xylem, in the phloem, laterally between these two tissues, and outward from the leaf.

Translocation of salts in xylem. Because of the evidence that has accumulated over the past three decades, there is little doubt that the salt accumulated in the xylem ducts of the root is carried upward with the transpiration stream. That salts move upward in the xylem tissues has been demonstrated in several ways. Ringing experiments by several investigators (12, 44, 48) have shown that the upward translocation of salts is unimpeded by removal of the phloem tissue. Relatively large amounts of dissolved salts have been detected in the xylem sap by direct analysis. If salts are carried upward in the transpiration stream, we should be able to observe an increase in salt uptake with an increase in transpiration rate. Arnon, Stout, and Sipos (1) made this observation with tomato plants. They found that radioactive phosphate traveled upward to the tip of the tomato plant much more rapidly under conditions favoring rapid transpiration (such as bright sunlight) than under less favorable conditions. Sutcliffe (57) showed that if transpiration by a leaf is inhibited by covering the leaf with a polyethylene bag, translocation of mineral salts to that particular leaf is reduced considerably.

Using radioactive tracers, Stout and Hoagland (56) obtained classical evidence that the pathway of upward translocation of salts is in the xylem tissue. They carefully separated the bark and xylem along a 9-inch length of willow stem. They then inserted a strip of impervious waxed paper between the xylem and the bark. The continuity of

the bark and the xylem tissue was undisturbed, and the plant was left intact. They allowed the willow to absorb radioactive potassium for a 5-hour period and then analyzed sections of the treated and intact areas of the stem for radioactive potassium.

The data shown in Figure 6–16 and Table 6–2 clearly show that potassium is translocated upward in the xylem tissue. Analyses of sections above and below the stripped area demonstrate that lateral interchange of potassium between the phloem and xylem takes place quite readily but that further translocation of potassium either upward or downward is retarded. If we assume that the strip of waxed paper inserted between the bark and xylem is completely impermeable to the labeled potassium being carried along in the transpiration stream, then we

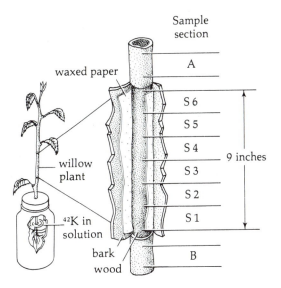

Figure 6–16. Method for detecting upward and lateral translocation of salts. Bark of willow plant is separated from wood by waxed paper; bark and wood are left intact. Willow is allowed to absorb ^{42}K. Intact areas are analyzed for radioactive potassium.

Section	Stripped Branch		Unstripped Branch	
	^{42}K in Bark (ppm)	^{42}K in Wood (ppm)	^{42}K in Bark (ppm)	^{42}K in Wood (ppm)
SA	53.0	47	64	56
S6	11.6	119		
S5	0.9	122		
S4	0.7	112	87	69
S3	0.3	98		
S2	0.3	108		
S1	20.0	113		
SB	84.0	58	74	67

Table 6–2. Results of experiment described in Figure 6–16. Source: *From P.R. Stout and D.R. Hoagland. 1939. Upward and lateral movement of salt in certain plants as indicated by radioactive isotopes of potassium, sodium and phosphorus absorbed by roots. Am. J. Bot. 26:320.*

must assume that some (although a minute amount of) translocation takes place in the phloem tissue. This assumption is based on the detection of small amounts of radioactivity in the bark along the stripped area. Stout and Hoagland's experiment demonstrates that the upward translocation of ions occurs normally in the xylem tissue and that a lateral interchange between the xylem, cambium, and phloem occurs quite readily. This interchange between the vascular tissue via the cambium has been shown also in cotton and bean plants (6, 42).

Lateral translocation of salts. In the Stout and Hoagland experiment, we noted that, in addition to the upward translocation of salts, there is also a lateral movement between vascular tissues. Generally, the xylem tissue is separated from the phloem tissue by a layer of living cells that constitutes the *cambial tissue.* The cambial tissue may regulate, to some extent, the amount of salt carried up in the transpiration stream. If the upward movement of salts were not regulated in some manner, certain areas of the plant would not be accommodated. The cambium is positioned in such a manner so that it is available, both metabolically and physically,

for regulation of the upward, lateral, and downward movement of salt. Biddulph (4) has suggested that the active accumulation of salt by the cambial cells may act as a deterrent against an "indiscriminate" sweep of salts upward in the transpiration stream.

The cambial tissue may discriminate among the different mineral salts carried in the transpiration stream. For example, if a particular element were present in high concentration in the phloem, and an equilibrium existed between the phloem and cambium, interference of the passage of that element in the transpiration stream would probably be negligible (4). However, if that element were present in low concentration in the phloem, the active accumulation of the element and its lateral translocation into the phloem would be enhanced.

Translocation of salts in phloem. The initial movement of salts in an upward direction occurs in the xylem tissue. However, as far back as 1935, Curtis (16) demonstrated that upward movement of mineral salts may occur in the phloem. Curtis demonstrated that stem tip growth is impeded if a ring of bark relatively high up on the stem is removed. This experiment appears to support the concept that upward translocation of mineral salts also occurs in the phloem tissue. However, because of the high position of the ring on the stem in Curtis's experiment, we must assume that the primary influence on stem tip growth was because of the blockage of salts moving out of the lower leaves and being transported upward in the phloem and not because of the root-absorbed salts. This assumption is based on the common observation that ringing a stem near the root level has no effect on salt nutrition.

Studies using radioactive tracers demonstrated the downward movement of salts in the phloem. Study of the outward movement of salts from the leaf showed that salts entering the main vascular stream from leaf sources move primarily in a downward direction in the phloem tissue (2, 42). Figure 6–17 and Table 6–3 provide data on the movement of salts in the phloem tissue. In support of Curtis's earlier observations, the experiment also noted the movement of salts in an upward direction. Table 6–3 also shows quite clearly that lateral transport

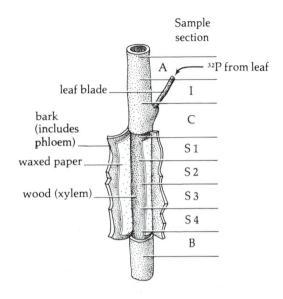

Figure 6–17. Method of detecting movement of ^{32}P downward. Bark immediately below leaf petiole of cotton plant is separated from wood by waxed paper; bark and wood are left intact. Phosphorus (^{32}P) is injected into leaf blade just above separated area of stem. One hour later, sections are analyzed for ^{32}P. See results in Table 7–2.

Section	Stripped Plant			Unstripped Plant		
	Bark		Wood	Bark		Wood
A		1.11				
I	0.485		0.100		0.444	
C		0.610				
S1	0.554		0.064	0.160		0.055
S2	0.332		0.004	0.103		0.063
S3	0.592		0.000	0.055		0.018
S4	0.228		0.004	0.026		0.007
B		0.653			0.152	

Table 6–3. Results of experiments described in Figure 6–17. Amount of ^{32}P detected in each section is given in milligrams.

Source: *From O. Biddulph and J. Markle. 1944. Translocation of radiophosphorus in the phloem of the cotton plant. Am. J. Bot. 31:65.*

between the vascular tissues takes place where the phloem and xylem are not separated. Both tissues may thus be involved with the upward translocation of mineral salts moving out from leaves.

There appears, then, to be a bidirectional movement of salts in the phloem tissue, which is thought of as a simultaneous movement in both directions in the same sieve elements. However, Crafts (14) suggested that movement of solutes (inorganic and organic) out of the leaf may occur in two different phloem channels, one toward the tip, the other toward the base of the plant. Evidence for bidirectional movement occurring in the same channel and in separate channels has been presented, and it is impossible at the present time to determine which theory is correct.

Outward movement of salts from leaves. Studies on the mineral nutrition of leaves of deciduous plants have shown that just prior to abscission there is a movement of mineral nutrients out of the leaf. Among the mineral nutrients moving out of the leaves are nitrogen, potassium, phosphorus, sulfur, chlorine, and under certain conditions, iron and magnesium. Those remaining include calcium, boron, manganese, and silicon (4). The withdrawal of mineral nutrients from leaves takes place primarily in the phloem tissue (see Figure 6–17 and Table 6–3).

A study of the path radioactive phosphorus will take when introduced to leaves at different levels on the plant has revealed that phosphorus from those leaves nearest the root system will mostly move downward toward the root, while phosphorus moving out of leaves positioned high on the plant will move mostly toward the apex (4, 5). Movement of mineral salts out of young, actively growing leaves is almost nonexistent; this characteristic greatly di-

minishes as the leaf matures. Younger leaves will often draw on the supply of mineral elements of the older more mature leaves. This phenomenon is especially noticeable when there is a deficiency of elements such as nitrogen and phosphorus, which are highly mobile in the plant. The deficiency symptoms will appear first on the lower leaves.

Circulation and Reutilization

Early work by Mason and Maskell (43, 44) suggested that minerals are taken up in the transpiration stream and moved to the leaves, and excess quantities are retranslocated downward in the phloem. The mineral salts could be laterally transported into the xylem tissue where upward translocation could take place again. Elements such as nitrogen, potassium, and phosphorus move readily in this circuit. Calcium ascends the stem but does not relocate in the phloem.

Biddulph and his colleagues (3, 5) have demonstrated that phosphorus is highly mobile in a plant and they have suggested the possibility of a continuous circulation of phosphorus. A given phosphorus atom, for example, may make several complete circuits in a plant in a single day (4). Phosphorus mobility is, perhaps, an essential feature of plant growth. Phosphorus is a necessary participant in such important metabolic schemes as photosynthesis, the synthesis of starch, glycolysis, and the synthesis of fats and proteins, and so on. Phosphorus is needed, therefore, at various points in the plant where any one of these processes is occurring. Biddulph (4) has suggested that a "pool" of phosphorus in a usable form is maintained throughout the plant in a relatively uniform concentration.

Sulfur is mobile in plants, but, because of its rapid uptake into metabolic compounds, it does not circulate in the plant like phosphorus. When radioactive sulfur is taken up by roots of a bean plant, it is rapidly translocated upward in the xylem tissue to the leaves. Within 24 hours, most of the labeled sulfur will be found in the younger leaves, the older more mature leaves having lost their sulfur to the young actively growing leaves (5). Since sulfur is a constituent of protein and protein synthesis is occurring to a much greater extent in the younger leaves compared to the older leaves, we can assume that movement to the younger leaves and metabolic capture of the sulfur at these sites is most probable. It has been suggested that sulfur makes one complete cycle before complete metabolic capture (4). Sulfur, then, is freely mobile in the plant but is rendered immobile rather quickly in metabolic reactions.

Radioactive calcium when taken up by bean roots is carried in the transpiration stream to the different areas of the plant. However, calcium is immobile in the phloem, and once delivered by the transpirational stream, it remains stationary (5).

The mobility of iron has been studied by Rediske and Biddulph (1, 49) in the red kidney bean plant, where it appears to be dependent primarily on the iron concentration of the plant tissues and secondarily on phosphorus availability and the pH of the nutrient medium. When the concentration of iron is low in the plant tissues, the mobility of iron injected into the phloem is highest. This mobility decreases with increased concentration of iron in the tissues. A pH of 4 in the nutrient solution gives high iron mobility. This mobility decreases as the pH increased to 7. A low phosphorus content in the nutrient solution favors the mobility of iron. High phosphorus concentrations in the plant tissues render iron immobile in the veins of the leaf.

In our discussion of the circulation of the mineral elements in the plant, we have touched upon four general directions of movement: upward, downward, lateral, and outward. The upward translocation of salts takes place primarily in the xylem tissue, although some upward movement does also take place in the phloem. Downward movement of the mineral elements takes place in the phloem tissue where upward movement also occurs. Movement of salts in the phloem tissue is generally spoken of as bidirectional. Lateral movement occurs between the xylem and the phloem, and this movement appears to be mediated by the cambium. Movement of salts out of leaves is a common occurrence, especially just prior to abscission, and occurs in the phloem tissue.

Keeping the above account in mind and considering the strong evidence supporting the different directions of salt movement in the plant, we can conclude that the theory that the circulation of mineral elements is a general phenomenon in plants is real and well documented.

Questions

6–1. Name several forms of passive diffusion that might take place in plants. Does passive diffusion account for appreciable ion accumulation in plants? Discuss your answer.

6–2. Define the following terms: active uptake, outer space, apparent free space, and carrier-ion complex.

6–3. Describe the details of a general model of the carrier concept. What is the significance of the activation of the carrier?

6–4. Describe the experimental evidence that supports the carrier-ion concept.

6–5. Describe two historical models that attempt to explain the active uptake of ions.

6–6. Explain and indicate the significance of the Nernst equation.

6–7. Indicate five major factors that influence salt absorption. Explain the possible major effect of each factor on salt absorption.

6–8. Where is the Casparian strip located in the root? How does it function in the uptake of ions and water?

6–9. Explain the method by which scientists are able to follow the path of elements (potassium for example) through the plant.

6–10. Are elements translocated through the phloem as well as the xylem? What is meant by mineral salt circulation in plants?

6–11. Suggest a major reason why one element may be highly mobile in plants and others are not. Is the mobility of an ion dependent on the plant or on the ion itself?

Suggested Readings

Baldwin, J.P. 1975. A quantitative analysis of the factors affecting plant nutrient uptake for some soils. *J. Soil Sci.* 26:195–206.

Carson, E.W., ed. 1974. *The Plant Root and Its Environment.* Charlottesville: University Press of Virginia.

Clarkson, D.T., and J.B. Hanson. 1980. The mineral nutrition of higher plants. *Ann. Rev. Plant Physiol.* 31:239–298.

Higinbotham, N. 1973. Electropotentials of plant cells. *Ann. Rev. Plant Physiol.* 24:25–46.

Hodges, T.K. 1973. Ion absorption by plant roots. *Adv. Agron.* 25:163–207.

Lüttage, U., and N. Higinbotham. 1979. *Transport in Plants.* New York: Springer-Verlag.

Mengel, K., and E.A. Kirkby. 1978. *Principles of Plant Nutrition.* Int. Potash Inst., eds. Bern: Der Bund.

Russell, R.S. 1977. *Root Function and the Soil.* New York: McGraw-Hill.

Torrey, J.G., and D.T. Clarkson, eds. 1975. *Development and Function of Roots.* New York: Academic Press.

Zimmerman, U., and J. Dainty. 1974. *Membrane Transport in Plants.* New York: Springer-Verlag.

Chapter 7

Functions of Essential Mineral Elements and Symptoms of Mineral Deficiency

Chrysanthemum plants grown in artificial medium. Plant on left received complete nutrient minus potassium, plants on right received complete nutrient plus potassium. *Courtesy of E.J. Holcomb, The Pennsylvania State University.*

In the previous two chapters, we discussed the occurrence, availability, absorption, and translocation of the essential mineral elements. We have studiously avoided any mention of the roles played by the mineral elements in the growth and development of the plant and of the deficiency symptoms that might accrue from a shortage of these elements. Since deficiency symptoms occur as a result of an essential function being retarded by the lack of an element necessary for that function, we will discuss these two aspects of mineral nutrition together.

Nitrogen

Function of Nitrogen

Perhaps nitrogen's most recognized role in the plant is its presence in the structure of the protein molecule. In addition, nitrogen is found in such important molecules as *purines, pyrimidines, porphyrins,* and *coenzymes.* Purines and pyrimidines are found in the nucleic acids RNA and DNA essential for protein synthesis. The porphyrin structure is found in such metabolically important compounds as the chlorophyll pigments and the cytochromes essential in photosynthesis and respiration. Coenzymes are essential to the function of many enzymes. Other compounds in the plant contain nitrogen (e.g., some vitamins), but since the above-cited molecules are often mentioned in connection with metabolic processes and plant growth, we will give them specific consideration in the next chapter.

Nitrogen Deficiency Symptoms

The most easily observed symptom of nitrogen deficiency is the yellowing (chloro-sis) of leaves due to a loss in chlorophyll. We usually notice this symptom first in the more mature leaves and last in the upper, more actively growing leaves. The nitrogen deficiency symptoms appear last in the younger leaves because of the high mobility of nitrogen in the plant. The younger leaves retain their nitrogen and, in addition, obtain nitrogen translocated from older leaves. Under severe conditions of nitrogen deficiency, the lowermost leaves on plants such as tobacco or bean will be dry and yellow and, in many cases, will abscise. Under these conditions, the topmost leaves are generally pale green in color.

One interesting characteristic of nitrogen deficiency found in many plants is the production of pigments other than chlorophyll when nitrogen is lacking. For example, in tomato plants, we can observe a purple coloring of the leaf petioles and veins caused by *anthocyanin* formation. We can also see this response to nitrogen deficiency on the stems of many plants.

If a plant is supplied high concentrations of nitrogen, there is a tendency to increased leaf cell number and cell size with an overall increase in leaf production (51, 57). We can assume from the above observations and from the fact that nitrogen is an essential constituent of protein that low nitrogen availability must cause a decrease in protein synthesis, which subsequently causes a decrease in cell size and especially cell division. Lutman (46) noted a decrease in leaf epidermal cell size due to nitrogen deficiency in millet and buckwheat.

Phosphorus

Function of Phosphorus

Phosphorus is found in plants as a constituent of *nucleic acids, phospholipids,* the *coenzymes NAD* and *NADP,* and, most important, as a constituent of *ATP* and other high-energy compounds. Phosphorus is found, of course, in other compounds of the plant, but these are considered most important. Heavy concentrations of phosphorus are found in the meristematic regions of actively growing plants, where it is involved in the synthesis of nucleoproteins. For example, not only is phosphorus found in the nucleic acid moiety of the nucleoprotein molecule, but it is also involved, through ATP, in the activation of amino acids for the synthesis of the protein moiety of this compound. Phospholipids, along with protein, are significant constituents of cell membranes. The coenzymes NAD and NADP are important in oxidation-reduction reactions in which hydrogen transfer takes place. Such essential plant processes as photosynthesis, respiration, nitrogen metabolism, carbohydrate metabolism, and fatty acid synthesis, to name a few, are dependent on the action of these coenzymes. The significance of ATP as an energy transfer compound is treated elsewhere in this book. There is no question as to the essentiality of phosphorus to the plant.

Phosphorus Deficiency Symptoms

Many of the symptoms of phosphorus deficiency can be confused with nitrogen deficiency, although the symptoms are not as pronounced as those found for nitrogen. Similar to nitrogen deficiency, phosphorus deficiency may cause premature leaf fall and purple or red anthocyanin pigmentation. Unlike plants lacking nitrogen, those lacking phosphorus may develop dead necrotic areas on the leaves, petioles, or fruits; they may have a general overall stunted appearance, and the leaves may have a characteristic dark to blue-green coloration. Because of

the high mobility of phosphorus in the plant and because of the tendency of younger leaves to deplete older leaves of mobile elements under deficiency conditions, the older leaves are usually the first to exhibit deficiency symptoms. Symptoms of zinc and phosphorus deficiencies may sometimes be confused. For example, lack of either one of these elements may cause a distortion in the shape of the leaves of some plants (31).

Lyon and Garcia (47, 48) performed anatomical studies on the stems of phosphorus deficient tomato plants. They found large amounts of pith and small amounts of vascular tissue. Central pith cells had disintegrated, and those that remained were large, succulent, and thin-walled, with abnormally large intercellular spaces. Phloem and xylem elements were thin-walled, and development of these vascular tissues was at a minimum.

A series of papers by Eaton (13, 14, 16) on phosphorus deficiency in sunflower, soybean, and black mustard demonstrated that a deficiency of this element causes an accumulation of carbohydrates.

Calcium

Function of Calcium

One well-known role played by calcium in the plant is as a constituent of cell walls in the form of *calcium pectate*. The middle lamella is composed primarily of calcium and magnesium pectates. The partial removal of calcium from the middle lamella with ethylenediaminetetraacetic acid (EDTA), a chelating agent, stimulates growth of the *Avena* coleoptile (4). Scientists assume that this stimulation is a result of increased plasticity caused by the removal of pectate-bound calcium. However, it may also be caused by an increase in cell permeability because of the removal of calcium.

Calcium is thought to be important in the formation of cell membranes and lipid structures. For example, the calcium salt of lecithin, a lipid compound, may be involved in the formation or organization of cell membranes (31). Calcium in small amounts is necessary for normal mitosis. In this respect, Hewitt (31) has suggested that calcium may be involved in chromatin or mitotic spindle organization. Abnormal mitosis may develop because of an effect of calcium deficiency on chromosome structure and stability. This suggestion is supported by the close correlation between calcium deficiency and chromosome abnormalities (18, 34, 68, 69) and by the suggestion that nucleoprotein particles are held together by divalent cations (49). A possible role for calcium as an activator of the enzyme phospholipase in cabbage leaves has been investigated (8). Calcium may also be an activator for the enzymes arginine kinase, adenosine triphosphatase, adenyl kinase, and potato apyrase (50).

Florell (20, 21) found that the number of mitochondria in wheat roots is reduced under calcium-deficiency conditions. Calcium deficiency in cotton plants results in increased levels of carbohydrates in the leaves and decreased levels in the stems and roots. Joham (38) interpreted this as a decrease in carbohydrate translocation due to calcium deficiency, an effect similar to that found in boron-deficient plants.

Calcium Deficiency

The easily observed symptoms of calcium deficiency are striking. Meristematic regions found at stem, leaf, and root tips are greatly affected and eventually die, thus ter-

minating growth in these organs. Roots may become short, stubby, and brown, as in calcium-deficient tomato plants (39). Chlorosis occurs along the margins of younger leaves, these areas usually becoming necrotic. Malformation or distortion of the younger leaves is also characteristic of calcium-deficient plants, a hooking of the leaf tip being the most easily detected symptom. Deficiency symptoms appear first in the younger leaves and the growing apices, probably as a consequence of the immobility of calcium in the plant.

Cell walls may become rigid or brittle in calcium-deficient plants (9, 39). Davis (9), in a study of calcium deficiency in *Pinus taeda*, demonstrated that cell enlargement, vacuolation, and differentiation occur closer to the shoot apex in deficient plants than in normal plants, an observation more recently made in tomato root tips by Kalra (39). Lutman (46) also observed vacuolation of cells occurring closer to the root apex of calcium-deficient rape and buckwheat plants.

Magnesium

Function of Magnesium

Two very essential roles played by magnesium in the plant is found in the important processes of *photosynthesis* and *carbohydrate metabolism*. Magnesium is a constituent of the chlorophyll molecule, without which photosynthesis would not occur. Many of the enzymes involved in carbohydrate metabolism require magnesium as an activator. Generally, ATP is involved in these reactions (see Table 7–1). Magnesium is also an activator for those enzymes involved in the synthesis of the nucleic acids (DNA, RNA) from nucleotide polyphosphates. Both the reactions mentioned above

Table 7–1. Some of the enzymes involved in carbohydrate metabolism that require Mg^{2+} as an activator.

Enzyme (trivial name)	Reaction
glucokinase	glucose + ATP $\longrightarrow$ glucose-6-P
fructokinase	fructose + ATP $\longrightarrow$ fructose-1-P
galactokinase	galactose + ATP $\longrightarrow$ galactose-1-P
hexokinase	hexose + ATP $\longrightarrow$ hexose-6-P
triosekinase	glyceraldehyde + ATP $\longrightarrow$ phosphoglyceraldehyde
gluconolactonase	6-phosphogluconolacton $\longrightarrow$ 6-phosphogluconate
6-phosphogluconic dehydrogenase	6-phosphogluconate $\longrightarrow$ ribulose-5-P
phosphopentokinase	ribulose-5-P + ATP $\longrightarrow$ ribulose-1,5-diP
enolase	2-phosphoglycerate + ATP $\longrightarrow$ phosphoenolpyruvate
pyruvic kinase	phosphoenolpyruvate + ADP $\longrightarrow$ pyruvate
carboxylase	pyruvate $\longrightarrow$ acetaldehyde
phosphoglyceric kinase	1,3-diphosphoglycerate + ADP $\longrightarrow$ 3-phosphoglycerate

and those requiring magnesium in carbohydrate metabolism involve phosphate transfer. Magnesium may participate in this type of group transfer as an intermediate carrier (55). Calvin (7) has stressed that ATP or ADP could become linked to the enzyme surface through a chelate complex involving the enzyme, magnesium, and the pyrophosphate group. In many cases, manganese can partially substitute for magnesium as an activator in the above enzyme systems. Furthermore, magnesium is necessary for full activity of the two principal CO_2 fixation enzymes, phosphoenolpyruvate carboxylase and ribulose-1,5-bisphosphate carboxylase.

Another function of magnesium has been postulated by T'so, Bonner, and Vinograd (70). They isolated ribosomal particles containing RNA, protein, and magnesium from homogenates of pea seedlings. They treated the particles with the chelating agent EDTA, which caused their dissociation into subunits. These investigators then suggested that magnesium binds these subunits together and that EDTA causes dissociation by its removal of the magnesium ion from the ribosomal particle. Thus magnesium may be assigned two roles in protein synthesis: as an activator in some of the enzyme systems involved in the synthesis of nucleic acids and as an important binding agent in ribosomal particles where protein synthesis takes place.

Magnesium Deficiency Symptoms

Since magnesium is a constituent of the chlorophyll molecule, the most common symptom of magnesium deficiency in green plants is extensive interveinal chlorosis of the leaves. Yellowing is apparent first in the basal leaves, and, as the deficiency becomes more acute, eventually reaches the younger leaves. The base-to-tip order of appearance of deficiency symptoms indicates that magnesium, like nitrogen and phosphorus, is mobile in the plant. Chlorosis is often followed by the appearance of anthocyanin pigments in the leaves. At a more acute stage in the deficiency—that is, following chlorosis and pigmentation—we may observe necrotic spotting.

Lyon and Garcia (47, 48) performed anatomical studies of tomato plants that were provided with an abundant or deficient supply of magnesium. An abundant supply of magnesium caused a depression of internal phloem development and an increase in size of parenchymatous cells adjacent to the endodermis. A deficient supply of magnesium caused more extensive chlorenchyma development, with the cells being smaller but greater in number and rather densely packed with chloroplasts. The investigators also observed smaller pith cells under deficient conditions.

Potassium

Function of Potassium

A deficiency in potassium affects processes such as respiration, photosynthesis, chlorophyll development, and water content of leaves. The best-known function of potassium is its role in stomatal opening and closing (see Chapter 5). The highest concentrations of potassium are found in the meristematic regions of the plant (55), an observation that seems in keeping with the findings of Webster and Varner (73, 74, 75) that potassium is essential as an activator for enzymes involved in the synthesis of certain peptide bonds. The accumulation of carbohydrates, often observed during the early stages of potassium deficiency, may be due to impaired protein synthesis (16)—that is, the carbon skeletons that would normally go into protein synthesis are accumu-

lated as carbohydrates. In addition to its role as an activator in protein metabolism, potassium also can act as an activator for several enzymes involved in carbohydrate metabolism. Apical dominance in several plants appears to be lacking or weak under potassium-deficient conditions (31). This may be caused by damage to the apical bud as a result of potassium deficiency.

Potassium Deficiency Symptoms

We can easily recognize the external symptoms of potassium deficiency on the leaves of the plant. A mottled chlorosis first occurs, followed by the development of necrotic areas at the tip and margin of the leaf. Because of the mobility of potassium, these symptoms generally appear first on the more mature leaves. There is a tendency for the leaf tip to curve downward and, as in the case of the French bean and potato, marginal regions may roll inward toward the upper surface (31). Generally, a plant deficient in potassium is stunted in growth with a pronounced shortening of the internodes.

Potassium deficiency in tomato plants causes disintegration of pith cells and results in an increase in the differentiation of secondary phloem parenchyma into sieve tubes and companion cells (47, 48).

Sulfur

Function of Sulfur

The content of sulfur in plants varies considerably and may reach a very high concentration, as observed by Gilbert (25) in the brassicaceus plants (members of the cabbage tribe of the mustard family). Its most obvious function is its participation in protein structure in the form of the sulfur-bearing amino acids cystine, cysteine, and methionine. Sulphur is taken up by the plant as the sulfate ion (SO_4^{2-}) and is subsequently reduced via an activation step involving the compound 3'-phosphoadenosine-5'-phosphosulfate (PAPS) and ATP. PAPS, first described by Robbins and Lipmann (62, 63), is synthesized in two distinct steps, an activation of sulfate by ATP and the enzyme sulfurylase to form adenosine-5'-phosphosulfate (APS), followed by the conversion of APS to PAPS by a specific kinase (3, 62, 63):

$$SO_4^{2-} + ATP \xrightarrow[Mg^{2+}?]{\text{sulfurylase}} APS + P\text{—}P$$

$$APS + ATP \xrightarrow[Mg^{2+}?]{\text{kinase}} PAPS + ADP$$

The activated sulfate is eventually reduced and incorporated into cystine, cysteine, and methionine, and finally into the protein structure.

(3'-phosphoadenosine-5'-phosphosulfate (PAPS)

When speaking of the function of sulfur in the plant, we must not forget the sulfer-bearing vitamins *biotin, thiamine,* and *coenzyme A.* Thus sulfur is involved in the metabolic activities of these vitamins and may also be found in sulfhydryl groups, which are present in many enzymes and are necessary for enzyme activity. Sulfur forms cross-links in the protein molecule and, in conjunction with the peptide and hydrogen bonding, acts to stabilize protein structure. It is a component of S-adenosyl-methionine, which is important in lignin and sterol biosynthesis. Also, sulfur is important in Fe-S proteins in photosynthesis, nitrogen metabolism, and ferredoxin synthesis.

Sulfur Deficiency Symptoms

The symptoms of sulfur deficiency resemble somewhat those of nitrogen deficiency. As in nitrogen-deficient plants, there is a general chlorosis, followed by the production of anthocyanin pigments in some species (15). Unlike nitrogen-deficient plants, sulfur-deficient plants show chlorosis of the younger leaves first. Under severe conditions, however, all of the leaves may undergo some loss of green color (25).

Hall and her colleagues (29) studied the ultrastructure of mesophyll chloroplasts in sulfur-deficient corn plants. They found that sulfur deficiency resulted in a marked decrease of stroma lamellae and an increase in grana stacking (see Figure 7–1). We also find increased grana stacking in nitrogen-deficient corn plants.

In a series of studies of sulfur deficiency in tomato, sunflower, black mustard, and soybean, Eaton (10, 11, 12, 15) found that starch, sucrose, and soluble nitrogen were accumulated under deficiency conditions but that reducing sugars were lower than normal. He suggested that the increase in soluble nitrogen resulted from an inhibition of protein synthesis and an increase in proteolytic activity.

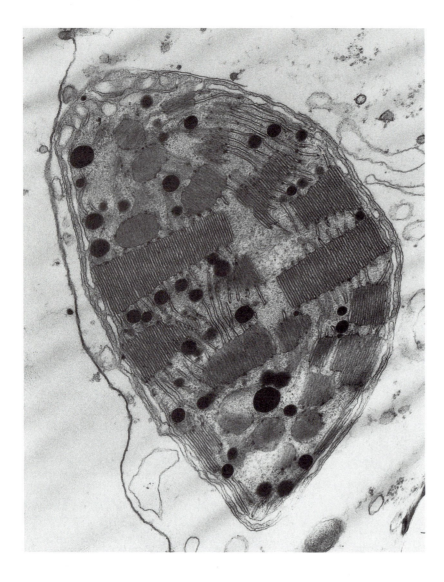

Figure 7–1. Electron micrograph of chloroplast from sulfur-deficient corn plant. Note extensive stacking of grana and marked reduction in stroma lamellae. Magnification 31,900×.

From J.D. Hall et al. 1972. Plant Physiol. 50:404. Photo courtesy of J.D. Hall, Robert Packer Hospital, Sayre, Pennsylvania.

Iron

Function of Iron

Iron has a number of important functions in the overall metabolism of the plant. Although iron is frequently taken up in the ferric state (Fe^{3+}), the *ferrous* state (Fe^{2+}) is generally accepted as the metabolically active form of iron in the plant. Iron is incorporated directly into the cytochromes, into compounds necessary to the electron transport system in mitochondria, and into ferredoxin. As we shall observe later, ferredoxin is indispensable to the light reactions of photosynthesis. Although iron is essential for the synthesis of chlorophyll, its chemical role in both the synthesis and degradation of chlorophyll is still uncertain (55). Iron is required in the synthesis of chloroplast proteins, possibly enzymes involved in chlorophyll synthesis (22). In a later chapter we will discuss the synthesis of chlorophyll and the involvement of protoporphyrin-9 as one of the intermediates in chlorophyll biosynthesis. This intermediate compound may represent a branch point in the biosynthesis of either cytochromes or chlorophyll. The synthetic path is dependent on which metal, magnesium or iron, is incorporated into the porphyrin structure (27). Price and Carell (60) found that the addition of iron to iron-deficient *Euglena* cells considerably increased the rate of chlorophyll synthesis (see Figure 7–2).

Iron has been identified as a component of various flavoproteins (metalloflavoproteins) active in biological oxidations. Iron is also found in the iron-porphyrin proteins, which include cytochromes, peroxidases, and catalases. We describe and discuss the functions of these enzymes in another section of this book.

Iron Deficiency Symptoms

The most easily observed symptom of iron deficiency in plants is extensive chlorosis in the leaves. The younger leaves are most affected, the more mature leaves sometimes show no chlorosis at all. Primarily, this is because of the relative immobility of iron in the plant. Thus the younger leaves cannot withdraw iron from the older leaves. One feature of iron-induced chlorosis is its characteristic interveinal nature, the surface of the leaf usually shows a fine reticulate network of green veins setting off chlorotic areas. Total chlorosis of the younger leaves is rare. However, secondary and tertiary

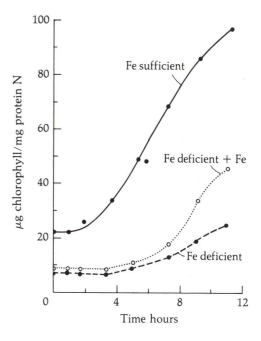

Figure 7–2. Time course of chlorophyll synthesis. Cells grown under low (50 footcandles) light intensity with $3 \times 10^{-5} M$ iron and low iron ($1.8 \times 10^{-7} M$). After harvest, deficient cells were resuspended in $10^{-3} M$ phosphate buffer, pH 6.0, with and without $3 \times 10^{-5} M$ iron added, and incubated under high light intensity; samples taken at various intervals for chlorophyll analysis.
From C.A. Price and E.F. Carell. 1964. Plant Physiol. *39:862.*

veins may undergo chlorosis under severe deficiency conditions.

Several investigators have made attempts to find a good correlation between iron deficiency and chlorophyll content. They have, however, had only limited success with this line of investigation. For example, some workers have found a good correlation between iron and chlorophyll content (36, 67, 72), others have found that chlorotic leaves may contain as much and even more iron than their normal counterparts (35, 44, 76). Jacobson and Oertli (37), in a study of iron deficiency in sunflowers, found that good correlation may be achieved if iron is supplied at a uniform rate. However, when they subjected a plant to a brief period of iron deficiency and then supplied it with adequate iron, they found no correlation between chlorophyll and iron, probably because of an enhanced iron uptake. They found that chlorosis is incompletely reversible in sunflower leaves. Thus if a chlorotic plant is restored to a normal iron supply, the chlorotic leaves of that plant are likely to accumulate as much or more iron than would be found under normal conditions. These investigators proposed that lack of iron may inhibit formation of the chloroplasts through inhibition

of protein synthesis, a fact that might explain incomplete recovery from chlorosis. Figure 7–3 illustrates the effect of iron deficiency in spinach leaves. The plastids are noticeably influenced in the iron-deficient plants as compared with those of the control plants.

Manganese

Function of Manganese

Manganese is an essential element in respiration and nitrogen metabolism; in both processes it functions as an *enzyme activator*. However, in many cases, especially

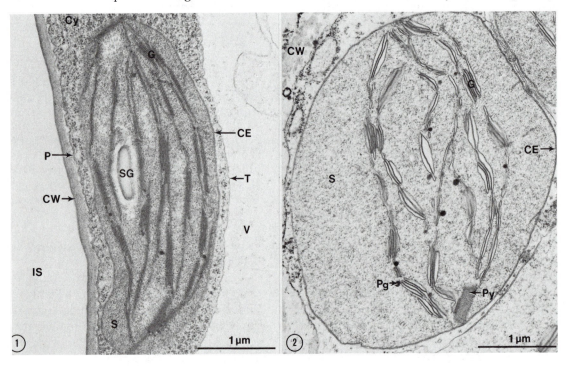

Figure 7–3. (1) Mesophyll cell chloroplast from normal spinach plant and (2) iron-deficient spinach plant. Abbreviations are: cell wall (CW), chloroplast envelope (CE), cytoplasm (Cy), granum (G), intercellular space (IS), plasma-lemma (P), plastoglobulin (Pg), phytoferritin (Py), starch granule (SG), stroma (S), tonoplast (T), and vacuole (V). Magnification 20,900×.

Courtesy of R. Rufner, Massachusetts Agricultural Experiment Station, University of Massachusetts.

with reactions in respiration, manganese can be replaced by other divalent cations, such as Mg^{2+}, Co^{2+}, Zn^{2+}, and Fe^{2+}. Magnesium is the most frequent replacer of manganese. Manganese appears to be essential, however, for some reactions in the metabolism of the plant. For example, malic dehydrogenase, an enzyme of the Krebs cycle, requires manganese as an activator. Another enzyme of the Krebs cycle, oxalosuccinic decarboxylase, requires the presence of manganese as an activator, although in this case, the manganese requirement may be partially substituted for by cobalt. From the very extensive work done on the enzymes of the Krebs cycle, we can draw the conclusion that manganese is the predominant metal ion of Krebs cycle reactions.

Scientists have known for sometime that manganese plays an important role in nitrate reduction (6, 14, 22). This role has been clarified to some extent. Manganese acts as an activator for the enzymes nitrite reductase and hydroxylamine reductase (53, 64). The preference of ammonia over nitrate as a nitrogen source by manganese-deficient cells (55) supports the above-mentioned analysis of the role of manganese. Manganese is also thought to be involved in the destruction or oxidation of indole-3-acetic acid (IAA), a natural auxin of plants (26, 41).

A decrease in the rate of photosynthesis in algae at an early stage of manganese deficiency suggests a direct role for manganese in photosynthesis (77). According to Eyster and colleagues (19), the sensitivity of chlorophyll to light destruction increases under conditions of manganese deficiency and leads ultimately to chlorosis in *Chlorella pyrenoidosa*. They found that oxygen evolution was suppressed under deficiency conditions. It appears from work on all the higher plants and on the alga *Ankistrodesmus braunii* that the site of manganese activity is

in the oxygen-producing step in photosynthesis (42, 43). More than likely manganese is involved in electron transfer from water to chlorophyll during the light reactions of photosynthesis. However, manganese deficiency does not seem to impair photoreduction in photosynthesis per se.

Manganese Deficiency Symptoms

Manganese deficiency is characterized by the appearance of chlorotic and necrotic spots in the interveinal areas of the leaf. These symptoms may appear first on the young leaves of some species, on other species they may appear first on the older leaves. We may observe a brown necrosis of the cotyledons of pea and bean seeds (30, 58). Manganese deficiency also appears to have a marked effect on the chloroplast. Eltinge (17) found that the chloroplasts of tomato leaves are the first part of the plant affected by manganese deficiency. The chloroplasts lose chlorophyll and starch grains, become yellow green in color, vacuolated, and granular, and finally disintegrate.

Copper

Function of Copper

There is little doubt as to the necessity of copper for normal plant metabolism. Copper acts as a component of phenolases, laccase, and ascorbic acid oxidase, and its role as a part of these enzymes probably represents an important function of copper in plants (55). Work by Neish (56) and by Green, McCarthy, and King (28) first suggested that copper may function in photosynthesis. For example, Neish found that the chloroplasts of clover contain most of the copper of the plant. In addition, Lousta-lot and others (45) found that CO_2 absorption is decreased in copper-deficient tung trees. The chloroplast possesses a copper-containing protein called plastocyanin that is essential as an electron carrier in photosynthesis. Also, plastid enzymes, namely, the phenolases, contain copper that is essential to their functioning.

Copper Deficiency Symptoms

The most easily recognized symptoms of copper deficiency are those found in exanthema and reclamation. *Exanthema* is a disease of fruit trees that is characterized by gummosis (gummy exudates), accompanied by dieback and glossy brownish blotches on leaves and fruit. *Reclamation* is a disease of cereals that occurs chiefly on newly reclaimed peat land and is characterized by chlorotic leaf tips and failure to set seeds. Copper deficiency causes a necrosis of the tip of young leaves that proceeds along the margin of the leaf and gives it a withered appearance. Under more severe conditions, the leaves may be lost, and the whole plant may appear wilted.

Zinc

Function of Zinc

Zinc may be involved in the biosynthesis of the plant auxin *indole-3-acetic acid* (IAA). Skoog (66) observed that there is a marked decrease in auxin content of zinc-deficient tomato plants and a significant increase in the IAA content when zinc is added to the deficient plants. Both of these responses (increase and decrease in auxin content) precede growth response to the absence or addition of zinc, thus suggesting that deficiency symptoms could, in part, be associated with the decrease in the plant auxin concentration. Later work demonstrated that the content of *tryptophan* parallels the content of auxin in the plant, both when zinc is deficient and when it is supplied to deficient plants. Scientists therefore concluded that zinc reduces auxin content through its involvement in the synthesis of tryptophan, a precursor of auxin (71). In support of this hypothesis, Nason (52) found that the activity of the enzyme tryptophan synthetase is low in zinc-deficient *Neurospora*. This enzyme catalyzes the reaction of the serine with indole to form tryptophan.

Zinc participates in the metabolism of plants as an activator of several enzymes. Carbonic anhydrase, an enzyme found in some marine plants but more common in animals, was the first zinc-containing enzyme to be discovered (40). This enzyme catalyzes the decomposition of carbonic acid to carbon dioxide and water. Other enzymes dependent on the presence of zinc are alcohol dehydrogenase and pyridine nucleotide dehydrogenases (32, 54). An accumulation of inorganic phosphorus in zinc-deficient tomato plants suggests that zinc may act as an activator for some phosphate transferring enzyme, such as hexose kinase or triosephosphate dehydrogenase. Another striking characteristic of zinc deficiency is the accumulation of soluble nitrogen compounds, such as amino acids and amides (59). We can assume from this observation that zinc must play an important role in protein synthesis.

Zinc Deficiency Symptoms

The first sign of zinc deficiency is an interveinal chlorosis of the older leaves, starting at the tips and margins. White necrotic spotting soon follows, as in cotton (5). Smaller leaves and shortened internodes, resulting in stunted growth, are characteristic of more severe zinc deficiency. The most easily recognized symptoms of zinc deficiency is the distorted appearance of the plant leaves. They are generally smaller in size, are distorted in shape and appearance, and may be clustered on short branches known as *rosettes*. The effect of zinc deficiency on leaves is sometimes referred to as "little leaf" disease. The absence of zinc also may have an adverse effect on the production of seeds in beans and peas and the development of fruit in citrus.

Boron

Function of Boron

Although symptoms of boron deficiency are striking, its role in plant metabolism has not, as yet, been ascertained. Gauch and Dugger (23, 24) have built a strong case for the involvement of boron in *carbohydrate transport* within the plant. They have drawn attention to the fact that the borate ion will complex readily with polyhydroxy compounds such as sugar. They propose that sugar is transported more readily across cell membranes as a borate complex. As an alternative proposal, they suggest that the borate ion may be associated with the cell membrane where it could complex with a sugar molecule and facilitate its passage across the membrane. Gauch and Dugger have also drawn attention to the fact that the common features of boron deficiency in plants are the death of stem and root tips and the abscission of flowers, areas of the plant high in metabolic activity. They suggest that symptoms of boron deficiency are actually symptoms of sugar deficiency.

Since areas of the plant that are high in metabolic activity also need higher quantities of sugar, these areas are the first to be affected under conditions of boron deficiency. The above-mentioned role of boron in sugar translocation has been strongly supported by experiments using ^{14}C-labeled sucrose (65). These experiments showed that the uptake and translocation of sugar is retarded in boron-deficient plants. Photosynthesis in the presence of $^{14}CO_2$ also supports the Gauch and Dugger theory of boron-facilitated translocation of sugars (65). The translocation of labeled photosynthate is much less efficient in boron-deficient plants.

Although many roles have been hypothesized for boron in plant metabolism, scientists generally accept its function in the translocation of sugar and role in DNA synthesis in meristems. Boron has been implicated in cellular differentiation and development, nitrogen metabolism, fertilization, active salt absorption, hormone metabolism, water relations, fat metabolism, phosphorus metabolism, and photosynthesis (55). However, convincing evidence of boron participation in these processes is still forthcoming. Indeed, we could argue that boron affects all of these processes only indirectly, through its effect on sugar translocation.

Boron Deficiency Symptoms

The first visible symptom of boron deficiency, which occurs only a few hours after boron withdrawal, is the death of the shoot tip because of its requirement for DNA synthesis. The growth of lateral shoots results, and the tips of these shoots also die. The leaves may have a thick coppery texture and sometimes curl and become quite brittle.

Flowers do not form and root growth is stunted. Storage or fleshy organs react in a striking manner to boron deficiency. A general disintegration of internal tissues results in abnormalities, such as heart rot in sugar beets, internal cork formation in apples, and water core development in turnips.

Molybdenum

Function of Molybdenum

Molybdenum has long been implicated in gaseous nitrogen fixation and nitrate assimilation. Chapter 8, which is devoted entirely to nitrogen metabolism, will cover the essential functions of molybdenum in nitrogen metabolism.

Several investigators have observed that molybdenum deficiency always leads to a drop in the concentration of ascorbic acid in the plant (1, 32). Hewitt (31), citing unpublished work by Hewitt and Hucklesby, drew attention to the fact that chloroplast disorganization occurs with the appearance of whiptail symptoms, a common disease of molybdenum deficiency. Evidence that molybdenum is involved in the phosphorus metabolism of the plant exists. However, we do not as yet have an explanation of the mechanism of action of molybdenum in phosphorus metabolism.

Molybdenum Deficiency Symptoms

Visible symptoms of molybdenum deficiency may start with chlorotic interveinal mottling of the lower leaves, followed by marginal necrosis and infolding of the leaves. Under more severe conditions, mottled areas may become necrotic and may cause the leaf to wilt. Flower formation is inhibited, and if flowers do form, they abscise before setting fruit.

One condition due to molybdenum deficiency, known as *whiptail*, is typically demonstrated in cauliflower plants. The leaves first show an interveinal mottling, and the leaf margins may become gray and flaccid and finally brown. The leaf tissues wither and leave only the midrib and a few small pieces of leaf blade, which gives the appearance of a whip or tail, hence the name.

Questions

7–1. What are the functions of each of the major elements in plants?

7–2. High nitrogen application to potato plants may often result in luxuriant growth of the stems and leaves but in production of small-sized tubers. What is the probable reason for this nitrogen effect?

7–3. In studying mineral deficiencies of plants, scientists have often observed that plants grown in salt solutions optimum for growth but lacking one element will show enhanced deficiency symptoms over plants grown in pure water. Explain the likely reason for this phenomenon.

7–4. Magnesium deficiency symptoms are often characterized by chlorosis and reduced photosynthetic rates. What are the reasons for these symptoms?

7–5. Symptoms of zinc deficiency are the distorted shape of leaves and their clustered appearance as rosettes. What is the probable reason for these deficiency symptoms?

7–6. When you observe a chloroplast from a plant deficient in iron, what structural changes due to the deficiency do you note?

7–7. Since a given deficiency symptom may not be characteristic of a given element, how might plant scientists determine which element is deficient and how might they then treat field crops exhibiting that deficiency?

7–8. How might you determine if apparent deficiency symtoms have been induced by pathogens or the absence of an element?

7–9. How might a scientist determine whether a plant is suffering from symptoms induced by the deficiency of an element or by toxicity due to an overabundance of an element?

7–10. Name several crops that might be grown without a requirement for the addition of nitrogen to the soil.

Suggested Readings

Chapin, F.S., III. 1980. The mineral nutrition of wild plants. *Ann. Rev. Ecol. Syst.* 11:233–260.

Clarkson, D.T., and J.B. Hanson. 1980. The mineral nutrition of higher plants. *Ann. Rev. Plant Physiol.* 31:239–298.

Hewitt, E.J., and T.A. Smith. 1975. *Plant Mineral Nutrition.* London: English Universities Press.

Rains, D.W. 1976. Mineral metabolism. In J. Bonner and J.E. Varner, eds., *Plant Biochemistry.* New York: Academic Press.

Sprague, H.B., ed. 1964. *Hunger Signs in Crops,* 3rd ed. New York: McKay.

Wallace, T. 1961. *The Diagnosis of Mineral Deficiencies in Plants,* 3rd ed. New York: Chemical Publishing.

Witham, F.H., D.F. Blaydes, and R.M. Devlin. 1971. *Experiments in Plant Physiology.* New York: Van Nostrand.

Chapter 8

Nitrogen Metabolism

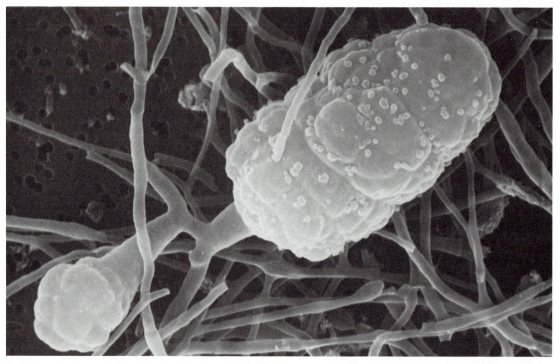

Scanning electron micrograph of actinomycete isolated by sucrose-density fractionation from nitrogen-fixing root nodules and cultivated in vitro. *Reprinted by permission from D. Baker, J.G. Torrey, and G.H. Kidd.* Nature 281:76. *Copyright © 1979 Macmillan Journals Limited. Photo courtesy of D. Baker and E. Seling.*

We will devote a complete chapter to the discussion of nitrogen metabolism. Yet one chapter is a small allotment when we consider the complexity and importance of this subject. With the exception of carbon, hydrogen, and oxygen, nitrogen is the most prevalent element in the living organism and is found in such essential compounds as proteins, nucleic acids, some of the plant growth regulators, and in many of the vitamins. As a component of these and many other compounds, nitrogen is involved in most of the biochemical reactions that compose life.

The large quantities of nitrogen present in the plant, the importance of nitrogen in the structure and metabolism of the plant, and the need of the plant for a continuous supply of nitrogen dramatically point out one of nature's most paradoxical situations. Since nitrogen composes 80 percent of the earth's atmosphere, the plant world may literally be said to be submerged in a sea of nitrogen, yet nitrogen in this form is unavailable to most plants. Indeed, nitrogen is one of the most inert elements and requires excessive temperatures and pressures in order to react with other elements or compounds. Although some forms of combined or fixed nitrogen may be contributed to the soil without the participation of living organisms (e.g., nitrogen oxides produced through electric discharge during lightning storms), a much greater quantity is fixed through the mediation of soil microorganisms. What, then, are the forms of nitrogen available to the plant, and how is atmospheric or molecular nitrogen converted to these forms? In the following pages we will discuss the forms and absorption of available nitrogen, the incorporation of reduced nitrogen into keto acids to form amino acids, the synthesis of protein, and finally the degradation of protein and amino acids.

Nitrogen Nutrition

With the exception of those microorganisms capable of fixing molecular nitrogen, plants absorb nitrogen in a fixed form from the soil. The forms of nitrogen available to the plant may be divided into four groups: *nitrate nitrogen, ammonia nitrogen, organic nitrogen,* and *molecular nitrogen.* Although most plants utilize the nitrate form of nitrogen, several plants can assimilate ammonia and certain forms of organic nitrogen. The utilization of molecular nitrogen is confined to a certain few groups found among the prokaryotic forms of plant life, including certain species of free-living bacteria (e.g., *Azotobacter, Clostridium*) and blue-green algae (e.g., *Anabaena, Nostoc*). However, the list of plant species capable of utilizing molecular nitrogen is growing every day.

Nitrate and Ammonia Nitrogen

The roots of most higher plants absorb nitrogen from the soil in the form of nitrate (NO_3^-). Nitrogen in this form, however, is not directly used by the plant but must be reduced to ammonia before it may be incorporated into the nitrogenous compounds of the plant. The reduction of nitrate to ammonia requires the energy of respiration. Thus the carbohydrates of the plant not only provide the carbon skeletons needed for the incorporation of ammonia but also provide, through their breakdown in respiration, the energy needed for nitrate reduction (5, 42). Many investigators have observed that under conditions of high nitrate reduction and assimilation in the dark, carbohydrate levels in the plant are significantly lowered. The lowering of carbohydrate levels under these conditions in the light is not as impressive because of the compensating effects of pho-

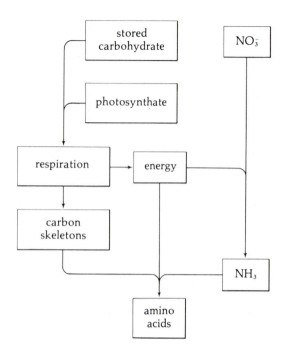

Figure 8–1. Relationship between carbohydrate status of plant and nitrate reduction and assimilation.

tosynthesis. Figure 8–1 shows the relationship between the carbohydrate status of the plant and nitrate reduction and assimilation.

The first step in nitrate reduction is the conversion of nitrate to nitrite (NO_2^-). Scientists originally provided evidence for this fact by identifying nitrite in plant tissues and by isolating nitrate reductase from soybean leaves and from *Neurospora* (12, 25). In addition, enzyme preparations from *Neurospora*, soybean leaves, and the blue-green alga *Anabaena cylindrica* appeared to contain nitrite reductase, an enzyme that catalyzes the reduction of nitrite to ammonia (25, 36).

Since the formation of nitrite from nitrate requires the transfer of two electrons to nitrate, scientists originally believed that the compound hyponitrite (HNO) was involved

as an intermediate in the transfer of electrons. Other observers, however, suggested that hyponitrite was not detected in plant tissues because of its great instability, which causes it to be converted to other compounds as quickly as it is formed (43). It seems clear now, however, that hyponitrite is not an intermediate in N_2 reduction (23).

Further, observers thought that another compound, hydroxylamine (NH_2OH) was an intermediate in the sequence leading from nitrate to ammonia. However, evidence has accumulated that rules out this compound also in the reduction of nitrate to ammonia (23). Thus the current idea is that the reaction proceeds as follows (the oxidation number for each compound is given under the formula name of that compound):

$$NO_3^- \xrightarrow[\substack{\text{nitrate} \\ \text{reductase}}]{2\ e^-} NO_2^- \xrightarrow[\substack{\text{nitrite} \\ \text{reductase} \\ \text{(plastid)}}]{6\ e^-} NH_3$$

nitrate	nitrite	ammonia
+5	+3	−3

Because of the widespread occurrence of the above intermediates in plants and the detection of enzymes, in different plant tissues, that catalyze the reduction, this inorganic sequence appears to be an important pathway for nitrate reduction in plants. We still need to determine whether or not nitrogen has to be reduced to the ammonia level before it will react with the organic compounds of the plant.

If we assume that nitrate must be reduced to ammonia before nitrogen can enter the metabolic system, then we should observe a more rapid assimilation of nitrogen when ammonia instead of nitrate is utilized as a nitrogen source. Investigations have shown that ammonia assimilation is indeed rapid compared to nitrate assimilation.

Healthy plants with an adequate supply of respirable carbohydrates will incorporate ammonia nitrogen so fast into the metabolic system that even during periods of high nitrogen uptake only traces of free ammonia may be found in the plant tissues (38). In contrast, relatively high amounts of free nitrate may be found in the plant tissues. As with nitrate reduction and assimilation, ammonia assimilation is dependent, in part, on the carbohydrate status of the plant. Because of the rapid assimilation of ammonia, however, the carbohydrate supplies of a plant utilizing ammonia as a sole nitrogen source may be depleted to a low point (26, 28, 40). For example, with the tomato plant a soft, succulent, nonfruitful, highly vegetative growth may result as a consequence of a high depletion of carbohydrate.

Nitrate and Nitrite Reductases

A discussion of the enzyme activity involved in each step of nitrate reduction is beyond the scope of this book. However, since an impressive amount of information has accumulated about the enzymes nitrate reductase and nitrite reductase, we will briefly discuss the nature of these enzymes and the cofactors involved in the reactions they catalyze.

Nitrate reductase is a metalloflavoprotein that catalyzes the reduction of nitrate to nitrite and was isolated in a highly purified form (12, 25). The enzyme system includes a reduced pyridine nucleotide (NADPH or NADH) as an electron donor, flavin adenine dinucleotide (FAD), and molybdenum. Electrons are passed from reduced pyridine nucleotide to FAD and produce reduced FAD ($FADH_2$). (We will examine these coenzymes and electron transfer in later chapters.) The electrons are in turn passed from $FADH_2$ to oxidized molybdenum and produce reduced molybdenum, which passes electrons to nitrate and reduces it to nitrite (see Figure 8–2) (26).

Nitrate reductase is an *inducible enzyme*. An inducible enzyme is distinguished from a *constitutive enzyme* (always present in the organism) in that it is only apparent in the presence of its particular substrate or inducer. The inducer for the formation of nitrate reductase seems to be nitrate in some systems, particularly enzyme preparations from higher plants. The data in Figure 8–3 clearly illustrates this point. In algae and other plants, however, the substrate induction of the enzyme is not clear and requires extensive study.

Effect of Light, CO_2, and Calcium on Nitrate Reductase

The presence of other factors such as light, CO_2, and calcium are also important in the formation of nitrate reductase. Several studies have shown that although the formation of nitrate reductase has been detected in a few cases in the dark, a much more efficient synthesis takes place when the plant is exposed to light (6, 15, 20). In-

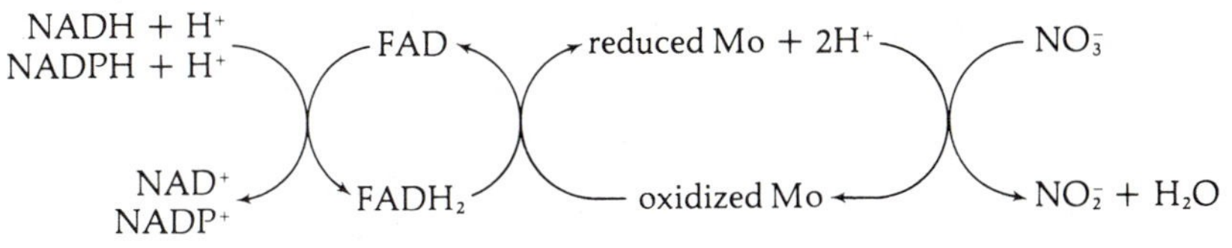

$$NADH + H^+ \quad NADPH + H^+ \qquad FAD \qquad \text{reduced Mo} + 2H^+ \qquad NO_3^-$$
$$NAD^+ \quad NADP^+ \qquad FADH_2 \qquad \text{oxidized Mo} \qquad NO_2^- + H_2O$$

Figure 8–2. Sequence of electron transport in nitrate reduction catalyzed by nitrate reductase.

From D.J.D. Nicholas and A. Nason. 1955. Plant Physiol. *30:135.*

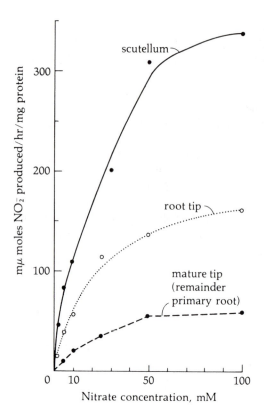

Figure 8–3. Influence of nitrate on level of nitrate reductase in maize seedling.

From W. Wallace. 1973. Plant Physiol. 52:191.

deed, Beevers and his colleagues (6) demonstrated with maize seedlings and radish cotyledons that the synthesis of nitrate reductase increased with increasing light intensity. Some investigators (20) believe that the light requirement reflects only the need for an active photosynthesis in the synthesis of this enzyme. This hypothesis is supported by the finding that when leaves of *Perilla* containing nitrate are illuminated in a CO_2-free atmosphere, formation of nitrate reductase cannot be detected (20). Nitrate reductase activity can be induced in darkness in green barley leaves supplied with nitrate, but the enzyme begins to disappear

after about 12 hours without light (41). This fact suggests that the role of light in nitrate reductase induction is to provide photosynthate for energy production (5). In support of this theory, Travis and Key (42) observed increased nitrate reductase activity in 3- and 8-day-old corn shoots, grown in the dark and exogenously supplied with glucose. Only photosystem I of photosynthesis seems to be necessary for nitrate reduction since the process promotes the reduction of $NADP^+$.

Paulsen and Harper (29) found that calcium-deficient wheat seedlings (*Triticum aestivum*) accumulated unusually high amounts of nitrite, which, in turn, caused the synthesis of nitrate reductase to be repressed. They suggested that the accumulation of nitrite was not due to any effect of calcium deficiency on nitrite reductase but to an inhibition of intracellular transport of nitrite caused by this deficiency. Nitrate reductase is found in the cytoplasm. Nitrite reductase, however, is located in the chloroplast (32). Calcium is a necessary factor in the structural integrity and functional performance of plant cell membranes (10). With the location of nitrite reductase and the effect of calcium in plant cell membranes in mind, we can see how the intracellular movement of nitrite into the chloroplast may be inhibited in calcium-deficient plants. This inhibition of movement could cause an accumulation of nitrite in the cytoplasm, which, in turn, would repress the synthesis of nitrate reductase.

Nitrite reductases have been isolated from both chlorophyllous tissues—where they reside in the chloroplasts—and from nonphotosynthetic tissues, such as tomato and barley roots and corn scutella (8, 32, 33). The chloroplast nitrite reductases operate with reduced ferredoxin, NADH, or NADPH as electron donors; the reductases from nonphotosynthetic tissues cannot ac-

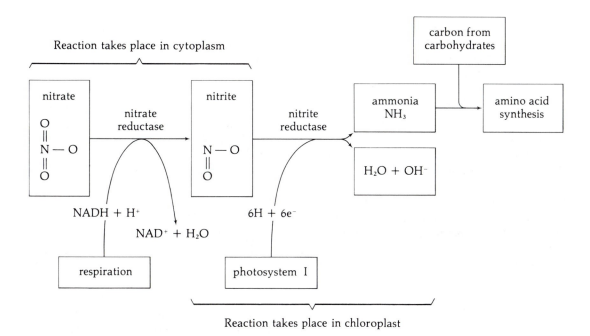

Figure 8–4. Generalized scheme for nitrate and nitrite reduction.

cept electrons directly from reduced pyridine nucleotides (8). In contrast, nitrite reductases of *Escherichia coli* accept electrons directly from the reduced pyridine nucleotides; in this way they are similar to the chloroplast nitrite reductases (21, 22). ATP and copper or iron or both also seem to be involved in nitrite reductase activity.

Figure 8–4 is a generalized scheme of nitrate reduction that clarifies the process. Although the diagram illustrates one of the major mechanisms in green plants, there are no doubt exceptions.

The nitrate reductase consists of the flavin protein (FAD) and molybdenum, which act as electron carriers from $NADH_2$ to the NO_3^- oxygens. Although $NADH_2$ is the normal electron donor in some plants, other donors such as $NADPH_2$, $FMNH_2$, and $FADH_2$ may function accordingly (23).

Nitrite reductase is also a metalloflavoprotein. Intermediates between NO_2 and

NH_3 were formerly believed to be bound to the enzyme. Currently, information by Murphy and others (24) suggests that the nitrite reductase is a single protein that catalyzes reduction of NO_2 to NH_3 directly from a reductant of photosystem I (see Chapter 13 for an explanation of this system). Reduced ferredoxin or reduced pyridine nucleotide may act as electron donors to nitrite reduction, and it appears that ATP is necessary for activity.

Organic Nitrogen

Many plants are capable of using organic as well as inorganic nitrogen as a nitrogen source for growth. Many of the amino acids and amides will provide available nitrogen for plant growth. Also, urea provides a good source of organic nitrogen. With a few exceptions, these compounds are the only organic nitrogen compounds capable of providing available nitrogen in the quantities needed to support normal plant growth. Much of the soil nitrogen is bound in organic form, primarily as proteins. The breakdown of proteins releases free amino acids. Amino acids may either be oxidized, thereby releasing their nitrogen in the form of ammonium, which is usually oxidized to nitrate before being absorbed by the plant, or the amino acids may be used directly by the plant. Many of the soil microorganisms can readily assimilate amino acids and compete with higher plants for this nitrogen source.

Investigators have given amino acid assimilation by intact plants only limited attention. However, a good deal of work has been accomplished on the assimilation of amino acids by plant tissues grown in aseptic cultures. Early work by White (48) demonstrated that certain amino acids can act as nitrogen sources for the growth of excised tomato roots. Since White's pioneer work, the uptake of amino acids by various plant tissue has been demonstrated.

Foliar application of urea

$$\underset{NH_2-\overset{\displaystyle O}{\overset{\displaystyle \|}{C}}-NH_2}{}$$

has proved to be a very effective method of relieving nitrogen shortages in many plants (19). The first step in the utilization of urea nitrogen is thought to be the rapid hydrolysis of urea by the enzyme urease to yield ammonia and carbon dioxide (27):

$$NH_2-\overset{\displaystyle O}{\overset{\displaystyle \|}{C}}-NH_2 \xrightarrow{\text{urease}} 2NH_3 + CO_2$$

Several investigators have suggested that urea might in some cases be assimilated directly, without prior hydrolysis to ammonia and carbon dioxide. One possible route for the incorporation of the intact urea molecule would be its condensation with ornithine (an amino acid) to form the amino acid arginine (7, 17, 46). However, convincing evidence for this pathway of urea incorporation has not been provided.

Molecular Nitrogen

By far the most abundant supply of nitrogen is found in the earth's crust, rocks, and sediment (17.5 to 18.4×10^{15} tons); the second largest reservoir of molecular nitrogen (N_2) is the atmosphere (3.5 to 4.0×10^{15} tons). However, only a relatively few plants are capable of *fixing*, or assimilating, this plentiful supply of nitrogen, and these plants are confined to the lower forms, such as certain groups of bacteria and blue-green algae. Although higher plants are not able to utilize molecular nitrogen directly, some are able to utilize it indirectly through the mediation of microorganisms in the soil. Before molecular nitrogen (N_2, or atmospheric nitrogen) can be used by most plants, it must be converted to nitrate (NO_3^-), ammonia (NH_3), or ammonium (NH_4^+, the cationic form of NH_3). The conversion of N_2 to NH_4^+ is accomplished by *asymbiotic nitrogen fixation* (nitrogen fixation by so-called free-living organisms, or organisms not in association with another). Molecular nitrogen may also be incorporated into amino acids by *symbiotic nitrogen fixation* (nitrogen fixation by an organism living in close association with another). Thus N_2 is made available to plants by nitrogen fixation, which is the reduction of N_2 to NH_4^+ and always appears to be carried on by prokaryotic organisms.

Asymbiotic Nitrogen Fixation

Nitrogen fixation by living organisms was recognized in the latter half of the nineteenth century. Jodin, in 1862, was able to determine a loss of atmospheric nitrogen and oxygen in a closed system containing a nonsterile solution and a source of carbon. In 1885 Berthelot demonstrated that the content of fixed nitrogen in nonsterile soil samples could be shown, by chemical analysis, to increase over a period of time. However, credit for actually showing that a living organism is involved in nitrogen fixation must be given to Winogradsky, who, in 1894, isolated the nitrogen-fixing anaerobic bacterium *Clostridium pastorianum*.

Two other, even more important, freeliving, nitrogen-fixing organisms were isolated by Beijerinck in 1901. In contrast to the anaerobic *C. pastorianum*, the two bacteria isolated by Beijerinck, *Azotobacter chroococcum* and *Azotobacter agile*, are aerobic. Since that time, several nitrogen-fixing species of *Azotobacter* have been found. Free nitrogen can also be fixed by a large number of blue-green algae. We will discuss briefly the requirements, inhibition, and biochemistry of molecular nitrogen fixation.

Environmental conditions necessary for nitrogen fixation. Other than those environmental conditions necessary for good growth, the process of nitrogen fixation places no special requirements on the organism. One possible exception may be the quantities of certain mineral elements needed for the most efficient nitrogen fixation. Many workers have established that the elements molybdenum, iron, and calcium are required in higher quantities when molecular nitrogen rather than ammonia nitrogen is used, thereby suggesting the participation of these elements in the nitrogen-fixing process. Figure 8–5 and Table 8–1 show the effects of different concentrations of these three elements on the growth of *Azotobacter vinelandii*.

The most extensive work on the requirements of higher levels of molybdenum, iron, and calcium for nitrogen fixation has been done with molybdenum. Wilson

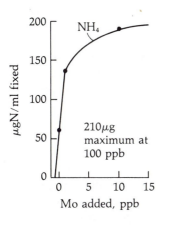
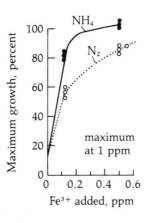
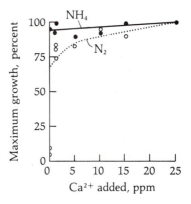

Figure 8–5. Effect of Mo, Fe^{3+}, and Ca^{2+} on growth of *Azotobacter vinelandii*. Molybdenum, iron, and calcium are required in higher quantities when molecular nitrogen rather than ammonia nitrogen is used.

From P.W. Wilson. 1958. Asymbiotic nitrogen fixation. In W. Ruhland, ed., Encyclopedia of Plant Physiology 8:9. *Berlin: Springer.*

Table 8–1. Requirement of molybdenum for molecular nitrogen fixation by *Azotobacter vinelandii.* All values are given as μg fixed N per millimeter.

| Experiment | N_2 | | NH_4^+ | |
	With Mo	Without Mo	With Mo	Without Mo
I	205	50	201	200
II	212	58	279	301

Source: *After R.G. Esposito as reported by P.W. Wilson (1958) in W. Ruhland, ed.,* Encyclopedia of Plant Physiology 8:9. *Berlin: Springer.*

(50) has pointed out that a requirement for molybdenum has been determined for every nitrogen-fixing organism studied.

Inhibition of nitrogen fixation. Inhibition of nitrogen fixation may be separated into three general situations: (1) inhibition of cellular metabolism, (2) inhibition with molecular hydrogen, and (3) inhibition with combined nitrogen. Since healthy growth is associated with nitrogen fixation, it is not surprising that inhibitors of cellular metabolism are also inhibitors of nitrogen fixation. A special case can be made for carbon monoxide (CO), an inhibitor of respiration. Nitrogen fixation appears to be far more sensitive to carbon monoxide poisoning than is respiration (57). This observation suggests that carbon monoxide may inhibit nitrogen fixation directly rather than indirectly through respiration.

Unlike carbon monoxide, molecular hydrogen acts as a specific inhibitor of nitrogen fixation. We mean by this that inhibition is only observed when the sole source of nitrogen is molecular nitrogen, not when other forms of combined nitrogen are applied (52, 53). Two explanations for this inhibition have been suggested: hydrogen may compete physically with nitrogen for an active site on the surface of some enzyme involved in nitrogen fixation, or inhibition may be related to the function of the enzyme *hydrogenase* in nitrogen fixation.

The second explanation has received the most attention since there is indirect evidence linking hydrogenase, an enzyme using molecular hydrogen as a substrate, with nitrogen fixation. For example, the hydrogenase content of *Azotobacter* and *Rhodospirillum* increases considerably when these organisms are fed molecular nitrogen instead of combined nitrogen (13, 14). When the alga *Chlorella pyrenoidosa* is transferred to an atmosphere of hydrogen, it develops an active hydrogenase (36, 37). The mediation of this enzyme in nitrite reduction has been demonstrated in *Chlorella* (37).

Generally, the fixation of nitrogen is inhibited by ammonia or compounds easily converted to ammonia, such as nitrate and nitrite. These compounds do not interfere with the mechanism of nitrogen fixation, but merely are preferred over molecular nitrogen as nitrogen sources. In other words, if both molecular nitrogen and combined nitrogen are present, the combined nitrogen will be used in preference to the molecular nitrogen. However, both forms of nitrogen may be used simultaneously and usually are.

Our knowledge of the pathway of nitrogen fixation is still rather meager. Experiments employing the stable isotope ^{15}N have shown beyond doubt, however, that ammonia occupies a key position in this pathway. The question of what intermediates are present between molecular nitrogen and ammonia has not been satisfactorily answered.

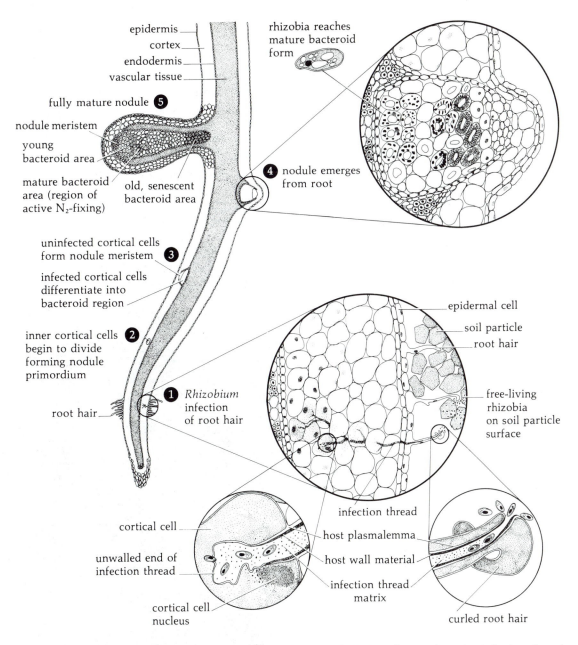

Figure 8–6. Penetration of rhizobia into root hair of legume plant. Root hair curls at tip, infection thread forms, and finally nodule forms.

Symbiotic Nitrogen Fixation

A relatively large group of plants, the legumes, obtain fixed nitrogen through a symbiotic association with soil bacteria of the genus *Rhizobium* (see Figure 8–6). In alder the symbiotic relationship is with certain species of the genus *Actinomyces*. In both cases neither organism alone is able to fix nitrogen. The symbiotic relationship between legumes and *Rhizobium* seems to be species-specific. When certain species of *Rhizobium* infect legumes, there is no evidence of nitrogen fixation. The actual site of nitrogen fixation is in the nodules formed in the roots of the legume plant as a result of the penetration of rhizobia. Through nodulation, the microorganisms provide the host plant with fixed (reduced) nitrogen, and the host plant provides the microorganisms with soluble carbohydrates.

Aside from the actual symbiotic fixation of nitrogen, the penetration of these bacteria and the resulting stimulation of root cell growth are interesting aspects of this association. Investigators commonly observe the accumulation of soil bacteria in the vicinity of plant roots, especially roots of legume plants. This accumulation probably occurs because plant roots excrete certain growth factors into the soil. Then the bacteria either penetrate the relatively soft root hair tip or invade damaged or broken root hairs and progress in an infection thread through the cortex tissue to the immediate area of the endodermis and pericycle. Cell divisions commence in the endodermis and pericycle area, and the nodule grows rapidly and pushes its way to the surface of the root. One rather remarkable observation, first made by Wipf and Cooper (55) in 1938, is that the nodule cells contain double the number of chromosomes found in the normal somatic cells of the plant. Wipf and Cooper, in a later study of nodule formation in pea and vetch (56), showed that successful nodule formation occurs only when the root nodule bacteria invade cells containing double the normal somatic complement of chromosomes. These cells are stimulated into meristematic activity by the invasion and form the nodule. If there are no cells with double the normal chromosome num-

ber in the area of the root penetrated by the infection thread, no nodule will form. Figure 8–7 shows clover roots with nodules, and Figure 8–8 is a scanning electron micrograph of infected autumn olive root nodules.

The factor or factors causing the profuse growth of the cells that form the nodules is at present unknown. Rhizobia are known to produce the plant hormone indole acetic acid (IAA). However, many other soil microorganisms are able to produce IAA but are not able to cause nodule formation.

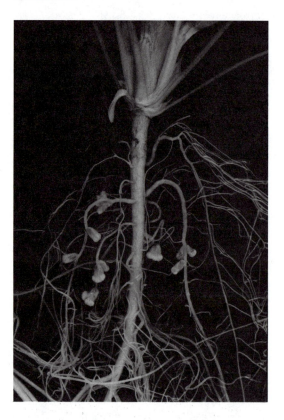

Figure 8–7. Nodules on clover roots.

Courtesy of B.W. Pennypacker and W.A.Kendall, The Pennsylvania State University, and USDA Regional Pasture Research Laboratory.

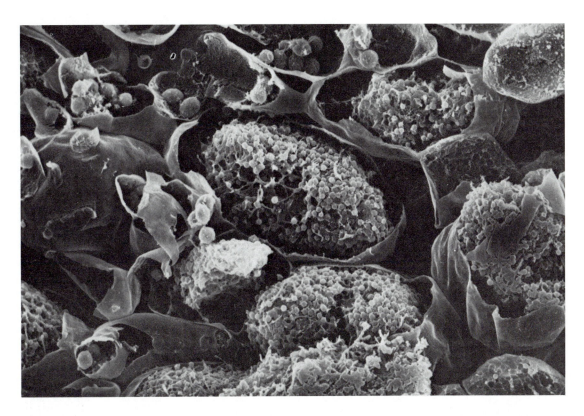

Figure 8–8. Scanning electron micrograph of actinomycete and infected cells of autumn olive root nodule. Magnification 650×.

From D. Baker, W. Necomb, J.G. Torrey. 1980. Characteri- *zation of an ineffective actinorhizal microsymbiont,* Frankia *sp. Eull (Actinomycetales). Can. J. Microbiol. 26:1072–89. Photo courtesy of D. Baker and E. Seling.*

Leghemoglobin and Mechanism of Symbiotic N_2 Fixation in Nodules

The dissection of a root nodule will reveal the presence of a red pigment that is remarkably similar in properties to the hemoglobin of red blood cells. The red pigment of the nodules is appropriately called leghemoglobin and appears to be a product of the rhizobium-legume complex, since the pigment is not present in either organism grown alone (3). Nodules that lack leghemoglobin are unable to fix nitrogen. Also,

numerous investigations (45) have shown a correlation between leghemoglobin concentration and rate of nitrogen fixation, which leads us to the conclusion that leghemoglobin and symbiotic nitrogen fixation are intimately related. Leghemoglobin is an oxygen carrier, the oxygen (O_2) is necessary for the electron transport chain of the rhizobium bacteroid (see Figure 8–9). Because of its very high affinity for oxygen, leghemoglobin provides oxygen to the root nodule bacteria quickly, even at very low levels of free oxygen (14). Observers also believe that the leghemoglobin keeps levels of molecular ox-

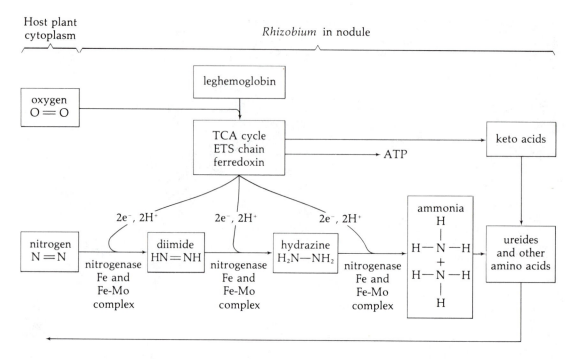

Figure 8–9. Major N_2 fixation reactions taking place in *Rhizobium* bacteroid as catalyzed by enzyme nitrogenase, an iron and iron-molybdenum protein complex.

ygen low in the bacteroid. This function is particularly important because nitrogenase is sensitive to O_2 and loses activity in its presence. Nodules that lack leghemoglobin may be unable to fix oxygen due to this condition.

Figure 8–9 illustrates the biochemistry of symbiotic nitrogen fixation. The reduction of nitrogen to ammonia is catalyzed by a complex of enzymes referred to as nitrogenase (30, 31). Certain micronutrients, such as iron, copper, cobalt, and molybdenum, appear to be essential. The iron requirement is accounted for by its presence in leghemoglobin. Copper is also required in leghemoglobin synthesis. Cobalt is an essential part of vitamin B_{12}, a compound that could be involved in the formation of leghe-

moglobin, possibly through the propionate pathway. The cobalt requirement has been demonstrated only in those plants capable of fixing molecular nitrogen (11). If combined nitrogen (e.g., nitrate or ammonium) is furnished to the nitrogen-fixing legume-symbiont system, there is no longer a requirement for cobalt (1, 2). Molybdenum functions alternately as an electron acceptor and donor in the reduction of nitrogen to ammonia.

As illustrated in Figure 8–9, N_2 is reduced to a diimide (HN=NH) (*imide* is any compound derived from ammonia by replacement of two hydrogen atoms), hydrazine (NH_2—NH_2), and then ammonia (NH_3). The N_2 reduction may be outlined as follows:

1. The electron and hydrogen appear to be donated through ferredoxin or other reducing agent of the electron transport system and Krebs cycle (see Chapters 13 and 16) of the bacteroid.

2. This bacteroid provides ATP by oxidative phosphorylation (see Chapter 16).

3. ATP is required in the transfer of electrons from an iron-protein complex via the FeMo of the nitrogenase system to the reduction process (54).

4. The Krebs cycle of the bacteroid produces the keto acids that are incorporated into reactions with NH_3 to form amino acids. Most of these amino acids are transported to the host.

5. The leghemoglobin functions to transport oxygen for the generation of ATP.

6. The nitrogenase enzyme, which is an iron-protein complex, mediates the transfer of electrons from ferredoxin on to an iron and molybdenum protein complex, where, possibly, N_2 reduction takes place.

The blue-green algae are also nitrogen fixers. Perhaps in the not to distant future, cropping systems will be made up of a culture of nitrogen-fixing algae and one of crop plants. Scientists at the University of California at Davis have already made significant advances in the cultivation of blue-green algae in cultures of rice.

Nitrogen Converters in Soil

Oxidation of ammonia to nitrate in the soil may occur through the mediation of two groups of bacteria: *Nitrosomonas* and *Nitrobacter*. The energy needed for the growth of these organisms is obtained through the oxidation of ammonia or of nitrite—in other words, *Nitrosomonas* and *Nitrobacter* are *autotrophic* bacteria and require only inorganic materials for growth. With only one major difference, this type of growth is similar to that found in green plants. In green plants, light provides the energy for growth; in the bacteria of nitrification, oxidation of ammonia or of nitrite provides this energy. Both of these organisms were isolated in 1891 by Winogradsky. He demonstrated that *Nitrosomonas* could convert ammonia only as far as nitrite and that *Nitrobacter* was needed for the further conversion of nitrite to nitrate. The conversion of ammonia to nitrite and then to nitrate is called *nitrification*.

$$NH_4^+ \xrightarrow{\text{Nitrosomonas}} NO_2^- \xrightarrow{\text{Nitrobacter}} NO_3^-$$

The conversion of nitrate to nitrous oxide (N_2O) and nitrogen gas also takes place through the mediation of a variety of soil organisms and is known as *denitrification*. The process of denitrification, which ends in the release of nitrogen gas to the atmosphere, completes nature's complex nitrogen cycle. Small amounts of fixed nitrogen are contributed to the soil from electrically produced nitrogen oxides, which are washed down from the atmosphere during rainstorms. Much greater amounts of fixed nitrogen are contributed by the molecular nitrogen–fixing organisms. The plants take up the fixed nitrogen and convert it to the many different organic nitrogen compounds of the plant. This organic nitrogen also contributes to the nitrogen development in animals since the animals are unable to convert inorganic nitrogen to organic nitrogen and must, therefore, ingest preformed organic nitrogen compounds as essential components of their diet. When ani-

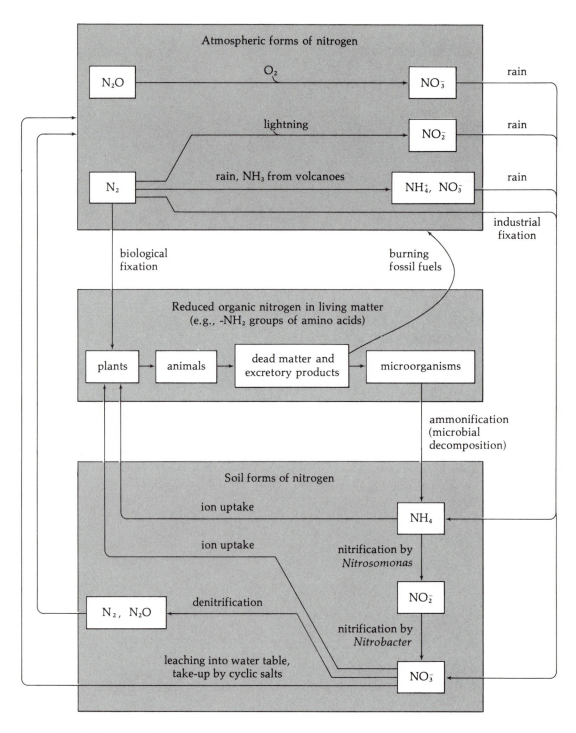

Figure 8–10. Nitrogen cycle.

mals and plants die, the organic nitrogen in them is returned to the soil, where, through microbial decomposition, ammonia is produced. The ammonia is rapidly converted to nitrate by the process of nitrification. Nitrate is then either available to the plant or is converted to nitrogen gas in the process of denitrification. Figure 8–10 provides an outline of this cycle.

Questions

8–1. Name the two enzymes involved in nitrogen reduction in some plants. What important factors are involved in the reduction of nitrate by one of these enzymes?

8–2. What is the major difference between an inducible and a constitutive enzyme?

8–3. Where in the plant cell is nitrate reductase located? Of what significance is its location to its activity?

8–4. Name the process involved in making nitrogen available to plants.

8–5. What are the benefits (to the plant and to the microorganisms) of the symbiotic relationship of rhizobia and legumes?

8–6. What is the role of leghemoglobin in nodules?

8–7. Outline the process of nitrogen fixation as it takes place in the bacteriod inhabiting a legume nodule.

8–8. In what other way besides in crop yields are legumes helpful to agriculture?

8–9. Explain the process of denitrification.

8–10. Is nitrogen mobile in plants? What major metabolic components require nitrogen for synthesis?

8–11. Name the microorganisms that are important in the nitrogen cycle. What are their specific roles?

Suggested Readings

Bauer, W.D. 1981. Infection of legumes by rhizobia. *Ann. Rev. Plant Physiol.* 32:407–449.

Brill, W.J. 1977. Biological nitrogen fixation. *Sci. Amer.* 236(3):68–81.

Burris, R.H. 1976. Nitrogen fixation. In J. Bonner and J.E. Varner, eds., *Plant Biochemistry*, 3rd ed. New York: Academic Press.

Guerrero, M.G., J.M. Vega, and M. Losada. 1981. The assimilatory nitrate-reducing system and its regulation. *Ann. Rev. Plant Physiol.* 32:169–204.

Hewitt, E.J., D.P. Hucklesby, and B.A. Notton. 1976. Nitrate metabolism. In J. Bonner and J.E. Varner, eds., *Plant Biochemistry*, 3rd ed. New York: Academic Press.

Mengel, K., and E.A. Kirkby. 1978. *Principles of Plant Nutrition.* Int. Potash Inst., eds. Bern: Der Bund.

Mortenson, L.E., and R.N.F. Thorneley. 1979. Structure and function of nitrogenase. *Ann. Rev. Biochem.* 48:387–418.

Phillips, D.A. 1980. Efficiency of symbiotic nitrogen fixation in legumes. *Ann. Rev. Plant Physiol.* 31:29–49.

Shanmugam, K.T., F. O'Gara, K. Andersen, and R.C. Valentine. 1978. Biological nitrogen fixation. *Ann. Rev. Plant Physiol.* 29:263–276.

Chapter 9

Proteins and Nucleic Acids

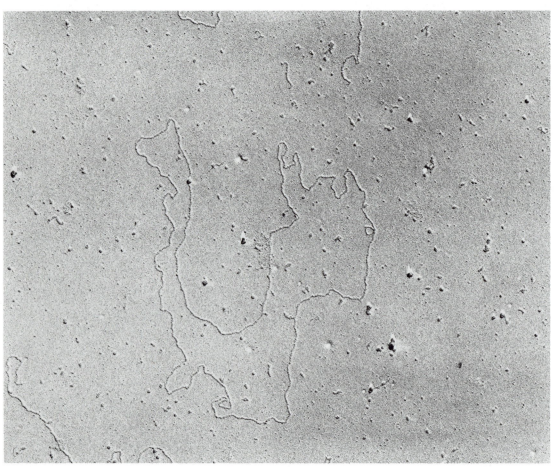

Electron micrograph of mitochondrial circular DNA molecule (mtDNA) from soybean (*Glycine max*). *From R.M. Synenki, C.S. Levings, III, and D.M. Shah. 1978.* Plant Physiol. *61:460.*

Within the last two decades it has become increasingly clear that the chemistry of nucleic acids, as expressed through the proteins, regulates the intricate biochemical properties of life and the dynamics of evolution. A most significant influence of proteins resides in the fact that many are functionally active as enzymes. The enzymes are vital for the rapid rate of biochemical reactions. Although many biochemical reactions will proceed to completion in the absence of enzymes, these reactions are extremely slow. Indeed, we could go so far as to say that enzymes and life are inseparable.

Two other important functions of proteins are as major natural *hydrogen ion buffers* and *structural components* of cells. Due to the ubiquitous nature and functions of proteins, investigators have studied them extensively. Certainly many important characteristics of proteins led scientists to the chemistry of the regulators of cellular information, the nucleic acids.

In this chapter we will consider the nucleic acids and proteins. Keep in mind, however, that we can present only the highlights concerning these nitrogen-containing chemicals here since the last decade has produced volumes of information on proteins and nucleic acids. We anticipate a similar compilation of new information within the decade to come.

Amino Acids and Amides

The acid hydrolysis of a protein molecule reveals that it is composed of smaller repeating units, the *amino acids*. With the exception of two secondary amino acids, the amino acids found in protein have the following general structure:

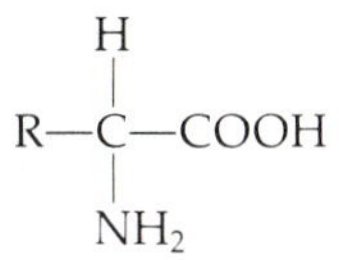

This structure depicts a *primary amino acid* in which the *amino group* (—NH_2) is attached to the carbon (α-carbon) adjacent to the *carboxyl group* (—COOH). Individual differences between the primary amino acids are found in the R-group, which may be quite different from one amino acid to the next. For example, the amino acids glycine, valine, and leucine have R-groups that are quite different. The structures for these amino acids are given with the R-groups circled.

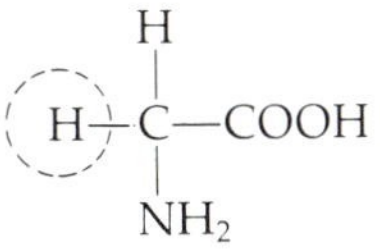

glycine

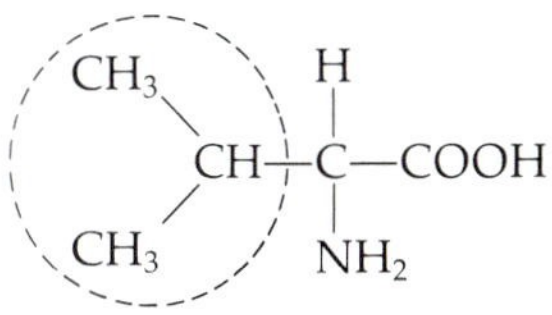

valine

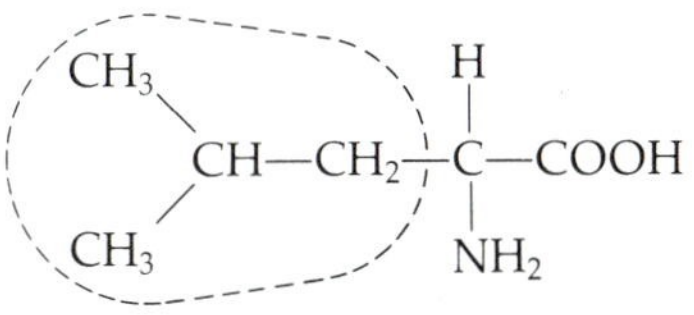

leucine

The amino acids found in plant protein as a result of extensive investigations by numerous workers are *glycine, alanine, valine, leucine, isoleucine, serine, threonine, phenylalanine, tyrosine, tryptophan, cysteine, methionine, proline, hydroxyproline, aspartic acid, glutamic acid, histidine, arginine,* and *lysine*. Table 9–1 illustrates the formulas of these amino acids.

Table 9–1. Amino acids found in plant proteins
and their chemical formulas.

Name	Formula	Type of Amino Acid
glycine	NH_2-CH_2-COOH	aliphatic
alanine	$CH_3-CH-COOH$ with NH_2	aliphatic
valine	CH_3, CH_3 $CH-CH-COOH$ with NH_2	aliphatic
leucine	CH_3, H_3C $CH-CH_2-CH-COOH$ with NH_2	aliphatic
isoleucine	$CH_3-CH_2-CH-CH-COOH$ with CH_3 NH_2	aliphatic
serine	$CH_2-CH-COOH$ with OH NH_2	aliphatic
threonine	$CH_3-CH-CH-COOH$ with OH NH_2	aliphatic
phenylalanine	$-CH_2-CH-COOH$ with NH_2	aromatic
tyrosine	$HO-$ $-CH_2-CH-COOH$ with NH_2	aromatic
tryptophan	$-CH_2-CH-COOH$ with NH_2; N–H	aromatic
cysteine	$HS-CH_2-CH-COOH$ with NH_2	S-containing

Table 9–1. (Continued)

Name	Formula	Type of Amino Acid
methionine	CH_3—S—CH_2—CH_2—CH—COOH (with NH_2 below CH)	S-containing
proline	(pyrrolidine ring with N—H, bearing —COOH)	secondary
hydroxyproline	(pyrrolidine ring with HO— substituent, N—H, bearing —COOH)	secondary
aspartic acid	HOOC—CH_2—CH—COOH (with NH_2 below CH)	acidic
glutamic acid	HOOC—CH_2—CH_2—CH—COOH (with NH_2 below CH)	acidic
histidine	(imidazole ring, N, NH) —CH_2—CH—COOH (with NH_2 below CH)	basic
arginine	H_2N—C—NH—CH_2—CH_2—CH_2—CH—COOH (with NH below C, NH_2 below CH)	basic
lysine	H_2N—CH_2—CH_2—CH_2—CH_2—CH—COOH (with NH_2 below CH)	basic

The proteins are the primary buffers in living systems due to the chemical properties of their amino acids. For a brief review of pH and buffers, see Appendix B. Depending on the pH of the solution, the alpha amino and alpha carboxyl functional groups of an amino acid may exist in one of the following forms:

$$\begin{array}{c} H \\ | \\ R-C-COOH \\ | \\ NH_3^+ \end{array} \xrightarrow{+H^+} \begin{array}{c} H \\ | \\ R-C-COO^- \\ | \\ NH_3^+ \end{array} \xrightarrow{+OH^-} \begin{array}{c} H \\ | \\ R-C-COO^- \\ | \\ NH_2 \end{array}$$

$(+)$ $(\pm)$ $(-)$

Zwitterion

As we can see, an amino acid might exist as a *Zwitterion*, or a molecule consisting of both a positive and negative charge. In this form, an amino acid is dipolar and is considered to be *amphoteric*, by which we mean that it can function as an acid or base. Further, the pH at which the *Zwitterion* form exists is usually referred to as the *isoelectric point*. Thus the isoelectric forms of amino acids exhibit a net zero charge and do not migrate when subjected to electrophoresis. In a solution very basic to the isoelectric point, an amino acid is an *anion* because of the predominating $-NH_2/-COO^-$ functional groups. Conversely, in solutions very acid (low pH) to its isoelectric point, an amino acid is a *cation* because of the predominating $-NH_3^+/COOH$ functional groups. We can easily visualize the tremendous buffering action of proteins when we consider the number of amino acids present.

Amino Acid Synthesis

Amino acids are generally considered to be the initial products of nitrogen assimilation. Evidence obtained by following the assimilation of inorganic nutrients containing ^{15}N has shown that, in most cases, the initial recipients of the nitrogen of these compounds are the free α-keto acids in the cytoplasm. The α-keto acid is similar to the amino acid with the exception of an oxygen instead of an amino group attached to the α-

carbon. We will discuss two ways by which nitrogen can be incorporated into α-keto acids.

Reductive Amination

Experiments with isotopically labeled nitrogen have shown that during the early stages of nitrogen assimilation, *glutamate* is by far the most-labeled compound. Confronted with this evidence, the investigator is very likely to conclude that there is a direct incorporation of ammonia into α-*ketoglutarate*, the corresponding keto acid of glutamate. The reaction is reversible and proceeds as follows:

$$\begin{array}{c} COOH \\ | \\ CH_2 \\ | \\ CH_2 \\ | \\ C=O \\ | \\ COOH \end{array} + NH_3 \rightleftharpoons \begin{array}{c} COOH \\ | \\ CH_2 \\ | \\ CH_2 \\ | \\ C=NH \\ | \\ COOH \end{array}$$

α-ketoglutarate   α-iminoglutarate

$$\xrightarrow[\substack{\text{glutamic} \\ \text{dehydrogenase}}]{NADH + H^+} \begin{array}{c} COOH \\ | \\ CH_2 \\ | \\ CH_2 \\ | \\ CH-NH_2 \\ | \\ COOH \end{array} + NAD^+$$

glutamate

The first reaction probably proceeds spontaneously, but the second reaction is catalyzed by the enzyme *glutamic dehydrogenase* and requires the presence of reduced nicotinamide-adenine-dinucleotide (NADH + H$^+$). Because of the central importance of glutamate in the synthesis of other amino acids and because of the high proportion of glutamate formed in this manner by the plant, this reaction is of utmost importance to the nitrogen metabolism of the plant. It represents, so to speak, a major "port of entry" into the metabolic system for inorganic nitrogen. The widespread occurrence of glutamic dehydrogenase in plants strongly supports the above statement.

Reductive amination as a means for the synthesis of amino acids other than glutamate is of limited importance. There is some indirect evidence for the direct amination of oxaloacetate and pyruvate to form aspartate and alanine, respectively.

We have, then, four ways by which ammonia nitrogen is incorporated into organic compounds to form amino acids:

$$\alpha\text{-ketoglutarate} + NH_3 \rightleftharpoons \text{glutamate}$$

$$\text{oxaloacetate} + NH_3 \rightleftharpoons \text{aspartate}$$

$$\text{fumarate} + NH_3 \rightleftharpoons \text{aspartate}$$

$$\text{pyruvate} + NH_3 \rightleftharpoons \text{alanine}$$

Of these four pathways, the amination of α-ketoglutarate appears to be the major reaction in the assimilation of nitrogen by plants.

Transamination

A very important reaction in amino acid synthesis is *transamination*, which involves the transfer of an amino group of an amino acid to the carbonyl group of a keto acid. When $^{15}NH_3$ is fed to plants, glutamic acid labeling is quite high in comparison with other amino acids and, therefore, suggests a key role for glutamate in this reaction. After inorganic nitrogen has gained entry primarily through the amination of α-ketoglutarate, the product, glutamate, is available for transamination reactions with keto acids to form the corresponding amino acids. The formation of 17 different amino acids takes place through transamination reactions with glutamate (13).

The enzymes catalyzing transamination reactions are called *transaminases*. When a specific transaminase is referred to, however, the substrate and product prefix the generic term. Thus the enzyme catalyzing the transfer of an amino group of glutamic acid (substrate) to the carbonyl group of oxaloacetate to form aspartate (product) is called *glutamic-aspartic transaminase*.

Although transamination reactions involving glutamic acid are by far the most prevalent in the plant, other transamination reactions have been found. For example, workers have found a transamination reaction in higher plants, involving aspartic acid and alanine. However, for the most part transamination reactions involve α-ketoglutarate or glutamate as an essential component (9).

Investigators have definitely established that transamination reactions involve the participation of *pyridoxal phosphate* or *pyridoxamine phosphate* as a co-enzyme. Apparently, pyridoxal phosphate, which is tightly bound to the enzyme, accepts an amino group from the amino acid to form pyridoxamine phosphate, thereby releasing the corresponding keto acid product. Pyridoxamine phosphate then passes the amino group to another keto acid, forming a new amino acid and regenerating pyridoxal phosphate. The reaction should proceed as follows:

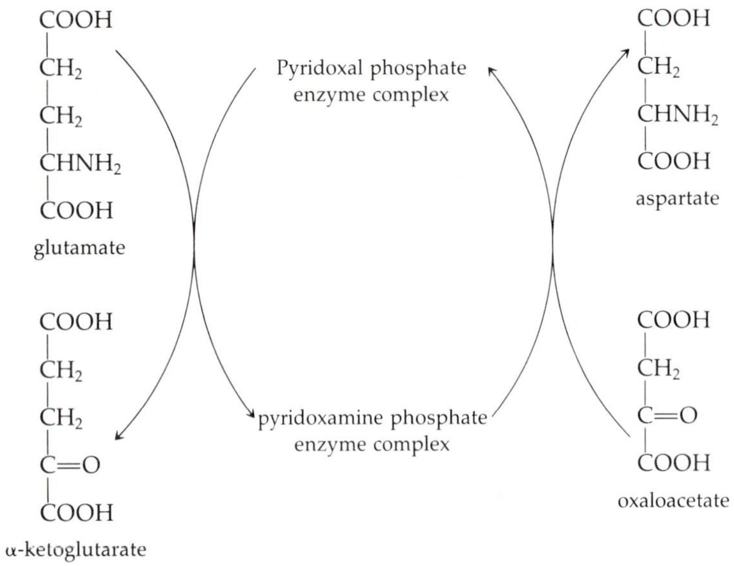

COOH
|
CH$_2$
|
CH$_2$
|
CHNH$_2$
|
COOH

glutamate

Pyridoxal phosphate
enzyme complex

COOH
|
CH$_2$
|
CHNH$_2$
|
COOH

aspartate

COOH
|
CH$_2$
|
CH$_2$
|
C=O
|
COOH

α-ketoglutarate

pyridoxamine phosphate
enzyme complex

COOH
|
CH$_2$
|
C=O
|
COOH

oxaloacetate

Before leaving the synthesis of amino acids for a discussion on proteins, we should mention the amides *asparagine* and *glutamine*. These compounds have been found in relatively high quantities in many plants and appear to function in the transport and storage of nitrogen. In the synthesis of glutamine, the hydroxyl group of one of the carboxyl groups of glutamic acid is replaced by an —NH$_2$ group. The enzyme catalyzing this reaction, *glutamine synthetase*, is activated by the metal cofactor, Mg^{2+}. In addition, ATP is required (see below).

The synthesis of asparagine from aspartate is thought to take place in the same manner and requires a metal activator and ATP. However, the enzyme *asparagine synthetase*, which would catalyze this reaction, has not as yet been isolated from plant tissues.

Proteins

Proteins are composed of repeating units, the amino acids. The amino acids are linked

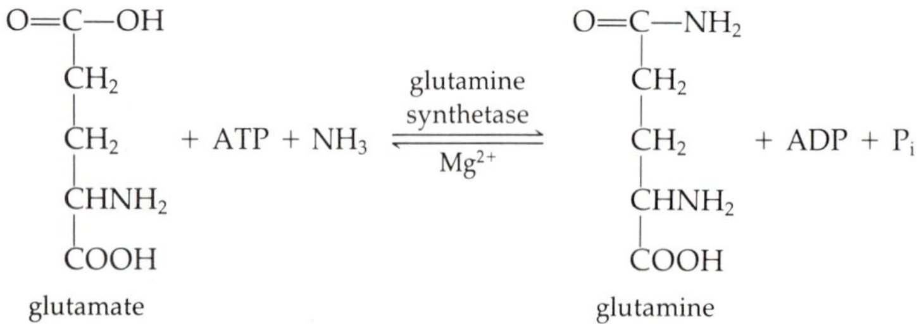

O=C—OH
|
CH$_2$
|
CH$_2$
|
CHNH$_2$
|
COOH

glutamate

+ ATP + NH$_3$ $\xrightleftharpoons[\text{Mg}^{2+}]{\text{glutamine synthetase}}$

O=C—NH$_2$
|
CH$_2$
|
CH$_2$
|
CHNH$_2$
|
COOH

glutamine

+ ADP + P$_i$

together by bonds connecting the carboxyl group of one amino acid with the amino group of another. This type of linkage, which is repeated many times in the protein molecule, is called the *peptide bond.* Each shaded area in the following diagram includes four atoms of the peptide bond:

A compound composed of two amino acids linked together by two peptide bonds is called a *dipeptide*—of three amino acids, a *tripeptide*, and so on. When a large number of amino acids are linked together in this fashion, the resulting compound is called a *polypeptide.* When we consider that a protein may be made up of twenty different amino acids, each one of which may be present many times and in different sequences, we get an idea of the complexity and size of the protein molecule. Proteins may range in size from the molecular weight of insulin, which is 6,000, to a molecular weight of several million.

Protein Structure

Primary structure. The biological properties of a protein molecule are related to its structure. The peptide bond and the definite sequence of amino acids give the protein its primary structure. Since many proteins contain more than one polypeptide chain, a connection between them other than by a peptide bond is a necessary feature of the protein molecule. The *disulfide* (—S—S—) *bond* between two cysteine molecules is important in this respect. Figure 9–1 illustrates the features of the primary structure of a small animal protein, beef insulin.

Evidence from numerous investigations indicates that peptide and disulfide bonds are not the only links involved in protein structure. For example, dissociation of many proteins may occur under mild condi-

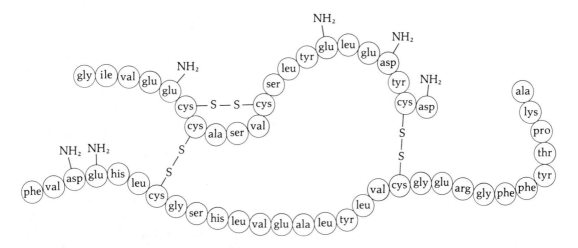

Figure 9–1. Composition and sequence of amino acids in beef insulin.

tions that would not disturb the peptide or disulfide bond.

Secondary structure. Polypeptide chains exhibit three major types of arrangement, or orientation: helical, pleated sheet, and random. These particular coil or spiral arrangements of the polypeptide chain constitute its secondary structure.

The α-*helix*, which is the most common helical arrangement, is maintained by an extensive assembly of hydrogen bonds throughout the chain. *Hydrogen bonds* are noncovalent links that occur as a result of the hydrogen atom sharing electrons with two oxygen atoms. Polypeptide strands also form other kinds of helices that are not as stable nor as common in proteins as the α-helix.

The *pleated sheet* orientation is formed when segments of the polypeptide chain are side by side and joined by hydrogen bonds in such a manner as to produce a zigzag appearance of the peptide backbone. Pleated sheets often take the appearance of chains running in opposite directions (*antiparallel*) or looped in such a manner that large portions of the chain run parallel (*parallel pleated sheets*).

With the so-called *random arrangement,* the polypeptide secondary structure may not appear to show geometrical order. This lack of geometry at the surface probably results from the folding of amino acid side chains rather than the folding of the peptide backbone. In addition to hydrogen bonds, salt bonds and van der Waals forces help maintain the helical structure.

Tertiary structure. With complete folding, or assumption of final form, by the chain, the secondary structure may take on its own specific final shape or pattern. This further folding of the various secondary structures

and random segments is referred to as the tertiary structure of the protein. The tertiary structure is maintained primarily by hydrogen bonding. Salt links and van der Waals forces are also involved. Unlike the secondary structure, the tertiary structure involves interactions of the R-groups. The secondary and tertiary structures of the protein molecule seem to be intimately associated with the molecule's biological function. Indeed, in many instances when the secondary and tertiary structures are disrupted, certain specific functions (e.g., enzyme activity) are irreversibly lost. This loss in activity may occur when proteins are exposed to relatively high temperatures, changes in pH, ultraviolet radiation, and so on. All of these conditions cause what is referred to as *denaturation.* Loss of many of the properties of the protein molecule—such as solubility, specific activity, crystallizability—follows its denaturation. In many cases, these properties cannot be recovered on return to normal conditions.

Quaternary structure. The quaternary structure of a protein molecule results from the association of two or more polypeptide chains. The final, three-dimensional shape (geometry) of the combined polypeptide chains (each is referred to as a subunit) as formed into a complete molecule constitutes its quaternary structure. Hydrogen bonds are again involved in holding subunits together in the maintenance of the quaternary structure. However, another kind of important interaction that holds the quaternary structure together is the association of hydrophobic groups that combine and exclude water. Many proteins that contain multiple subunits seem to be constructed and held together by the hydrophobic side chain interactions among subunits. Figure 9–2 shows the different kinds of bonds that may occur in the protein molecule.

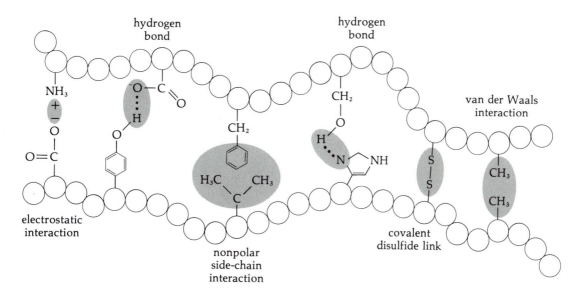

Figure 9–2. Some bonds found in protein molecules: electrostatic interaction, hydrogen bonding between tyrosine residues and carboxylate groups on side chains, interaction of nonpolar side chains caused by mutual repulsion of solvents, and van der Waals interactions.

Protein Classification

Because of the similar general structure found in the many different proteins, it is easy to separate them from other nitrogenous compounds into their own general group. However, because of this similarity it is difficult to establish a classification for proteins themselves. Indeed, many consider as unsatisfactory the classification that exists at the present time. A classification based on specific structural characteristics is next to impossible because of our scanty knowledge of the secondary and tertiary structures of proteins. Therefore, a classification has been attempted that is based in part on solubility properties and in part on known chemical and physical differences.

Simple proteins. Simple proteins are compounds that, on hydrolysis, yield only amino acids. Classification of the simple proteins is primarily based on solubility properties. The simple proteins can be divided into six major groups: *albumins, globulins, glutelins, prolamines, histones,* and *protamines.*

1. *Albumins.* The albumins are soluble in water and in dilute salt solutions. They can be coagulated by exposure to heat. The β-amylase of barley is a good example of an albumin.

2. *Globulins.* The globulins are either insoluble or sparingly soluble in water. They are soluble in dilute salt solutions. The globulins will also coagulate when exposed to heat. Many examples of globulins may be found in the storage proteins of seeds.

3. *Glutelins.* The glutelins are insoluble in neutral solutions but are soluble in weak acid or basic solutions. These

proteins are chiefly found in the cereal grains. Glutenin is an example of a glutelin protein from wheat. Another example would be oryzenin from rice.

4. *Prolamines.* The prolamines are insoluble in water, but are soluble in 70 to 80 percent ethanol. They are not soluble in absolute ethanol. On hydrolysis, these proteins yield relatively large quantities of proline and ammonia—thus the term prolamine. Examples of plant prolamines are zein of maize, gliadin of wheat and rye, and hordein of barley.

5. *Histones.* The histones are rich in basic amino acids, such as arginine and lysine, and are soluble in water. These have been found in cell nuclei and may be associated with nucleic acids.

6. *Protamines.* The protamines, like the histones, are rich in basic amino acids and are soluble in water. Also, like the histones, they are found in the nucleus and are probably associated with the nucleic acids. The amino acids tyrosine and tryptophan are not found in these proteins. Also, the protamines contain no sulfur.

Conjugated proteins. The conjugated protein is associated with a nonamino acid component that may be referred to as a *prosthetic group.* We can divide the conjugated proteins into five major groups: *nucleoproteins, glycoproteins, lipoproteins, chromoproteins,* and *metalloproteins.* From the terms used to describe the different groups, we can see that the conjugated proteins are named in accordance with their associated prosthetic group.

1. *Nucleoproteins.* On hydrolysis, the nucleoproteins yield a simple protein plus a nucleic acid.

2. *Glycoproteins.* As the name implies, glycoproteins are proteins containing small amounts of carbohydrates as prosthetic groups. Some of the proteins of the cell membrane may be glycoproteins.

3. *Lipoproteins.* The lipoproteins are generally insoluble in water and are common components of membranes.

4. *Chromoproteins.* The chromoproteins comprise a diverse group of compounds, which includes flavoproteins, biliproteins, (phycobilins and phytochrome), carotenoid proteins, chlorophyll proteins, and hemoglobins. The one property they all have in common is a prosthetic group that is a pigment.

5. *Metalloproteins.* Many of the enzymes belong to the metalloprotein group. These enzymes require a metal as an activator. We will encounter this particular type of protein frequently in the discussion of the respiratory enzymes.

Nucleic Acids

Before discussing the subject of protein synthesis, we must acquaint ourselves with the nucleic acids: *ribose nucleic acid* (RNA) and *deoxyribose nucleic acid* (DNA). The nucleic acids are large polymeric molecules composed of repeating units called *nucleotides,* which, in turn, are composed of three components: a *purine* or *pyrimidine* base, a *pentose* or *deoxypentose sugar,* and *phosphoric acid.* The nucleotides are bound together by sugar-phosphate linkages (see Figure 9–3).

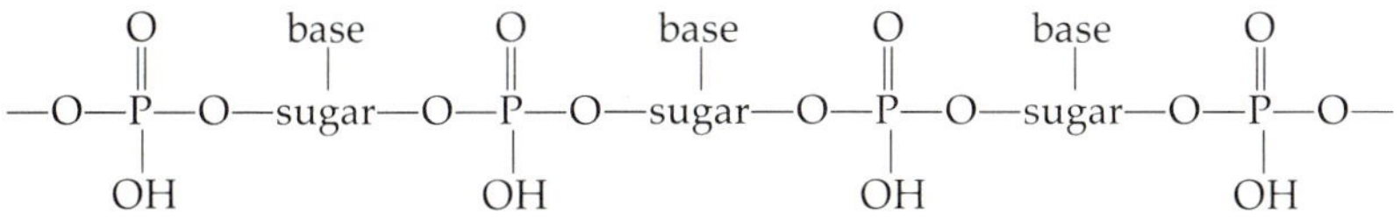

Figure 9–3. Sequence of DNA molecule showing sugar-phosphate linkages, purine nucleotide, adenylic acid, and pyrimidine nucleotide, cytidylic acid.

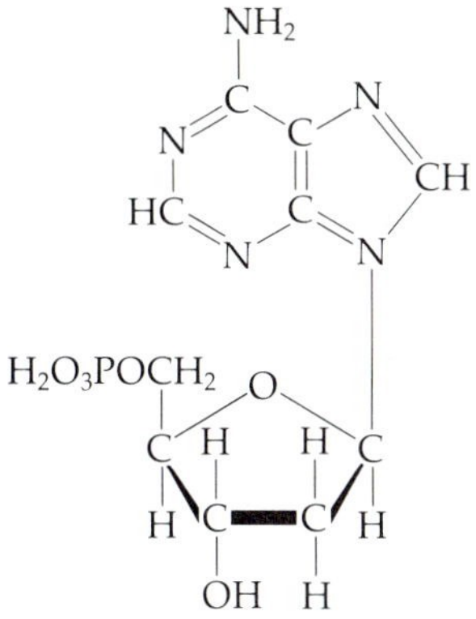

sugar-phosphate linkages

purine nucleotide,
adenylic acid

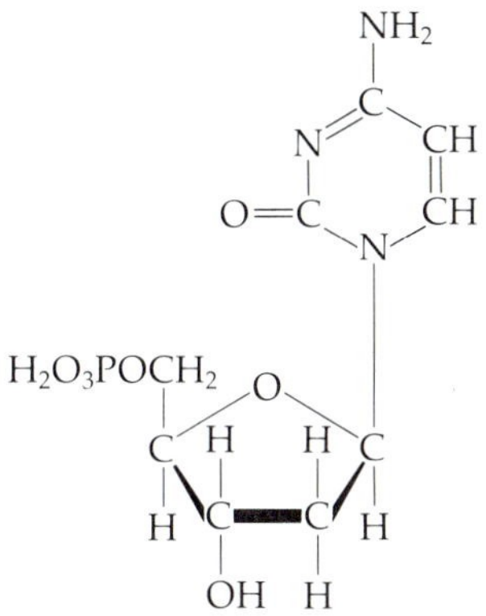

pyramidine nucleotide,
cytidylic acid

Separation of the nucleic acids into two large groups is determined by the sugar component present, a factor that has also influenced the naming of these groups. Thus ribose nucleic acid (RNA) contains ribose, and deoxyribose nucleic acid (DNA) contains deoxyribose. The difference between these two sugars can be found in the second carbon (see below).

Numerous investigations have unequivocally established that two types of nitrogenous bases occur in nucleic acids—the purines and pyrimidines. *Adenine* and

guanine are the two purines commonly found, while *thymine, cytosine,* and *uracil* are the pyrimidines that are generally present. As with the sugar moiety, here also there is a difference between DNA and RNA. The pyrimidine thymine is found only in DNA, while the pyrimidine uracil is confined to the RNA molecule. The structures of the different major nitrogenous bases found in the nucleic acids are shown on page 182.

DNA is found in chromosomes, in plastids, and in mitochondria. RNA is found in chromosomes, nucleoli, cytoplasm

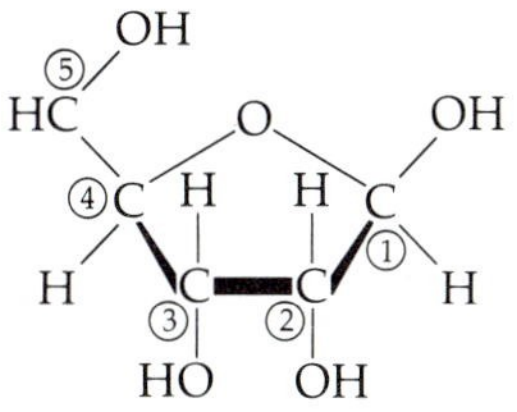

D-ribose

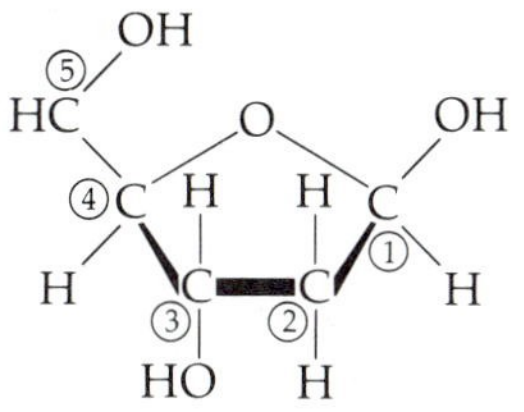

2-deoxy-D-ribose

adenine

guanine

thymine

cytosine

uracil

(as transfer RNA), ribosomes (ribosomal and messenger RNA), plastids, and mitochondria. The biological functions of DNA and RNA are the transmission of hereditary characteristics and the biosynthesis of proteins. DNA is primarily associated with the transmission of genetic information, RNA is directly connected with protein synthesis.

The molecular structure and nucleotide sequence of the nucleic acids has been studied intensively. Evidence accumulated from X-ray diffraction studies indicates that the DNA molecule is a double helical structure; the two strands are entwined and complementary, as shown in Figure 9–4 (12). The two strands are linked together through hydrogen bonding between base pairs. Chemical analyses of the DNA molecule indicate that there is a 1:1 relationship between adenine and thymine and between

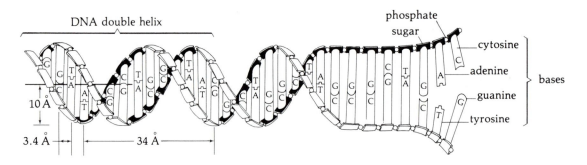

Figure   9–4. Line drawing of double helix structure of DNA.

Reprinted by permission of the publisher and Professor M.H.F. Wilkins, The University of London King's College,

from M. Feughelman et al., Molecular structure of DNA and nucleoprotein, Nature *175:834–838. Copyright © 1955 Macmillan Journals Limited.*

guanine and cytosine. This observation and others suggest that base pairing between the two helical strands occurs between purines and pyrimidines, not between two purines or between two pyrimidines. However, the ratio of adenine-thymine to guanine-cytosine can differ from one DNA molecule to another. The DNA molecule is a self-replicating molecule. Presumably, under the appropriate conditions and with the necessary enzymes, the two chains unwind from the double helix, draw from a base pool, and duplicate each other.

RNA is also a helical structure (single strand) and is composed of a sequence of nucleotides spaced in much the same manner as those of DNA. Uracil replaces thymine in the RNA molecule. However, just as adenine pairs with thymine in DNA, so also does it pair with uracil in RNA. Three types of RNA have been identified that vary in size and function. The largest RNA is found in ribosomes and is generally referred to as *ribosomal RNA* (rRNA). *Messenger RNA* (mRNA) is much smaller but is still of considerable size; in electron micrographs mRNA can be identified as long fibrous molecules, with several attached ribosomes. The whole complex is called a *polysome* or *polyribosome*. Finally, small molecules of RNA, called transfer RNA (tRNA), have been identified and are associated with protein synthesis.

Transcription

DNA directs the synthesis of RNA essentially by acting as a template. In the process, DNA uncoils and the strands separate between the bases, thereby leaving single strands of DNA. At this point each strand consists of repeating nucleotides. If complementary nucleotides, which contain deoxyribose and which come from the cell nucleotide reservoir (referred to as a "pool"), join one of the open DNA strands by forming base pairs, another double-stranded DNA molecule results. However, if bases of ribose nucleotides complementary to the DNA strand are polymerized (catalyzed by RNA polymerase) off the template, a molecule of RNA is formed in which the sequence of nucleotides is complementary to that of the DNA template strand. Since RNA is formed from the base sequence pattern of DNA, the process is sometimes referred to as *DNA-dependent RNA synthesis*. Further, the DNA is said to be transcribed into a complementary RNA chain. Therefore, when we refer to the process from the standpoint of informational transfer, we call it *transcription*. Once synthesized, an mRNA molecule may leave the nucleus via pores in the nuclear membrane and may become associated with ribosomes in the cytoplasm.

Although little is known of the process, rRNA is formed in the nucleolus before being released into the cytoplasm. Details of tRNA synthesis are still unknown.

The sequence of bases in the mRNA molecule is important since it represents a pattern that controls the synthesis of proteins. Plant proteins are composed of at least 20 different amino acids. A pattern of 1 base to 1 amino acid would account for only 4 different amino acids and combinations of 2 bases with 1 amino acid would account for only 16 different amino acids. However, with a *triplet* code, 64 combinations of bases are possible, which more than adequately account for the 20 different amino acids found in plants. The triplet groups of sequentially arranged bases on the mRNA molecule are called *codons*, and each codon represents the code for a specific amino acid. For example, the triplet sequence UUU

Table 9–2. Assignment of specific amino acids to 61 out of 64 possible codons. Remaining three codons are called nonsense codons, or triplets, because they do not code for any amino acid.

First Letter	Second Letter U		C		A		G		Third Letter
U	UUU UUC	phenylalanine	UCU UCC	serine	UAU UAC	tyrosine	UGU UGC	cysteine	U C
	UUA UUG	leucine	UCA UCG		UAA UAG		UGA UGG	tryptophan	A G
C	CUU CUC CUA CUG	leucine	CCU CCC CCA CCG	proline	CAU CAC	histidine	CGU CGC CGA CGG	arginine	U C A G
					CAA CAG	glutamine			
A	AUU AUC AUA	isoleucine	ACU ACC ACA	threonine	AAU AAC	asparagine	AGU AGC	serine	U C A
	AUG	methionine	ACG		AAA AAG	lysine	AGA AGG	arginine	G
G	GUU GUC GUA GUG	valine	GCU GCC GCA GCG	alanine	GAU GAC	aspartic acid	GGU GGC GGA GGG	glycine	U C A G
					GAA GAG	glutamic acid			

(three uracil nucleotides) serves as the template for a molecule of the amino acid phenylalanine. The 64 possible codons and the amino acids they will code for are given in Table 9–2. Note that three of the codons—UGA, UAA, and UAG—do not code for any amino acid. These are called *nonsense triplets;* their function in protein synthesis may be to mark the ending of one protein and the beginning of another. We will discuss the significance of the triplet code later.

Translation (Protein Synthesis)

Activation of amino acids. Amino acid activation, the first step in protein synthesis, consists of the selection of specific amino acids

from a heterogenous pool in the cytoplasm. The selection of amino acids is accomplished through the aid of highly specific enzymes, each amino acid having at least one activating enzyme. In the presence of ATP, the activating enzyme catalyzes the formation of an energy-rich, enzyme-bound amino acid adenylate (E—AA—AMP) and the release of pyrophosphate.

$$O—\overset{\overset{O}{\|}}{P}—O—\overset{\overset{O}{\|}}{P}—O$$
$$\overset{|}{O} \qquad \overset{|}{O}$$

Amino acid–tRNA complex. The activation of an amino acid is followed by its attachment to transfer RNA (tRNA), which is a relatively small molecule containing anywhere from 70 to 100 nucleotides. A specific tRNA molecule for each amino acid (2, 10), leads us to assume that the transfer of amino acids from the enzyme-bound active complex to tRNA is an additive rather than a competitive process. The point of attachment between tRNA and the activated amino acid is at the second or third carbon of the ribose sugar of a terminal adenylic acid.

Polypeptide formation. Once formed, messenger RNA (mRNA) becomes associated with ribosomes in the cytoplasm to form a polysome. Amino acids are conveyed to the polysomes by tRNA, the amino acid being attached to one end of the tRNA molecule. At the other end of the tRNA molecule is a triplet of nucleotides, or an *anticodon*, which is the complement of the messenger codon for that amino acid. For example, for the messenger codon UUG, the tRNA anticodon would be AAC (the anticodon for leucine). The codon and anticodon lock rapidly into place by complementary attraction, as they do in other nucleic acid activities such as DNA replication and mRNA transcription. When the anticodons of a number of tRNA molecules are locked in place, the amino acids at their opposite ends are lined up in sequence for polypeptide formation. The ribosomes are thought to move along the mRNA molecule from one end to the other and to join the amino acids in peptide linkages with the aid of specific enzymes (see Figure 9–5). Thus the role of tRNA appears to be to convey amino acids to the mRNA-ribosome complex (polysome) and to hold them in place according to the pattern dictated by RNA codons. The pattern is fixed, in turn, by the DNA on which it was transcribed. The actual joining of the amino acids is under the control of protein-synthesizing enzymes.

The above-described mechanism does not take into account any interaction between the amino acids and their respective

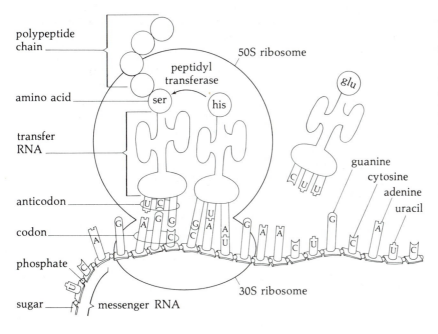

polypeptide chain

50S ribosome

peptidyl transferase

amino acid

glu

ser

his

transfer RNA

guanine
cytosine
adenine
uracil

anticodon

U C

codon

G A G G G A A G A A

phosphate

30S ribosome

sugar

messenger RNA

Figure 9–5. Polypeptide formation. Ribosomes move from one end of mRNA molecule to another as amino acids from cytoplasm form peptide linkages by enzymatically catalyzed reactions.

codons at the ribosomal surface. Recent work by Hendry and his colleagues suggests a different interpretation. Working with CPK space-filling models, Hendry and Witham (7) showed that it is chemically feasible for specific amino acids to react with the first two bases of their respective codons. These investigators showed that the R-groups of amino acids may be recognized by the first two bases of specific codons. Thus rather than being passively carried and held in position by tRNA molecules for peptide synthesis, amino acids may play an active role in protein synthesis through their own chemical stereospecificity.

Protein Degradation

Protein metabolism in plants is in a continuous state of flux between synthesis and breakdown. Observers have found proteolytic enzymes, such as *protease* and *peptidase*, in various plant organs. This finding suggests that the activity of these enzymes may, in part, control protein degradation.

Investigators have studied protein degradation principally in germinating seeds and in detached leaves. During germination, a massive breakdown of storage protein occurs in the cotyledon or endosperm, paralleled by a rapid synthesis of protein by the embryo. Observers have also found an accumulation of amino acids and amides in the embryo. Apparently, the physiological factors leading to germination set in motion the degradation of storage protein, the migration of the products of this degradation (amino acids) to the embryo, and the synthesis of new protein from these amino acids.

Studies of nitrogen metabolism during the germination of peas (3) and barley (5) show that storage proteins are among the

first compounds to disappear. Development of oat and barley embryos is retarded when they are excised from their storage parts and grown in a nutrient medium. Development of the embryos is revived to some extent if amino acids are added to the culture medium (4). In a study of the protein metabolism of intact and excised corn embryos, Oaks and Beevers (9) produced evidence suggesting that the large quantities of preformed amino acids being transferred from the endosperm to the developing embryo restrict the synthesis of new amino acids within the embryo. When the corn embryo is excised from its storage parts and grown on a nutrient medium containing glucose and inorganic nitrogen, the level of protein nitrogen is considerably less than that of the intact embryo grown for the same period of time. This finding suggests that the embryo possesses only a limited ability to incorporate inorganic nitrogen and synthesize new amino acids. However, the ability to incorporate inorganic nitrogen and synthesize new amino acids and proteins was found to develop in the excised embryo that was grown for a period of time on a medium low in soluble amino acids (9).

The mechanisms of protein synthesis and especially protein breakdown are not clearly understood and represent a challenge of vital importance to the scientist specifically and to humankind in general. A study of protein metabolism has a practical as well as an academic interest. For example, an understanding of the mechanisms involved in the synthesis and breakdown of proteins in the plants may enable the scientist to block or retard protein breakdown and possibly raise protein levels in plants used for food. This development would certainly be welcomed by those countries of the world whose populations live on a diet consisting almost wholly of carbohydrates.

Perhaps protein levels in crop plants will be almost completely under the control of humans in the future.

Questions

9–1. List the major roles of amino acids in plants.

9–2. Provide the structure of a primary amino acid. What is the major structural difference between primary and secondary amino acids? What is meant by the R-group of an amino acid?

9–3. What makes an amino acid basic or acidic?

9–4. Define *Zwitterion*, isoelectric point, amphoteric, and primary structure.

9–5. Name some of the other components of cells that are acids or bases. What are the significances of acids and bases in biological systems?

9–6. Why is transamination an important process in amino acid metabolism?

9–7. Are proteins translocated throughout the plant? Explain the occurrence of proteins in separate plant parts.

9–8. Describe the structure of a peptide bond. What role does it play in protein synthesis?

9–9. Explain the composition of the primary, secondary, tertiary, and quaternary structures of a protein.

9–10. List three major roles of proteins in plants.

9–11. Amino acids and proteins are said to be amphoteric. What major function of proteins in plants is due to this property?

9–12. What is the pH of a solution in which the hydrogen ion concentration is 0.00001 N, 10^{-9}, or .001 N?

9–13. Let us assume you found the pH of plant cells to vary with time from 4.5 to 6.6. To find the appropriate average of pH over that time interval, would you simply add the two volumes and divide by two? Explain.

9–14. How are proteins classified? Name some so-called simple proteins and conjugated proteins.

9–15. Where are proteins primarily synthesized in the cell?

9–16. Starting with the bases, what is the hierarchy of organization of structural components in RNA and in DNA?

9–17. Where are nucleic acids (DNA or RNA) located in the cell? What tissues of the plant might you expect to be relatively high in nucleic acids or nucleic acid derivatives?

9–18. What are the biological functions of DNA and RNA? Do proteins affect the functioning of nucleic acids? Explain.

9–19. Where in the structure of DNA might other biochemicals react?

9–20. Describe the accepted mechanism for the transcription and translocation of biochemical information as expressed through the process of protein synthesis.

9–21. Would you expect a protein to be highly water soluble and physiologically active at its isoelectric point?

Suggested Readings

Bedbrook, J.R., and R. Kolodner. 1979. The structure of chloroplast DNA. *Ann. Rev. Plant Physiol.* 30:593–620.

Bohinski, R.C. 1979. *Modern Concepts in Biochemistry*, 3rd ed. Boston: Allyn and Bacon.

Flavell, R. 1980. The molecular characterization and organization of plant chromosomal DNA sequences. *Ann. Rev. Plant Physiol.* 31:569–596.

Howell, S.H. 1982. Plant molecular vehicles: potential vectors for introducing foreign DNA into plants. *Ann. Rev. Plant Physiol.* 33: 609–650.

Key, J.L. 1976. Nucleic acid metabolism. In J. Bonner and J.E. Varner, eds., *Plant Biochemistry*, 3rd ed. New York: Academic Press.

Lehninger, A.L. 1982. *Principles of Biochemistry*. New York: Worth.

McGilvery, R.W., with G. Goldstein. 1979. *Biochemistry: A Functional Approach*. Philadelphia: Saunders.

Chapter 10

Enzymes

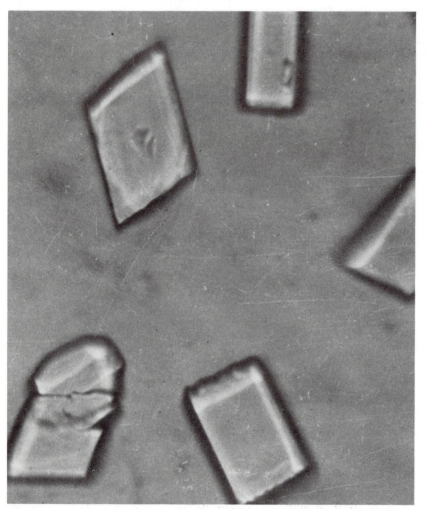

Micrograph of amylase crystals isolated from *Aspergillus oryzae.* Magnification 300×.
Courtesy of J.H. Pazur, Professor of Biochemistry, The Pennsylvania State University.

The dynamic state of the biochemistry of living systems is for the most part under the regulation of enzymes. An enzyme is partly or entirely a protein that can tremendously increase the efficiency of a biochemical reaction and is generally specific for that reaction. As with inorganic catalysts, the final products of the reaction are not affected by the enzyme. Although a biochemical reaction will proceed to completion in the absence of an enzyme, the process would be extremely slow—so slow, in fact, as to make life as we know it impossible.

The use of enzymes by humans for practical purposes can be traced back to the ancient Greeks, who used the action of enzymes in the process of fermentation to produce wine. Other practices in which the activity of enzymes is a necessary feature and which have a long history are the making of cheese and vinegar and the leavening of bread. During the drive to improve the quality and production of these products (especially wine), indirect knowledge of enzyme activity was gathered, and this knowledge led eventually to the recognition of living cells as essential participants. Much of the credit for this work must be given to Louis Pasteur, a great French scientist. In all of this early work, living intact cells, not enzymes per se, were held responsible for the activity observed. However, a significant breakthrough in the study of enzymes was made when Buchner discovered, in 1897, that the juice of ground and pressed yeast cells could ferment sugar. Buchner observed that living yeast cells contributed some factor that was able to catalyze the fermentation of sugar in a cell-free environment.

The next significant breakthrough in the study of enzymes was the isolation of the enzyme urease by Sumner (4) in 1926, and his discovery that enzymes are proteins. The observation that enzymes are proteins was greeted with much skepticism. But, with the isolation of several other enzymes that were unequivocally shown to be protein in nature, this skepticism died, and it is now universally accepted that enzymes are proteins.

Nature of Enzymes

Enzymes, being organic catalysts, have many of the properties of inorganic catalysts and thus can be characterized as follows:

1. Enzymes are active in extremely small amounts—that is, in a biochemical reaction only a small amount of enzyme is necessary to convert a large amount of substrate to product. The two terms *substrate* and *product* signify the starting and ending material of a reaction, respectively. The number of moles of substrate converted per minute by 1 mole of enzyme is called the *turnover number* of the enzyme. We can see a dramatic example of the variance among the activities of enzymes in different biochemical reactions when we compare their turnover numbers. This number may vary from 100 to over 3,000,000.

2. True catalysts remain unaffected by the reaction they catalyze. Enzymes under stable conditions approach this property of the ideal catalyst very closely. Because of the protein nature of enzymes, however, their activities are confined to narrow ranges of temperature, pH, and so on. Under conditions other than those considered optimum, an enzyme is a relatively unstable compound and may be affected by the reaction it catalyzes.

3. Although an enzyme considerably hastens the completion of a reaction, it will not affect the equilibrium of that reaction. The reversible reactions usually present in

the living system would proceed toward equilibrium at a very slow rate in the absence of enzymes. However, an enzyme will speed a reaction in either direction—that is, it will bring about the equilibrium of that reaction at a much faster rate.

4. Catalytic action is specific. Enzymes exhibit specificity for the reactions they catalyze—that is, an enzyme that will catalyze one reaction may not catalyze another. This specificity is very strong for some enzymes and rather general in others. Nevertheless, the property of specificity remains one of the most important properties of enzymes.

Catalytic Action

How does an enzyme influence the rate of a reaction? Perhaps we can answer

this question best by describing what happens when a substance is spontaneously converted to another substance, first in the absence of an enzyme and then in the presence of an enzyme (see Figure 10–1). In ordinary chemical reactions, the beginning molecule (*substrate*) must be changed into one or more different unstable forms before it can be transformed to the product. We might refer to this unstable form as the *intermediate form* (transition state form). The intermediate form is usually at a higher energy level than is the starting molecule. Therefore, the probability is low that the intermediate form will occur unless additional energy is provided to the reaction. Thus all reactions, even spontaneous ones, require "activation" of the starting molecule before conversion to product can take place. Without a catalyst, most biological reactions

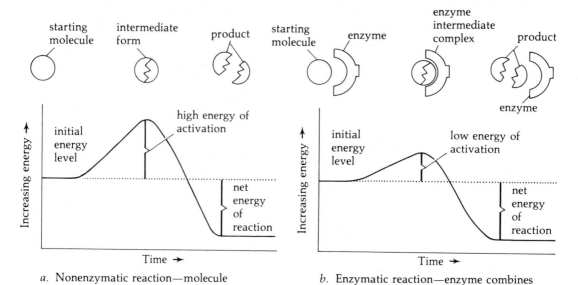

a. Nonenzymatic reaction—molecule assumes unstable intermediate form before conversion to product.

b. Enzymatic reaction—enzyme combines with substrate, unstable intermediate forms, product conversion occurs.

Figure 10–1. Enzyme action. Activation energy for formation of intermediate unstable form is lower for enzyme-catalyzed reaction. In presence or absence of enzyme, net energy of reaction is the same.

would occur very slowly and be rate limited by the formation of an intermediate compound that requires a high *energy of activation*. One way to overcome the energy of activation barrier is to supply energy (heat). With an increase in temperature, a greater number of starting molecules will gain sufficient energy of activation to the intermediate form, and the intermediate forms will convert spontaneously to the product.

In enzymatically catalyzed reactions, the enzyme combines with the starting molecule in such a manner as to produce a change in the conformation of the substrate to that of an intermediate form. Because to form the enzyme-intermediate complex requires less energy than it does to activate molecules by heat, we can accurately say that the enzyme lowers the energy of activation required of the substrate and thus increases the rate of the unstable intermediate formed and hence the final product. If the energy of activation required for the formation of the enzyme-intermediate complex is low, many more starting molecules can participate in the reaction than would be the case if the enzyme were absent. For example, in the absence of the enzyme catalase, the energy of activation for the decomposition of hydrogen peroxide (H_2O_2) is 18,000 cal/mole; but in the enzyme's presence, it is 6,400 cal/mole. When the energy of activation of a reaction is lowered, it is lowered for the backward as well as the forward reaction—in other words, an enzyme will speed a reaction to its equilibrium. Figure 10–1 illustrates these principles.

Another simplification of the foregoing discussion is that when two reactants are involved, enzymes may bring them together in a particular geometric conformation at specific sites that nonenzymatic molecules do not have. These sites consist of specific spatial arrangements of bonding groups that complement the conformation of the substrate and increase catalytic effectiveness tremendously.

Specificity and Enzyme-Substrate Complex

The specificity of enzymes is an important feature of the metabolism of living systems. We should think of it in terms of the enzyme binding a particular compound because the amino acid residues in the enzyme site conform, or fit, to it in a complementary manner. The catalytic property of an enzyme is confined to one of a group of related reactions. For example, the enzyme urease is highly specific for the substrate urea:

$$\text{urea} + H_2O \xrightleftharpoons{\text{urease}}$$
$$\text{carbon dioxide} + \text{ammonia}$$

In contrast, some enzymes are less restricted, and their specificity may be confined to a certain chemical linkage. Some esterases, for example, can act on the ester link between different fatty acids and alcohols without making a distinction among the various ester linkages. Nevertheless, esterases are specific in that they catalyze the hydrolysis of only ester links. They do not catalyze the hydrolysis of other types of chemical linkages nor do they catalyze oxidation reactions, decarboxylation reactions, or other nonhydrolytic reactions.

Studies on the kinetics of enzyme (E) action indicate that enzymes combine with their substrates (S) before yielding the products (P) of the reactions they catalyze. In other words, the enzyme and substrate form a *complex* before transformation of the substrate can occur.

$$E + S \rightleftharpoons ES \longrightarrow E + P$$

Enzymes have active sites with which a substrate molecule can form an intimate association. If we visualize enzymes that have active sites and that are surrounded by numerous substrate molecules, which are very small by comparison, we can immediately see that random collisions play an important part in enzyme-substrate reactions. Since the major portion of the enzyme molecule is devoid of active sites, we would expect many collisions between substrate and enzyme molecules before an active collision takes place. However, if enough substrate molecules are present, the active sites on an enzyme may become completely occupied, and at this point the rate of reaction will be at a maximum—all other factors being held constant. The enzyme-substrate complex offers a good explanation of the specificity of enzymes. Apparently, active sites are specifically shaped within the numerous folds of the enzyme molecule.

Prosthetic Groups, Cofactors, and Coenzymes

Many enzymes have, in addition to their protein structure, an attached nonprotein group. Proteins (in this case enzymes) with attached nonprotein groups are called *conjugated proteins*. Proteins or enzymes of this type may be thought of as consisting of two parts: an *apoenzyme,* composed only of amino acids (protein), and a nonamino acid *prosthetic group*. We can see a good example of this type of complex in enzymes that require a certain metal for activity. We refer to the metal as an *inorganic cofactor* (by earlier terminology, an activator). Investigators have shown definite correlations between the catalytic properties of some enzymes and their association with various metal components. Indeed, the separation of the apoenzyme from its metal component usu-

ally results in complete loss of activity. Restoration of the metal to the apoenzyme returns activity. Many of the enzymes associated with glycolysis require metal cofactors. Some metals known to be cofactors of enzyme systems are copper, iron, manganese, zinc, calcium, potassium, and cobalt.

In contrast to the metal-requiring enzymes, some enzymes require an association with certain organic substances for activity. These organic prosthetic groups are sometimes an integral part of the enzyme and do not dissociate from the protein (apoenzyme) portion. Certain organic substances, however, dissociate readily and may act as prosthetic groups with an array of apoenzymes. These organic prosthetic groups are called *coenzymes*.

During enzymatic activity, a coenzyme generally acts as a donor or acceptor of atoms that are added to or removed from the substrate. Many coenzymes are very important in oxidation-reduction reactions. Further, under laboratory conditions they can be separated easily from the protein portion and when this occurs, the catalytic properties of the enzyme are greatly reduced. Some of the coenzymes that have been identified are nicotinamide adenine dinucleotide (NAD), nicotinamide adenine dinucleotide phosphate (NADP), coenzyme A (CoA), flavin mononucleotide (FMN), and flavin adenine dinucleotide (FAD). Most of the coenzymes are composed of *vitamins*, which, as many of us know, are organic compounds that are synthesized in plants but not in mammals.

Nomenclature and Classification of Enzymes

Enzymes are generally named according to the substrate they complex with or the type of reaction they catalyze. The usual practice is to add the suffix -ase to the name of the substrate involved. Thus the enzymes arginase and tyrosinase are so named because of their substrates arginine and tyrosine. Enzymes may also be grouped under a more general term describing a certain group of compounds with which they react. Thus we have lipases, proteinases, carbohydrases, and so on. Finally, enzymes may be named according to the type of reaction they catalyze. Some examples are hydrolases, oxidases, carboxylases, and phosphorylases. Unfortunately, some of the older nomenclature still persists, and occasionally, we run into a name of an enzyme that has no bearing on the reaction it catalyzes. However, this is more often the exception than the rule.

In order to work with the enormous number of enzymes active in metabolism, some system of classification is necessary. The older system of classification, based on type of chemical reaction catalyzed, is still widely used, but serious students of biochemistry should learn the *numbering system* for transferases, hydrolases, isomerases, and others since "enzymes by the numbers," as recommended by the Commission on Enzymes of the International Union of Biochemistry, are required in most professional biochemical and chemical research publications and abstracts. We recommend that students interested in enzymes consult the most recent textbooks, some of which are listed at the end of this chapter. For purposes of convenience and introduction, we will use the trivial names of the enzymes in this book.

Hydrolytic Enzymes

Hydrolytic enzymes catalyze the addition of water across a specific bond of the substrate. The classification of this type of

enzyme as hydrolytic is an arbitrary one. Since most hydrolytic reactions are reversible, the hydrolytic enzyme could just as well be called a condensation or synthetic enzyme.

$$RCO—OR' \xrightleftharpoons{HOH} RCOOH + R'OH$$

Some examples of hydrolytic enzymes are esterases, carbohydrases, and proteases.

Oxidation-Reduction Enzymes

Oxidation-reduction enzymes catalyze the removal or addition of hydrogen, oxygen, or electrons from or to the substrate, which is oxidized or reduced in the process.

$$RH_2 + A \longrightarrow R + AH_2$$
(removal of hydrogen)
$$RO + \tfrac{1}{2}O_2 \longrightarrow RO_2$$
(addition of oxygen)
$$R^{2+} \longrightarrow R^{3+} + e^-$$
(removal of electron)

These enzymes occupy a major position in cellular metabolism. Because of their importance, we will cover their function in metabolism in more detail in a later chapter. Examples of oxidation-reduction enzymes are the dehydrogenases and oxidases.

Phosphorylases

The phosphorylases catalyze the reversible phosphorolytic cleavage of a specific bond on a substrate. The best-known phosphorylases are those that catalyze the addition of the elements of phosphoric acid to the $\alpha(1,4)$ glycosidic linkages of starch and glycogen.

$$starch + phosphate \rightleftharpoons glucose-1-phosphate$$

The activity of these enzymes is somewhat analogous to that of the hydrolytic enzymes, except that the elements of phosphoric acid instead of water are added.

Transferases

Transferases catalyze the transfer of a group from a donor molecule to an acceptor molecule. This group of enzymes is very large and includes such enzymes as transglycosidases, transpeptidases, transaminases, transmethylases, and transacylases. Probably the best-known example of a transferase is the enzyme glutamic-aspartic transaminase. This enzyme catalyzes the transfer of an amino group from glutamic acid to oxaloacetic acid to form aspartic acid.

| glutamic acid | oxaloacetic acid | α-ketoglutaric acid | aspartic acid |

Carboxylases

The carboxylases catalyze the removal or addition of carbon dioxide. An example of an enzyme that removes CO_2 is glutamic decarboxylase. This enzyme catalyzes the removal of CO_2 from glutamic acid to yield γ-aminobutyric acid.

glutamic acid $\rightleftharpoons$ γ-aminobutyric acid + CO_2

An example of an enzyme that catalyzes the addition of CO_2 is ribulose bisphosphate carboxylase. This enzyme is important in photosynthesis, during which it catalyzes the carboxylation of ribulose-1,5-bisphosphate. We will discuss this reaction in more detail in a later chapter.

Isomerases

Isomerases catalyze the interconversion of aldose and ketose sugars. For example, the interconversion of glucose-6-phosphate with fructose-6-phosphate is catalyzed by the enzyme phosphoglucoisomerase.

glucose-6-phosphate   fructose-6-phosphate

Epimerases

Epimerases are enzymes that catalyze the conversion of a sugar or a sugar derivative to its epimer. Epimers are molecules that differ only in the configuration of a single carbon atom. The conversion of a molecule to its epimer is called *epimerization*. An example of an epimerization is the reversible conversion of xylulose-5-phosphate to ribulose-5-phosphate.

xylulose-5-phosphate   ribulose-5-phosphate

Distribution of Enzymes in Plant Cells

The development, in relatively recent years, of techniques enabling scientists to study enzyme systems outside of the living cell has given us a good picture of the distribution and action of enzymes within the architecture of the cell. Unicellular organisms such as yeasts, bacteria, and algae—because of their high protein content and less complex structures—have proved to be excellent sources of enzymes for this type of study. Also, the physiological functions of certain parts of the cell are good guides to the location of enzymes involved in these functions. For example, ribosomes function chiefly in the process of protein synthesis. Thus the enzymes involved in translation should be located on the surface or in the very immediate vicinity of the ribosomes.

Many of the enzymes of cellular metabolism are associated with the organelles of the cell. Perhaps the highest concentrations of enzymes may be found in mitochondria and chloroplasts. In the Krebs cycle, all of the enzymes necessary for the complete oxidation of pyruvate to CO_2 and H_2O are present in the mitochondria. Included are the enzymes necessary for electron transfer to oxygen with the formation of H_2O. The passage of electrons from the intermediates of the Krebs cycle to oxygen occurs via the electron transport system and results in the formation of ATP.

The chloroplast is even more remarkable for the diverse array of enzymes that are within it. The enzymes necessary for the CO_2 fixation reactions of photosynthesis are present in the stroma of the chloroplast. Cytochromes have been located in chloroplasts and, as in mitochondria, their activity is important to the production of ATP. In addition, the enzymes necessary for the synthesis of the pigments of the chloroplasts (chlorophylls, carotenoids, and so on) are present.

The enzyme deoxyribonuclease is located in the nucleus. This enzyme catalyzes the hydrolytic cleavage of DNA. The *ground phase* of the cytoplasm (cytoplasm not containing organelles) abounds in enzymes. The enzymes of glycolysis and the hexose monophosphate pathway are located in the cytoplasm, and various hydrolytic enzymes and phosphorylases are also present.

Besides enzymes associated with specific areas of the cell, there are enzymes that we may consider extracellular. These enzymes function in the extracellular digestion and transport of nutrients into the cell. For example, some bacteria utilize proteins and polysaccharides as nutrients. These molecules are very large and complex and could not possibly penetrate the cell membrane. However, bacteria excrete enzymes that reduce these large molecules to smaller molecules that are able to penetrate the cell.

A certain degree of compartmentalization of enzymes takes place within the cell, which, in many cases, affords a better association of enzyme and substrate and results in a more efficient system. The compartmentalization of enzymes reaches a high degree in the mitochondrion and the chloroplast. However, even the cytoplasm is thoroughly partitioned by the endoplasmic reticulum and other organelles (peroxisomes and other microbodies), thereby suggesting that metabolites as well as enzymes are compartmentalized.

Factors Affecting Enzyme Activity

Like all chemical reactions, an enzyme-catalyzed reaction is susceptible to external conditions. Because of their protein nature, however, enzymes are unusually sensitive to the fluctuating influences of their immediate environment. Thus the substrate concentration, enzyme concentration, temperature, and pH affect the rate of an enzyme-catalyzed reaction because they all affect the active site of the enzyme and the formation of the enzyme-substrate complex.

Substrate Concentration

The formation of an enzyme-substrate complex precedes transformation of the substrate; therefore, we can clearly describe the effect of substrate concentration on the rate of an enzyme-catalyzed reaction. Usually, the enzyme molecule is much larger than its substrate. Consider, then, a giant enzyme molecule, surrounded by a relatively low concentration of substrate molecules, some near and some far from the active sites on the enzyme. Because of the low concentration of substrate molecules, the active sites on the enzyme molecule may not be occupied. In addition, when an occupied site is vacated, there may be a brief interval before the site is contacted by another substrate molecule. Obviously, under these conditions, the enzyme is not working at maximum efficiency. An increase in the substrate concentration will bring about an increase in the number of molecules in the immediate vicinity of the enzyme's active sites, and will result in an increase in the chance of a substrate molecule coming in contact with an active site. Therefore, at a constant enzyme concentration, an increase in substrate concentration will result in an increase of the rate of an enzyme-catalyzed reaction. When the substrate concentration is increased to the point of "saturating" the active sites, the enzyme is said to be working at maximum efficiency, all other factors being constant. Further increase in sub-

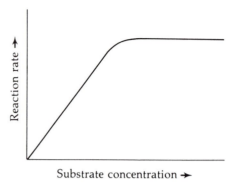

Figure 10–2. Typical effect of substrate concentration on rate of enzyme-catalyzed reaction.

strate concentration will have no effect on the rate of reaction. Figure 10–2 illustrates these relationships.

Enzyme Concentration

When we consider the effect of substrate concentration on an enzyme-catalyzed reaction, we should also be able to understand why an increase in enzyme concentration would cause an increase in the rate of reaction. Suppose at a specific enzyme concentration, we have saturated the active sites with substrate molecules and no longer can effect an increase in the rate of reaction by adding more substrate. If we now increase the enzyme concentration, we, in effect, increase the number of available active sites and thus the chance of reactive contact between enzyme and substrate.

Generally, when we measure the activity of an enzyme, we use low concentrations of enzymes in high concentrations of substrate. Under these circumstances, there would be maximum enzyme activity no matter what enzyme concentration is used, as long as the enzyme concentration is sufficiently low for continuous contact between active sites and substrate molecules. In this situation, we can observe that the rate of reaction is directly proportional to enzyme concentration (see Figure 10–2). We should not lose sight of the fact, however, that if the substrate concentration were relatively low, an increase in enzyme concentration would cause a rise in the rate of reaction up to a point, and then the rate of reaction would remain constant. In other words, increasing the enzyme concentration would have the same effect on the rate of reaction as would increasing the substrate concentration (see Figure 10–3).

Figures 10–2 and 10–3 illustrate how the kinetics of reactions (the study of the rates of reactions or the speed with which they occur) can be analyzed. These rates of reactions can be evaluated according to their *kinetic order* (degree to which the rate is dependent on reactant concentrations). For example, we call the straight-line portions of the plots in Figures 10–2 and 10–3 *first order* because the reaction rate is directly proportional to the concentration of the substrate

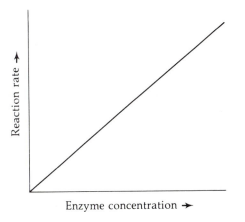

Figure 10–3. Typical effect of enzyme concentration on rate of reaction. Substrate concentration is sufficiently high to allow for active sites to be continuously occupied.

or enzyme. After the rate levels off with the slope change (Figure 10–2), we refer to the rate as *zero order* because the reaction rate is independent of the reactant (substrate) concentration and indicates that the enzyme is working at maximum efficiency.

Temperature

As with any chemical reaction, an enzyme-catalyzed reaction is influenced by temperature. However, the protein nature of enzymes causes them to be particularly sensitive to temperature changes and confines their activity to a much narrower temperature range than would be encountered in most ordinary chemical reactions. At 0°C the rate of an enzyme-catalyzed reaction is practically zero. As the temperature increases, the reaction rate increases more or less steadily. Generally, the reaction rate increases on the average 2.5 times for every 10°C rise in temperature up to 25°C. Two factors are involved here: (1) increase in the kinetic energy of both the substrate and enzyme molecules and (2) increase in the chance of collision between enzyme and substrate molecules as a result of their greater agitation by higher temperatures.

As 30°C is approached, however, factors leading to enzyme denaturation become more apparent. The complex molecular structure of an enzyme is an essential factor in its catalytic activity. This structure is held in its unique pattern by numerous, weak links called hydrogen bonds. Because of increased thermal activity, these bonds are stretched and finally broken as the temperature increases. The rupture of one hydrogen bond makes the rupture of the next easier, and so on, until the integrity of the enzyme structure can no longer be maintained and catalytic properties are com-

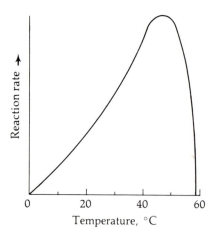

Figure 10–4. Typical effect of temperature on enzyme-catalyzed reaction.

pletely lost. Collapse of the enzyme structure caused by an increase in temperature or by other factors generally results in a gross loss of conformation of the protein, which is known as *denaturation* (i.e., loss of natural properties). The loss of catalytic properties is rather abrupt. In typical cases it begins at about 35°C and is complete as 60°C is approached (see Figure 10–4).

We must also consider a time factor when discussing the effect of temperature on enzyme activity. In Figure 10–4 we can see that the reaction rate is approaching maximum at 45°C. However, at this temperature, destruction of the essential structure of the enzyme molecule is also taking place. If the reaction occurs for any length of time at this temperature, there will be a gradual falling off in activity.

Hydrogen Ion Concentration (pH)

Changes in pH can also cause denaturation of the enzyme molecule and result in a loss of activity. However, this does not appear to be the major effect of pH on en-

zyme-catalyzed reactions. Typically, an enzyme has an optimum pH; a shift to the alkaline or acid side of the pH scale causes a drop in activity. Characteristically, proteins possess many ionic groups, which may either be charged or uncharged depending on the hydrogen ion concentration of their immediate environment. If these ionic groups happen to be functional groups, say as part of an active site, and the formation of the enzyme-substrate complex is dependent on their ionic state, we can easily see why a change in pH can cause a change in enzyme activity. In addition, if the substrate is an electrolyte, as is often the case, then its ionic state is also affected by changes in pH. Thus if the ionic state of the substrate is an important factor in the reaction, the reaction rate will be influenced by any change in the ionic state of the substrate caused by a change in pH. Other conditions being equal, the highest efficiency in an enzyme-catalyzed reaction can be expected at the pH level that leaves the greatest number of molecules in the proper ionic state. From this statement, we can infer that different enzymes have different optimal levels of pH, as Figure 10–5 illustrates.

Inhibitors

Since enzymes are proteins, they possess a variety of functional groups capable of interacting with numerous other compounds. Interaction of an enzyme with substances other than the normal substrate in many instances leads to an alteration of structure essential to catalytic activity. If this alteration occurs, there is a loss in catalytic efficiency or complete inactivation of the enzyme. Enzyme inhibitors may be either competitive or noncompetitive. Some biochemists refer to a third category of inhibitors known as uncompetitive inhibitors,

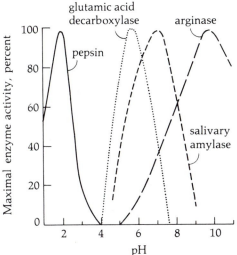

Figure 10–5. Effect of pH on activity of pepsin, glutamic acid decarboxylase, salivary amylase, and arginase.

From *J.S. Fruton and S. Simmonds. 1959. General Biochemistry. New York: Wiley.*

which are only subtly different from noncompetitive inhibitors. We will restrict our discussion to competitive and noncompetitive inhibitors.

Competitive inhibitors. Structural analogs of a substrate molecule may, in some cases, occupy active sites on an enzyme that are normally occupied by the substrate. The new complex formed is reversible and inactive

with respect to the formation of products. In other words, these structural analogs compete with the normal substrate molecule for active sites on the enzyme. Substances that act in this manner are called *competitive inhibitors*, and their inhibition of enzyme activity is called *competitive inhibition.*

$$E + S \rightleftharpoons ES \longrightarrow E + P$$
$$\text{(enzyme + substrate)}$$

or

$$E + I \rightleftharpoons EI \quad \text{(enzyme + inhibitor)}$$

Competitive inhibition can be overcome by increasing the concentration of substrate until all the active sites are occupied by substrate molecules.

One of the classic examples of competitive inhibition is the inhibition by malonic acid of the enzyme succinic dehydrogenase, which catalyzes the conversion of succinic acid to fumaric acid. The inhibitor, malonic acid, closely resembles in chemical structure the normal substrate, succinic acid, and as a result is able to occupy active sites normally occupied by succinic acid. Malonic acid is a competitive inihibitor since inhibition can be overcome by increasing the concentration of succinic acid (see Figure 10–6).

Noncompetitive inhibitors. In contrast to competitive inhibitors, noncompetitive inhibi-

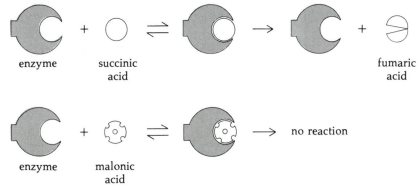

enzyme   succinic acid   fumaric acid

enzyme   malonic acid   no reaction

Figure 10–6. Competitive inhibition. Malonic acid is similar in structure to succinic acid and can occupy active sites on enzyme.

tors do not compete with the substrate per se for active sites on the enzyme surface. As a result, noncompetitive inhibition cannot be completely overcome by the addition of more substrate. Generally, in noncompetitive inhibition, the inhibitor reacts either with parts of the enzyme not involved in catalytic activity or with the enzyme-substrate complex. The relationship of the enzyme, substrate, and noncompetitive inhibitor may be as follows:

$$E \begin{cases} + I \rightleftharpoons EI & \text{(enzyme + inhibitor)} \\ + I \longrightarrow EI & \text{(enzyme + inhibitor)} \end{cases}$$

or

$$E \begin{cases} + I + S \longrightarrow EIS & \text{(enzyme + inhibitor + substrate)} \\ + S \longrightarrow ES \longrightarrow P + E & \text{(enzyme-substrate interaction only)} \end{cases}$$

In the enzyme-inhibitor relationship, inhibition is often caused by a modification of the enzyme structure, which destroys the ability of the enzyme and substrate to interact. In the enzyme-inhibitor-substrate relationship, the inhibitor renders the enzyme-substrate complex inactive.

See the Suggested Readings list for some current biochemistry books that provide an in-depth treatment of enzyme kinetics, inhibitors, allosterism, and enzyme regulation since these subjects are beyond the scope of this book.

Questions

10–1. What is an enzyme, structurally and functionally?

10–2. Although the term *activation* is not customarily used when referring to the substrate, what might substrate activation mean?

10–3. The term *activation* is usually applied to enzyme activation. What does enzyme activation mean? Does the term relate to energy of activation? Explain.

10–4. Explain the terms *energy of activation, intermediate* (transition state) *form,* and *product* as they pertain to enzyme-catalyzed reactions.

10–5. How does an enzyme operate to lower the energy of activation of a given reaction. Does the substrate play a role in its own transition to an intermediate form during an enzyme-catalyzed reaction?

10–6. Are enzymes specific for only specific compounds? Explain.

10–7. Indicate those events in an enzyme-catalyzed reaction that might potentially limit the rate of product formation.

10–8. Define inorganic cofactor, prosthetic group, apoenzyme and coenzyme, and conjugated proteins.

10–9. Name several coenzymes. What is the role of coenzymes? Give an example in your explanation. What is the relationship of vitamins to coenzymes.

10–10. The older system of classification of enzymes is based on the type of chemical reactions catalyzed. Provide some examples (by name) of enzyme reactions. What is the basis for the newer means of classification of enzymes? Consult textbooks in biochemistry, if necessary.

10–11. What is the degree of compartmentalization of enzymes in plant cells?

10–12. Describe those factors that influence the rate of enzyme-catalyzed reactions.

Suggested Readings

Bohinski, R.C. 1979. *Modern Concepts in Biochemistry*, 3rd ed. Boston: Allyn and Bacon.

Lehninger, A.L. 1982. *Principles of Biochemistry*. New York: Worth.

McGilvery, R.W., with G. Goldstein. 1979. *Biochemistry: A Functional Approach*. Philadelphia: Saunders.

Metzler, E.D. 1977. *Biochemistry*. New York: Academic Press.

Preiss, J., and T. Kosuge. 1976. Regulation of enzyme activity in metabolic pathways. In J. Bonner and J.E. Varner, eds., *Plant Biochemistry*, 3rd ed. New York: Academic Press.

Smith, H., ed. 1977. *The Molecular Biology of Plant Cells*. Berkeley: University of California Press.

Stryer, L. 1981. *Biochemistry*, 2nd ed. San Francisco: Freeman.

Chapter 11

Carbohydrates

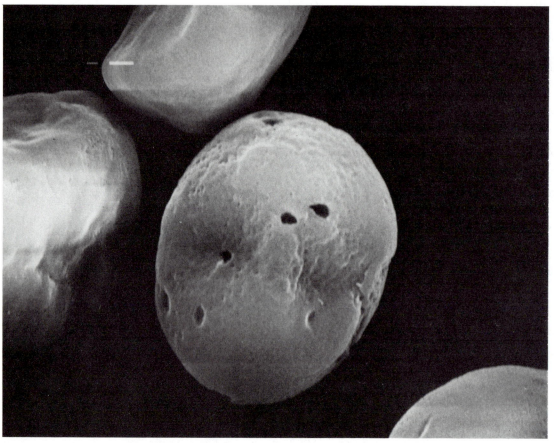

Scanning electron micrograph of starch granules from endosperm cells of maize. Magnification 6,500×. *Courtesy of C.D. Boyer, The Pennsylvania State University.*

As the term implies, the carbohydrates are a group of organic compounds containing the elements carbon, hydrogen, and oxygen, generally in the ratio of $1:2:1$. However, the definition of this group has been broadened to include compounds containing nitrogen and sulfur and compounds that do not conform to a strict $1:2:1$ ratio of carbon, hydrogen, and oxygen. Carbohydrates, therefore, are considered to be *polyhydroxy-aldehydes* or *polyhydroxyketones* and their derivatives.

The carbohydrates are important to the plant in several ways. First, they represent a means for the storage of the energy that is converted from light, in the process of photosynthesis—a function of the utmost importance to animals as well as plants. Second, they are important constituents of the supporting tissues that enable the plant to achieve erect growth, in some cases to the height of 400 feet. Third, they provide the carbon skeletons for the organic compounds that make up the plant.

Classification

We may divide the carbohydrates roughly into three categories: the *monosaccharides*, the *oligosaccharides*, and the *polysaccharides*. The first group, the monosaccharides, are the least complex of the carbohydrates and do not yield simpler carbohydrates on hydrolysis. If we were to adhere to the original definition of carbohydrates (hydrates of carbon), then we would have to consider two-carbon compounds such as formaldehyde and acetic acid among the carbohydrates. However, some of the chemical and physical properties associated with the carbohydrates are lacking in these compounds. Monosaccharides are the building units for the more complex oligosaccharides and po-

lysaccharides. The oligosaccharides are also relatively simple; they are made up of two or more monosaccharides held together by *glycosidic linkages* (covalent bonds between sugars). The polysaccharides, in contrast, are complex molecules of high molecular weight composed of a large number of monosaccharides joined through glycosidic linkages. The border line between oligosaccharides and polysaccharides is ill defined. We may call a large oligosaccharide, a polysaccharide, or a small polysaccharide, an oligosaccharide.

Monosaccharides

The simplest soluble carbohydrates are the three-carbon compounds *glyceraldehyde* and *dihydroxyacetone*.

$$
\begin{array}{ll}
& \text{H} \\
& | \\
\text{①} & \text{C=O} \\
& | \\
\text{②} & \text{CHOH} \\
& | \\
\text{③} & \text{CH}_2\text{OH} \\
& \text{glyceraldehyde}
\end{array}
\qquad
\begin{array}{ll}
\text{①} & \text{CH}_2\text{OH} \\
& | \\
\text{②} & \text{C=O} \\
& | \\
\text{③} & \text{CH}_2\text{OH} \\
& \text{dihydroxyacetone}
\end{array}
$$

A consideration of the above compounds will help us with the general terminology used to describe sugars. For example, we classify the monosaccharides in accordance with the number of carbons present. Thus we call glyceraldehyde and dihydroxyacetone *trioses*. Note also that on these compounds one of the carbons bears a carbonyl oxygen, on the first carbon of glyceraldehyde to form an *aldehyde group*, and on the second carbon of dihydroxyacetone to form a *ketone group*. Thus we can differentiate between the two trioses by calling glyceraldehyde an *aldose* and dihydroxyacetone a *ketose*. The aldehyde group and the ketone group are known as *reducing groups* because of their ready oxidation by certain compounds, which are themselves reduced in the reaction. Sugars possessing these groups are called *reducing sugars*.

Pentoses. The pentoses are five-carbon sugars, which are rarely found dissolved in the free state in the cell cytoplasm. However, they are quite abundantly found as constituents of some of the more complex carbohydrates of the plant. Thus D-*xylose* and L-*arabinose* occur in plants as constituents of *xylans* and *arabans*, respectively—large polysaccharides with a structural function in the cell wall.

| D-xylose | L-arabinose |

In addition to xylose and arabinose, the five-carbon sugars D-*ribose* and 2-*deoxy-D-ribose* are also commonly found in plants as components of the nucleic acids. Certain coenzymes important in hydrogen and group transfer reactions have D-ribose as a component of their structure. Note the close

| D-ribose | 2-deoxy-D-ribose |

similarity between D-ribose and 2-deoxy-D-ribose. These pentoses differ only in the substituents about the second carbon. In place of a hydroxyl group, 2-deoxy-ribose has a hydrogen atom (deoxy means "without oxygen"). These sugars are part of the nucleotide structure of the nucleic acids.

Hexoses. The hexoses are six-carbon sugars. Four hexoses—D-*glucose, D-fructose, D-mannose,* and D-*galactose*—are commonly found in most plants, either as a component of some more complex carbohydrate or free in the cell. Generally, glucose and fructose are the only hexoses found dissolved in the free form.

We can see immediately that there appears to be very little difference in the structures of the hexoses. In the first three, the only differences are on the first and second carbons. Fructose differs from both glucose and mannose in that it is a ketose and the other two are aldoses. However, the last four carbons are identical in these three compounds. Galactose differs from glucose only in the position of the hydroxyl group on carbon number 4.

The hexoses characteristically contain several *asymmetric* carbons (containing four different substituents), thus allowing for the existence of several diastereoisomers that differ in physical, chemical, and biological properties and that are known by different names, such as glucose, mannose, galactose, and so on. However, these sugars may also have mirror images known as *enantiomers*, which are identical in all physical properties except optical rotation. We mean by optical rotation that a plane of polarized light transmitted by pure solutions of these mirror compounds will be rotated either to the left (levorotatory) or to the right (dextrorotatory) depending on what mirror image is present. Conventionally, the italic letter *d*

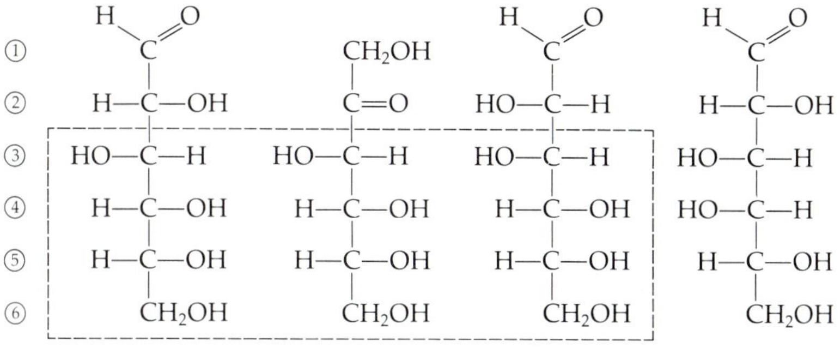

①
H O
  C

② H—C—OH

③ HO—C—H

④ H—C—OH

⑤ H—C—OH

⑥ CH₂OH

D-glucose

CH₂OH

C=O

HO—C—H

H—C—OH

H—C—OH

CH₂OH

D-fructose

H O
  C

HO—C—H

HO—C—H

H—C—OH

H—C—OH

CH₂OH

D-mannose

H O
  C

H—C—OH

HO—C—H

HO—C—H

H—C—OH

CH₂OH

D-galactose

or the plus sign (+) is inserted before the name of a sugar for rotation to the right and the italic letter *l* or the minus sign (−) for rotation to the left. Thus we have *d*(+)-glucose and *l*(−)-glucose.

Although the use of *d* or *l* (+ or −) tells something about the optical properties of sugars, it does not give any information on the configuration around the centers of asymmetry in the molecule. A system has been devised based on configurational rather than optical properties, and the key atom generally used in this system is the highest numbered asymmetric carbon. In the hexoses, this is carbon number 5 and the hydroxyl group of this carbon is said to be either in the D position or in the L position. When we put the structure of a sugar on paper, we write the hydroxyl of carbon 5 of a D-hexose to the right of the carbon chain. In an L-hexose, we write the hydroxyl to the left of the carbon chain, as shown in the formulas for glucose and fructose. Practically all of the sugars found in the plant are of the D configuration. However, the rarely found L-galactose is a constituent of agar.

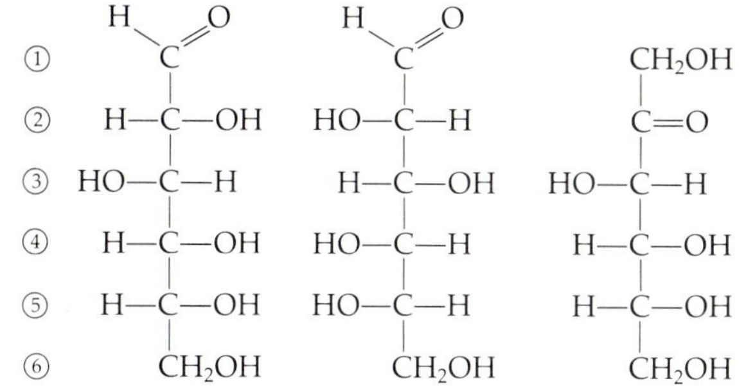

Ring structure. So far in our discussion of carbohydrates, we have considered sugars only as straight-chain structures when, in

①
H O
  C

② H—C—OH

③ HO—C—H

④ H—C—OH

⑤ H—C—OH

⑥ CH₂OH

D(+)-glucose

H O
  C

HO—C—H

H—C—OH

HO—C—H

HO—C—H

CH₂OH

L(−)-glucose

CH₂OH

C=O

HO—C—H

H—C—OH

H—C—OH

CH₂OH

D(−)-fructose

fact, carbohydrates are largely found in *cyclic* or *ring* forms. In the carbon chain of glucose, there are four centers of asymmetry (carbons 2, 3, 4, 5). If carbons 1 and 5 come in close proximity of each other, however, as might occur in solution, an oxygen bridge may form between these carbons and result in the formation of a hydroxyl group on carbon 1. A new center of asymmetry around carbon 1 is thus created, and the glucose molecule now has five instead of four asymmetrical carbons. The newly formed hydroxyl group may be in either the α or β position on carbon 1, thus adding another feature to our classification of the carbohydrates.

The sugars (β-D-glucose and α-D-glucose) are presented in the Haworth (stereochemical) configuration on page 206.

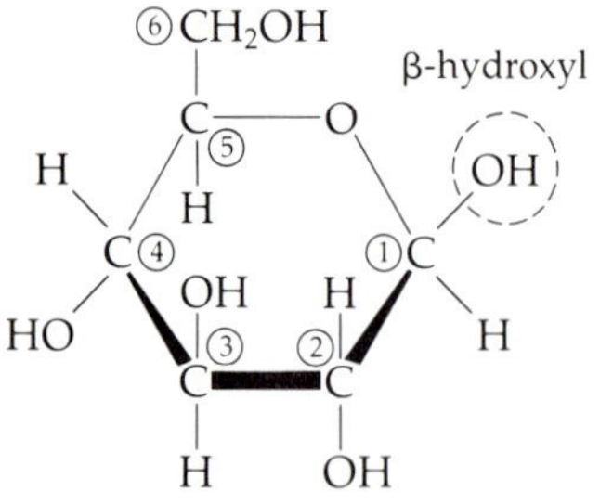

β-D-glucose

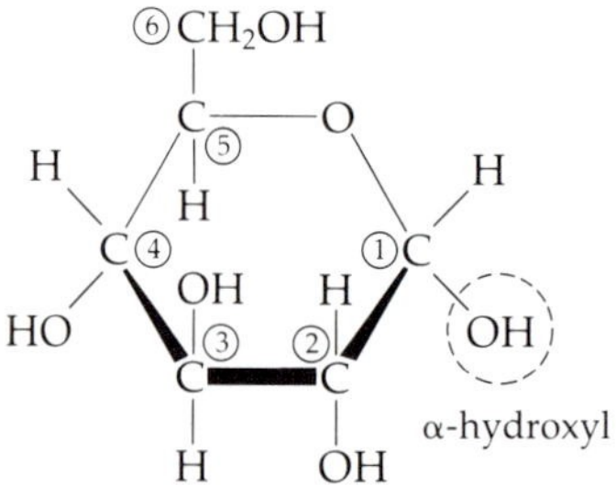

α-D-glucose

Although α- and β-D-glucose are very similar in structure, they are different in their physical, chemical, and biological properties. For example, β-D-glucose units make up the structure of *cellulose*, a polysaccharide of the cell wall. Structural support is the obvious function here. And α-D-glucose units make up the structure of the polysaccharide *starch*. Starch is the most common storage material of plants.

Branched chain. Two-branched chain monosaccharides of natural origin have been found in plants. One is a five-carbon sugar called *apinose*, and the other, a six-carbon sugar called *hamamelose* (22). Apinose has been found in parsley and arrowwood as a constituent of at least three different glycosides. Studies have shown that the occurrence of apinose in plants is widespread, and in some cases it has been found in large quantities. Other plants that contain apinose include duckweed, oleander, and eelgrass.

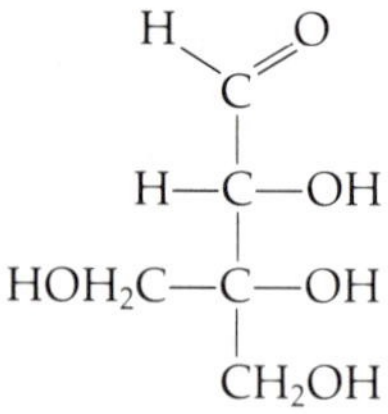

apinose

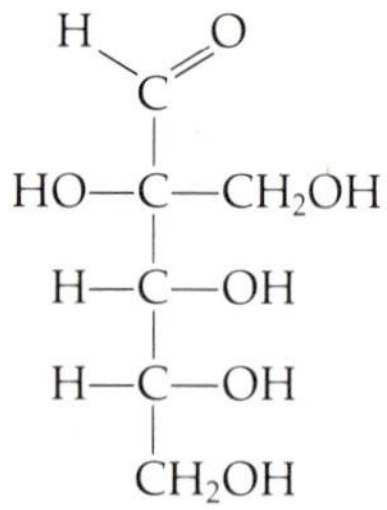

hamamelose

Hamamelose was first discovered in the bark of witch hazel, where it occurs in combination with tannin. The widespread occurrence of hamamelose in higher plants, especially in the *Primula* species, has been shown in studies by Scherpenberg, Grobner, and Kandler (29) and by Sellmair and Kandler (31).

Oligosaccharides

The oligosaccharides are generally classified according to the number of monosaccharide units found in their structures. Therefore, if two monosaccharide units make up a sugar, it is called a *disaccharide*; if three, a *trisaccharide*; if four, a *tetrasaccharide*; and so on.

The principle disaccharide of higher plants is *sucrose*, a condensation product of glucose and fructose—that is, in the formation of sucrose, glucose and fructose are linked together, a combination that results in the elimination of water. Sucrose is common table sugar and is of commercial value to humans. Thus we value highly plants such as the sugar cane and the sugar beet, which produce large quantities of this sugar.

Although glucose and fructose, which make up the structure of sucrose, are reducing sugars, sucrose itself is not. The reducing groups of both the simple sugars are

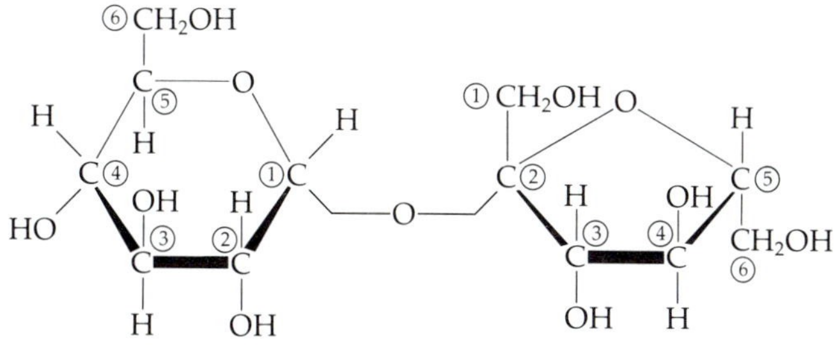

sucrose

involved in the bond that links them together to form sucrose—that is, the oxygen bridge between the two monosaccharides occurs between carbon 1 of glucose and carbon 2 of fructose and results in the elimination of the free carbonyl groups of both these sugars. We should also note from the structure of sucrose that fructose occurs as a five-membered ring (furanose ring) as compared to glucose, which occurs as a six-membered ring (pyranose ring).

Sucrose is the principal form in which carbohydrates are transported in higher plants. In recent years, the use of radioactive materials has clearly demonstrated this fact. A plant undergoing photosynthesis in an atmosphere of radioactive carbon dioxide will show that translocation of this radioactive carbon out of the leaf after it has been assimilated will be primarily in the form of sucrose.

Other disaccharides of importance are usually products of the partial degradation of polysaccharides, such as starch and cellulose. Therefore, a partial degradation of starch may yield the disaccharide *maltose*, a compound composed of two molecules of D-glucose joined together in a α-1,4-linkage. The numbers refer to the carbons involved in the link between the two glucose molecules. Partial degradation of cellulose or lignin may yield the disaccharide *cellobiose*, a compound composed of two D-glucose mol-

ecules joined together by a β-1,4-linkage. In contrast to sucrose, both maltose and cellobiose are reducing sugars.

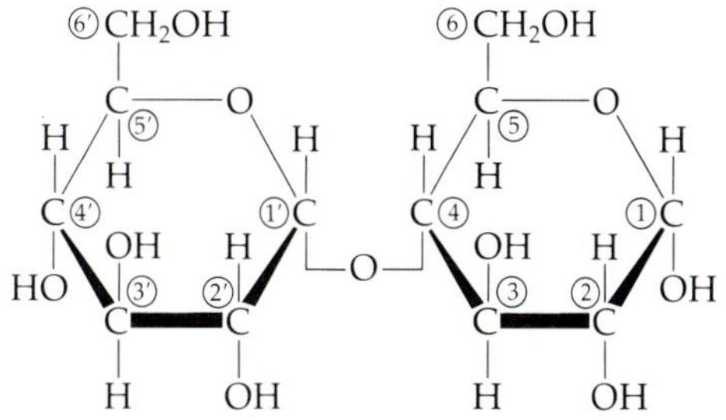

maltose (α-1,4-linkage)

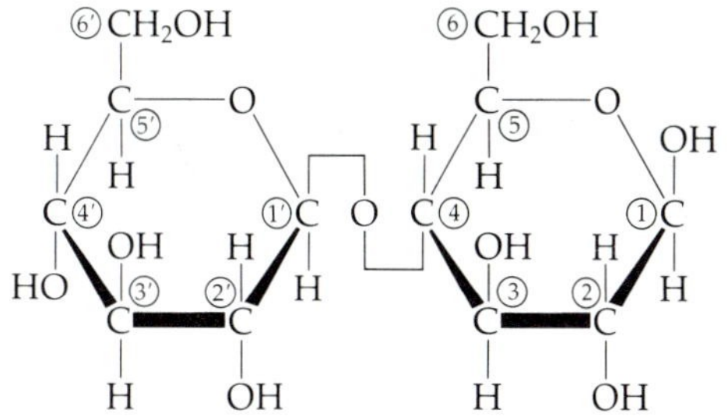

cellobiose (β-1,4-linkage)

Naturally occurring trisaccharides, such as *gentianose* and *raffinose*, have been found in many plants (22). On hydrolysis, gentianose yields two molecules of glucose and one of fructose. Hydrolysis of raffinose yields glucose, fructose, and galactose. Both gentianose and raffinose are nonreducing sugars. Small amounts of raffinose are found in the leaves of many plants, much larger amounts are found in storage organs such as seeds where it accumulates during maturation and is consumed during germination (22). Loss of water by plant tissues (as in seed formation) appears to be accompanied by an increasd rate of raffinose synthesis. Zimmerman (39, 40) found the tetrasaccharide *stachyose* in several tree species. On hydrolysis, stachyose yields glucose, fructose, and two molecules of galactose. Webb and Burley (35) made the interesting observation that instead of sucrose, stachyose is the principal carbohydrate transported in *Fraxinus americana, Cucurbita pepo,* and *Verbascum thapsus.*

Polysaccharides

Unlike the oligosaccharides, polysaccharides are high molecular weight polymers consisting of monosaccharide-repeat-ing units (monomers). In many cases, the simple sugars produced by a plant are converted to polysaccharides. The two most common polysaccharides of plants are starch, a storage product of plants, and cellulose, a structural polysaccharide, which makes up the greater part of the plant cell wall. In the lower plants, such as the algae, bacteria, and fungi, other polysaccharides of structural and nutrient function are found in addition to cellulose and starch.

Starch is a compound of high molecular weight, which, on complete hydrolysis, yields only α-D-glucose molecules. Cellulose also has a high molecular weight and, on complete hydrolysis, yields β-D-glucose molecules. Both of these compounds and polysaccharides in general (with several exceptions) differ from monosaccharides and oligosaccharides in being insoluble in water and lacking sweetness. The molecular structures of starch and cellulose are as shown below.

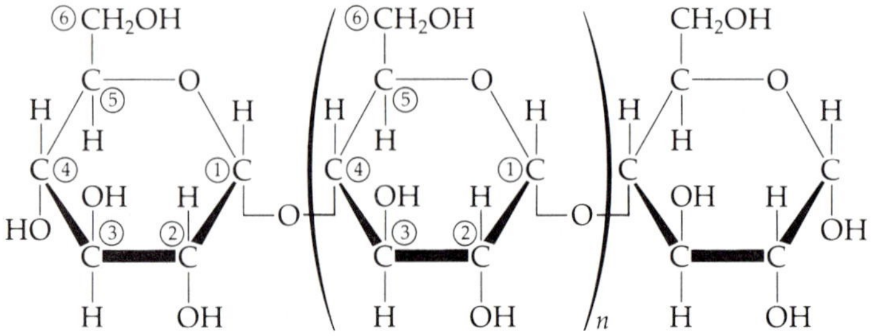

starch

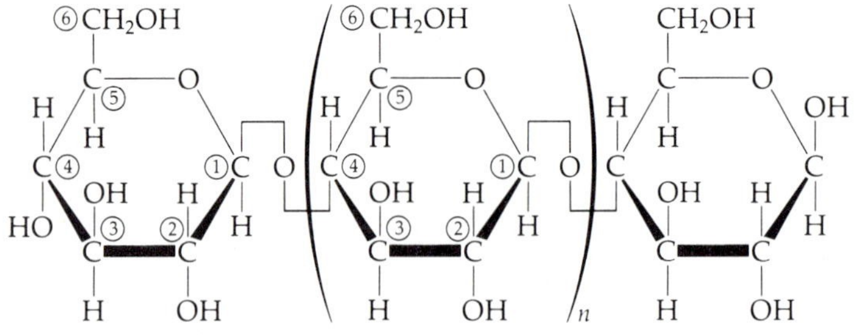

cellulose

Starch. Much of the sugar produced in photosynthesis is converted to starch, which is formed in plastids as starch grains. Starch grains are very prevalent in storage organs, such as seeds, tubers, bulbs, where they functon as reserve nutriment for the growth and development of the plant. Starch grains differ in shape and size from plant to plant and are large enough to be distinguished microscopically.

Although we generally think of starch as a straight-chain polymer of glucose units, it is actually composed of two polysaccharides—*amylose* and *amylopectin*. Both of these polysaccharides yield α-D-*glucose* units on hydrolysis. However, amylose is a straight-chain polymer of glucose units; and amylopectin is a branched molecule. Only α-1,4-links are found in the amylose molecule. In contrast, amylopectin has α-1,6-links and sometimes α-1,3-links, in addition to α-1,4-links. Because of its more complex structure, amylopectin is less soluble in water than is amylose and we may partially separate it from amylose by allowing starch to stand in water for prolonged periods of time. The blue-black color that occurs when we add iodine to starch is due to amylose. The amylopectins give a red to purple color with iodine. A representation below of the amylopectin molecule shows the α-1,4-linkages and α-1,6-linkages.

Cellulose. Cellulose is a straight-chain polymeric molecule of high molecular weight composed of D-glucose units bound together with β-1,4-links. It is a fundamental component of the cell wall and, as such, is the most abundant natural product in the world. When the primary wall is formed on new cells, it is composed of approximately 20 percent cellulose; the remainder consists

amylopectin

of noncellulosic polysaccharides and a small amount of protein. As the cells mature and new wall material is deposited to form secondary walls, the cell wall becomes impregnated with noncarbohydrate materials such as lignin, suberin, or cutin. Cellulose composes about 43 percent of the secondary wall.

Cellulose is a relatively inert material and is completely degraded only under the most strenuous chemical treatment. For example, it may be hydrolyzed to glucose when it is treated with concentrated sulfuric or hydrochloric acid or with concentrated sodium hydroxide. Cellulose is insoluble in water but may be dissolved in ammoniacal solutions of cupric salts. Because of the lack of digestive enzymes such as cellulase, it is of no direct nutritive value to humans. In certain organisms—ruminants, some bacteria, termites, and certain protozoa—cellulose is digested to glucose and has excellent nutritive value. However, due to the lack of appropriate enzymes in plants and the fact that cellulose has certain properties, it is excellent for structural purposes. Although we generally think of the structural value of cellulose to the plant, we should also consider its structural value to humans. Since well before the "dawn of history," the properties of cellulose have served humans well, especially in the tools they fashioned and in the structures they built to shelter them-

selves from their environment. Indeed, not only is cellulose the most abundant organic compound in the world, but it is also one of the most valued compounds.

Pectic compounds. Three general types of pectic substances have been observed in plants: *pectic acid* and two derivatives of pectic acid called *pectin* and *protopectin*. Pectic substances are found most abundantly in the middle lamella between cell walls, usually in the form of calcium or magnesium salts of pectic acid. However, pectin and protopectin are also present. Pure pectic acid is an unbranched molecule consisting of about 100 D-galacturonic acid residues bound together by α-1,4-linkages. On complete hydrolysis, pectic acid releases galacturonic acid molecules. Galacturonic acid differs from galactose only in carbon 6, which is a carboxyl group (—COOH) rather than a carbinol group (—CH$_2$OH). Pectic acid is soluble in water and may be precipitated by calcium ions.

Pectin very closely resembles pectic acid, the only difference is in the esterification of many of the carboxyl groups with methyl groups. Pectin will form a colloidal suspension in water that will "set," or form a gel, on the addition of small concentrations of alcohol or high concentrations of sugar. The ability of pectin to form a gel

pectic acid

Component	Red Maple (Acer rubrum)	White Birch (Betula papyrifera)	Balsam Fir (Abies balsamea)
cellulose	45	42	42
lignin	24	19	29
glucuronoxylan	25	35	—
glucomannan	4	3	—
arabinoglucuronoxylan	—	—	9
galactoglucomannan	—	—	18
pectin, starch	2	1	2

Table 11–1. Chemical composition of wood from two angiosperms and one gymnosperm. All values in percent of extractive-free wood.

Source: *From T.E. Timell. 1965. In W.A. Coté, Jr., ed., Cellular Ultrastructure of Woody Plants. Syracuse, N.Y.: Syracuse University Press. Reprinted by permission.*

makes it commercially valuable for the manufacture of food jellies.

The term protopectin is reserved for all *insoluble* pectic substances. Because of the instability of protopectin, this compound has not been isolated effectively. As a result, not much is known about the structure and composition of protopectin, although it is thought to be a much larger molecule than either pectic acid or pectin. It is accumulated in large quantities in some fruits, such as the apple and pear. During the ripening of the fruit, protopectin is converted into the more soluble substances— pectin and pectic acid.

Although α-1,4-linked galacturonic acid residues account for most pectic substances, polymerized nonuronide sugars also appear to be present in small amounts. Nonuronide sugars that have been isolated from hydrolyzed pectic substances include D-galactose, L-arabinose, L-rhamnose, D-glucose, 2-O-methyl-L-fucose, and 2-O-methyl-L-xylose (9, 38).

Pentosans. Polymers of five-carbon sugars are also found in plants. Two pentosans commonly found are *xylan* and *araban,* which on hydrolysis yield xylose and arabinose, respectively. Xylan is the pentosan most commonly present in plants and is an important constituent of the cell wall matrix. Xylans generally are relatively small unbranched polymers, composed of D-xylose units bound together by β-1,4-links. Also, within the xylan structure other sugar units (e.g., L-arabinose) and sugar acid units (e.g., glucuronic acid) may be found.

Araban is also thought to be a relatively small polymer, composed chiefly of L-arabinose units bound together by α-1,5-links. Although L-arabinose is the chief sugar present, other sugars such as D-xylose may also be present. Table 11–1 gives the chemical composition of wood from two angiosperms and one gymnosperm.

Synthesis and Degradation of Sucrose

Sucrose Synthesis

Sucrose synthesis, at least in higher plants, appears to involve the participation of *uridine diphosphate glucose* (UDPG), a compound first discovered in yeast cells (8). The enzyme *sucrose synthetase* catalyzes the transfer of glucose from UDPG to fructose. In a somewhat similar reaction, the transfer of glucose from UDPG to fructose-6-phosphate is catalyzed by the enzyme *sucrose phosphate synthetase*. These reactions are shown here:

$$\text{UDPG + fructose} \overset{\text{sucrose synthetase}}{\rightleftharpoons} \text{UDP + sucrose}$$

UDPG + fructose-6-phosphate

$$\overset{\text{sucrose phosphate synthetase}}{\rightleftharpoons} \text{UDP + sucrose phosphate}$$

The sucrose phosphate formed in the second reaction can be hydrolyzed by a phosphatase enzyme to yield sucrose. Whether sucrose is synthesized simultaneously in the plant by these pathways is not yet clear. However, sucrose synthetase and sucrose phosphate synthetase activity has been observed in many plants. The existing evidence suggests that UDPG is an essential feature of the biosynthesis of sucrose in higher plants.

Sucrose Degradation

The enzyme *invertase* catalyzes the hydrolysis of sucrose, thereby yielding glucose and fructose.

$$\text{sucrose + H}_2\text{O} \overset{\text{invertase}}{\longrightarrow} \text{glucose + fructose}$$

This reaction is thought to be unidirectional; thus the hydrolysis goes almost to completion. The fact that invertase has been isolated from a variety of plant tissues suggests that the main route of sucrose degradation in plants may be through the activity of this enzyme.

It is interesting to note that gibberellic acid, a plant growth regulator, has been shown to promote invertase synthesis in several different plant growth systems (11, 18, 25). We will discuss the role of gibberellic acid in the physiology of plants in a later chapter.

Synthesis and Degradation of Starch

Starch Synthesis

The study of starch metabolism in the plant cell has developed into a complex and interesting subject. One general conclusion that we can draw from the numerous studies on this subject is that the synthesis and degradation of starch is under the regulation of a variety of enzymes, some of which have both a synthetic and degradative function, depending on immediate conditions at the site of action.

Hanes (16) detected the presence of *starch phosphorylase* in potato and pea plants and demonstrated its activity in vitro. He found that in the presence of this enzyme and glucose-1-phosphate, a polymer of glucose molecules could be formed. Also required is a primer molecule (acceptor), composed of anywhere from 3 (maltotriose) to an optimal number of 20 glucose residues strung together in α-1,4-linkages.

$$n(\text{glucose-1-phosphate}) + \text{acceptor}$$
$$\overset{\text{starch phosphorylase}}{\rightleftharpoons} \text{amylose} + n(\text{P}_i)$$

The glucose of glucose-1-phosphate is added to the nonreducing end of the primer molecule to form an α-1,4-link at that point. Thus the enzyme starch phosphorylase catalyzes the addition of glucose units one by one to the nonreducing end of a primer molecule building a straight-chain amylose molecule (see Figure 11–1).

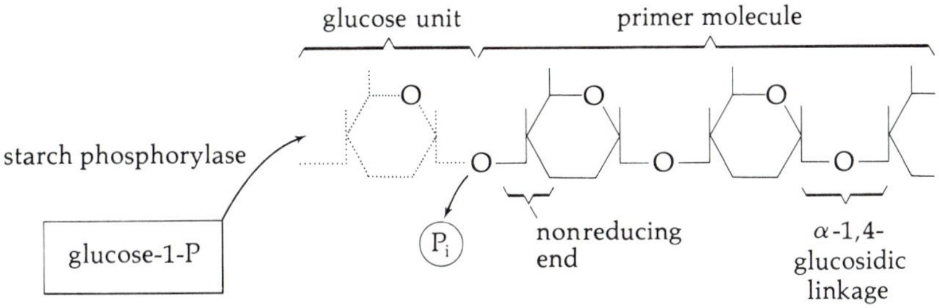

Figure 11–1. Synthesis of amylose molecule by addition of glucose units to nonreducing end of primary molecule. Reaction is catalyzed by starch phosphorylase.

We may also consider starch phosphorylase primarily a degradative enzyme (19). In the presence of inorganic phosphate, starch phosphorylase can catalyze the phosphorolytic cleavage of the α-1,4-link of amylose to form glucose-1-phosphate molecules, a process known as *phosphorolysis*. Phosphorolysis differs from hydrolysis in that it involves the elements of phosphoric acid instead of water. High concentrations of inorganic phosphate and high pH favor phosphorolysis; lower pH and lower concentrations of inorganic phosphate favor the synthetic action. Starch phosphorylase has been isolated from a number of plants and appears to have universal distribution (36).

Another enzyme capable of forming α-1,4-links by the addition of glucose units onto a primer molecule is *UDPG transglycosylase.* This enzyme was first detected in bean, corn, and potato where it was shown to catalyze the transfer of glucose from UDPG to an acceptor or primer molecule. The primer molecule could be maltose, maltotriose (3 glucose units), maltotetrose (4 glucose units), or even a starch molecule (27). When starch is used as the primer molecule, glucose units can be added to either amylose or amylopectin. Thus UDPG transglycosylase apparently requires the presence of at least one α-1,4-link, such as would be found in maltose, and catalyzes the formation of additional α-1,4-linkages.

$$\text{UDPG} + \text{acceptor} \xrightleftharpoons[]{\overset{\displaystyle \text{UDPG}}{\text{transglycosylase}}} \text{UDP} + \alpha\text{-1,4-glucosyl-acceptor}$$

Sucrose may function as a glucose donor in starch synthesis. Akazawa, Minamikawa, and Murata (1) found that the incubation of sucrose-^{14}C with starch granules, sucrose synthetase, and uridine diphosphate (UDP) resulted in a significant amount of the label being transferred to starch. They proposed that the glucose of sucrose is first transferred to UDP to form UDPG, as a result of the reversal of sucrose synthesis. Then the glucose transferred to UDPG is in turn transferred to starch. This scheme explains how a continual supply of UDPG can be maintained for the synthesis of starch.

Some findings suggest that UDPG may play only a secondary role in starch synthesis. For example, Murata and his colleagues (23, 24) demonstrated that adenosine disphosphate glucose (ADPG) is utilized much more efficiently in starch synthesis than is UDPG. These findings are supported by the discovery of ADPG as a natural compound in rice (23, 24). In view of the above comments, UDPG transglycosylase might more appropriately be called *amylose synthetase.* In the scheme for starch synthesis presented in Figure 11–2, ADPG could be substituted for UDPG.

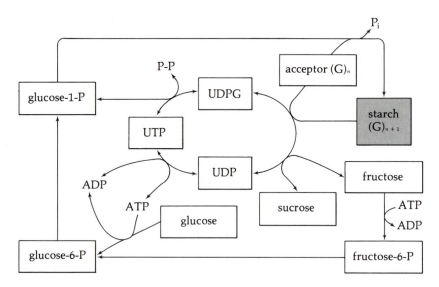

Figure 11–2. Starch synthesis.

From T. Akazawa et al. 1964. Plant Physiol. *39:371.*

Still another enzyme has been found that will catalyze the formation of α-1,4-glycosidic links. This enzyme, called *D-enzyme,* was first discovered in potato by Peat, Whelan, and Rees (26) and was shown to catalyze the reversible transfer of two or more glucose units from malto-dextrins (α-1,4-linked glucose chains of more than two units) to a variety of acceptors. If we consider one molecule of maltotriose as a substrate and another as an acceptor, then D-enzyme would catalyze the formation of malto-pentose. The malto-dextrins are added onto the nonreducing end of the acceptor molecule (see below).

Walker and Whelan (34) showed that if the glucose that accumulates in the above reaction is removed by some other meta-

bolic reaction, amylose chains of significant length can be built up by the D-enzyme. For example, glucose can be phosphorylated if hexokinase and ATP are present.

Starch phosphorylase, UDPG transglycosylase (amylose synthetase), and D-enzyme all catalyze the formation of α-1,4-glycosidic links. However, the starch-molecule also contains α-1,6-glycosidic links at its branching points. Potato extracts contain an enzyme (*Q-enzyme*) capable of forming an amylopectin type molecule, using amylose as a substrate. Q-enzyme was first isolated from potato extract by Baum and Gilbert (3). Q-enzyme is thought to catalyze the transfer of small chains of glucose units from an amylose type molecule, which we will call the donor molecule, to an acceptor molecule of at least four α-1,4-linked glucose units. The small chains being transferred are "tacked" onto the carbon 6 of one of the glucose units of the acceptor molecule to form α-1,6-glycosidic linkages.

Starch is probably synthesized as a result of the simultaneous activity of Q-enzyme and one or more of the enzymes known to catalyze the formation of α-1,4-linkages. However, this supposition has not been demonstrated, and the question of how amylose and amylopectin can be synthesized together in the same starch granule has not been answered. Indeed, the incubation of Q-enzyme with starch phosphorylase in the same reaction mixture only results in a mixture of branched polysaccharides, not in the individual synthesis of amylose and amylopectin. Perhaps they are synthesized at different sites on the granule.

It is interesting to note that in at least one plant (sweet corn) a glycogen-type polysaccharide (phytoglycogen), as well as amylose and amylopectin (36), is synthesized. Phytoglycogen, like animal glycogen, has a higher degree of branching than amylopectin and contains numerous interchain linkages. Since none of the enzymes previously discussed is able to debranch phytoglycogen, it seems likely that sweet corn possesses additional enzymes to perform this function (19).

Starch Degradation

The α- and β-*amylases* are of primary importance to the degradation of starch. The amylases have been found in a wide variety of plants and represent the best means for the mobilization of carbohydrate reserves in the plant. The amylases are hydrolytic enzymes that catalyze the addition of the elements of water to the α-1,4-glycosidic linkage.

β-amylase, which is found most abundantly in seeds, has been isolated from several plants. Incubation of this enzyme with amylose will result in the complete degradation of the amylose molecule to maltose. Starting at the nonreducing end of an amylose molecule with an even number of glucose units, β-amylase successively removes maltose units until the molecule is completely degraded to maltose. However, if the amylase molecule happens to be composed of an odd number of glucose units, hydrolysis with β-amylase will result in the formation of maltose and one maltotriose molecule. The maltotriose represents the terminal three glucose units on the reducing end of the amylose molecule. If the molecule happens to be amylopectin, then β-amylase can start at the nonreducing end of each branch and successively remove maltose units to within two glucose units of the α-1,6-linkages (see Figure 11–3).

donor (optimum 40 units)   acceptor

Q-enzyme

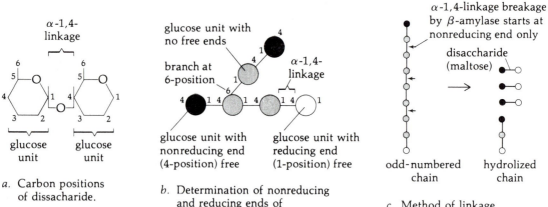

a. Carbon positions of dissacharide.

b. Determination of nonreducing and reducing ends of polysaccharide chain.

c. Method of linkage breakage by β-amylase.

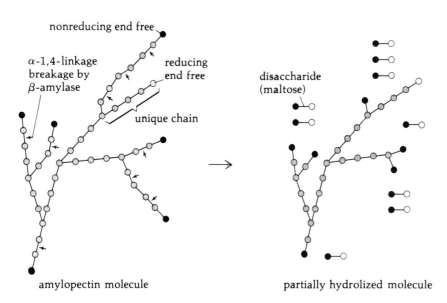

d. Hydrolysis of amylopectin by β-amylase.

Figure 11–3. Hydrolysis of starch by β-amylase.

A study of the activity of the α- and β-amylases will show that their mode of action is quite different. β-amylase removes maltose units one by one from the nonreducing end of a chain of glucose units; α-amylase attacks at random any α-1,4-link on the starch molecule—that is, α-amylase may hydrolyze α-1,4-links at either end or in the middle of the molecule. If a branched chain is attacked, all α-1,6-links to within three units of the α-1,4-link can be hydrolyzed. The products of α-amylase activity on starch are a variety of oligosaccharides or dextrins (see Figure 11–4).

α-1,4-linkage breakage
by α-amylase
proceeds at random

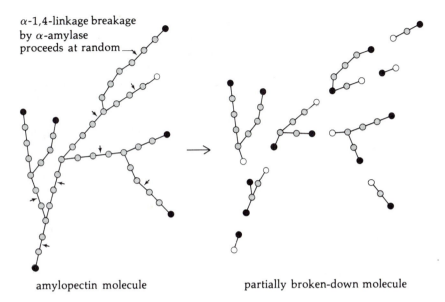

amylopectin molecule   partially broken-down molecule

a. Action of α-amylase on amylopectin.

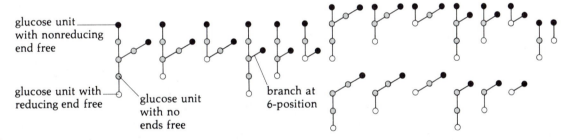

glucose unit
with nonreducing
end free

glucose unit with
reducing end free   glucose unit   branch at
                           with no   6-position
                           ends free

b. Possible structures of limit dextrins from amylopectin breakdown.

Figure 11–4. Action of α-amylases on amylopectin.
From P. Bernfield. 1951. Adv. Enzymol. 12:379.

In addition to the activity of the α- and β-amylases, the enzyme starch phosphorylase can also degrade starch by a phosphorolytic cleavage of α-1,4-glycosidic linkages. In our discussion of the degradation of starch, we have covered the hydrolysis and phosphorolysis of the α-1,4-link by the α- and β-amylases and starch phosphorylase. Total breakdown of the amylopectin molecule is not accomplished by these enzymes because of the presence of the α-1,6-linkages. However, *R-enzyme* isolated from broad bean and potato (17) and *isoamylase* isolated from yeast (20) are both capable of catalyzing the hydrolysis of α-1,6-linkages. Both enzymes are specific for the α-1,6-linkage and do not catalyze the hydrolysis of α-1,4-linkages. As might be expected, the activity of the α- and β-amylases on amylopectin is increased considerably in the presence of either R-enzyme or isoamylase. With the exception of glucose-1-phosphate, the product of starch phosphorylase activity, the simplest degradative product formed as a consequence of the activity of the above enzymes on starch, is maltose. However, maltose is not a form of sugar that is readily available to the plant. This problem is solved by the almost universal distribution in plants of the enzyme *maltase*. Maltase, which is often found in association with amylases (15), catalyzes the hydrolysis of the glucosidic bond of maltose to yield molecules of glucose.

We have, therefore, a general overall picture of the synthesis and degradation of starch, starting with glucose and ending with glucose. As noted, several enzymes are involved in the metabolism of starch, and the harmonious activity of a number of them together is required to either build up or break down a starch molecule.

Synthesis and Degradation of Cellulose

Synthesis

Unlike the metabolism of starch, our knowledge of the metabolism of cellulose is very limited. Most of the information about cellulose synthesis comes from studies on cellulose-producing bacteria of the genus *Acetobacter*. When radioactive ^{14}C-labeled carbohydrate intermediates such as glucose are fed to cultures of *Acetobacter*, the labeled carbon can eventually be found in cellulose. Carbon sources other than glucose can also be utilized as intermediates in cellulose synthesis (6), thereby suggesting the cooperation of a complex of enzymes in this process. In other words, when carbohydrates other than glucose are fed to *Acetobacter* (e.g., mannitol, glycerol), the enzymes necessary for the conversion of these carbo-

hydrates to glucose have to act before the carbon of these compounds can be incorporated into cellulose.

When *Acetobacter acetigenum* is fed lactic acid labeled in the carboxyl carbon (—^{14}COOH), the label is carried over into cellulose. The symmetrical distribution of the label in the cellulose molecule suggests that the glucose units of cellulose arise by the fusion of two three-carbon compounds (5).

$$
\begin{array}{ccc}
\text{C} & & \text{C} \\
| & & | \\
\text{C} & & \text{C} \\
| & & | \\
\text{C} & & \text{C} \\
+ & \longrightarrow & | \longrightarrow \text{cellulose} \\
\text{C} & & \text{C} \\
| & & | \\
\text{C} & & \text{C} \\
| & & | \\
\text{C} & & \text{C}
\end{array}
$$

Evidence has accumulated suggesting that although glucose does not undergo any prior cleavage, phosphorylation of the glucose molecule might be necessary before incorporation into cellulose is possible (30).

Some very interesting work has been done on the possible participation of UDPG in cellulose synthesis as well as in starch synthesis. Glaser (14) found that cell-free enzyme preparations from *A. xylinum* could synthesize cellulose in the presence of glucose-labeled UDPG. However, substitution of ^{14}C-labeled glucose for UDPG produces negative results. The synthesis of cellulose in a UDPG system is considerably enhanced by the addition of an acceptor molecule (cellodextrins) to the mixture.

UDPG + acceptor $\longrightarrow$
    UDP + β-1,4-glucosyl-acceptor

Of even greater significance, Brummond and Gibbons (7) demonstrated that a cell-free enzyme preparation from *Lupinus albus* (a higher plant) is able to synthesize cellulose from UDPG. At least in some instances, the synthesis of cellulose appears to be analogous to that of starch. More work needs to be done on this aspect of cellulose synthesis, but the UDPG theory does present a promising mechanism for the incorporation of glucose into the cellulose chain (32).

Degradation

Needless to say, an essential feature of the environment of this world is the degradation of cellulose. If this degradation were not possible, we would literally be "covered" with dead plant material, and there would be an appreciable depletion of atmospheric CO_2. However, nature has provided us with a variety of lower forms of life capable of degrading cellulose, among which certain bacteria and fungi are most important.

According to available evidence, the enzymatic hydrolysis of cellulose can be considered a random attack on the β-1,4-linkage. The cellulose molecule is reduced

to cellodextrins and eventually to cellobiose, a disaccharide composed of two glucose units. The enzymes involved in the random hydrolysis of cellulose to cellobiose have not as yet been characterized, but have been grouped under the generic term *cellulase*.

The β-1,4-link of cellobiose can be hydrolyzed by the enzyme cellobiase.

cellulose $\xrightarrow{\text{cellulases}}$ cellodextrins $\longrightarrow$ cellobiose $\xrightarrow{\text{cellobiase}}$ glucose

Synthesis and Degradation of Pectic Substances

The primary pathway for the synthesis of pectic substances is generally thought to be through the mediation of UDPG. This supposition is supported by the observation that both glucose and galactose are good substrates for the synthesis of pectic acid and that UDPG and UDP-galactose are readily interconverted. Figure 11–5 shows a possible pathway by which pectic acid might be synthesized. From this pathway we can see where either glucose or galactose can enter into the synthesis of pectic acid.

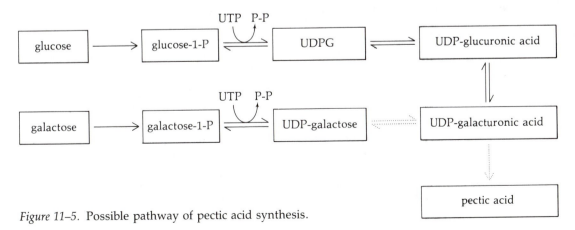

Figure 11–5. Possible pathway of pectic acid synthesis.

All of the reactions indicated have been demonstrated in plants, except for the incorporation of galacturonic acid from UDP-galacturonic acid into the pectic acid chain. However, this last step seems to be a logical assumption, particularly in view of the participation of UDPG in the synthesis of other polysaccharides, such as starch and cellulose. Methyl groups, which are found in pectic substances esterified to the carboxyl group of the galacturonic acid units, are most likely contributed by *methionine* through *S-adenosylmethionine*. The compound S-adenosylmethionine has been demonstrated to be active in the transfer of methyl groups.

Hydrolysis of the α-1,4-linkage of pectic substances is catalyzed by the enzyme *pectin polygalacturonase*. Enzymatic hydrolysis of the methyl ester bonds of pectin is catalyzed by *pectin methyl esterase*.

inulin (~35 fructose units)

Inulin

Before leaving a discussion of the carbohydrates, we should mention the reserve material *inulin*, which is found predominantly in the compositae plants. Particularly good sources are the tubers of dahlia, chicory, and the Jerusalem artichoke. Inulin is thought to be an unbranched polymer of about 35 fructose units joined by β-2,1-linkages. On hydrolysis, however, inulin yields a small amount of glucose. Two glucose units are thought to be in the inulin molecule, one somewhere in the center and the other at the reducing end of the chain to give a sucrose-type linkage. The following molecular structure depicts only repeating units of fructose residues; the two glucose units are left out for the sake of simplicity.

Available evidence suggests that inulin is synthesized by the transfer of the fructose part of the sucrose molecule to an acceptor molecule:

glucose-fructose + glucose-(fructose)$_n$ $\rightleftharpoons$
glucose-(fructose)$_n$-fructose + glucose

Enzymes capable of the hydrolysis of the β-2,1-linkages of inulin have been found in the Jerusalem artichoke (11). These enzymes seem to function in the mobilization of the inulin that is utilized during the sprouting of the artichoke tuber.

Other short chain polysaccharides, composed predominantly of β-2,6-linked fructose units, have been uncovered in the

Graminal family. These polysaccharides are called *levans* and, like inulin, they are terminated at one end by a sucrose residue.

Questions

11–1. What is a good explanation for the term *carbohydrate*?

11–2. Name the three major categories used to describe the carbohydrates. What is the basis of each category? Given an example of each.

11–3. What is the Haworth configuration of a sugar?

11–4. To what major category of carbohydrates do the two sugars α-D-glucose and β-D-glucose belong? In which polysaccharides are they found and how do these polysaccharides differ from one another in chemistry and in function and location in plant cells?

11–5. Name some of the oligosaccharides that are commonly found in plants. Are they translocated?

11–6. Name the chemical that is the basic component of pectic compounds. Is this compound a carbohydrate? Where on the molecule does calcium react?

11–7. Name some of the chemical components of wood.

11–8. Name some of the enzymes involved in starch synthesis. Where are these enzymes located in the plant cell?

11–9. Indicate the characteristics of the reactions that are mediated by α- and β-amylases.

11–10. What are cellulases? Do they result in the formation of glucose directly or are there other enzymes involved?

Suggested Readings

Akazawa, T. 1976. Polysaccharides. In J. Bonner and J.E. Varner, eds., *Plant Biochemistry*, 3rd ed. New York: Academic Press.

Bohinski, R.C. 1979. *Modern Concepts in Biochemistry*, 3rd ed. Boston: Allyn and Bacon.

Dennis, D.T., and J.A. Miernyk. 1982. Compartmentation of nonphotosynthetic carbohydrate metabolism. *Ann. Rev. Plant Physiol.* 33:27–50.

Gander, J.E. 1976. Mono- and oligosaccharides. In J. Bonner and J.E. Varner, eds., *Plant Biochemistry*, 3rd ed. New York: Academic Press.

Lehninger, A.L. 1982. *Principles of Biochemistry.* New York: Worth.

Lüttge, U., and N. Higinbotham. 1979. *Transport in Plants.* New York: Springer-Verlag.

McGilvery, R.W., with G. Goldstein. 1979. *Biochemistry: A Functional Approach.* Philadelphia: Saunders.

Preiss, J. 1982. Regulation of the biosynthesis and degradation of starch. *Ann. Rev. Plant Physiol.* 33:431–454.

Robinson, T. 1967. *The Organic Constituents of Higher Plants*, 2nd ed. Minneapolis, Minn.: Burgess Publishing.

Shannon, J.C. 1978. Physiological factors affecting starch accumulation in corn kernels. *Thirty-Third Ann. Corn and Sorghum Res. Conf.*

White, A., P. Handler, E.L. Smith, R.L. Hill, and I.R. Lehman. 1978. *Principles of Biochemistry*, 6th ed. New York: McGraw-Hill.

Chapter 12

Pigments and Structure of Photosynthetic Apparatus

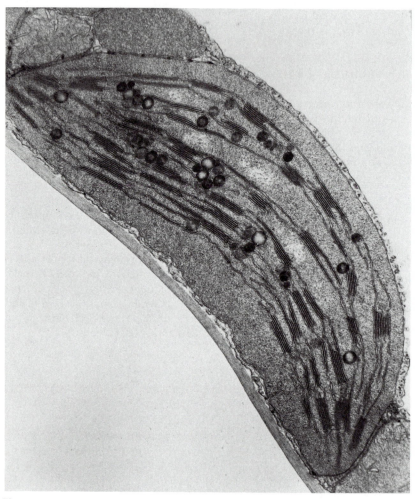

Electron micrograph of chloroplast from alfalfa (*Medicago sativa*) leaf. Magnification 22,800×. *Courtesy of R. Rufner, Massachusetts Agricultural Experiment Station, University of Massachusetts.*

The continued existence of plants as viable reproducing organisms depends on their efficiency in capturing, transforming, translocating, storing, and consuming energy. The source of energy that sustains our biosphere is the sun. Green plants, through their elaborate photosynthetic apparatus, absorb visible light energy and "package" it as bond energy and reducing power in the form of the "high-energy" compound *adenosine triphosphate* (ATP) and the coenzyme reduced *nicotinamide adenine dinucleotide phosphate* (NADPH+H$^+$). These compounds drive the reactions that are involved in the fixation of carbon dioxide into carbohydrates. The carbohydrates, in turn, provide a cellular source of energy and of starting materials for the synthesis of lipid, protein, and other plant products.

Although we most often associate the process of photosynthesis with the production of carbohydrates, the fundamental significance of the process is that the photosynthetic apparatus, equipped with an intricate array of membranes and pigments, traps light energy and effects its transduction (conversion) to chemical energy. We will begin our study of photosynthesis with the pigments involved and their associated photosynthetic membrane structure—the chloroplast.

Pigments Involved in Photosynthesis

It is hard to conceive of life originating or existing without the absorption and conversion of radiant energy to chemical energy. As Glass (23) stated, "life is a photochemical phenomenon." The compounds most important in this conversion of light energy to chemical energy are the pigments that exist within the chloroplasts, or chromatophores,

of plants. Light initiates the process of photosynthesis through these chemicals and organelles.

Chlorophyll Pigments

Chlorophylls, the green pigments of plants, are the most important pigments active in the photosynthetic process. We can distinguish at least nine types: chlorophylls *a*, *b*, *c*, *d*, and *e*; bacteriochlorophylls *a* and *b*; and the chlorobium chlorophylls 650 and 660 (2, 14).

Chlorophylls *a* and *b* are the best known and most abundant and are found in all autotrophic organisms except pigment-containing bacteria. Chlorophyll *b* is also absent from the blue-green, brown, and red algae. Chlorophyll *a* usually appears blue green, chlorophyll *b* is yellow green. Chlorophylls *c*, *d*, and *e* are found only in algae and in combination with chlorophyll *a*. Bacteriochlorophylls *a* and *b* and the chlorobium chlorophylls are pigments found in photosynthetic bacteria.

The chlorophyll *a* molecule has a *cyclic tetrapyrrolic structure* (*porphyrin*), with an isocyclic ring containing a *magnesium atom* at its center. The *phytol chain* of the chlorophyll molecule extends from one of the pyrrole rings. The empirical formula of the chlorophyll molecule is $C_{55}H_{72}O_5N_4Mg$ and its molecular structure is shown in Figure 12–1. The phytol chain, which is esterified with the carboxyl group on the seventh carbon of the porphyrin is a long hydrophobic chain that contains one double bond. The phytol chain is thought to be produced via the same metabolic pathway as are the carotenoids and may be a derivative of vitamin A. The phytol chain appears to extend into the interior of the chloroplast's membranes and to interact with other hydrophobic lipid

Figure 12–1. The chlorophyll *a* and *b* molecule.

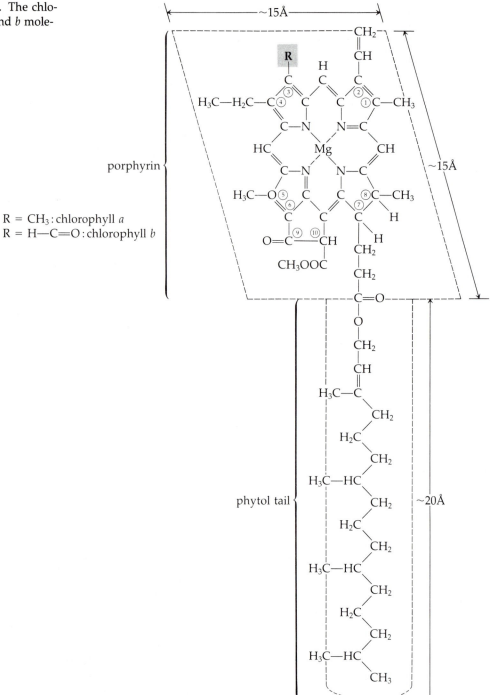

R = CH₃ : chlorophyll *a*
R = H—C=O : chlorophyll *b*

molecules. The difference between chlorophyll *a* and chlorophyll *b* is found at the third carbon. Chlorophyll *a* has a methyl group (CH_3) attached to the third carbon, and chlorophyll *b* has an aldehyde (HC=O) attached to the third carbon.

In addition to their minor differences in molecular structure, chlorophylls *a* and *b* exhibit different absorption spectra. An *absorption spectrum* is a measure of the extent to which a given substance absorbs the light of different colors or wavelengths. It is a function of the relationship between absorption—which may be expressed as absorbance (A), optical density (OD), absorption, specific absorption, and extinction—and wavelength expressed in nanometers (nm). Since an absorption spectrum depends on the unique electronic structure and hence light absorption characteristics of a compound, scientists use it to identify specific substances with a great deal of accuracy.

Figure 12–2 illustrates the absorption spectra for chlorophyll *a* and *b*, as determined with a *spectrophotometer*. Both chlorophyll *a* and *b* show absorption maxima in the blue-violet region and orange-red region of the visible spectrum. Notice the minimum absorption of the green and yellow wavelengths (500 to 600 nm). The absorption spectrum of chlorophyll provides indirect evidence of the wavelengths of light that are absorbed for the process of photosynthesis. The absorption maxima of these most important photosynthetic pigments should give a hint as to the quality of light most effective in the process of photosynthesis.

The absorption spectra we have just discussed are of extracts of chlorophyll—that is, chlorophyll in an organic solvent. The absorption spectra of chlorophyll in vivo may be different. Indeed, the absorp-

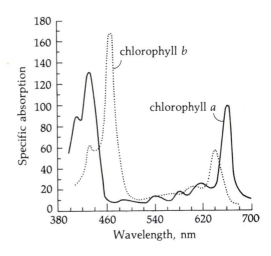

Figure 12–2. Absorption spectra of ether extracts of chlorophylls *a* and *b*.

Reprinted from Botanical Gazette 102:463 by F. Zscheile and C. Comar by permission of The University of Chicago Press. Copyright 1941 The University of Chicago Press.

tion spectra of the chlorophylls differ slightly when in different solvents. Often the wavelength positions of the peaks may vary by a few nanometers with chlorophyll extracted from different species.

Chlorophyll Synthesis

How the chlorophylls and iron porphyrins are synthesized within the living cell has not, as yet, been conclusively shown. However, studies by many investigators on the metabolism of heme and chlorophyll and on the biosynthesis of porphyrins have uncovered most of the steps involved in the syntheses of these most important compounds. There is general agreement that succinyl CoA, a Krebs cycle intermediate, and the amino acid glycine initiate the biosynthetic pathway that leads to chlorophyll formation. The condensation of these two compounds leads to the formation of the unsta-

$$\begin{array}{c}
\text{COOH} \\
| \\
\text{CH}_2 \\
| \\
\text{NH}_2
\end{array}
\;+\;
\begin{array}{c}
\text{COOH} \\
| \\
\text{CH}_2 \\
| \\
\text{CH}_2 \\
| \\
\text{O}{=}\text{C}{-}\text{S}{-}\text{CoA}
\end{array}
\;\longrightarrow\;
\text{HOOC}{-}\overset{\overset{\displaystyle \text{NH}_2}{|}}{\text{CH}}{-}\underset{\underset{\displaystyle \text{O}}{\|}}{\text{C}}{-}\text{CH}_2{-}\text{CH}_2{-}\text{COOH}$$

glycine   succinyl CoA   α-amino-β-ketoadipic acid

pyridoxal phosphate

CO_2 ↙   δ-aminolevulinic acid synthetase

$$\text{H}_2\text{C}{-}\overset{\overset{\displaystyle \text{NH}_2}{|}}{\underset{\underset{\displaystyle \text{O}}{\|}}{\text{C}}}{-}\text{CH}_2{-}\text{CH}_2{-}\text{COOH}$$

δ-aminolevulinic acid

ble α-amino-β-ketoadipic acid, which, on decarboxylation, yields δ-aminolevulinic acid. This reaction requires the presence of the cofactor pyridoxal phosphate and is catalyzed by the enzyme δ-aminolevulinic acid synthetase (21, 31). There is strong evidence that the synthesis of δ-aminolevulinic acid is mediated by light.

In the presence of the enzyme δ-aminolevulinic acid dehydrase (as yet not isolated from plant material), two molecules of δ-aminolevulinic acid condense to form porphobilinogen, a monopyrrole. Two molecules of water are lost in the reaction. Scientists have demonstrated this condensation reaction in extracts of chlorella and spinach (25) and in etiolated barley (26) or bean leaves (33).

The enzymes uroporphyrinogen synthetase and uroporphyrinogen III cosynthetase catalyze the formation of uroporphyrinogen III from four molecules of porphobilinogen. Observers have detected both enzymes and uroporphyrinogen III in a variety of plant material (9).

Uroporphyrinogen decarboxylase catalyzes the decarboxylation of the four acetic acid substituents of uroporphyrinogen III to yield coproporphyrinogen III. Under aerobic conditions and in the presence of coproporphyrinogen oxidative decarboxylase, protoporphyrinogen IX is formed from coproporphyrinogen III. Oxidation of protoporphyrinogen IX results in the formation of protoporphyrin IX, which then incorporates magnesium to form Mg-protoporphyrin IX. The enzyme Mg-protoporphyrin methyl esterase catalyzes the addition of a methyl group to Mg-protoporphyrin IX to form Mg-protoporphyrin IX monomethyl ester. The methyl group donor in this reaction is thought to be S-adenosyl methionine.

Next in the reaction sequence leading to the biosynthesis of chlorophyll is the conversion of Mg-protoporphyrin IX monomethyl ester to protochlorophyllide. No chlorophyll is found in angiosperm seedlings that have been allowed to germinate and develop in the dark (etiolated). The

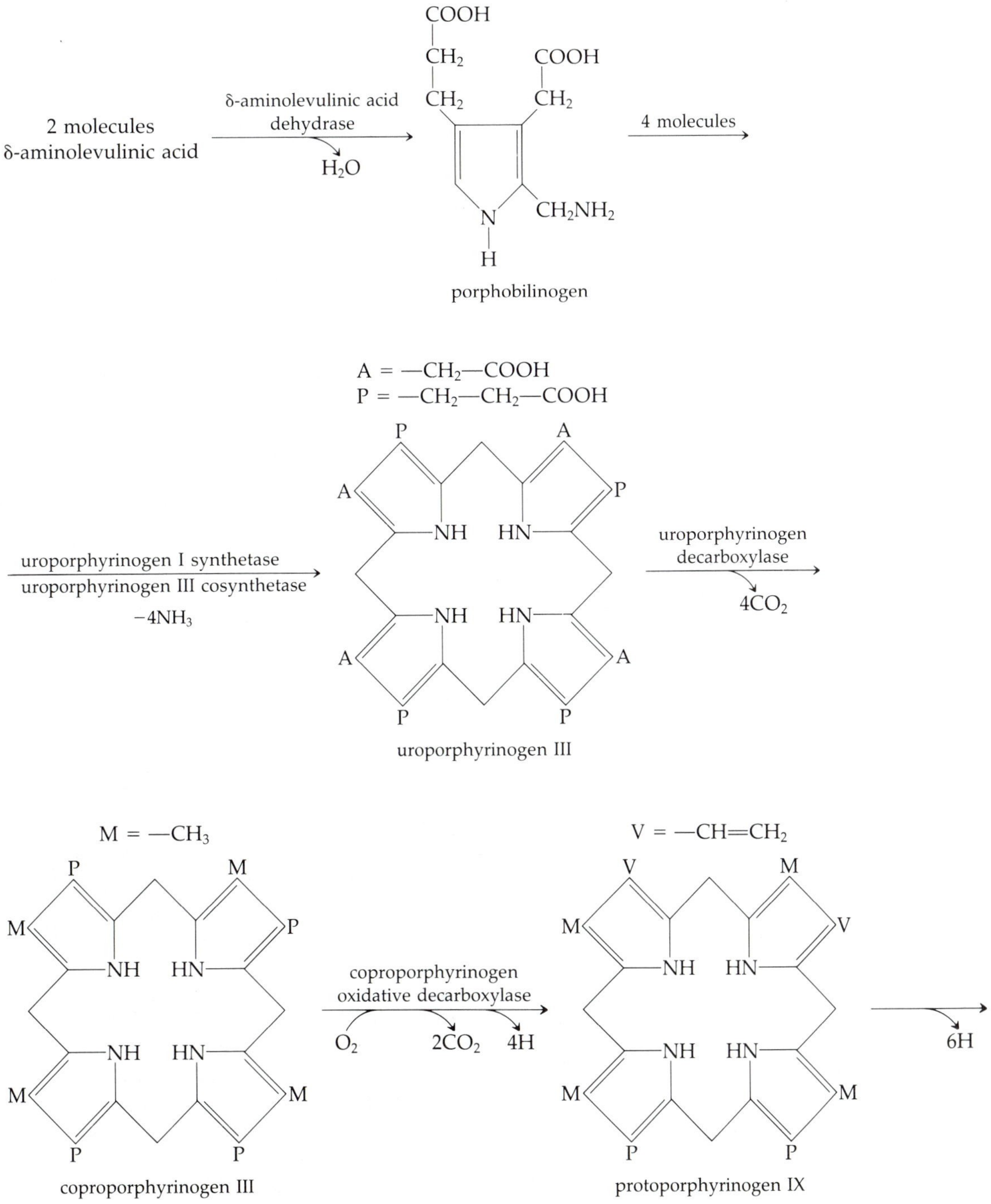

COOH
CH₂
CH₂

COOH
CH₂

2 molecules
δ-aminolevulinic acid

δ-aminolevulinic acid
dehydrase

H_2O

CH₂NH₂

N
H

4 molecules

porphobilinogen

A = —CH₂—COOH
P = —CH₂—CH₂—COOH

uroporphyrinogen I synthetase
uroporphyrinogen III cosynthetase

$-4NH_3$

uroporphyrinogen III

uroporphyrinogen
decarboxylase

$4CO_2$

M = —CH₃

coproporphyrinogen III

coproporphyrinogen
oxidative decarboxylase

O_2 $2CO_2$ 4H

V = —CH=CH₂

protoporphyrinogen IX

6H

protoporphyrin IX

$$\xrightarrow{\text{Mg}} \text{Mg-protoporphyrin IX} \xrightarrow[\substack{\text{Mg-protoporphyrin} \\ \text{methyl esterase}}]{\substack{\text{S-adenosyl} \\ \text{methionine}}}$$

Mg-protoporphyrin IX
monomethyl ester

prominent color exhibited by the seedlings under these conditions is yellow due to the presence of carotenoids. However, detectable amounts of protochlorophyllide and protochlorophyll are also present and, together with the carotenoids, give the etiolated seedling a yellow-green color.

Protochlorophyll is formed by the addition of a phytol group to protochlorophyllide. The immediate precursor of chlorophyll *a* was once thought to be protochlorophyll. However, convincing evidence now suggests that the immediate precursor of chlorophyll *a* is *chlorophyllide a*. When we subject the etiolated angiosperm seedling to light, protochlorophyllide is reduced to form chlorophyllide *a* (1, 19, 20, 42, 57). Note that in this photoreduction, hydrogen atoms are added to carbon atoms 7 and 8. The final step in chlorophyll *a* synthesis is activated by the enzyme chlorophyllase, which catalyzes the esterification of a phytol group to chlorophyllide *a* to form chlorophyll *a*. Although it is not conclusively proven, most investigators believe that chlorophyll *b* is formed from chlorophyll *a* (7, 9, 53).

The light requirement for the conversion of protochlorophyllide to chlorophyllide *a* appears to be absolute in angiosperms. In gymnosperms, some ferns, and

many algae, however, chlorophyll can be synthesized completely in the dark solely through enzymatic activity. Work by Sudyina (56) on several gymnosperms suggests that the biosynthetic pathway for chlorophyll formation is the same in the dark as in the light. However, we must remember that some workers have suggested that the formation of δ-aminolevulinic acid is *light mediated*. Perhaps a difference between dark chlorophyll synthesis and light-mediated chlorophyll synthesis can be found at this early part of the biosynthetic pathway (14).

There is a close relationship between the chlorophylls and the metal-porphyrins of the living cell, the blood pigment heme,

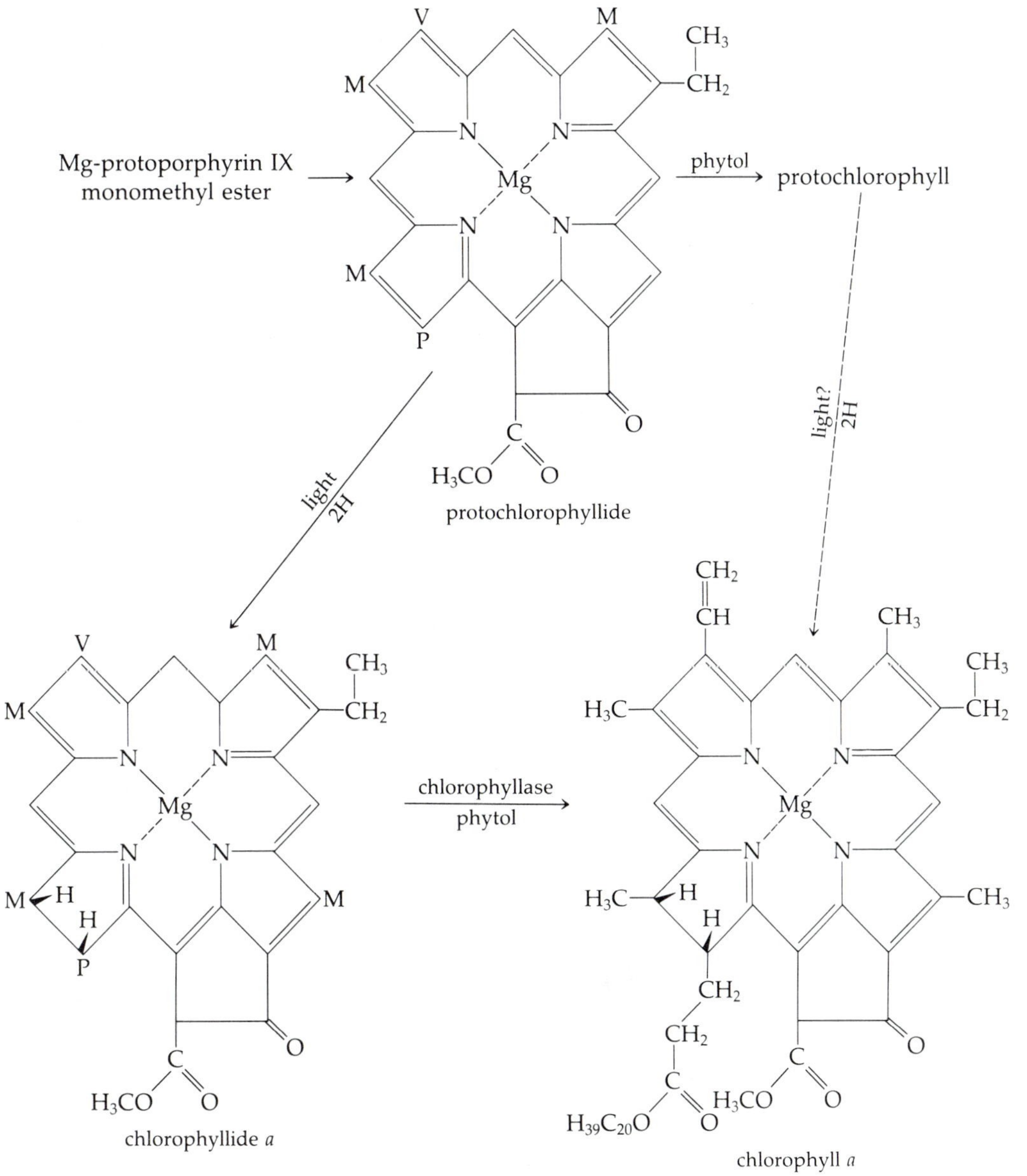

Mg-protoporphyrin IX monomethyl ester $\longrightarrow$

$\xrightarrow{\text{phytol}}$ protochlorophyll

protochlorophyllide

chlorophyllide *a*

$\xrightarrow[\text{phytol}]{\text{chlorophyllase}}$

chlorophyll *a*

and the cytochromes. The major difference between the chlorophylls and these other compounds is that chlorophyll has magnesium and a phytol tail, and the cytochromes and hemes have iron and no phytol tail. However, all of these compounds appear to originate along the same biochemical pathways.

Carotenoid Pigments

Carotenoids are lipid compounds that are distributed widely in both animals and plants and range in color from yellow to purple. Carotenoids are present in variable concentrations in nearly all higher plants and in many microorganisms, including red and green algae, photosynthetic bacteria, and fungi (24). *Carotene*, the first carotenoid to be named, was isolated from carrot root tissue by Wackenroder in 1831. However, it was not until after 1925 that several investigators—notably, Karrer, Jucker, Lederer, Kuhn, and Zechmeister—definitely established the structures of some of the carotenoids.

We can consider the natural carotenoids as derivatives of *lycopene*, a red pigment found in tomatoes and in many other plants. Lycopene is a highly unsaturated, straight-chain hydrocarbon composed of two identical units joined by a double bond between carbon atoms 15 and 15′. The empirical formula is $C_{40}H_{51}$. Each half of the molecule is probably derived from four isoprene units; isoprene has the formula $CH_2 = C(CH_3) — CH = CH_2$. Thus carotenoids are composed of eight isoprene-like residues. Molecular structures of three carotenoids are shown below.

The major carotenoid found in plant tissues is the orange-yellow pigment β-carotene, which is generally accompanied by varying amounts (0 to 35 percent) of α-carotene (41). The chemical difference between α-carotene and β-carotene is that β-carotene consists of the two β-ionone rings and α-

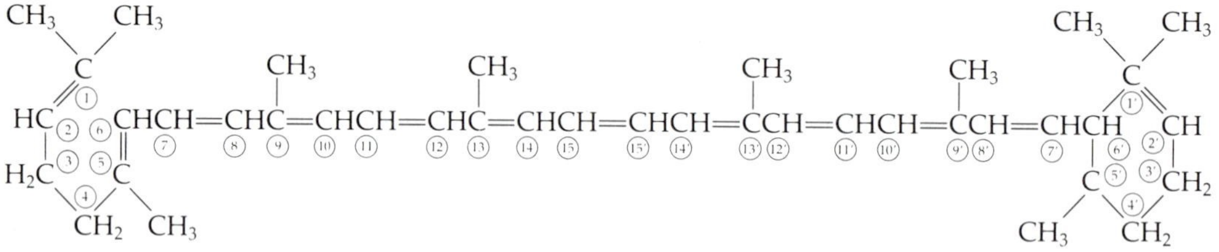

lycopene

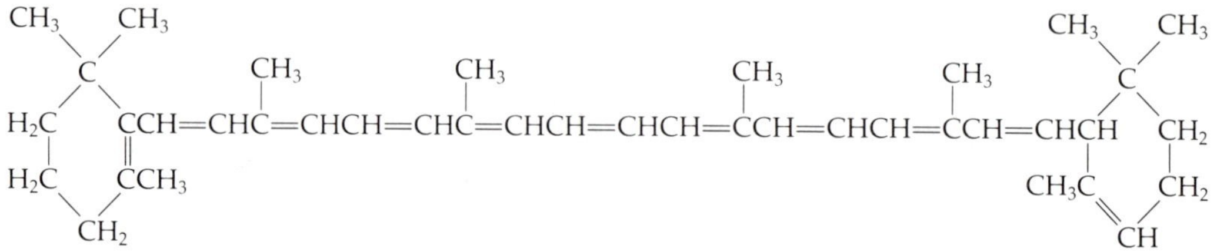

α-carotene

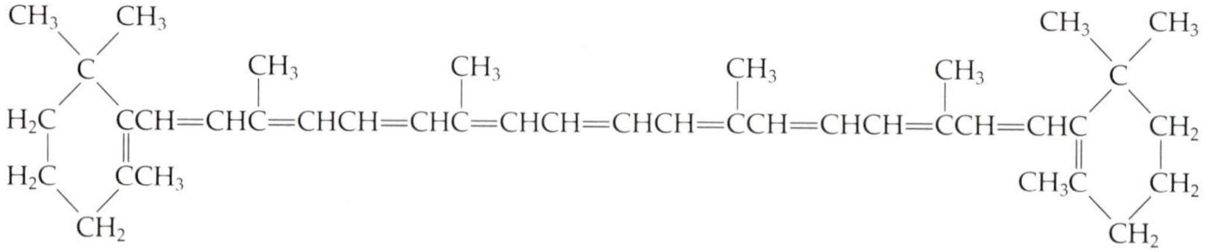

β-carotene

R = unsaturated side chain (see α-carotene for example)

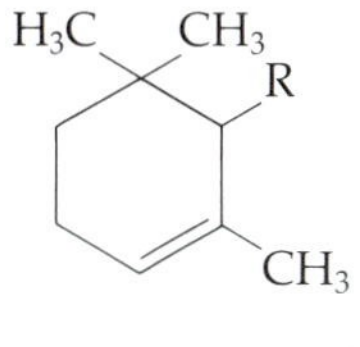

α-ionone ring

R = side chain (see β-carotene for example)

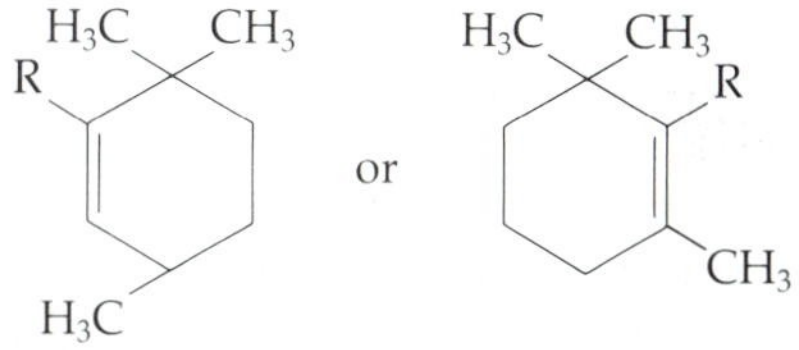

β-ionone ring

carotene consist of one α- and one β-ionone ring. The basic structural configuration of each ring is shown above.

Hydrogen carotenoids (i.e., carotenoids that consist exclusively of carbon and hydrogen) are termed carotenes, and carotenoids that contain oxygen are called *xanthophylls*. Generally, the common names used to describe the different carotene species end with -ene, and those used to describe the different xanthophyll species end with -in. The xanthophylls are more abundant in nature than are the carotenes, and in growing leaves the concentration of xanthophylls may exceed that of carotenes by about 2:1 (24). Table 12–1 shows the major xanthophylls found in green leaves.

Carotenoids, like chlorophyll, are located in the chloroplast and in the chromatophore (13, 60, 61) as water-insoluble protein complexes. The specific orientation of the carotenoids in relation to the chlorophylls within the lamellar system of the chloroplast must be an important aspect of the photosynthetic process.

Probable Role of Carotenoids in Plants

As we might expect, most studies of the physiological role of carotenoids have centered around their relationship with vitamin A and animal nutrition. In recent years scientists have directed considerable attention to the possible role of carotenoids in plants. At least two probable roles of carotenoids are evident: (1) they protect against the photooxidation of chlorophyll and (2) they absorb and transfer light energy to chlorophyll *a*.

Protection against Photooxidation of Chlorophyll

The blue-green mutant of *Rhodopseudomonas spheroides* is practically devoid of carotenoids and appears to be vulnerable to chlorophyll-catalyzed photooxidation in the presence of oxygen. However, *R. spheroides* will grow and photosynthesize under an-

Pigment	Structure	Relative Amount (% of total)	
cryptoxanthin	3-hydroxy-β-carotene	4	Table 12–1. Major xanthophylls found in green leaves.
lutein	3,3-dihydroxy-α-carotene	40	Source: *Data from Goodwin (1960). Reprinted with permission.*
zeaxanthin	3,3-dihydroxy-β-carotene	2	
violaxanthin	5,6,5',6'-diepoxyzeaxanthin	34	
neoxanthin	$C_{40}H_{56}O_4$ (exact structure unknown)	19	

aerobic conditions (50). A pale green mutant of *Chlamydomonas* has been found to be almost completely lacking in carotenoid pigments. As we might expect, this mutant must be grown exclusively in darkness. It dies when grown in light (50). In fact, a few of the so-called chlorophyll mutants (lacking chlorophyll) are actually carotenoid mutants (lacking carotenoids); this phenomenon suggests that carotenoids protect against chlorophyll breakdown.

Protection by carotenoids against the photooxidation of chlorophyll apparently also occurs in higher plants. For example, when we expose the corn mutant white seedling-3, a plant devoid of carotenoids (36), to light under aerobic conditions, it synthesizes chlorophyll. However, if we prolong illumination, the chlorophyll is destroyed, thereby suggesting that the mutant is capable of forming chlorophyll but that the chlorophyll breaks down (34). That this chlorophyll breakdown is possibly due to photooxidation is shown by the fact that no chlorophyll loss occurs when we illuminate the mutant in a nitrogen atmosphere. The protective action of carotenoids against chlorophyll photooxidation has also been demonstrated with sunflower mutants.

Many investigators think that carotenoids protect against chlorophyll photooxidation by acting as preferred substrates in photosensitized oxidations. This suggestion was made by Calvin (12) in 1955 and later by Sistrom, Griffiths, and Stanier (54) in 1956. They suggested that carotenoids are oxidized through the light-catalyzed formation of epoxides across double bonds and are then reduced by a dark, enzyme-catalyzed reaction. Lundegårh (39) provided evidence suggesting the possibility of a light-catalyzed conversion of carotenes to xanthophylls in isolated spinach chloroplasts.

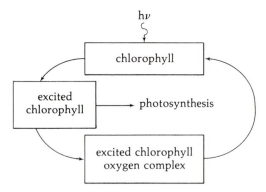

a. Deactivation of excited chlorophyll-oxygen complexes

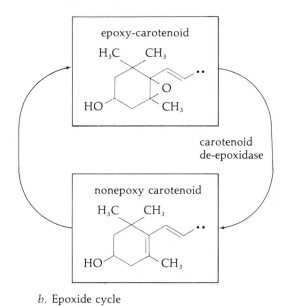

b. Epoxide cycle

Figure 12–3. Epoxide cycle and deactivation of excited chlorophyll-oxygen complexes.

Bamji and Krinsky (3) demonstrated in *Euglena gracilis* a dark reductive deepoxidation of an epoxy-carotenoid to form a nonepoxy-carotenoid. The reaction is catalyzed by the enzyme *carotenoid deepoxidase*. The reverse reaction, the re-formation of an epoxy-carotenoid, was also found to occur in *E. gracilis* under the combined influence of light and molecular oxygen. Consideration of these results led Krinsky (35) to propose that an epoxide cycle exists and that the function of this cycle is to protect against the photooxidation of chlorophyll (see Figure 12–3). Chlorophyll that is activated by the absorption of light is generally returned to its original state as a result of its participation in photosynthesis. However, activated chlorophyll can and sometimes does combine with molecular oxygen, a complex that can lead to its photooxidation. The chlorophyll-oxygen complex can be inactivated by a nonepoxy-carotenoid, which is itself oxidized to form its epoxide derivative. The protective nonepoxy-carotenoid could then be regenerated from its epoxy derivative through a dark reaction catalyzed by carotenoid deepoxidase.

Transfer of Energy to Chlorophyll

Because of the presence of carotenoids in all photosynthetic tissue, we might anticipate their role in photosynthesis. However, this role must be secondary since tissues rich in carotenoids and devoid of chlorophyll do not photosynthesize. Light energy absorbed by carotenoids appears to be transferred to chlorophyll *a* (or bacteriochlorophyll *a*) and utilized there in the photosynthetic process. Observers have obtained strong evidence of this supposition in tests that demonstrate that absorption of light by carotenoids results in the fluorescence of chlorophyll. Figure 12–4 illustrates the absorption spectrum of β-carotene.

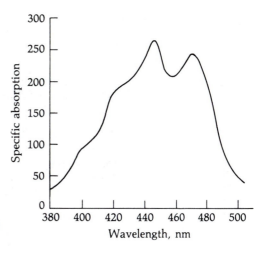

Figure 12–4. Absorption spectrum of β-carotene in hexane.

From F. Zscheile et al. 1942. Plant Physiol. *17:331.*

Phycobilins

The red and blue *biliproteins*, called *phycoerythrins* and *phycocyanins*, respectively, are found in algae and in photosynthetic bacteria. The chromophore moiety of the biliprotein, termed a *phycobilin*, is strongly attached to the associated protein and makes the study of phycobilins in the pure state very difficult. Consequently, most of our information on these pigments comes from studies of the pigment-protein complex.

The absorption spectra of the phycobilins are most interesting when we consider that the phycobilins are active in the transfer of light energy to chlorophyll for utilization in the process of photosynthesis. From the absorption spectra given in Figures 12–5 and 12–6, we can see that the phycocyanins and phycoerythrins effectively absorb light over a range of wavelengths that cannot be absorbed by chlorophyll. This is one of the

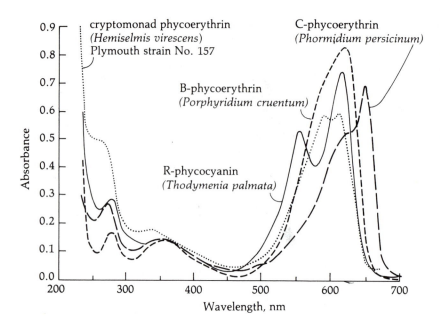

Figure 12–5. Absorption spectra of aqueous solutions of phycocyanins (pH 6 to 7).

From R.A. Lewin, ed. 1962. Physiology and Biochemistry of Algae. New York: Academic Press. Reprinted by permission.

reasons we consider carotenoids and phycobilins to be active in absorbing light energy that is utilized in photosynthesis. Hence we refer to the carotenoids and phycobilins as *accessory pigments* primarily because their role is indirect—that is, the energy they absorb is transferred to chlorophyll before it is active in photosynthesis. From the standpoint of energy transfer, these accessory pigments exhibit fluorescence peaks that

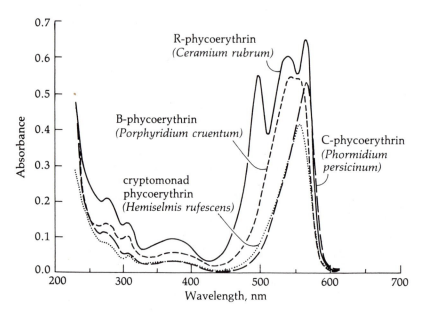

Figure 12–6. Absorption spectra of aqueous solutions of phycoerythrins (pH 6 to 7).

From R.A. Lewin, ed. 1962. Physiology and Biochemistry of Algae. New York: Academic Press. Reprinted by permission.

overlap. This overlapping is important to the energy transfer system of the photosynthetic apparatus.

We can obtain experimental evidence of the participation of other pigments as well as chlorophyll in light capture by comparing the absorption spectrum of a plant tissue, cell, or pigment with the action spectrum of photosynthesis. An *action spectrum* is the measure of the efficiency of a process (believed to be mediated by light) induced by light of different wavelengths but of the same intensity. In other words, it is actually a plot of the extent of a response (such as photosynthesis) against different wavelengths of light. A comparison of an action spectrum with the absorption spectrum of a pigment indicates whether or not the pigment is involved in the response. For example, the absorption spectrum of chlorophyll is very similar to the action spectrum of photosynthesis in most plant tissues. The absorption spectrum as compared with an action spectrum of a cell or tissue also provides information as to whether or not other pigments besides chlorophyll are involved in a light-stimulated process. Figure 12–7 provides a comparison of such an absorption and action spectrum.

There is some question as to the actual location of the phycobilins. Red algae contain chloroplasts, and cells of the blue-green algae and the algal division Cryptophyta contain a free lamellar structure (i.e., unconfined by a chloroplast membrane). Giraud (22) suggested that the phycobilins are located in the matrix of the chloroplast. Electron microscope analyses by Bogorad and others (8, 11) of the lamellar structure of several mutants of *Cyanidium caldarium*—some of which lacked phycobilins—appear to support Giraud's observations. These workers found that the lamellae of the wild type were no thicker than the lamellae of those mutants lacking phycobilins, a finding that suggests that these pigments are not involved in the lamellar structure.

Work by Gantt and Conti (18) with the red alga *Porphyridium cruentum* strongly suggests that the phycobilins are not found free in the matrix but are attached to the chloroplast lamellae. In this organism, phycoerythrin is confined to small granules attached to the lamellae in a very regular arrangement. Although the granules could be mistaken for ribosomes, they are larger than ribosomes, do not exhibit the characteristic arrangement of membrane-bound ribo-

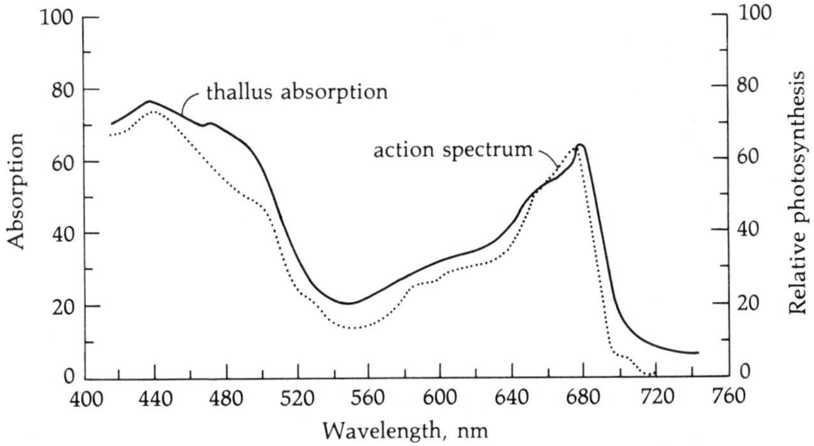

Figure 12–7. Action and absorption spectra for thallus of green algae *Ulva taeniata*. Photosynthesis is active from 480 to 500 nm, suggesting some energy transfer from carotenoids to chlorophyll.

Reproduced from F. Haxo and L. Blinks. The Journal of General Physiology, 1950, 33:389 by copyright permission of The Rockefeller University Press.

somes, and are resistant to ribonuclease extraction. Phycoerythrin granules have also been observed in other red algae.

Gantt and Conti (16, 17, 18) found that in algae in which phycocyanin is the predominant phycobilin, the granules are replaced by "loosely segmented rows, or cords, consisting of thin disks." They observed this type of arrangement, for example, in *P. aerugineum*, an alga having phycocyanin as its predominant phycobilin. These granules, or disks, have been well studied and are referred to as *phycobilisomes*.

Chloroplasts

The process of photosynthesis takes place from start to completion in the chloroplast, a cytoplasmic organelle of surprisingly complex architecture. Chloroplasts are found in green algae, bryophytes, and vascular plants. They are absent in the fungi, blue-green algae, and photosynthetic bacteria. In the latter, however, the pigments are contained in membranous structures termed *chromatophores*. In the blue-green algae, the

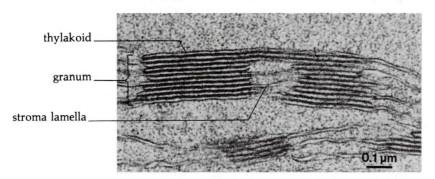

a.

b.

Figure 12–8. The chloroplast. (a) Electron micrograph of two grana and associated stroma thylakoids from mesophyll chloroplast of alfalfa leaf. Magnification 67,400×. (b) Three-dimensional structure of grana and plastid lemellae. Note waferlike structure of grana thylakoids.

Courtesy of R. Rufner, Massachusetts Agricultural Experiment Station, The University of Massachusetts.

pigments are found on or within membranes.

Chloroplast Structure

Chloroplasts are easily observable with the light microscope as organelles of varying sizes (4 to 6 μm in diameter) and shape (round to oval). However, their fine structure can only be distinguished by the electron microscope (see Figure 12–8).

To review briefly, the contents of chloroplasts are bounded by a double-membrane system, or envelope. These membranes are continuous with no perforations or regularly attached particles (50). Scientists have obtained evidence that the envelope, as is the case with most membranes, is differentially permeable. For example, Mudrack (43) observed that chloroplasts in *Agapanthus umbellatus* were plasmolyzed and deplasmolyzed in the same sense as the plasmalemma of the cell when exposed to solutions of different osmotic potentials.

In higher plant chloroplasts, the grana lamellae are highly ordered and contain most of the biochemical machinery for the photochemical reactions of photosynthesis. In cross section the grana lamellae are shown to be paired or form saclike structures, which are termed thylakoids. Some of the thylakoids may be connected to other grana by membranes in the stroma. These connecting links between grana are the *stroma lamellae* (see Chapter 1).

The chloroplasts of algae usually lack grana, and in the blue-green algae the lamellae are found in the cytoplasm with no confining envelope. In such cases, the pigments of the chloroplast are found evenly distributed on or within the lamellae. The phycobilins are confined to phycobilisomes attached to the lamellae (see Figure 12–9). In higher plants, however, the chloroplast pigments are confined to the membranes. Granules, lipid droplets, starch grains, and vesicles, in addition to the lamellae system, may be found in the stroma. Eye spots and pyrenoid bodies, often found in algae cells, are also found in the matrix.

In the past, the basic photosynthetic unit was thought to be the quantosome, a spherical particle embedded in the phospholipoprotein of thylakoid lamellae. We mention it here only because some modern textbooks still refer to it as an important phososynthetic structure. This entity, however, is not a functioning photosynthetic unit. It does not seem to be large enough to contain the necessary components for complete photosynthesis nor is it an autonomous photosynthetic unit in the organism (in vivo) or outside of living cells (in vitro).

Chloroplast Origin and Formation

The current evidence strongly indicates that plastids do not arise *de novo* (newly assembled) but arise from preexisting plastids, termed *proplastids*. Chromoplasts and leucoplasts arise from the division of proplastids transmitted maternally, from one generation to the next, in the cytoplasm of egg cells. During the growth of immature plant tissues (leaves, roots), the production of plastids seems to keep pace through division with those of the dividing cell.

We can deduce that existing plastids replicate by noting that plastids that do not produce chlorophyll give rise to plastids also devoid of chlorophyll. Furthermore, scientists have observed the division of mature chloroplasts in higher plants (47). Plastids also contain circular DNA and RNA dissimilar to that of the cell's genome; this

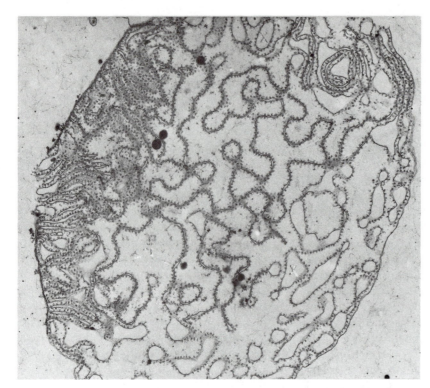

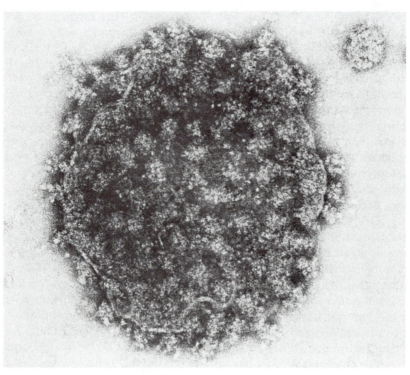

Figure 12–9. Section of chloroplast of *Porphyridium cruentum* (top). Phycobilisomes are attached to "outer" side of each lamella and are absent from chloroplast envelope. Magnification 15,700×. Granules from *P. cruentum* chloroplast fragment (bottom), greatly enlarged, showing small but distinct subunits. Magnification 164,000×. *From E. Gantt and S.F. Conti. 1967. Brookhaven Symp. Biol. 19:393. Photo courtesy of E. Gantt, Smithsonian Radiation Biology Laboratory.*

fact suggests that plastids have some but not complete informational autonomy and control.

The similarities between the two cytoplasmic particles—chloroplasts and mitochondria—are surprising. Both are basically lipoprotein complexes, both contain respiratory enzymes and pigments, both produce ATP, both increase in the cell, both are enclosed by a double membrane, both contain an internal lamellar system, and specific RNA and DNA are found in both. These similarities suggest a close relationship between these organelles and provide the basis for speculation by some scientists that mitochondria and chloroplasts evolved in primitive cells from endosymbionts (self-replicating organisms).

Von Wettstein (58), with the aid of the electron microscope, was probably the first to follow the development of the chloroplast from the early proplastid stage to the fully mature chloroplast. The stages, as illustrated in Figure 12–10, are as follows: (1) early proplastids give rise to vesicles, which appear to arise from a bleb derived from the inner membrane of the proplastid, (2) the vesicles attach to each other and arrange themselves in layers, (3) further fusion and growth in surface area of the newly formed lamellar disks takes place, (4) characteristics of the lamellae are easily distinguished at this point, (5) multiplication of the lamellae to form a more or less continuous lamellar system takes place, and (6) differentiation of grana occurs.

Not all chloroplasts arise directly from proplastids. For example, many of the plastids present in leaf mesophyll cells arise from the division of mature plastids as the leaf expands. Further, the production of mesophyll cell chloroplasts is regulated by light, with the red and blue wavelengths being the most effective.

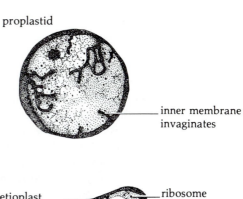

proplastid

inner membrane invaginates

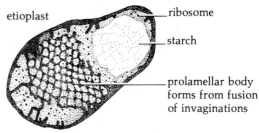

etioplast

ribosome

starch

prolamellar body forms from fusion of invaginations

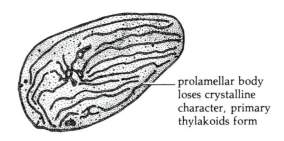

prolamellar body loses crystalline character, primary thylakoids form

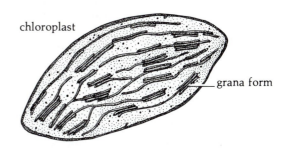

chloroplast

grana form

Figure 12–10. Chloroplast development.

Lamellar System and Chlorophyll Formation

When angiosperms are grown in darkness, the plants become very elongated and pale yellow; the color is due to the presence of carotenoids and the striking absence of chlorophyll. These *etiolated* (pale and weak) plants contain proplastids termed *etioplasts*, in which the carotenoids are concentrated in the envelope membranes and in the usually sparse stroma lamellae. In addition, the internal lamellae are usually arranged in a meshwork of canals called the *prolamellar body* (also found in proplastids of light-grown plants). The etioplast does not contain grana since light and subsequent chlorophyll synthesis are required for lamellae formation. Upon exposure to light, however, the prolamellar body disappears, chlorophyll synthesis is initiated, and grana formation takes place. We have known for sometime that the presence of chlorophyll is necessary for lamellar formation and that the presence or absence of carotenoids, unlike chlorophyll, has no noticeable effect on the structural organization of the chloroplast (50).

By obtaining absorption spectra of the leaf tissue before and after exposure to light, we may spectrophotometrically follow the initial changes that take place when we illuminate etiolated plants (10). As Figure 12–11 illustrates, protochlorophyllide present in the etiolated plant is photoreduced to chlorophyllide, the immediate precursor of chlorophyll. Scientists think that the photoreduction of protochlorophyllide is closely associated with certain early morphological changes in lamellar formation and arrangement (10, 59).

The investigations of Gassman and Bogorad (19, 20) showed that light not only causes the photoreduction of protochlo-

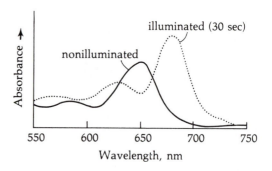

Figure 12–11. Absorption spectrum of etiolated maize leaf before and after 30 seconds of illumination. Absorption maximum (approximately 650 nm) before illumination is due to protochlorophyllide-protein complex (holochrome). On illumination, protochlorophyllide is reduced to chlorophyllide.

From L. Bogorad. 1967. *Chloroplast structure and development.* In A. San Pietro, F.A. Greer, and T.J. Army, eds., Harvesting the Sun—Photosynthesis in Plant Life. *New York: Academic Press. Reprinted by permission.*

rophyllide but that it also enhances the synthesis of the chlorophyll precursor, δ-aminolevulinic acid (ALA). In this regard we may make the following pertinent observations: (1) Etiolated barley leaves supplied with ALA produce up to ten times more protochlorophyllide than do control leaves (26, 27); this fact suggests that there is a paucity of ALA in the etiolated leaf. (2) Very little protochlorophyllide accumulates in etiolated bean leaves that are treated with chloramphenicol or puromycin (inhibitors of protein synthesis) prior to illumination. (3) Administration of ALA to chloramphenicol- or puromycin-treated leaves partially overcomes the effect of these inhibitors. These observations indicate that protein synthesis (possibly specific enzymes for ALA synthesis) requires illumination.

Chloroplast DNA and RNA

Within recent years it has become increasingly apparent that chloroplasts possess the genetic information for coding proteins such as ribulose-1,5-bisphosphate carboxylase. DNA and RNA found in the matrix of the chloroplast (28, 51, 62) exhibit base sequences different from those of the nucleus and appear to code for some but not all of the chloroplast proteins. Circular DNA isolated from chloroplasts appears to replicate within the plastid, especially at times preceding chloroplast division. Not only has RNA been extracted from chloroplasts (40) but the presence of ribosomes has been shown by electron microscope studies (30). See Figure 12–12 for an electron micrograph of an alfalfa leaf chloroplast with clearly defined chloroplast ribosomes.

Since the chloroplast contains specific DNA and RNA and is self-replicating, we might be inclined to view the chloroplast as a partially independent organelle. Chloroplasts have the machinery for some protein synthesis. Observations relating to the synthesis of proteins in plastids include RNA polymerase synthesis (10), RNA synthesis, and the incorporation of amino acids into proteins (45). Indeed, when we illuminate etiolated bean plants, the protein content of the leaf chloroplasts doubles in 48 hours (25). However, studies by Kirk (32) show that many chloroplast proteins are coded by nuclear DNA. Thus the chloroplast organelle is not entirely independent of the cell. Isolated chloroplasts do not live in culture because they require many proteins from the cytoplasm.

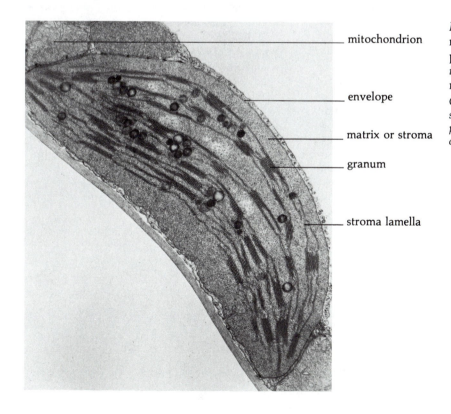

mitochondrion

envelope

matrix or stroma

granum

stroma lamella

Figure 12–12. Electron micrograph of chloroplast from alfalfa (*Medicago sativa*) leaf. Magnification 17,400×.

Courtesy of R. Rufner, Massachusetts Agricultural Experiment Station, University of Massachusetts.

Questions

12–1. What are the basic chemical components comprising the chlorophyll molecule? Where does the synthesis of chlorophyll take place in a plant cell?

12–2. Diagram and name the basic components of lycopene. To what group of pigments does lycopene belong? Is lycopene located in the vacuoles or plastids of cells comprising the tomato berry?

12–3. What is the significance of the α-and β-ionone rings found in some of the carotenoid pigments?

12–4. List two possible roles of carotenoids in plants.

12–5. What is considered to be the major function of phycobilins?

12–6. What is an action spectrum and an absorption spectrum? How are they employed in plant physiology?

12–7. How is the membranous organization of a chloroplast related to its function?

12–8. Describe the meanings of the terms *etioplast* and *prolamellar body.*

12–9. List some of the changes that take place when etiolated plants are illuminated. Is there any significance to the functional economy of plants through the etiolation process? Explain.

12–10. List the observations or facts that suggest that chloroplastids evolved from autonomous organisms. Is it germane to the problems of plant physiology that scientists be concerned with the origin of chloroplasts? Why or why not?

Boasson, R., W.M. Laetsch, and I. Price. 1972. The etioplast-chloroplast transformation in tobacco: correlation of ultrastructure, replication and chlorophyll synthesis. *Am. J. Bot.* 59:217–223.

Ellis, J. 1981. Chloroplast proteins: synthesis, transport and assembly. *Ann. Rev. Plant Physiol.* 32:111–137.

Haupt, W. 1982. Light-mediated movement of chloroplasts. *Ann. Rev. Plant Physiol.* 33:205–233.

Heber, U., and H.W. Heldt. 1981. The chloroplast envelope: structure, function, and role in leaf metabolism. *Ann. Rev. Plant Physiol.* 32:139–168.

Kirk, J.T.O., and R.A.E. Tilney-Bassett. 1978. *The Plastids: Their Chemistry, Structure, Growth and Inheritance.* New York: Elsevier North-Holland.

Lehninger, A.L. 1982. *Principles of Biochemistry.* New York: Worth.

Newcomb, E.H. 1967. Fine Structure of protein-storing plastids in bean root tips. *J. Cell Biol.* 33:143–163.

Park, R.B. 1976. The Chloroplast. In J. Bonner and J.E. Varner, eds., *Plant Biochemistry* 3rd ed. New York: Academic Press.

Possingham, J.V. 1980. Plastid replication and development in the life cycle of higher plants. *Ann. Rev. Plant Physiol.* 31:113–129.

Raven, P.H., R.F. Evert, and H. Curtis. 1981. *Biology of Plants,* 3rd ed. New York: Worth.

Suggested Readings

Bailey, J.L., J.P. Thornber, and A.G. Shyborn. 1966. The chemical nature of chloroplast lamellae. In T.W. Goodwin, ed., *Biochemistry of Chloroplasts.* New York: Academic Press.

Barber, J. 1982. Influence of surface charges on thylakoid structure and function. *Ann. Rev. Plant Physiol.* 33:261–295.

Chapter 13

Electron Transport and Phosphorylation Reactions of Photosynthesis

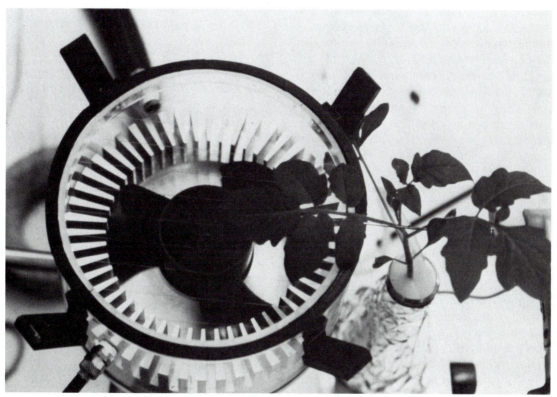

Tomato plant leaves in leaf chamber, used to measure carbon dioxide uptake and photosynthesis. *Courtesy of R.N. Arteca, The Pennsylvania State University.*

The absorption of light energy by chlorophyll induces a rearrangement of the molecule's electronic structure to a highly unstable configuration (excited state); the structure returns to normal configuration (ground state) in approximately 10^{-9} seconds or less. Yet, this seemingly rapid process, termed *photochemical excitation* in chloroplasts, is directly responsible for the oxidative change of water (*photooxidation*), the reduction of nicotinamide adenine denucleotide phosphate, $NADP^+$ to NADPH (*photoreduction*), and phosphorylation of adenosine diphosphate to adenosine triphosphate (*photophosphorylation*). This combined effect of the photochemical reactions (light reactions) is truly a unique aspect of photosynthesis and the primary source of all biochemical energy.

ATP and NADPH are utilized in the reactions of carbon dioxide fixation that are commonly referred to as the "dark reactions" of photosynthesis. In many plants, carbon dioxide is almost exclusively fixed in the chloroplast by its combination with ribulose-1,5-bisphosphate to produce 3-phosphoglyceric acid. The subsequent conversions of 3-phosphoglyceric acid to phosphorylated sugars via a series of reactions are known collectively as the *Calvin-Benson pathway*. In other plants, the so-called C_4 plants and those with crassulacean acid metabolism (CAM)—which we will discuss later—carbon dioxide reacts with phosphoenolpyruvic acid to form a four-carbon product. The utilization of these products as architectural components and a source of energy is indeed one of the most fascinating aspects of plant physiology, the details of which have been clarified only relatively recently. We will begin with a look at the history of the concept of photosynthesis.

History

When we consider the importance, indeed the absolute necessity, of photosynthesis to life, it is surprising what little attention the process attracted before the eighteenth century. By this time, agriculture had existed for more than 10,000 years, and practical discussions of crop production had been written at least 2,000 years before (22). The early Greeks taught that the plant obtained its food directly from the earth, which had converted plant and animal debris to a form readily absorbed by roots. The obvious increase in crop production as a result of the addition of plant and animal matter to the soil gave credence to this theory and left it practically uncontested until the eighteenth century.

In the early 1600s Van Helmont performed a simple but nevertheless important experiment. He planted a willow seedling weighing 2 kg in a tub of carefully weighed soil and followed its growth for five years. He provided the willow with only rainwater. At the end of the five year period, the willow weighed 75 kg, and the soil had lost only a few grams in dry weight. Thus Van Helmont concluded that it was water and not soil that contributed to the growth of the plant. We know today that those few grams of material taken up from the soil were of vital importance and, in fact, essential to growth.

It was left to Woodward in 1699 to state that plants required more than water. After growing sprigs of mint in various samples of water (including rainwater, river water, Hyde Park drainage water), he reached the following conclusion:

Vegetables are not formed of water but of a certain peculiar terrestrial matter. It has been shown that there is considerable quantity of this matter contained in rain, spring and river water; that the greatest part of the fluid mass that ascends up into plants does not settle there but passes through their pores and exhales up into the atmosphere; that a great part of the terrestrial matter, mixed with water, passes up into the plant along with it and that the plant is more or less augmented in proportion as the water contains a greater or less quantity of that matter; from all of which we may reasonably infer, that earth, and not water, is the matter that constitutes vegetables.*

Since the chemistry of carbon dioxide was, as yet, unknown, ignorance of its role in the growth of plants is understandable. Nevertheless it is surprising how little attention was paid to the role of light in plant growth. Hales, often referred to as the "father of plant physiology," may have had some insight into this role when he wrote in 1727:

Plants very probably draw through their leaves some part of their nourishment from the air . . . may not light, also by freely entering surfaces of leaves and flowers, contribute much to ennobling the principles of vegetables.

Studies by Priestley in 1772 were concerned only with the gas exchange that accompanies the process of photosynthesis.

* This and the following two quotations are from W. Loomis, 1960. Historical introduction. In W. Ruhland, ed., *Encyclopedia of Plant Physiology* 5, Part I: 85–114. Berlin: Springer.

Priestley wrote that air "contaminated" by the burning of a candle could not support life in a mouse. He noted, however, that if a sprig of mint were grown in this same air, the air could be rendered "pure," enabling it to support a living mouse again. He also observed that the sprigs of mint flourished in the so-called "contaminated" air.

Although Priestley did recognize the difference in gas exchange between animals and plants when he concluded that

. . . plants, instead of affecting the air in the same manner with animal respiration, reverse the effects of breathing and tend to keep the atmosphere sweet and wholesome when it has become noxious in consequence of animals either living and breathing or dying and putrefying in it[,]

he did not recognize the role of either carbon dioxide or light in photosynthesis.

However, Ingenhousz, a contemporary of Priestley, reported in 1779 that plants purified the air only in the presence of light. He also wrote that only the green parts of the plant produced the purifying agent (oxygen), while nongreen tissues contaminated the air. Thus Ingenhousz recognized the participation of chlorophyll and light in the photosynthetic process.

Although Priestley had flirted with the idea of CO_2 absorption and utilization when he noted that plants "thrived in a most surprising manner" in the "putrid air" in which a mouse had died and partially decayed, he did not establish that fixed air (CO_2) was responsible for this effect.

It was left to Senebier in the years 1782 through 1788 to prove the importance of fixed air and to recognize that the production of oxygen by plants is dependent on the presence of carbon dioxide. Actually, it was not until Lavoisier's study of the composition of carbon dioxide in 1796 that Ingenhousz suggested that this compound is an important source of carbon for plants.

In 1804 de Saussure (11) published work to which we can trace the history of so many of the physiological functions of plants. He agreed with Ingenhousz that two types of gas exchange occur, one in the light and another in darkness, and that green tissues alone carry on the process of carbon dioxide absorption and oxygen evolution in light. He also recognized, to a limited extent, the participation of water in photosynthesis.

Mayer's establishment in 1842 of the law of the conservation of energy was a giant step toward the elucidation of energy transfer in photosynthesis. Mayer stated that the ultimate source of the energy utilized in both plants and animals is the sun and that this light energy, when absorbed by plants, is converted to chemical energy in the process of photosynthesis.

Despite the great efforts of these brilliant men, the mechanism of photosynthesis still remained a mystery until 1905 when Blackman, an English plant physiologist, surprised the scientific world by demonstrating that photosynthesis is not only a photochemical reaction but also a biochemical reaction. As we know today, the photochemical reaction, or light reaction, is exceedingly rapid and requires light energy. In contrast, the biochemical reaction, or dark reaction (carbon dioxide fixation), proceeds at a relatively slow rate. However, since these so-called dark reactions can proceed in light or darkness, we can more correctly refer to them as the *carbon dioxide fixation reactions*. Although Blackman's contribution was remarkable for that time, there was still very little known about the

nature of the light and dark reactions of photosynthesis. It was not until thirty-two years after Blackman's discovery that some solid information on the nature of the light reaction was provided.

In 1937 Hill, an English biochemist, demonstrated that isolated chloroplasts in the presence of light, water, and a suitable hydrogen acceptor evolve oxygen in the absence of carbon dioxide (16). The significance of Hill's experiments was that they provided evidence that the evolution of oxygen is a consequence of photochemical reactions. They also indicated the importance of oxidation-reduction reactions to photosynthesis. We now know that the O_2 of photosynthesis comes from water and not from carbon dioxide.

Origin of Oxygen in Photosynthesis

In the comparative biochemical studies of van Niel (32), we can see some of the initial steps that led to our modern concept of photosynthesis. Van Niel established that CO_2 reduction by photosynthetic bacteria requires the simultaneous oxidation of a substrate (hydrogen donor) from the growth medium. He also noted that in bacterial photosynthesis, CO_2 assimilation is not accompanied by O_2 evolution and ceases to occur when the supply of hydrogen-donor substrate is exhausted. A variety of compounds can be utilized as hydrogen-donor substrates by the many different forms of photosynthetic bacteria that exist. Some of these compounds are organic, such as simple alcohols and organic acids; others are inorganic, such as hydrogen sulfide, thiosulfate, and molecular hydrogen.

Carbon dioxide assimilation by green sulfur bacteria requires the presence of hydrogen sulfide (H_2S) as a hydrogen source. One of the products of this reaction is molecular sulfur. By comparison, algal and higher plant photosynthesis requires the presence of H_2O as a hydrogen source. Molecular oxygen is one of the products of this reaction. The following equations illustrate the two types of photosynthesis.

$$2H_2S + CO_2 \xrightarrow[\text{light}]{\substack{\text{green} \\ \text{sulfur bacteria}}} 2S + (CH_2O) + H_2O$$

$$2H_2O + CO_2 \xrightarrow[\text{light}]{\substack{\text{algae and} \\ \text{higher plants}}} O_2 + (CH_2O) + H_2O$$

The apparent similarity between bacterial and higher plant photosynthesis prompted van Niel to propose a general formula for photosynthesis:

$$2H_2A + CO_2 \xrightarrow{\text{light}} 2A + (CH_2O) + H_2O$$

Two very important points are implied in van Niel's observations of photosynthesis (33): (1) the O_2 evolved in higher plant photosynthesis comes from H_2O and not from CO_2 and (2) the actual assimilation of CO_2 is not light dependent. The photochemical act in this case provides the energy necessary to transfer the hydrogen needed for the reductive steps in CO_2 assimilation.

That H_2O provides the sole source for O_2 evolved in algal and higher plant photosynthesis is strongly supported by isotopic studies using the heavy isotope of oxygen (^{18}O). If photosynthesis proceeds in the presence of $H_2^{18}O$ and normal CO_2, molecular oxygen containing the heavy isotope is evolved:

$$2H_2^{18}O + CO_2 \xrightarrow[\text{chloroplasts}]{\text{light}} {}^{18}O_2 + (CH_2O) + H_2O$$

In contrast, if photosynthesis takes place in the presence of normal H_2O and $C^{18}O_2$, normal molecular oxygen is evolved:

$$2H_2O + C^{18}O_2 \xrightarrow[\text{chloroplasts}]{\text{light}} O_2 + (CH_2{}^{18}O) + H_2{}^{18}O$$

Even stronger support is given by the Hill reaction. This reaction demonstrates that isolated chloroplasts can evolve oxygen, provided they are supplied light, water, and a suitable hydrogen acceptor. The presence of H_2O and the absence of CO_2, coupled with the strongly supported assumption that the Hill reaction simulates the light reaction of photosynthesis, make a very good case for H_2O being the sole source for the O_2 evolved in photosynthesis. From this discussion we can assume with reasonable certainty that water provides the hydrogen necessary for the reductive steps leading to the assimilation of CO_2.

Nature of Light

Toward the middle of the seventeenth century, it was generally believed that light consisted of a stream of minute particles (corpuscles) that were emitted by light sources such as the sun or the flame of a candle. These minute particles penetrated transparent material and were reflected from the surfaces of opaque material. This widely accepted explanation of the nature of light was spoken of as the "corpuscular theory."

In 1670 Huygens noted that the laws of reflection and refraction could better be explained on the basis of a wave theory than on the corpuscular theory. However, the wave theory was not immediately accepted, and not until experiments by Fresnel and Young in 1827 was the corpuscular theory found to be inadequate. In addition, Maxwell showed that an oscillating electrical circuit could radiate electromagnetic waves. The velocity of the propagation of these waves was found to be 3×10^{10} cm/sec. This velocity was remarkably close to the measured velocity of the propagation of light, and developed a strong case for the existence of light as electromagnetic waves of very short wavelength. The problem appeared to be solved. However, one puzzling observation seemed to contradict the wave theory of light—namely, the phenomenon of photoelectric emission (ejection of electrons from a conductor by light incident on its surface). Any change in wavelength of the incident radiation covering a limited spectral region produces changes in the distribution of photoelectron kinetic energies. But, if the wavelength remains fixed, the distribution of electron energies also remains fixed. This phenomenon is true even if the intensity of the incident radiation is increased or decreased. Also there is a direct proportional relationship between the number of electrons and the radiation intensity. With this evidence, Einstein (12) postulated that the energy in a light beam, instead of being distributed through space in the electric and magnetic fields of an electromagnetic wave, is concentrated in small particles called *photons*. Thus, because of the quantization of electromagnetic radiation by Einstein, people were led to consider the photon as a type of particle, thereby giving some validity to the corpuscular theory. However, since the photon was considered to have a frequency and the energy of a photon was believed proportional to its fre-

quency, some of the wave theory was retained. In order to understand the nature of light, scientists had to appreciate its *dual wave-particle* characteristics.

The wavelengths of light that are capable of affecting plant growth lie between 0.00003 and 0.00009 cm (see Figure 13–1). The use of such large units of measure to describe the length of a light wave would obviously be an awkward nuisance. Therefore, scientists studying light effects on plants employ fractions of a meter. Wavelengths are expressed in much smaller units such as the nanometer (1 nm = 10^{-9} m = 10^{-7} cm).

The energy in a quantum may be determined from the wavelength of radiation, the energy being greater the shorter the wavelength. This is shown in Planck's law:

$$q \text{ (quantum)} = h\nu = \frac{hc}{\lambda}$$

where h is Planck's constant (6.624×10^{-27} erg sec), ν is the frequency of light in waves per second, c is the velocity of light (2.998×10^{10} cm/sec); and λ is the wavelength expressed in centimeters. When a chlorophyll molecule absorbs a light quantum (photon), the molecule is excited—that is, it is brought from its normal ground state to an excited state (higher energy level). Not all quanta are able to lift chlorophyll to a higher state of energy. First the light has to be absorbed, and then the quantum absorbed has to have a sufficient amount of energy to do the job. According to Einstein's law of photochemical equivalence, only one molecule or atom can be excited or activated by one quantum—that is, one quantum of light, regardless of its energy level, will only activate one molecule. We usually consider the amount of energy absorbed by a mole of a substance rather than the energy absorbed by one

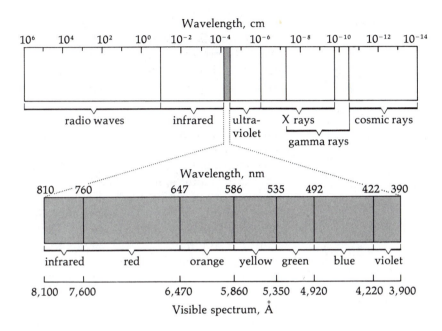

Figure 13–1. Electromagnetic spectrum.

molecule. Therefore, we would need N quanta (N = Avogadro number = 6.02×10^{23}) to excite 1 mole (N molecules) of substance. We can say that N quanta is 1 mole of quanta (1 einstein). A mole of quanta is known as a photochemical equivalent, and the energy (E) contained in a photochemical equivalent may be calculated from the following equation.

$$E = Nh\nu$$

Now, if we substitute c/λ for ν we get:

$$E = \frac{Nhc}{\lambda}$$

$$E = \frac{(6.02 \times 10^{23})(6.624 \times 10^{-27})(2.998 \times 10^{10})}{\lambda}$$
$$\text{erg/mole}$$

$$E = \frac{1.197 \times 10^8}{\lambda} \text{ erg/mole}$$

and if we convert erg to calories (1 erg = 0.239×10^{-7} cal),

$$E = \frac{2.86}{\lambda} \text{ cal/mole}$$

At this point, we have the wavelength in centimeters, which should be converted to angstroms to get:

$$E = \frac{2.86 \times 10^8}{\lambda} \text{ cal/mole}$$

With the above equation we can obtain the photochemical equivalent in calories per mole for any wavelength, for example:

400 nm = 71,500 cal/mole
500 nm = 57,200 cal/mole
600 nm = 47,667 cal/mole

In this manner we can determine the

amount of energy that is being absorbed at any one wavelength.

Free Radicals

Experimenters often mention free radicals in connection with the photosynthetic process. *Free radicals* are atoms or molecules containing an unpaired electron and are produced when bonds are broken symmetrically in homolytic reactions. In such reactions electron pairs are divided, one electron going to each nucleus. If a free radical contains only one unpaired electron, it is called a monoradical, and if it contains two unpaired electrons, it is called a biradical. The existence of a biradical can be demonstrated by the irradiation of ethylene. In fact, a free biradical is almost always produced when a double bond between two carbon atoms is changed to a single bond.

$$H_2C{=}CH_2 \xrightarrow{\text{light}} H_2\overset{\cdot}{C}{-}\overset{\cdot}{C}H_2$$

Electrons pair because only two electrons can occupy the same energy state, and these two electrons must have spins, or angular momentum, about their axis, in opposite directions. This is known as the *Pauli exclusion principle.* The electron has been observed to have an intrinsic magnetic moment and can be pictured as a spinning, charged body that sets up its own magnetic field. All electrons possess the same intrinsic spin, which is characterized by a spin quantum number s, the magnitude of which is always $\frac{1}{2}$. If we define this spin in relation to a magnetic field direction, we can assign a spin of $+\frac{1}{2}$ or $-\frac{1}{2}$ to an electron. According to Pauli's principle, two electrons in the same orbit would have opposite spins, thus neutralizing their magnetic momenta. The resulting spin would be zero $(+\frac{1}{2} - \frac{1}{2} = 0)$. For example, in a helium atom the spins of its

two electrons in the ground state are in opposite directions, and the total spin of the atom is zero. This state is known as the singlet state because the total spin can only have the single value zero in a given direction.

In free radicals the spin of an unpaired electron is not compensated for by a partner electron's spin in the opposite direction and so the resulting spin is $+\frac{1}{2}$ or $-\frac{1}{2}$. In a biradical the resulting spin would be $+1$ or -1. Because of a resulting spin other than zero, free radicals act as paramagnetic substances—that is, when attracted by a magnet, they assume a position parallel to that of a magnetic force.

These properties of free radicals make them very useful in the detection of photobiological processes. In 1945, Zavoisky discovered *electron spin resonance* (ESR) absorption, which initiated the development of spectrophotometers capable of detecting the presence of unpaired electrons. Figure 13–2 illustrates the principle of magnetic resonance absorption measurements.

The ESR phenonemon is related to the intrinsic magnetic moment arising from the spin of the electron. When an unpaired electron is placed between the poles of a magnet, the electron, generating its own mag-

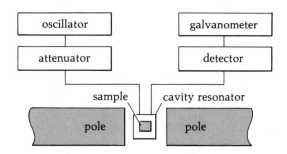

Figure 13–2. Principle of magnetic resonance absorption measurements.
From P.W. Selwood. 1956. Magnetochemistry. *New York: Interscience Publishers.*

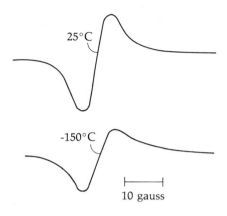

Figure 13–3. Electron spin resonance of whole spinach chloroplasts at 25°C and at −150°C.

From M. Calvin. 1959. The photochemical apparatus—its structure and function. Brookhaven Symp. Biol. 11:160.

netic field, will orient itself with or against the external field. A significant difference in energy exists between these two fields. The separation between these two energy levels is a function of the external magnetic field. Thus we may create an energy difference of our own choosing by merely adjusting the strength of the external magnetic field. We may calculate the energy involved from the following equation:

$$\Delta E = h\nu = gBH$$

where ΔE is the energy difference; h is Planck's constant; ν is the frequency; B is a constant, the Bohr magneton (0.927×10^{-20} erg/gauss); and H is the magnetic field strength in gauss. The interaction between the magnetic moment of the electron and the external magnetic field is given by g, which has a value of 2.0023. Interaction between the electron spin and the electron's orbital angular momentum may cause the g value to deviate slightly from the 2.0023 value.

ESR measurements have been made on a variety of biological materials, such as illuminated chloroplasts, heme proteins, bacterial cells, and oxidation-reduction systems. Figure 13–3 gives an example of ESR measurements of illuminated whole spinach chloroplasts. Note that temperature does not have a significant effect on the photo-induced signals, thereby suggesting the lack of enzyme participation.

Light Absorption by Chlorophyll and Transfer of Energy

Not all of the pigment molecules absorb light or are activated at once. Light energy absorbed by one pigment molecule is thought to be transferred through many other pigment molecules before reaching its site of action. This transfer of light energy may be from one chlorophyll *a* molecule to another, from chlorophyll *b* to chlorophyll *a*, from carotenoids to chlorophyll *a*, or from phycobilins to chlorophyll *a* (14).

To gain a general understanding of how light energy can be absorbed and transferred from molecule to molecule, we must first consider the excited states of molecules (see Figure 13–4). In the ground state, the spins of paired electrons are in opposite directions (Pauli's exclusion principle), the total spin being zero. If a pigment molecule absorbs a quantum of blue light, the electron may exist (through transition from one excited state to another) for a very short time at a higher energy level (excited singlet state), returning to ground state in about 10^{-9} sec. Similarly, red absorption will raise the electron to a lower energy singlet state, with a return to ground state after 10^{-9} sec. It is also possible that absorption of another quantum of light soon after the first will raise the molecule to a second excited state,

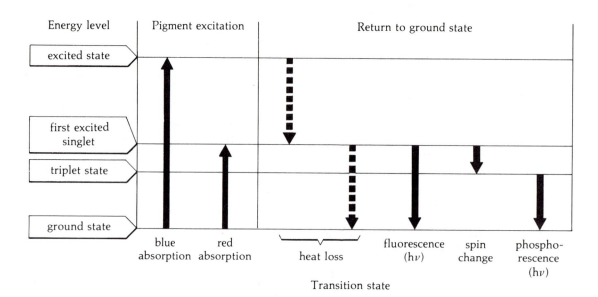

Figure 13–4. Diagram depicting absorption of light and resulting different energy levels due to electronic changes. When electron returns to ground state, there is an accompanying release of energy as heat or light (*hν*).

with a return to the first excited state or to ground level through different transition states. The light energy absorbed in both instances does not just disappear but has to reappear as radiation or in some other form. Thus, with the return of the electron from the excited singlet state to the ground state, the energy previously absorbed can be given off as radiation energy. This phenomenon is known as *fluorescence* because it takes place almost immediately after the absorbing compound is irradiated (chlorophyll exhibits red light fluorescence in solution). As we would expect, the fluorescence of chlorophyll is temperature independent.

There is always the possibility that the electron, brought to a higher energy level (excited singlet state) by the absorption of a quantum of light, may have its spin reversed. Since two electrons cannot exist at the same energy level with parallel spins, this excited electron cannot return to its companion. The electron is said to be "trapped" at a high energy level, and the electron is in what is called the *triplet state*, which, due to transitions and a slight loss of energy, is at a lower energy level than the excited singlet state. However, the electron can have its spin changed again, return from the triplet state to the ground state, and give off its excess energy in the form of radiation. This process is termed *phosphorescence* and is also temperature independent.

In essence, the major difference between fluorescence and phosphorescence is the amount of time for each process to occur after the initial absorption of a quantum of sufficient light energy. Also, since the half-life of the singlet state is approximately 10^{-9} seconds and that of the triplet is 10^{-3} seconds or longer, the triplet half-life seems most appropriate as the energy level responsible for electron transfer in the reduction of a chemical acceptor and the initiation of the electron flow for ATP production. But, surprisingly, it is the singlet state energy that is used in photosynthesis.

Electronic energy transfer between accessory pigments and chlorophyll molecules takes place by resonance. It is a wave resonance, however, which may be visualized as the wave action created when a stone is thrown into a pool of water. Accordingly, before the resonance transfer can take place, the energy donor has to be fluorescent at frequencies the energy acceptor can absorb—that is, the fluorescence spectrum of the energy donor must overlap the absorption spectrum of the energy acceptor. The phycobilins, carotenoids, and chlorophylls exhibit the necessary overlapping fluorescent and absorption characteristics. That is *not* to say, however, that energy transfer takes place by fluorescence absorbance. In addition, the molecules must be packed together relatively closely (100 nm and less) for good resonance energy transfer to take place. The molecular arrangement of the chloroplast is such that the pigment molecules are packed together sufficiently closely for this phenomenon to take place.

Emerson Effect

Light energy absorbed by accessory pigments is transferred to chlorophyll *a* before it becomes active in photosynthesis. While studying the photosynthetic role of accessory pigments in algae, several investigators, working independently, observed a curious phenomenon. They found that light absorbed directly by chlorophyll *a* was less efficient in photosynthesis than light absorbed by the accessory pigments (phycocyanin in blue-green algae and both phycocyanin and phycoerythrin in red algae). The absorption and action spectra of the red alga, *Porphyra nereocystis*, shown in Figure 13–5, clearly illustrate this phenomenon. The same effect was observed in measurements of chlorophyll *a* fluorescence. Light absorbed by the phycobilins promotes more efficient chlorophyll *a* fluorescence than light absorbed directly by chlorophyll *a*. One explanation for this seemingly contradictory observation was that chlorophyll *a* exists in two forms, a photosynthetically active fluorescent form and an inactive nonfluorescent form. Observers thought that

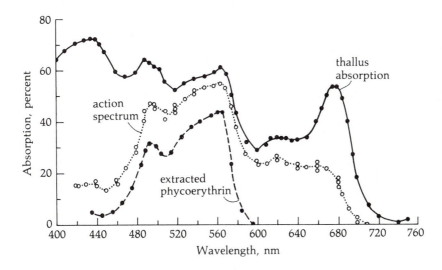

Figure 13–5. Absorption and action spectra of red alga *Porphyra nereocytis*. Note conspicuous lack of activity in 675 and 680 nm regions, although thallus spectrum shows definite absorption peak over that range.

From L.R. Blinks. 1964. In A.C. Giese, ed., Photophysiology. New York: Academic Press. Reprinted by permission.

light energy absorbed by the phycobilins was preferentially transferred to the fluorescent form of chlorophyll *a*. This explanation was subsequently proven wrong, but it did show that there were different forms of the photosystem units.

With the use of monochromatic light of different wavelengths, Emerson made precise measurements of the quantum yield (the number of oxygen molecules released per light quanta absorbed) of photosynthesis over the visible spectrum (13). He observed a significant decrease in quantum yield at wavelengths greater than 680 nm, an area of the spectrum occupied by the red absorption band of chlorophyll *a*. Because of its location in the red area of the spectrum, the decrease in quantum yield noted by Emerson is generally referred to as the *red drop*. The discovery of the red drop added another chapter to the mystery concerning the activity of chlorophyll *a* in photosynthesis.

Emerson and his co-workers soon discovered that the efficiency of photosynthesis at wavelengths exceeding 680 nm can be restored by a simultaneous application of a shorter wavelength. The effect of the two superimposed beams of light on the rate of photosynthesis exceeds the sum effect of both beams of light used separately. This photosynthetic enhancement is referred to as the *Emerson effect*.

Two Pigment Systems

In the late 1950s and early 1960s, the Emerson effect received a great deal of attention. It became increasingly apparent that photosynthesis requires (the interaction of) two distinct groups of functioning pigments termed *photosystems*. In addition, numerous

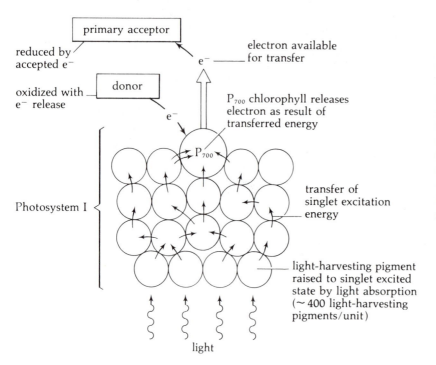

reduced by / accepted e^-

primary acceptor

oxidized with e^- release

donor

Photosystem I {

P_{700}

electron available for transfer

e^-

P_{700} chlorophyll releases electron as result of transferred energy

e^-

transfer of singlet excitation energy

light-harvesting pigment raised to singlet excited state by light absorption (~ 400 light-harvesting pigments/unit)

light

Figure 13–6. Harvesting of light by chlorophyll. Absorption of light quantum by chlorophyll molecule raises that molecule to singlet excited state. Light quantum, in form of singlet excitation energy, migrates from molecule to molecule by resonance. Excitation of P_{700} results.

spectral analyses of chlorophyll *a* in vivo show that the greater part of the chloroplast chlorophyll *a* exists in two forms, one form with an absorption maximum at 673 nm (Chl *a* 673) and the other with an absorption maximum at 683 nm (Chl *a* 683) (9). A long wave-absorbing chlorophyll discovered by Kok (20) is also present, but in much smaller amounts than Chl *a* 673 and Chl *a* 683. This chlorophyll, called P_{700}, has an absorption maximum at 703 nm and is another form of chlorophyll *a* (10).

The photochemical phase of photosynthesis includes two separate photosystems. Photosystem I is rich in chlorophyll *a* and contains carotenoids and less chlorophyll *b* than does photosystem II. In both photosys-

tems most of the pigments operate to harvest light energy and transfer it, possibly by resonance, to chlorophyll *a* molecules located at photochemically active reactive centers termed *traps*. The active center pigment for photosystem I consists of chlorophyll *a*, which absorbs at 703 nm and is called P_{700}. The chlorophyll *a* collecting pigment at the reactive center of photosystem II exhibits an absorption peak at 682 nm and is termed P_{680}. The chlorophyll *a* molecules (donor molecules) reduce specific electron acceptors (A) and become oxidized themselves. The electron carriers that are thus reduced initiate electron flow and the conversion of light energy to chemical energy (*transduction*) (see Figures 13–6 and 13–7).

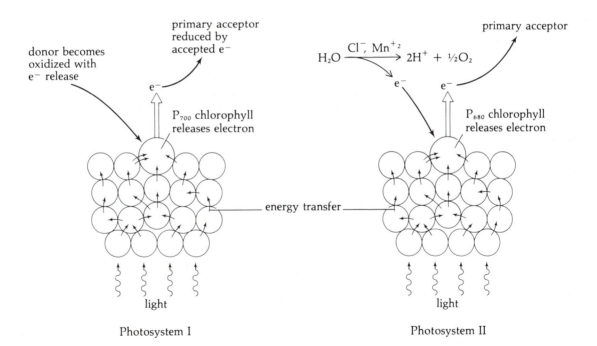

Photosystem I   Photosystem II

Figure 13–7. Photoexcitation of P_{700} causes release of electron to primary acceptor. P_{700} is oxidized and acceptor is reduced. Light quanta absorbed by photosystem I migrate from molecule to molecule by resonance transfer. Acceptor is reduced by electrons flowing from photoexcited photosystem II (P_{680}). Electrons released by photosystem II are regained from water.

Photosynthetic Unit

Early workers in photosynthesis believed that the absorption and conversion of light energy required the presence of intact chloroplasts. In the last fifteen years, however, several investigators have demonstrated the Hill reaction with extremely small chloroplast fragments, thereby suggesting the possibility that the chloroplast may be composed of numerous, minute photosynthetic units. A *photosynthetic unit* is defined as the smallest group of collaborating pigment molecules necessary to effect a photochemical act—that is, the absorption and migration of a light quantum to a trapping center where it promotes the release of an electron. We can think of the basic photosynthetic unit as a collection of about 400 light-harvesting chlorophyll molecules and a trapping center. The tight arrangement of chlorophyll molecules in grana allows for excellent energy migration by resonance transfer. These tightly packed light-harvesting chlorophyll molecules are sometimes called *antennae chlorophyll*. A light quantum absorbed by a single antenna chlorophyll will migrate from one molecule to another until it dissipates into heat, into fluorescence, or into chemical work (ATP and NADPH formation).

A chlorophyll molecule that absorbs a light quantum is raised to a singlet excited state. The light quantum remains in the form of singlet excitation energy for approximately 10^{-9} seconds, thereby leaving very little time for this excess energy to do chemical work. However, the migration of singlet excitation energy between closely packed molecules is very efficient (about 1,000 molecules in 10^{-12} seconds) and not entirely random (21).

Quantum migration occurs preferentially from a pigment with a shorter (higher energy) wave absorption to a pigment with a longer (lower energy) wave absorption band. In photosystem I the long wave–absorbing, or long-wave–collecting, pigment is P_{700}. In photosystem II the absorbing, or collecting, pigment is P_{680}. Once these pigments are excited, they can reduce their electron acceptors, which are otherwise difficult to reduce but which, when once reduced, will pass electrons on to another molecule.

Oxidation reduction reactions involving organic compounds occur within the photosynthetic apparatus. Also, the presence of free radicals in photosynthesizing chloroplasts indicates the existence of one or more electron transport systems.

Production of ATP and NADPH

Having discussed some of the aspects of photochemical reactions, we are now ready to construct a scheme for photosynthesis. We should ask ourselves whether this scheme should be confined solely to the chloroplast or whether the complete cell should be considered. For over a hundred years, photosynthesis was known to be associated with the chloroplast, but it was not known whether photosynthesis was completely confined to this cytoplasmic particle. In fact, for a good many years scientists thought that only the light reaction occurred in the chloroplast and that CO_2 reduction occurred in the cell cytoplasm. In 1954 experimenters observed that isolated chloroplasts, under suitable experimental conditions, could assimilate carbon dioxide. Therefore, the enzymes involved in CO_2 reduction and the *assimilatory power* (reducing power) NADPH needed to accomplish this assimilation must be present, and perhaps produced, within the confines of the chloroplast.

Photosynthetic Phosphorylation

With the discovery that CO_2 fixation takes place in isolated chloroplasts came the realization that the chloroplast contains the required enzymes and produces the ATP necessary for CO_2 fixation and carbohydrate production. Arnon and others (4, 5) demonstrated that isolated chloroplasts produce ATP in the presence of light. They termed the process of ATP synthesis in chloroplasts *photosynthetic phosphorylation* or *photophosphorylation*. Elucidation of this process revealed the hitherto unknown fact that mitochondria are not the only cytoplasmic organelles involved in ATP formation. The formation of most ATP in mitochondria takes place by means of a process known as *oxidative phosphorylation*.

Also, ATP formation in chloroplasts differs from that in mitochondria in that it is independent of respiratory oxidations. The independence of photosynthetic phosphorylation from molecular oxygen is demonstrated in Figure 13–8. What is truly significant here is that light energy is used in the formation of ATP—that is, light energy is converted to chemical energy.

But ATP is only one of the necessary requirements for carbohydrate production. A reductant must be formed in photosynthesis that will provide the hydrogens or electrons. As far back as 1951, Arnon (2) demonstrated that isolated chloroplasts are capable of reducing pyridine nucleotides when these chloroplasts are exposed to light. The photochemical reaction has to be coupled with an enzyme system capable of utilizing the reduced pyridine nucleotide as quickly as it is formed. Observers found that NADPH is the reduced pyridine nucleotide active in photosynthesis (6). In the presence of H_2O, ADP, and orthophosphate (P_i), substrate amounts of NADP were re-

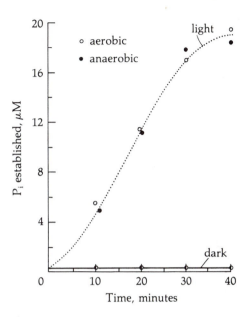

Figure 13–8. Incorporation of inorganic phosphate to form ATP by broken chloroplasts. Note light dependence and oxygen independence of photosynthetic phosphorylation.

From D. Arnon. 1959. The photochemical apparatus—its structure and function. Brookhaven Symp. Biol. 11:181.

duced, accompanied by the evolution of oxygen in accordance with the equation:

$$2\ ADP + 2\ P_i + 2\ NADP + 4\ H_2O$$

$$\xrightarrow[\text{energy}]{\text{light}}\bigg|\text{chloroplasts}$$

$$2\ ATP + O_2 + 2\ NADPH + 2\ H_2O$$

As this equation and Figure 13–9 show, the evolution of 1 mole of O_2 is accompanied by the reduction of 2 moles of NADP and esterification of 2 moles of orthophosphate. Together, ATP (see Figure 13–10) and NADPH (see Figure 13–11) provide the energy and reducing power for CO_2 fixation and reduction. In bacterial photosynthesis NADH is utilized instead of NADPH (34).

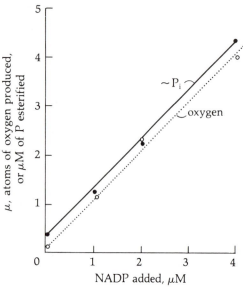

Figure 13–9. Incorporation of inorganic phosphate by isolated chloroplasts to form ATP in presence of different concentrations of NADP. Note linear agreement between amount of NADP supplied and amount of inorganic phosphate taken up; also note evolution of oxygen parallels phosphate uptake.

From D. Arnon. 1959. The photochemical apparatus—its structure and function. Brookhaven Symp. Biol. 11:181.

Z-Scheme: Electron Transport and Photophosphorylation

The Z-scheme (so named because of its shape, see Figure 13–12) illustrates electron transport and the production of NADPH and ATP in chloroplasts. It is a composite of numerous research findings and is therefore subject to change and different interpretations. Although we will not be able to cover all the speculations and details of the photochemical events as they relate to this scheme, the following description, based on much of the previous discussions, will clarify the major concepts. We must keep in mind, however, that not all scientists agree on the details nor sequence of all the intermediates.

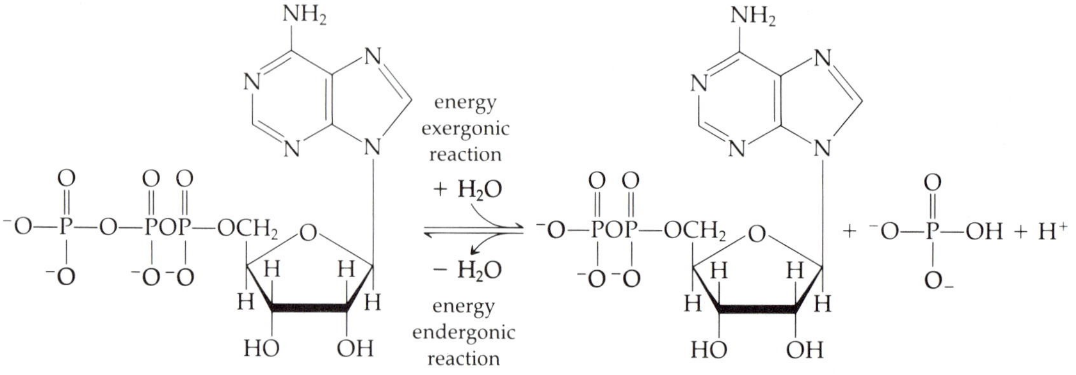

Figure 13–10. Relationship of adenosine diphosphate (ADP) and adenosine triphosphate (ATP). ATP has higher energy content than ADP. In process of ATP conversion to ADP, energy is released and utilized in various ways by organism. Note production of inorganic phosphate (P_i) and proton (H^+). In living systems ATP functions as major source of chemical energy.

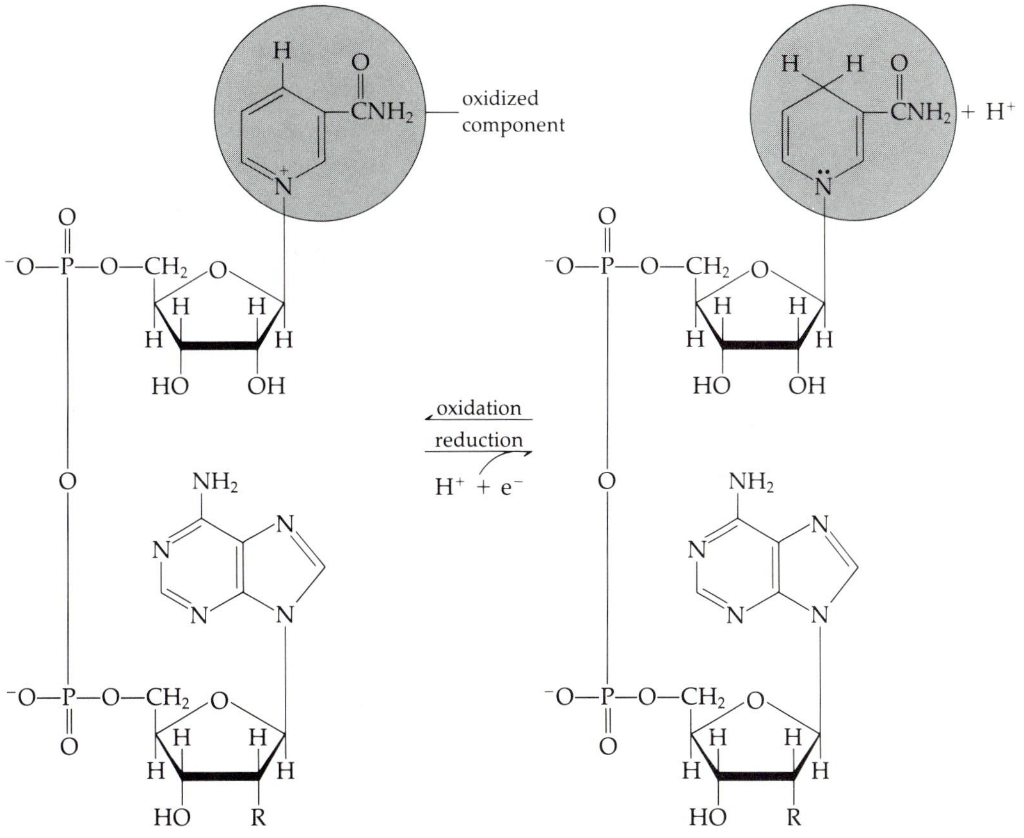

R equals OH; nicotinamide adenine dinucleotide (NAD⁺) $\longrightarrow$ NADH

R equals OPO₃H₂; nicotinamide adenine dinucleotide (NADP⁺) $\longrightarrow$ NADPH

Figure 13–11. Structures of nicotinamide adenine dinucleotide phosphate (NADP) and nicotinamide adenine dinucleotide (NAD). NAD differs from NADP in that NADP has an additional phosphate on the sugar at second carbon position. These coenzymes are important in oxidative reduction reactions of photosynthesis (NADP) and respiration (NAD). NADPH + H⁺ (NADPH) is important in fixation and reduction of CO_2.

Noncyclic Photophosphorylation

The primary flow of electrons within a given granum thylakoid may be initiated almost simultaneously for each photosystem through integrated (coupled) reactions and photolysis of water, which provide the necessary electron flow to produce ATP and NADPH. This integration of the two photosystems is most commonly referred to as *noncyclic photophosphorylation* to describe one means of ATP production in chloroplasts. We may also use the term *noncyclic electron*

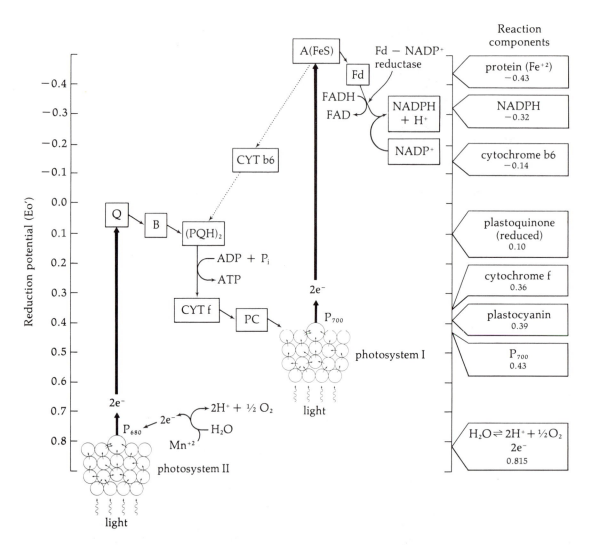

Figure 13–12. Photoinduced electron transport (Z-scheme) in photosynthesis showing cyclic and noncyclic photophosphorylation. Abbreviations are: plastoquinone (PQ), cytochrome b_6 (CYT b_6), cytochrome f (CYT f), plastocyanin (PC), iron-sulfur protein acceptor (A[FeS]), ferredoxin (Fe), flavine adenine dinucleotide (FAD), and FAD reduced (FADH).

transport to refer to the manner of electron flow during the process.

As Figure 13–12 illustrates, after excitation of P_{700}, the "trap" chlorophyll of photosystem I, the electrons are passed on to an unknown primary electron acceptor, believed to be an iron-sulfur protein and des-

ignated A(FeS). The electrons are then passed to ferredoxin and ultimately to $NADP^+$, with the formation of NADPH + H^+. For convenience we refer to the reduced form of NADP as NADPH, although the actual situation is NADPH + H^+. The transfer of electrons (to $NADP^+$) creates an *electron debit* (commonly referred to as a "hole") in photosystem I. However, this deficit is made up by the excitation of P_{680} of photosystem II, subsequent photoejection of electrons and their transport to P_{700} through a system of carriers—Q, B, plastoquinone (PQ), cytochrome f (CYT f), and plastocyanin (PC). At this point Q and B are unidentified compounds.

As illustrated in Figure 13–12, plastoquinone shuttles protons and passes electrons to cytochrome f. At this point ATP is produced (see Figures 13–12 and 13–13). The "hole" created in photosystem II is filled by electrons that are derived from the splitting (photolysis) of water. Thus the passage of electrons requires both photosystems and results in the synthesis of ATP and NADPH. In other words, the electrons are drained off into the production of NADPH and ATP, and result in the Z-scheme or noncyclic photophosphorylation.

Cyclic Photophosphorylation

Theoretically, one way of excluding noncyclic photophosphorylation is to illuminate chloroplasts with wavelengths of light greater than 680 nm. Under these conditions only photosystem I is activated and electrons are not removed from H_2O, as illustrated by the lack of O_2 evolution under these circumstances. When the flow of electrons from H_2O is stopped, noncyclic photophosphorylation is also stopped and, as a consequence, CO_2 assimilation is retarded. With CO_2 assimilation retarded, oxidized NADP is no longer available as an electron acceptor. Activation of photosystem I by wavelengths of light greater than 680 nm causes electrons to flow from P_{700} to A(FeS). When electrons are not passed to $NADP^+$, they may be lost to cytochrome b_6. Cytochrome b_6 will, in turn, pass electrons back

to P_{700} via cytochrome f and plastocyanin (see Figure 13–12). There is some evidence that plastoquinone instead of cytochrome b_6 may act as the primary acceptor of electrons from A(FeS). This possibility is quite likely because plastoquinone is necessary for proton transport across the thylakoid membrane for the generation of ATP.

Although some schemes show the synthesis of ATP in this cyclic transport system as theoretically possible at two locations—synthesis of ATP between A(FeS) and cytochrome b_6 and between cytochrome b_6 and cytochrome f—it is not very likely without plastoquinone mediation. Nevertheless, the term cyclic photophosphorylation is used to denote the cycling of electrons from the donor (excited P_{700} system) to an acceptor (possibly FeS) and back to the P_{700} trap with some generation of ATP. If cyclic photophosphorylation does indeed operate appreciably in certain organisms, it only produces limited ATP. Also, the exact route of the electrons is not clear as indicated in Figure 13–12 by the dotted lines.

Primary Electron Acceptors and Donors

Before we go further in our discussion of photosynthetic phosphorylation, let us take a closer look at photosynthetic NADP reduction. In the late 1950s scientists thought that the reduction of $NADP^+$ was associated with a soluble protein factor found in chloroplasts. Arnon and his colleagues (6) observed that this protein preferentially reduced $NADP^+$ with the evolution of stoichiometric amounts of oxygen. They called it the NADP-reducing factor. The NADP-reducing factor was purified and named photosynthetic pyridine nucleotide reductase (PPNR), since its catalytic activity was only apparent when chloroplasts were illuminated (27). In 1962 the true nature of PPNR was uncovered. Tagawa and Arnon (31) recognized that PPNR is one of a family of nonheme, nonflavin, iron-containing proteins that is universally present in chloroplasts. We use the generic term *ferredoxin* to describe these proteins. Scientists have isolated proteins of the ferredoxin family from the chloroplasts of a variety of plants and have assigned them various functions. What we now call ferredoxin has been called methaemoglobin-reducing factor, NADP-reducing factor, photosynthetic pyridine nucleotide reductase (PPNR), heme-reducing factor, and red enzyme.

Before the discovery of ferredoxin, $NADP^+$ was thought to be the initial electron acceptor of the photosynthetic light reaction. However, neither $NADP^+$ nor ferredoxin is believed to be the primary acceptor of electrons from P_{700}. There is evidence that suggests the existence of an intermediate of—an iron-sulfur protein acceptor, A(FeS)—between ferredoxin and photosystem I.

In earlier Z-schemes, plastoquinone was designated as the primary electron acceptor from P_{680}. Some scientists question whether the oxidation-reduction, or redox, potential of plastoquinone is sufficiently high for the compound to be a primary acceptor from P_{680}. Since quinones are abundant in chloroplasts, there are other likely candidates. Q in Figure 13–13 stands for the unknown primary acceptor that quenches the fluorescence of chlorophyll *a*. Plastoquinone is reduced by transfer of electrons from Q through B, the latter being a secondary unidentified acceptor that is associated with a photosystem II membrane protein.

The reduced plastoquinone is oxidized by the transfer of an electron to cytochrome f. Either cytochrome f or plastocyanin (Cu-containing protein) is the immediate electron donor to photooxidized P_{700}. Both compounds are found associated with the photosynthetic tissues of algae and higher plants and both compounds have redox potentials close to that of P_{700} (about 0.43 volts). However, there is some indication that plastocyanin is located closer than cytochrome f to the photoreaction center (P_{700}) of photosystem I. Therefore, plastocyanin is considered the immediate electron donor to photooxidized P_{700}. Cytochrome f, in this case, would transfer electrons to plastocyanin.

Proposed Mechanisms of ATP Formation

Electron flow and the phosphorylation of ADP to ATP and H_2O are distinct processes that are coupled, or have energy transferred from one to the other, by some common reactant. Evidence for this coupling is based on several observations: (1) In the presence of uncoupling agents, ATP formation can be inhibited but electron transport continues and often shows an increase in rate. When the uncoupler is removed, ATP formation resumes in pace with electron transport. (2) When electron transport is impeded or blocked by certain herbicides, such as the triazines, triazinones, biscarbamates, and diuron (3-[3,4-dichlorophenyl]-1,1-dimethylurea), phosphorylation is also inhibited. (3) Scientists have commonly observed simultaneous oxidation of NADPH (NADH in respiration) and FADH with ATP formation.

Although scientists have studied the coupling of electron flow and phosphorylation intensively, they have not as yet completely elucidated the mechanisms. However, the major "working" hypotheses that experimenters have proposed are *conformational coupling, chemical coupling*, and *chemiosmotic coupling*.

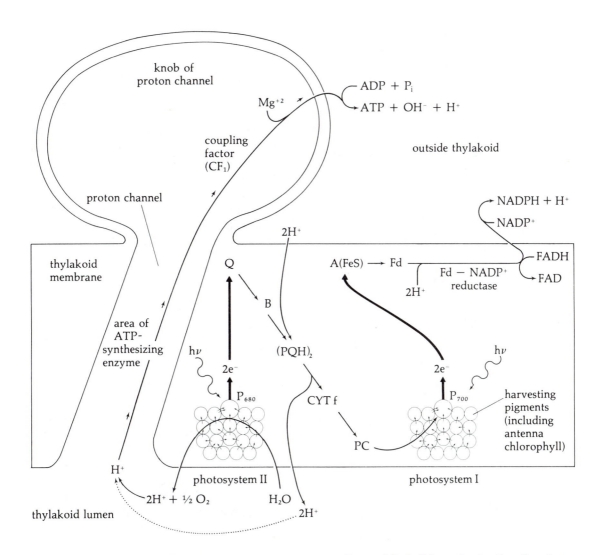

Figure 13–13. Granum thylakoid membrane illustrating location of photophosphorylation and coupling of electron flow to ATP production according to Mitchell hypothesis. Coupling factor (CF) is believed to be ATPase.

Conformational Coupling

Conformational coupling is premised by the idea that the membranes of the mitochondria or chloroplast thylakoids undergo structural changes and that these changes presumably induce high-energy states, or conformations, that favor the release of energy for the ATPase-catalyzed production of ATP. We should note here that although ATPase normally catalyzes ATP decomposition to ADP and inorganic phosphate (P_i), it will work in the reverse direction when sufficient energy is available. Electron micrographs illustrate differences in structure of membranes (mostly mitochondrial membranes) during organelle activity. But we lack further unequivocal evidence demonstrating a correlation between activated membranes and ATP production.

Chemical Coupling

Another hypothesis, developed in the 1960s, suggests that an unknown coupling protein might act as an energy transfer agent between electron transport and ATP formation. According to this idea, a coupling factor (CF), believed to be a protein, initially forms a high-energy CF complex with one of the electron carriers, a participant at the site of phosphorylation along the electron transport chain. The formation of the CF-carrier complex involves an endergonic reaction provided with energy released during electron transfer. The CF-carrier complex then enters into an exchange reaction in which inorganic phosphate (P_i) exchanges with the electron carrier to form a high-energy phosphorylated coupling factor (CF-P) complex. The CF-P complex then releases the high-energy phosphate to ADP, thereby forming ATP. Thus, according to the chemical coupling hypothesis, the endergonic formation of ATP is accomplished via a coupling factor that transfers electron energy promoted by light (photosynthesis) or oxidation of organic chemicals (respiration). Although this hypothesis is consistent with the effects of inhibitors and uncoupling agents, it does not have as strong a following as does the chemiosmotic coupling hypothesis.

Chemiosmotic Coupling

This hypothesis is the most widely accepted explanation for oxidative phosphorylation in mitochondria and has recently gained importance as an explanation for photophosphorylation in thylakoid membranes.

In 1961, after observing that hydrogen ions are actively released from "respiring" mitochondria at the expense of energy derived from the electron transport process, Mitchell (23, 24) proposed the idea of chemiosmotic coupling. He suggested that a concentration gradient of protons is established across the mitochondrial membrane because there is an accumulation of hydrogen on one side of the mitochondrial membrane. The proton accumulation is necessary for energy transfer to the endergonic ADP phosphorylation process. These ideas were later extrapolated to ATP production in chloroplasts, mainly through the work of Jagendorf (19). Jagendorf demonstrated that a pH gradient across the thylakoid membrane stimulated ATP production when chloroplasts were maintained in darkness. Further, under "normal" light conditions, he demonstrated that an H^+ concentration gradient is established in actively photosynthesizing chloroplasts. As Figure 13–13 illustrates, the electron transport carriers are located in the granum membrane. The photolysis of water is depicted as taking place in the thylakoid interior (see Figure 13–12, also). ATP and NADPH are produced on the stroma side surface of the thylakoid.

An important aspect of the model (Figure 13–13) is the "mobility" of plastoquinone. This carrier presumably transfers electrons to cytochrome f and in addition picks up H^+ ions on the outside and releases protons to the thylakoid channel. The transfer of protons to the inside and the production of protons from the photolysis of water incurs a buildup of protons inside and a pH gradient across the thylakoid membrane to the outside (stroma side), where the hydrogen concentration is relatively low. The membrane itself is not permeable to protons concentrated on the channel side, which represent a source of energy, much like water behind a dam. It is believed that protons flow from the inside to the stroma side of the membrane through special pathways of CF (stalks) that terminate as knobs at the outer (stroma side) surface. These stalks and knobs are the sites of photophosphorylation. The proton flow along the gradient provides the necessary energy for the following reaction:

$$ADP + P_i \xrightarrow{\text{ATPase}}$$

$$ATP + H_2O + 8{,}000 \text{ cal/mole}$$

The proton flow and phosphorylation are thought to be brought together (coupled) by the activity of the enzyme ATPase (also called a coupling factor). As mentioned previously, ATPase is associated with the destruction of ATP, but it will operate in the reverse situation as long as sufficient energy is supplied (in this case from the proton flow).

As indicated in Figure 13–13, for every two electrons passing through the transport system, two protons are transported by reduced plastoquinone, a water molecule is photolyzed, and four protons are accumulated. Theoretically, one molecule of ATP is produced for every three protons passing through the CF.

The light reaction phase of photosynthesis illustrated in Figure 13–13 may be summarized by the following equation, which represents the photochemical, photophosphorylation, photoreduction, and photooxidation (splitting of water) events:

$$2H_2O + 2NADP^+ + (ADP)_n$$

$$+ (P_i)_n \xrightarrow[\text{($h\nu$)}_n]{\text{chloroplasts}} (ATP)_n$$

$$+ 2NADPH + 2H^+ + O_2$$

The summary equation also indicates that the stoichiometry of the overall equation is not exact, particularly for ATP production and the quanta required. We do not know the number of ATP molecules produced per oxygen molecule liberated. Some investigators claim 2 molecules of ATP are produced for every oxygen molecule liberated, and others maintain 4.

Observers still do not agree on how much light energy is required to produce sufficient energy to fix one molecule of CO_2 into a sugar phosphate. Warburg suggested in 1922 that 4 quanta were sufficient. However, many scientists do not consider this figure very realistic. In view of Einstein's law of photochemical equivalence (i.e., 1 quantum is required to excite 1 electron), an efficiency of 100 percent is not very likely. Therefore, many plant scientists feel that at least 8 and maybe more quanta (8 quanta for a 4-electron process) are necessary for 50 percent or less efficiency. In line with the preceding discussion, 8 to 12 quanta (photons) appear to be necessary to produce the NADPH and ATP sufficient for CO_2 fixation. Approximately 2 NADPH and 3 ATP molecules are required to incorporate 1 molecule of CO_2 into a sugar phosphate.

Questions

13–1. Describe the early contributions of the following men to an understanding of photosynthesis: Van Helmont, Woodward, Priestley, Ingenhousz, de Saussure, Mayer, Blackman, and Hill.

13–2. What is the source of oxygen evolved during photosynthesis?

13–3. Describe Planck's law and Einstein's law of photochemical equivalence. What do they tell us about light absorption by chlorophyll?

13–4. Describe how we can visualize the excitation of a pigment (e.g., chlorophyll) by an understanding of Pauli's exclusion principle.

13–5. On what basis can we explain the Emerson photosynthetic enhancement effect?

13–6. What wavelengths of light appear to be optimum for promoting the photosynthetic process? List facts to support your answer.

13–7. What are some of the similarities and differences between oxidative phosphorylation and photophosphorylation?

13–8. What is the Z-scheme of photosynthesis? What are the products of the photoreaction and how are some used in the process of CO_2 fixation?

13–9. The transfer of electrons during the photochemical reactions may create an electron debit, or "hole," in photosystem I. What does this mean and how is the deficit eliminated?

13–10. Explain the current thinking concerning the mechanism by which ATP is generated in the grana thylakoids.

13–11. In the absence of CO_2, the chlorophyll of a green leaf may fluoresce. In the presence of CO_2, we do not observe this phenomenon. Provide an explanation for these observations.

13–12. How much light energy is required to produce the chemical energy to fix one molecule of CO_2 into a sugar phosphate? Use additional references to discuss the answer.

13–13. At certain times carbon dioxide levels in a greenhouse atmosphere may be relatively high, at other times the amount is so low that photosynthesis is limited. How can these fluctuating conditions be explained?

13–14. What procedures may be employed in greenhouses to provide adequate levels of carbon dioxide for photosynthesis?

13–15. What is the role of Cl^- in photosynthesis? What other elements are directly involved in the light reactions?

Suggested Readings

Anderson, J.M. 1975. The molecular organization of chloroplast thylakoids. *Biochim. Biophys. Acta* 416:191–235.

Barber, J. 1982. Influence of surface charges on thylakoid structure and function. *Ann. Rev. Plant Physiol.* 33:261–295.

Bearden, A.J., and R. Malkin. 1975. Primary photochemical reactions in chloroplast photosynthesis. *Q. Rev. Biophys.* 7:131–177.

Bearden, A.J., and R. Malkin. 1977. Chloroplast photosynthesis: the reaction center of photosystem I. *Brookhaven Symp. Biol.* 28:247–266.

Blankenship, R.E., and W.W. Parson. 1978. The photochemical electron transfer reactions of photosynthetic bacteria and plants. *Ann. Rev. Biochem.* 47:635–653.

Bolton, J.R. 1978. Primary electron acceptors. In R.K. Clayton and W.R. Sistrom, eds., *The Photosynthetic Bacteria.* New York: Plenum Publishing.

Dutton, P.L., R.C. Prince, D.M. Tiede, K. Petty, K.J. Kaufmann, T.L. Netzel, and P.M. Rentzepis. 1977. Electron transfer in the photosynthetic reaction center. *Brookhaven Symp. Biol.* 28:213–327.

Fajer, J., M.S. Davis, A. Forman, V.V. Klimov, E. Dolon, and B. Ke. 1980. Primary electron acceptor in plant photosynthesis. *J. Am. Chem. Soc.* 102:7143–7145.

Feher, G., and M.Y. Okamura. 1978. Chemical composition and properties of reaction centers. In R.K. Clayton and W.R. Sistrom, eds., New York: Plenum Publishing.

Lehninger, A.L. 1982. *Principles of Biochemistry.* New York: Worth.

Malkin, R. 1982. Photosystem I. *Ann. Rev. Plant Physiol.* 33:455–479.

Malkin, R., and A.J. Bearden. 1979. Iron-sulfur centers of the chloroplast membrane. *Coord. Chem. Rev.* 28:1–22.

Metzler, D.E. 1977. *Biochemistry.* New York: Academic Press.

Stryer, L. 1981. *Biochemistry,* 2nd ed. San Francisco: Freeman.

White, A., P. Handler, E.L. Smith, R.L. Hill, and I.R. Lehman. 1978. *Principles of Biochemistry,* 6th ed. New York: McGraw-Hill.

Chapter 14

Carbon Dioxide Fixation and Reduction

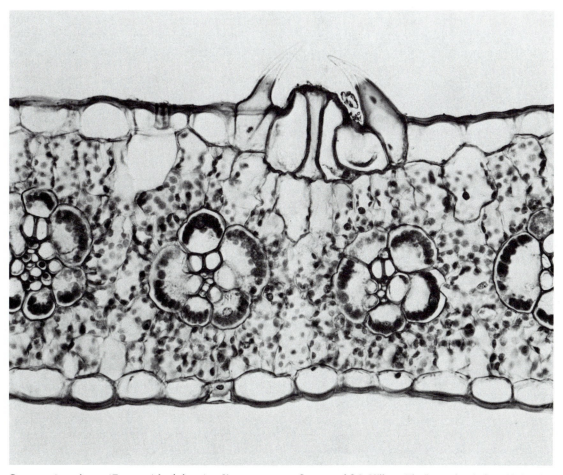

Cross section of corn (*Zea mays*) leaf showing *Kranz* anatomy. *Courtesy of C.J. Hillson, The Pennsylvania State University.*

With the production of ATP and reduced NADP from the photochemical reactions, the fixation and consequent reduction of CO_2 to carbohydrates follow. To Liebig belongs the credit for the first theory on carbon reduction in photosynthesis; he suggested that plant acids are compounds intermediate in reduction between CO_2 and sugars. However, he did not provide experimental evidence in support of this theory, which he developed from observation. For example, ripening fruit is at first sour and later becomes sweet.

Baeyer (1), in 1870, offered the first really strong opposition to Liebig's theory when he proposed that CO_2 is first reduced to formaldehyde, followed by condensation of the formaldehyde molecules to form sugars. The relatively simple formaldehyde theory received strong following, although very little experimental evidence was given in support of it. Indeed, formaldehyde, even at very low concentrations, is toxic to many plants. Paechnatz (36) found that *Elodea*, *Chlorella*, and *Tropaeolum* were not capable of utilizing formaldehyde for the formation of sugar. In fact, she found that concentrations of formaldehyde as low as 0.003 percent were toxic to both respiration and photosynthesis.

Radioactive Tracers

Let us now turn to the early studies of Calvin and his associates. Obviously, the "path of carbon in photosynthesis" was not to be found solely by the proposition of a theory, but rather by careful laboratory experimentation in which each product was carefully analyzed and unequivocally proven a participant in the overall sequence leading to the reduction of CO_2 to sugar. To make such a careful analysis presented immense problems because of the dual role of many of the enzyme systems that are involved in respiration and photosynthesis. Because of the constant mixing of the intermediates of photosynthesis and respiration, it was just about impossible to pinpoint what compound belonged to what system. The methods and instrumentation that existed at that time presented no solution to this complex problem. What was needed was a method of "tagging" compounds in timed experiments with living, photosynthesizing organisms and of placing these compounds in the correct sequence of synthesis.

Using radioactive carbon dioxide was the first step toward solving the problem (42, 43, 44). Experimenters found that fixation of radioactive carbon dioxide ($^{11}CO_2$) by barley leaves and *Chlorella* took place not only in the light but also in the dark. However, the dark fixation of CO_2 occurred only when the leaves were exposed to short periods of darkness. After 3 hours of darkness, no fixation of CO_2 occurred in the barley leaves. Early workers were unsuccessful in their attempts to identify the initial products of photosynthesis, but they did establish that in these products a carboxyl group was present and that this group contained most of the radioactivity. Because of the short half-life of ^{11}C (22 minutes), the pioneer work of these investigations was limited to very brief analytical procedures. This obstacle was overcome with the identification of another radioactive isotope of carbon, ^{14}C, a beta ray emitter with a half-life of 5,000 years (43, 44).

Work with the radioactive tracing of the assimilation of CO_2 in photosynthesis practically came to a standstill during World War II. After the war, however, work with

$^{14}CO_2$ picked up and gathered momentum. Finally, Calvin and Benson (12) produced their remarkable work—the mapping and identification of the intermediates involved in the assimilation of CO_2 in photosynthesis.

Radioautograph

In addition to the use of the radioisotope ^{14}C, experimenters employed a combination of paper chromatography and radioautography. Paper chromatography provides a means for good separation of small quantities of intermediates from very complex mixtures. Radioautography enables investigators to identify on a chromatogram those compounds that are radioactive and therefore involved in the photosynthetic assimilation of $^{14}CO_2$. The complete chromatogram is exposed to sensitized photographic film, which will develop spots on areas in contact with radioactive spots on the chromatogram. Quantitative evaluation of the concentrations of each metabolite present may be obtained by simultaneously exposing a compound with a known amount of ^{14}C and then comparing the relative densities. Figure 14–1 presents a radioautograph from an experiment on photosynthesis.

Type of Plants Used

Calvin and co-workers selected *Chlorella* and *Scenedesmus* for study. These green algae are particularly adapted to studies dealing with the assimilation of carbon dioxide. They are small, unicellular, and can easily be maintained under laboratory con-

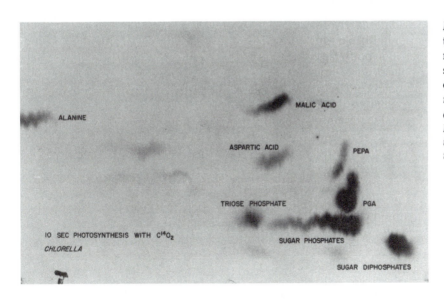

Figure 14–1. Radioautograph showing some metabolites of photosynthesis after 10-second exposure of *Chlorella* to $^{14}CO_2$.

Courtesy of J.A. Bassham, Lawrence Berkeley Laboratory, University of California, Berkeley.

ditions. In addition, they can readily be grown in clonal cultures, thus permitting work with large populations and, consequently, minimizing individual variation. Most important, however, is the fact that a great deal of work has been published about the physiology of these two organisms. Application of this knowledge to the culturing of these organisms made possible the use of uniform and reproducible biological material, a factor that is essential to any detailed metabolic study.

Sequence of Product Formation

Experimenters had to solve one more problem. They had to find a method that allowed for very short times of exposure to $^{14}CO_2$ in order to limit the labeling of compounds to the first few steps of the carbon assimilation pathway. They solved this problem in a simple and ingenious way. A suspension of algae (*Chlorella* or *Scenedesmus*) was allowed to photosynthesize under conditions of constant temperature and illumination in a transparent reservoir. Carbon dioxide was bubbled into the reservoir at an above optimum concentration for photosynthesis. Under these circumstances steady-state conditions for CO_2 fixation are presumably reached. The algal cells were then forced through a narrow, transparent tube into a beaker of boiling methanol, thereby instantly terminating all metabolic activity. Photosynthesis continued in the tube as it did in the reservoir. The length of the tube and the time needed for an algal suspension to traverse the tube were known. Therefore, when $^{14}CO_2$ was injected into the tube at certain, specific points, the time of exposure of algae to radioactive carbon could be calculated. The time of exposure varied from 1

to 15 seconds. The alcohol extract was subjected to the analytical procedures just described. The incorporation of radiocarbon was found to be linear with time of exposure, thereby suggesting steady-state conditions. Figure 14–2 is a schematic representation of the apparatus used by Calvin and his co-workers.

After an exposure of only 5 seconds to $^{14}CO_2$, most of the radioactive carbon is found in 3-phosphoglyceric acid (3-PGA), a three-carbon compound. Moreover, most of the radiocarbon is located in the carboxyl group of this compound. When time of exposure to $^{14}CO_2$ is increased to 30 to 90 seconds, most of the isotopic carbon is found in hexose phosphates as well as in 3-PGA.

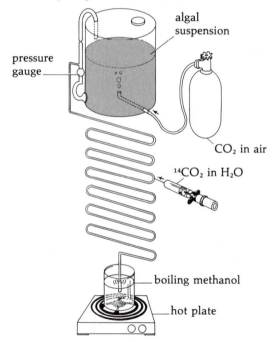

Figure 14–2. Flow-through system for short-time exposure to $^{14}CO_2$.

Reprinted with permission from J.A. Bassham et al. 1954. J. Am. Chem. Soc. 76:1760. Copyright by the American Chemical Society.

Since the carbons 3 and 4 of the hexose phosphates contained most of the radioactivity, it is reasonable to assume they arose from 3-PGA via 3-phosphoglyceraldehyde, fructose-1,6-diphosphate, glucose-6-phosphate, and glucose-1-phosphate. From glucose-1-phosphate both starch and sucrose can be synthesized directly. NADPH is the reductant of 3-PGA to 3-phosphoglyceraldehyde in photosynthesis.

Although fructose-1,6-diphosphate derived from the Calvin-Benson cycle is symmetrically labeled, glucose phosphates formed in photosynthesis are asymmetrically labeled (16, 21). The asymmetrical distribution of radiocarbon in these compounds argues against a head-to-head condensation of two symmetrically labeled triosephosphates as a reaction leading to their formation, although fructose-1,6-diphosphate appears to be so derived. The asymmetrical distribution of radiocarbon in glucose formed during photosynthesis is known as the *Gibbs effect*. It suggests that the two halves of glucose are derived from different pools of triose and that fructose is not the precursor of glucose.

Initial Acceptor of Carbon Dioxide

What compound or compounds give rise to 3-PGA—that is, what compound is the initial acceptor of the carbon dioxide molecule? Calvin and Benson obtained evidence that the five-carbon compound *ribulose-1,5-bisphosphate* (RuBP) is the initial acceptor of the CO_2 molecule. It is now well established that RuBP is carboxylated and then cleaved enzymatically to form two 3-PGA molecules. The enzyme that is required for this reaction is *ribulose bisphosphate carboxylase*, probably one of the most abundant enzymes in photosynthetic tissues. Experimenters found stronger evidence for RuBP being the initial CO_2 acceptor when they studied the distribution of radioactive carbon under light and dark conditions. A change from light to dark produced significant changes in the concentrations of 3-PGA and RuBP. There was a marked increase in 3-PGA and a decrease in RuBP. Figure 14–3 shows this relationship.

Presumably, a steady-state condition exists when the cells are illuminated—that is, 3-PGA and RuBP are continually being formed and broken down. However, when the light is turned off, there is a sharp increase in 3-PGA. This phenomenon suggests that the carboxylation by which 3-PGA is formed does not directly require any of the ATP or NADPH produced in the light reactions of photosynthesis. But the reactions by whch 3-PGA is reduced to 3-phosphoglyceraldehyde show a definite dependence on ATP and NADPH. Since the chemicals are found in the cell in only very small amounts, we presume that they are used up very quickly when the light is turned off. Therefore, 3-PGA would continue to be formed until the supply of CO_2 acceptor (RuBP) is used up. However, the reaction by which 3-PGA is used up would cease very soon after the lights are turned off. With the increase in 3-PGA, there is a rapid decrease in RuBP, thereby pointing to that compound as the initial acceptor of the CO_2 molecule.

Calvin-Benson Pathway

Through the determination of the relative concentrations of radioactive carbon in the different hexoses, pentoses, heptuloses, and so on produced in the algae under different conditions of illumination, Calvin

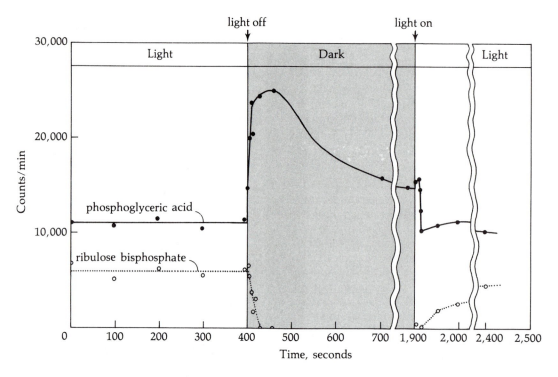

Figure 14–3. Effect of presence or absence of light on concentration of 3-PGA and RuBP.

From J.A. Bassham and M. Calvin, The Path of Carbon in Photosynthesis, © *1957. By permission of Prentice-Hall, Inc., Englewood Cliffs, New Jersey.*

and his co-workers were able to map a metabolic path of carbon assimilation that was hitherto unknown (see Figure 14–4).

As indicated in Figure 14–4, each molecule of ribulose-1,5-bisphosphate fixes one molecule of carbon dioxide with the addition of water, thereby resulting in the formation of two molecules of 3-phosphoglyceric acid (3-PGA). The conversion of two molecules of 3-PGA to 1,3-diphosphoglyceric acid requires 2 ATP molecules coming from the light reactions. Another molecule of ATP, derived from the light reactions is required for the conversion of ribulose-5-phosphate to RuBP. Two molecules of 3-phosphoglyceraldehyde are produced from the two 3-PGA molecules, and this reaction

requires two molecules of NADPH produced by the light reactions. Thus the fixation and reduction of one molecule of CO_2 requires three molecules of ATP and two of NADPH, coming from the photochemical reactions.

Figure 14–4. Calvin-Benson cycle. Enzymes involved are: (1) ribulose bisphosphate carboxylase, (2) 3-phosphoglyceric acid kinase, (3) 3-phosphoglyceraldehyde dehydrogenase, (4) triose phosphate isomerase, (5) aldolase, (6) transketolase, (7) aldolase, (8) fructose-1,6-diphosphatase, (9) sedoheptalose-1,7-diphosphatase, (10) transketolase, (11) ribose phosphate isomerase, (12) ribulose phosphate epimerase, (13) ribulose phosphate kinase.

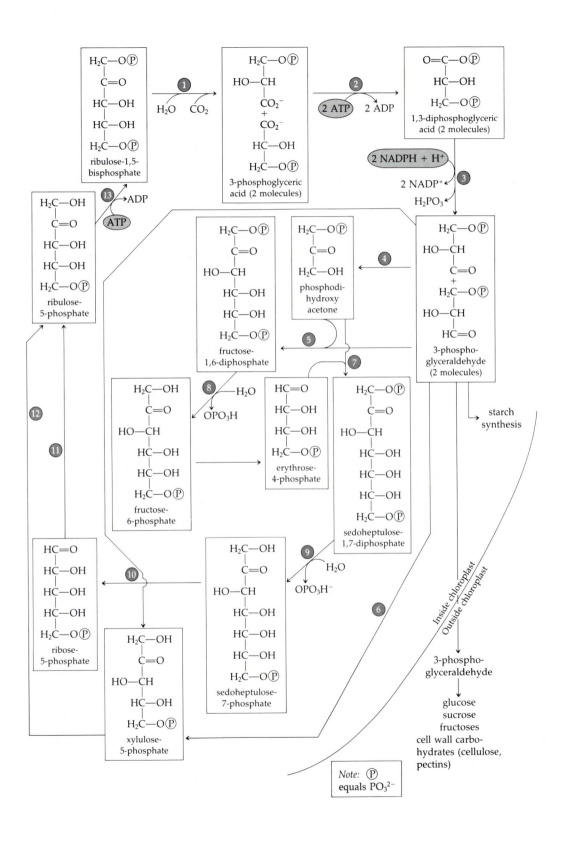

The compound 3-phosphoglyceraldehyde holds a pivotal position in the cycle. It may be transported out of the chloroplast and converted into various hexoses, including glucose, sucrose, fructosans, and cell wall carbohydrates. It may also be diverted to starch synthesis within the plastid via 1-hexose phosphates, or it may move into the metabolic pool.

For every six molecules of 3-phosphoglyceraldehyde produced at the expense of 9 ATP, 6 NADP, and 3 CO_2, one molecule enters the metabolic pool as a net gain in energy and substance to the metabolic system. The five remaining molecules are interconverted through reactions, producing various phosphorylated sugars required for the synthesis of three molecules of ribulose-5-phosphate. This latter compound reacts with ATP, from the light reactions, to form additional RuBP, which accepts CO_2 and initiates the cycle.

For some time after the discovery of the intermediates and the formulation of the scheme, the Calvin-Benson pathway appeared to be the primary method by which plants fix CO_2. Also, plants that utilize primarily RuBP to fix CO_2, which results in the formation of the three-carbon compound 3-PGA, are called C_3 *plants*. But, as we shall see, a vast number of plants fix carbon dioxide in other ways.

C_4 Plants and Carbon Dioxide Fixation (Hatch-Slack Pathway)

In some plants, particularly tropical plants, the predominant ^{14}C-labeled compounds formed after very short periods of photosynthesis in the presence of [14]CO_2 are *malic* and *aspartic acids* (17, 18, 23). A very small percentage of labeled 3-PGA can be recovered, thus indicating that this compound is not the initial product of CO_2 fixation. In addition, ribulose bisphosphate carboxylase, the enzyme that catalyzes the carboxylation of RuBP, is neither active nor concentrated in the mesophyll cells of these plants. The enzyme that catalyzes the formation of phosphoenolpyruvic acid (PEP) from pyruvic acid and ATP (pyruvate phosphate kinase) is found in relatively high quantities (48). The significance of this enzyme is that it mediates the accumulation of PEP, a compound that can be carboxylated to form oxaloacetic acid.

The initial work of Kortschak, Hartt, and Barr (23), which demonstrated that rapidly photosynthesizing sugar cane plants fix CO_2 into aspartic and malic acids, was confirmed by Hatch and Slack (17, 18). Most important, Hatch and Slack were able to detect the relatively unstable ^{14}C-labeled oxaloacetic acid as the first carboxylation product of PEP. They then proposed a new pathway of CO_2 fixation via the carboxylation of PEP. Because the products are four-carbon compounds (oxaloacetic, malic, and aspartic acids), plants exhibiting this pathway (sometimes termed the Hatch-Slack pathway) are referred to as C_4 *plants* (see Figure 14–5).

The anatomy of the leaves of C_4 plants is particularly interesting and diagnostic of C_4 metabolism. Typically, unlike the leaves of C_3 plants, which exclusively fix CO_2 via the Calvin-Benson pathway, the leaves of C_4 plants are characterized by a sheath of parenchyma cells that are radially arranged around each vascular bundle. The vascular bundle, in turn, is enclosed by loosely packed, "spongy" mesophyll cells. This closed, vascular bundle sheath arrangement is known as *Kranz anatomy* ("wreath," in German) and is characteristic of C_4 plants

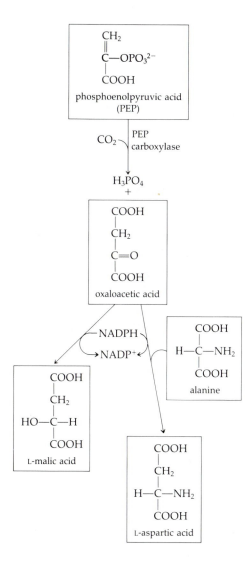

Figure 14–5. Hatch-Slack pathway.

such as sugar cane, sorghum, maize, various grasses originating in the tropics, and, indeed, many other species of plants (see Figure 14–6).

The leaf cells of C_4 plants possess two types of chloroplasts. Within the bundle sheath cells are large chloroplasts that usually lack grana and contain numerous starch grains. The mesophyll cells of the leaf contain smaller chloroplasts that have well-defined grana but do not accumulate starch (see Figure 14–7).

The mesophyll cells of C_4 plants exhibit high activity of phosphoenolpyruvate (PEP) carboxylase, which catalyzes the fixation of CO_2 with PEP to form oxaloacetic acid. Conversely, the bundle sheath cells exhibit high RuBP carboxylase and the other enzymes of the Calvin-Benson cycle. It is now evident that the leaves of C_4 plants are compartmentalized and exhibit a division of labor with respect to the fixation of CO_2 into C_4 acids (mesophyll cell chloroplasts) and the subsequent formation of phosphorylated sugars and starch (bundle sheath chloroplasts). Figure 14–8 illustrates the anatomical and reaction sequence relationships in C_4 plants. We should note that Figure 14–8 illustrates the pathways of synthesis and decarboxylation of the carbon fixation products, malate and aspartate. Plants produce only one or the other as the primary product of the Hatch-Slack pathway. One product of fixation is transported from the mesophyll cells to the bundle sheath chloroplasts where decarboxylation takes place. The liberated CO_2 from the decarboxylation reaction is incorporated into the Calvin-Benson cycle, with the subsequent production of phosphorylated sugars, sucrose, and starch. It is a curious phenomenon that C_4 plants are as highly productive photosynthetically (production of phosphorylated sugars) in view of the low affinity of RuBP

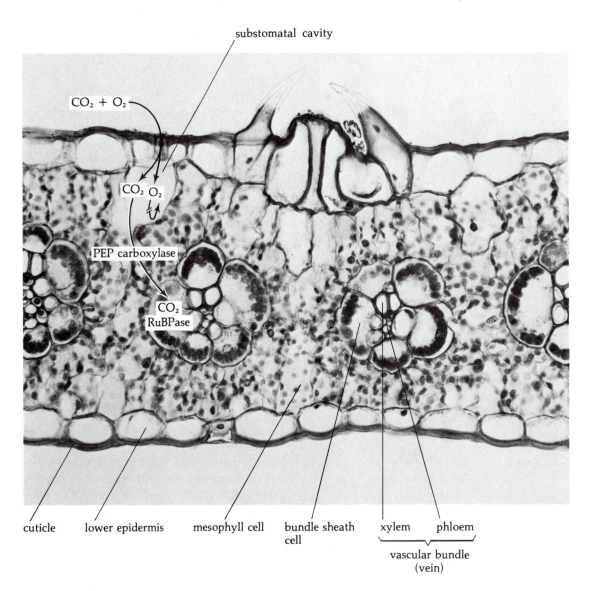

Figure 14–6. Cross section of corn (*Zea mays*) leaf illustrating typical *Kranz* anatomy, with tightly packed parenchyma cells of each vascular bundle sheath evident. Bundle sheath of each vein is not exposed to atmosphere. Stomata are between veins, thereby contributing to reduced photorespiration.

Courtesy of C.J. Hillson, The Pennsylvania State University.

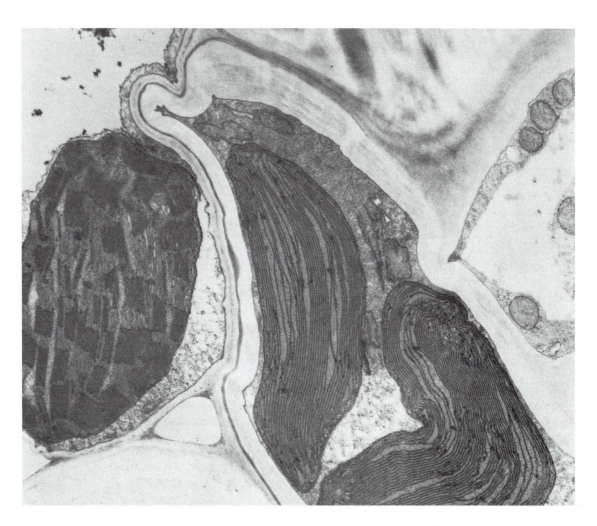

Figure 14–7. Section of sugar cane leaf showing bundle sheath cell chloroplast (right) and mesophyll cell chloroplast (left). In mesophyll chloroplast, note abundance of grana. Magnification 24,500×.

Photo courtesy of W.M. Laetsch, University of California, Berkeley.

carboxylase for its substrate. But CO_2 is concentrated as C_4 acids due to the activity of the PEP carboxylase, which in effect maintains high pool levels of CO_2. These pool levels more than compensate for the low substrate affinity of RuBP. Further, the localization of the Calvin-Benson pathway en- zymes accounts for the production of starch in the bundle sheath parenchyma. Such compartmentation provides a highly favorable and efficient situation for carbohydrate interconversions, and it provides a port of entry for the loading and transport of sucrose in the phloem.

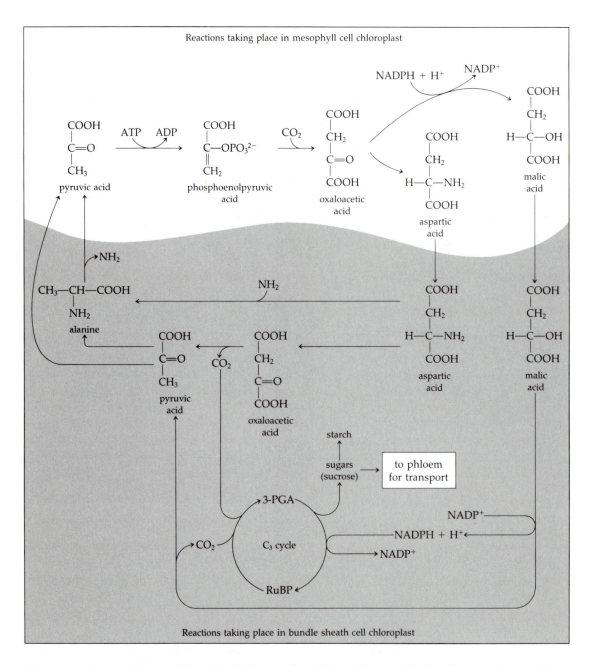

Figure 14–8. Two pathways of C$_4$ cycle of photosynthesis depending on whether plant produces malic or aspartic acid.

Crassulacean Acid Metabolism

Plants such as *Kalanchoë, Agave,* and *Sedum,* which grow in acid habitats, have fleshy stems and leaves with low transpiration rates and are commonly referred to as succulents. Many succulents are C_4 plants that fix CO_2 into malic acid. However, they do not possess the *Kranz* anatomy.

Before they discovered the C_4 metabolism in sugar cane, scientists knew that some nonsucculents, and principally species of the succulent Crassulaceae family, fixed CO_2, with accompanying acidification (C_4 acid formation). Hence the process was named *Crassulacean acid metabolism* (CAM). Unlike other C_4 plants, CAM plants fix CO_2 at night because their stomata are open at night and closed during the day. Thus stomatal behavior and the reduced rate of stomatal transpiration due to nighttime environmental factors account for the low transpiration rates of CAM plants and their survival in dessert and other arid habitats. As Laetsch (26) pointed out, CAM plants have a low ratio of surface area to volume, which is an important structural character for water retention but not necessarily for efficient gas exchange. Lack of water was probably the greatest selective pressure for the evolution of an efficient water-conservation mechanism. C_4 plants, however, exist in habitats with alternate periods of drought and rain. Again, Laetsch points out that we can think of the *Kranz* anatomy as a structural compromise between water economy and efficient CO_2 fixation.

We should note again that CO_2 fixation occurs at night (acidification), with carbohydrate formation taking place during the day (deacidification), apparently within the mesophyll cells. In other words, CAM plants do not exhibit the structural compartmentation (C_3 and C_4 cycles taking place in different cells) of conventional C_4 plants. Perhaps in the C_4 habit, the compartmentation is coupled with rapid growth rates that are essential for the successful competition of C_4 plants with mesophytes, particularly during times of water availability. Further inspection of C_3, C_4, and CAM metabolism reveals some very striking similarities and differences among the three types (see Table 14–1). These features influence photosynthesis.

Factors Affecting Photosynthesis

Photosynthesis, like any other physiochemical process, is affected by the conditions of the environment in which it occurs. Students of the concept of the *three cardinal points,* a theory introduced by Sachs in 1880, were the first to consider the dependence of photosynthesis on external factors. According to this concept, there is a minimum, optimum, and maximum for each factor in relation to photosynthesis. For example, any species has a minimum temperature below which no photosynthesis takes place, an optimum temperature at which the highest rate takes place, and a maximum temperature above which no photosynthesis will take place. Figure 14–9 shows these relationships graphically.

However, on application of this theory, most investigators have been confronted with fluctuating optimums. Scientists may find that the optimum concentration for CO_2 in one experiment changes in another experiment without having realized that the second experiment may have been carried out under different

Table 14–1. Photosynthetic features of C_3, C_4, and CAM plants.

Features	C_3	C_4	CAM
Kranz anatomy	No	Yes	No
CO_2 acceptor	RuBP	PEP	PEP
CO_2 fixation product	3-PGA	Oxaloacetic acid C_4 acids	Oxaloacetic acid and other C_4 acids
Carboxylase	RuBP carboxylase	PEP carboxylase; RuBP carboxylase	PEP carboxylase; RuBP carboxylase
CO_2 fixation*	Light*	Light*	Darkness: C_4 cycle; light: C_3 cycle
O_2 inhibition of photosynthesis	Yes	No	Yes
Chloroplasts	One structure	Two structures	?
Photorespiration	High	Low (bundle sheath cells only)	Very low
Transpiration	High	Low	Very low
Productivity	Low to high	High	Low to high
CO_2 compensation point	High (25–100 ppm)	Low (0–10 ppm)	Low (0–5 ppm)
Temperature (30–40°C) effect on CO_2 uptake	Inhibits	Promotes	Promotes

Although CO_2 fixation can take place in darkness, the amount of CO_2 fixed is greater in the light because of the availability of ATP and NADPH from the light reactions and because the stomata are open and facilitate gas exchange.

light and temperature conditions. Obviously, experimenters cannot treat the external factors affecting photosynthesis individually but have to treat them in relation to one another.

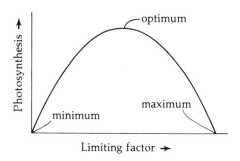

Figure 14–9. Concept of three cardinal points.

Matters remained as such until early in the twentieth century when Blackman proposed the *principle of limiting factors.* The origin of this theory goes back twenty years before the concept of the three cardinal points. Blackman's principle of limiting factors is actually a modification of Liebig's *law of the minimum,* which states that the rate of a process controlled by several factors is only as rapid as the slowest factor permits. At lower concentrations of the limiting factor, a proportional relationship often appears to exist between rate and the quantity of the limiting factor present, but at higher concentrations this is not so. Blackman's contribution was in the discovery that the effect of external factors on the rate of photosynthesis can be measured individually within certain limits—that is, an approximation of the effect of these factors can be obtained. After studying the details of the photochemical and CO_2 fixation reactions, we could expect the factors of light, oxygen, carbon dioxide, temperature, water, and nutrients to influence considerably the process of photosynthesis.

Light

The plant is capable of using only a very small portion of the incident electromagnetic radiation that falls on a leaf or the radiation that is absorbed by the pigment complex of the leaf. Each pigment has its own absorption spectrum, which is usually represented by a curve showing the amount of light absorbed at each wavelength. If we examine the absorption spectra for the major pigments of a leaf (chlorophylls *a* and *b* and β-carotene), we can readily see why most leaves are green in color. The chlorophylls absorb heavily in the blue and red regions of the spectrum, and β-carotene absorbs mostly in the blue region. Most of the light reflected, then, is in the green region, thereby giving the leaf a green color.

Studies by Billings and Morris (5) on the amount of light reflected by a leaf show that peak reflectance is found at about 550 nm and, at this wavelength, about 15 percent of the incident light is reflected (see Figure 14–10). A sharp rise in percentage of reflectance starts at 675 nm and reaches a plateau at 725 nm. About 50 percent of the incident light is reflected at this plateau. In general, the reflectance properties of most green leaves are about the same. However, the environment of the leaf and its surface characteristics influence the amount of reflectance. For example, we find greater leaf reflectance (up to 26.6 percent at 550 nm) in

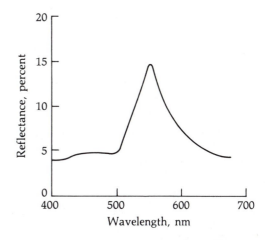

Figure 14–10. Percentage of reflectance from leaves of lilac (*Syringa vulgaris*).
From W. Billings and R. Morris. 1951. Am. J. Bot. 38:327.

environments in which there is greater exposure to light (such as desert localities).

As we might expect, we find the greatest absorption in the thicker leaves with, of course, a lower percentage of transmitted light (light that passes completely through the leaf) as compared to thinner leaves. The average green leaf will transmit only about 10 percent of the incident infrared-free white light (38, 47). Leaves, in general, are almost transparent to infrared and far-red radiation (40). Consequently, workers have found that the average leaf transmits anywhere from 25 to 35 percent of the incident sunlight, which includes infrared radiation.

A direct relationship can be demonstrated between the rate of photosynthesis and the intensity of light, provided that no other factor is limiting. If we plot the rate of photosynthesis against light intensity, this direct relationship is shown at the lower light intensities. As we increase the intensity of light, however, the photosynthetic rate falls off because of some other limiting factor or because of the destructive effects of high light intensity. Also, the point of saturation may be reached, at which time the rate of photosynthesis will remain stationary. Figure 14–11 shows the relationship between photosynthetic rate and light intensity at different temperature levels.

Experimenters take most measurements of the rate of photosynthesis at different light intensities under controlled laboratory conditions. When they study this relationship in the field under natural conditions, they must take many variables into consideration. For example, on bright sunny days, the CO_2 concentration of the atmosphere, not light intensity, is usually the limiting factor. However, on cloudy days, light may be the limiting factor (see Figure 14–12).

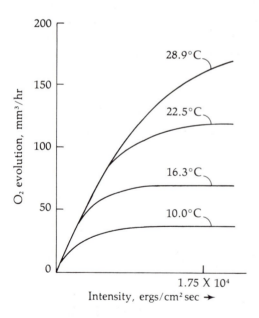

Figure 14–11. Effect of increasing light intensity on rate of photosynthesis by *Chlorella* at different temperatures.

From E. Wassink et al. 1937. Enzymologia 5:100.

Another variable experimenters must consider is the shading effect of one species on another or even the shading effect of outer leaves on the inner leaves of a tree. As we have pointed out, leaves are almost transparent to infrared radiation, a factor that allows the understory of a forest, for example, to receive light a great deal richer in the longer wavelengths. And, of course, the intensity of light reaching the forest floor is greatly diminished, thereby making light a limiting factor under these conditions. Heinicke and Childers (19) studied the rate of photosynthesis by an apple tree under natural conditions. They found that the rate steadily increased with light intensity up to about full sunlight, even though saturation intensity for a single exposed leaf was a good deal lower. For example, about one-fourth of full summer sunlight (2,500 to 3,000 footcandles) is all that is needed for maximum photosynthesis in a single, normally exposed leaf of corn (54). Undoubtedly, the need for higher light intensities for maximum photosynthesis of an entire tree is due to partial illumination of the inner leaves.

Approximately 90 to 95 percent of all light absorbed by a leaf is lost as heat. The remainder is utilized for the photochemical reactions. Plants differ, however, in respect to the radiant energy required to balance photosynthesis exactly with respiration. The intensity of light in which the CO_2 utilized in photosynthesis is equal to that liberated by respiration is called the *light compensation point*. The light compensation point is different for each species and must be appreciably exceeded for a plant to survive, grow, and develop.

Optimum, or saturation, intensities may vary considerably for different species. Some plants grow very well in shaded habitats (*shade plants*), others require exposure to

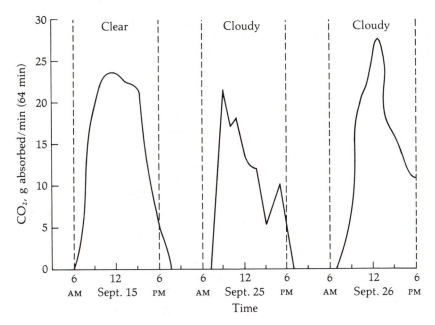

Figure 14–12. Diurnal assimilation of CO_2 by alfalfa on 3 days. September 15 was cloudless, September 25 and 26 were cloudy.

Reprinted by permission from M.D. Thomas and G.R. Hill. 1949. In J. Franck and W.E. Loomis, eds., Photosynthesis in Plants. *Ames: Iowa State University Press.*

full sunlight (*sun plants*). In contrast with sun plants (many crop plants), shade plants usually have very low light compensation points and photosynthesize at higher rates under low light intensity, with the photosystems seemingly saturated at relatively lower intensities than those of sun plants. Some plants adapt to shade—*Pinus taeda,* for example (6). Young seedlings of this species can become shade adapted when they grow under the canopy of older trees, whereas older seedlings and young trees of that species are unable to survive under the same conditions.

The leaves of shade trees exhibit morphological and anatomical features that differ from the leaves of those growing in open sunlight. As we would expect, plants growing under a forest canopy have leaves that tend to be thinner, have more surface area, and contain more chlorophyll than those of sun plants. Shade plants tend to exhibit elongated stems and growth orientation to-

ward the light. It is interesting to note that C_4 plants, most of which are sun plants or have high saturation points of their photosynthetic light-harvesting system, show very high photosynthetic rates under appropriate light conditions. Conversely, C_3 plants tend to exhibit photosynthetic rate saturation prior to one-half the intensity of full sunlight. Although we do not know the reasons for these differences, we can speculate, based on our understanding of the C_3 and C_4 physiology. We might expect that photosynthetic efficiency, among other characteristics of C_4 plants, is related to very high light-harvesting saturation points. One viewpoint, however, is that saturation of light harvesting is not the primary distinguishing feature of efficiency but rather it is the size ratio of the reaction center to light-harvesting efficiency or the photosynthetic unit size (PSU). Photosynthetic size is small in C_4 and sun plants but larger in C_3 and shade plants.

When the intensity of light incident on leaves is increased beyond a certain point, the chlorophyll within is subject to photooxidation. This phenomenon, sometimes referred to as *solarization*, is dependent on O_2 and is clearly evident soon after shade plants are placed in full sunlight; the leaves become chlorotic and die. One explanation is that many more chlorophyll molecules than can possibly be utilized become excited and in the presence of O_2 are readily subject to oxidation (22, 49, 58).

In addition to O_2 the presence of carotenoids and CO_2 influences the extent of photooxidation taking place. CO_2 will tend to inhibit the process. With high concentrations of CO_2, the photooxidative consumption of O_2 occurs at much higher light intensities than it does at lower CO_2 levels (20). Carotenoids play a protective role. Some observers (15) suggest that they act as antioxidants (react preferentially with activated O_2). Carotenoids may also absorb the light energy and somehow divert it from chlorophyll through heat dissipation. In fact, in high light intensity more chlorophyll triplets appear to be formed, and the energy is transferred to carotene triplet states. Therefore, the carotene provides a channel to dispose of excess energy absorbed by chlorophyll. While carotenoids are important in chlorophyll protection, many shade plants are still not sufficiently protected to survive in full sunlight.

Oxygen and Inhibition of Photosynthesis and Photorespiration

In the early 1920s the renowned German biochemist Warburg reported his observation that O_2 inhibited photosynthesis. Although the inhibition of photosynthesis by O_2 was first discovered in algae, the phenomenon is widespread in land plants. In fact, the concentration of O_2 in the atmosphere is inhibitory to photosynthesis. The early work of McAlister and Myers (28) showed the effect of high and low concentrations of O_2 on the process of photosynthesis (see Figure 14–13).

The *Warburg effect*, or the inhibition of photosynthesis by high O_2 levels, was not understood until the late 1960s, even though observers had made several general suggestions that were surprisingly close to the truth years earlier. One suggestion was that O_2, as a necessary component of respiration, favored a more rapid respiratory rate and allowed the process of respiration to compete favorably for intermediates common to photosynthesis and respiration. A second suggestion was that O_2 might compete with CO_2 for hydrogen and become reduced in place of CO_2 (15).

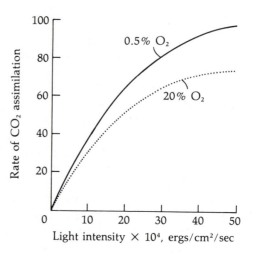

Figure 14–13. Effect of oxygen concentration on photosynthesis of wheat plants at different light intensities.

From E. McAlister and J. Myers. 1940. Smithsonian Miscellaneous Collections 99, no. 6.

With these speculations, the solution to the problem was close at hand in the early 1950s. However, even when the rate of respiration in some plants was known to be greater in light than in darkness, this fact was not readily accepted. Additional work in the late 1960s and 1970s revealed that the rate of respiration, as measured by O_2 consumption or CO_2 liberation from the leaves of C_3 plants, was often as much as two times greater in light than in darkness. Further investigations showed that this "light respiration" is similar to true aerobic respiration, which is found in many plants and animals and which is also characterized by O_2 uptake and CO_2 liberation, but that in "light respiration" no energy liberation occurs (no ATP production from phosphorylated sugars). This form of respiration is termed *photorespiration* because of its similarity to true respiration with regard to the gas-exchange process.

At high light intensity, high O_2, and high temperature, the mesophyll cells in the leaves of all C_3 plants exhibit high rates of photorespiration. C_4 plants, however, exhibit low rates of photorespiration, even though this process has been detected in the bundle sheath cells of C_4 plants, where the Calvin-Benson pathway is operative.

In 1971 Orgen and Bowes (35) provided an explanation of the Warburg effect by showing that the site of action of O_2 was ribulose bisphosphate carboxylase and substrate ribulose 1,5-bisphosphate. In the presence of O_2, the enzyme operates as an oxygenase catalyzing the oxidation (addition of O_2) of RuBP to phosphoglycolic acid (see Figure 14–14). O_2 therefore competes for the RuBP with CO_2, thus causing the net effect of a reduced rate of CO_2 fixation and diminution of phosphorylated sugar synthesis. The entire process takes place in three organelles: chloroplasts, peroxisomes,

and mitochondria. In the chloroplast that contains the Calvin-Benson enzymes, CO_2 is normally fixed by its combination with RuBP to form two molecules of 3-phosphoglyceric acid (3-PGA). As the light intensity and temperature increase, the rate of CO_2 fixation increases proportionately. If CO_2 is in high concentration, higher light intensities will be required before saturation. However, light intensities may be attained that promote the oxidation of each molecule of RuBP to a molecule of phosphoglycolic acid and one molecule of 3-PGA. This oxidation is catalyzed by ribulose bisphosphate carboxylase and results in the formation of only one 3-PGA molecule for each O_2 molecule fixed (no net gain in 3-PGA) rather than the two produced with each CO_2 molecule fixed.

The phosphoglycolic acid is then converted to glycolic acid (by a phosphatase reaction), and the glycolic acid translocates to a peroxisome. Within this ctyoplasmic organelle, glycolic acid is oxidized (the enzyme is glycolic acid oxidase) to produce glyoxylic acid and hydrogen peroxide (H_2O_2). The glyoxylic acid is converted to the amino acid glycine with the liberation of O_2. Two glycine molecules react to produce serine and CO_2. In a mitochondrion the serine is metabolized to carbohydrates or incorporated into proteins. It may also be recycled to the chloroplast, thereby saving three of the four carbons produced through phosphoglycolic acid.

From the foregoing consideration of photorespiration, we now understand why the competition between molecular O_2 and CO_2 accounts for the inhibition of photosynthesis in C_3 plants. However, in C_4 plants, even though the bundle sheath cell chloroplasts are susceptible to O_2 inhibition, the C_4 acid pools (malic or aspartic acids) provide sufficient CO_2 to decrease the O_2 com-

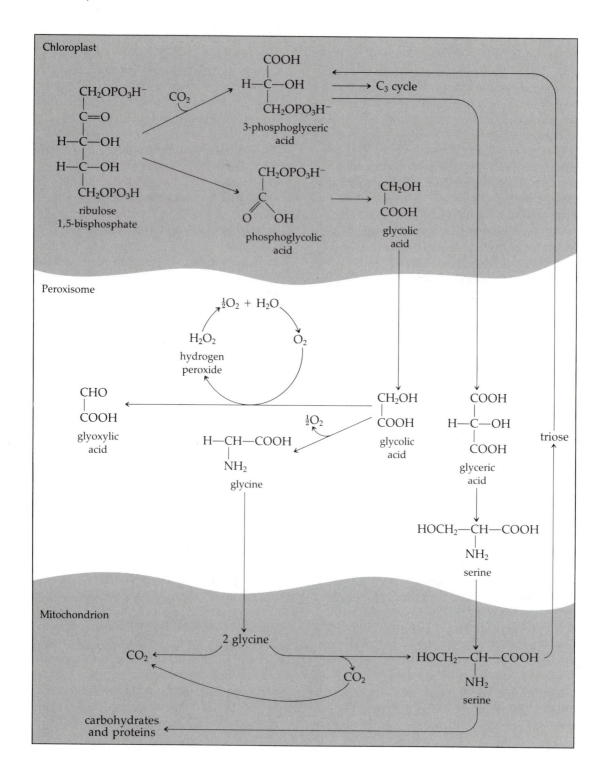

petition, with minimal photorespiration in the bundle sheath cell area.

Photorespiration is essentially nonproductive photosynthetically because the carbon is used to regenerate RuBP, with no gain in carbohydrate synthesis for respiration or storage. At relatively high light intensities, photosynthesis in C_3 plants is said to be saturated because the rate of photorespiration is equal to the net rate of CO_2 fixation. At higher intensities, photorespiration predominates, with an appreciable loss in CO_2 fixation. It is interesting to note that the CO_2 release from photorespiring cells is a reflection of the amount of CO_2 not fixed due to O_2 fixation. The actual CO_2 product of photorespiration, however, is the CO_2 released in the production of serine from two glycines. Paradoxically, the high light intensities are effective in ATP and NADPH synthesis through photophosphorylation and reduction. The ATP and NADPH are required to regenerate the RuBP for photorespiration as well as CO_2 fixation.

The serine produced may be incorporated into protein or converted to carbohydrates via its direct conversion to 3-PGA. Actually, 3-PGA may be produced from three-carbon acids as well as from glycolic acid, glyoxylic acid, glycine, and serine. The formation of 3-PGA from these latter compounds is referred to as the *glycolate pathway*.

Some plant physiologists speculate that photorespiration evolved as a pathway in response to phosphoglycolate accumulation and as a regulating mechanism for the levels of phosphorylated sugars. It may also be important as a mechanism for the intracellular transport and interconversions of

carbohydrates and nitrogenous compounds (glycolate to glycine to serine to PGA). However, currently we do not know what the function of photorespiration is in plants.

Carbon Dioxide

The concentration of CO_2 in the air is small, about three parts in 10,000 or 0.03 percent by volume. This amount is relatively constant and provides a steady and adequate supply of CO_2 to the plant world. Since the plant population, which utilizes a good deal more CO_2 than it gives off, far exceeds the animal population of the world, why does the CO_2 concentration of the atmosphere remain relatively constant? Obviously, CO_2 is also supplied to the atmosphere from sources other than animal respiration.

Carbon dioxide reservoir. The largest single contribution of CO_2 is made by bacteria found in the soil, in fresh water, and in the ocean. The oxidation or decay of organic matter in every location is brought about by these organisms and is a process by which most of the carbon trapped in organic material is released as CO_2. There is little doubt that this source of CO_2 alone exceeds that of all animal respiration. In the oceans and in fresh water accumulations, CO_2 is found primarily as dissolved carbonic acid (H_2CO_3). Certainly, water is one of the primary reservoirs of CO_2.

Another CO_2 source, of lesser importance but still of significance, is the combustion of fuels, which results in the liberation into the atmosphere of hundreds of thousands of tons of CO_2 annually. It would not be surprising if the air over industrial centers and cities is significantly higher in CO_2 concentration.

Figure 14–14. Scheme showing pathway involved in photorespiration in chloroplasts, peroxisomes, and mitochondria.

During the Carboniferous Age, about 300,000,000 years ago, conditions for plant growth were the best they have ever been in the earth's history. The world resembled a large greenhouse of high humidity and high CO_2 concentration. The CO_2 concentration of the atmosphere at that time was probably a good deal higher than the concentration found today. Because of the vast amount of photosynthesis that took place during this period, millions of tons of carbon were trapped and stored within plant tissues. Large quantities of this plant material accumulated under the mud and water of swamps where conditions prevented decay, and eventually they formed the giant coal beds and oil pools of today.

By far the most important factor in the stabilization of the atmospheric CO_2 concentration is the ocean water, which represents immense stores of CO_2 in many different forms. Much of the oceanic CO_2 is available for photosynthesis by plants. The respiration of marine plants and animals releases CO_2 into the water. Part of the CO_2 trapped by marine plants in photosynthesis is released either by the respiration of the organism that consumes it or when the organism dies and decays. Also, the lime for shells of many different marine animals is obtained from the conversion of calcium bicarbonate ($Ca[HCO_3]_2$) to calcium carbonate ($CaCO_3$). Half of the CO_2 tied up in calcium bicarbonate is released in this reaction. Certain marine animals whose shells are made of calcium phosphate carry the above reaction still further and release all of the CO_2 held by the calcium. Today, nearly three-fourths of the earth's surface is covered with ocean water, and this vast body of water is estimated to contain eighty times as much carbon as the atmosphere in forms available to plants. Indeed, a theory has been proposed that the CO_2 concentrations of the at-

mosphere and of the oceans are maintained in a dynamic equilibrium, a drop in the CO_2 concentration of the atmosphere being compensated for by a release of CO_2 from the ocean (29). Of course, an increase in atmospheric CO_2 concentration results in dissolution of CO_2 in the oceans. This equilibrium is probably the primary factor in the stabilization of the CO_2 concentration of the atmosphere. Volcanoes and mineral springs also release CO_2 into the atmosphere, but their contribution is insignificant.

Forces other than photosynthesis can achieve the reduction of the CO_2 concentration of the atmosphere. For example, in the weathering or decomposition of feldspars, CO_2 is utilized and eventually tied up in an unusable chemical form. The chemical decomposition of orthoclase, one mineral of the feldspar class, is as follows:

$$KAlSi_3O_8 + H_2O + CO_2 \longrightarrow$$
orthoclase
$$clay + SiO_2 + K_2CO_3$$
silica   potassium carbonate

The overall cycle of CO_2 in nature thus involves a complex maze of reactions (see Figure 14–15).

Carbon dioxide and plants. Scientists have studied the diffusion of CO_2 through the stomatal pores of a leaf for more than sixty years. Indeed, the stomata seem to be the primary path of entrance of CO_2 into the leaves. Stomatal behavior is much more important in regulating CO_2 availability to the leaves than in regulating O_2 penetration. O_2 penetrates the cuticle of leaves readily, CO_2 is essentially impeded. Therefore the opening and closing of the stomata have an important effect on the regulation of photo-

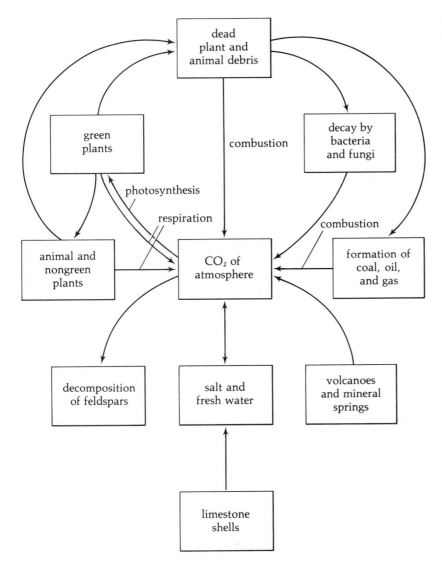

Figure 14–15. Carbon cycle in nature.

synthetic activity, particularly in C_3 plants, which incorporate CO_2 directly into phosphorylated sugar intermediates. A correlation between CO_2 concentration and rate of photosynthesis was reported in the quantitative studies of Kreusler (24, 25), Brown and Escombe (9), and Pantanelli (37). These investigators observed that there is an increase in the rate of photosynthesis when there is an increase in the concentration of CO_2 at three different light intensities. Figure 14–16 shows the effect CO_2 concentration has on the rate of photosynthesis at different light intensities.

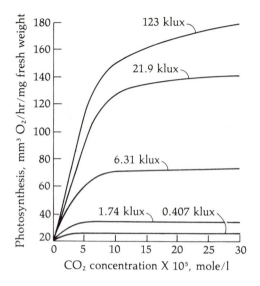

Figure 14–16. Effect of CO_2 concentration on rate of photosynthesis at different light intensities. Klux equals 1,000 meter candles.

From E. Smith. 1938. J. Gen. Physiol. 22:21.

At a given low concentration of CO_2 and nonlimiting light intensity, the photosynthetic rate of a given plant will be equal to the total amount of respiration (true respiration plus photorespiration). The atmospheric concentration of CO_2 under which photosynthesis just compensates for respiration is referred to as the *CO_2 compensation point.* The CO_2 compensation point is reached when the amount of CO_2 uptake is equal to that generated at a nonlimiting light intensity. Apparent photosynthesis under these conditions is zero. In C_3 plants the CO_2 compensation point is usually much higher (25 to 100 ppm CO_2) than it is in C_4 plants (less than 5 ppm). The obvious implication is that C_4 plants have high CO_2 levels in the bundle sheath chloroplasts and high pool levels of CO_2 in the mesophyll cells. The CO_2 in C_4 plant leaves is distributed as organic acids that maintain high pool levels of CO_2. The amount of free CO_2

in the mesophyll of C_3 plants is not as high because C_3 plants have no such mechanism of CO_2 fixation. Significantly high levels of CO_2 inhibit photorespiration in C_4 plants because at high concentrations, CO_2 competes better than O_2 for the RuBP carboxylase active site and is fixed at a greater rate than O_2.

Although normally the concentration of CO_2 in the atmosphere is relatively constant, there are examples where deviation from the average 0.03 percent is considerable. Undoubtedly, in areas of concentrated photosynthesis, such as above a forest canopy or immediately above a dense maize or wheat field, the CO_2 concentration is significantly diminished during the daylight hours. Verduin and Loomis (54) found that the CO_2 concentration at a height of 100 cm over a maize field dropped from an average high of 0.0675 percent at night to a low of 0.045 percent in the morning. This study not only demonstrates how rapidly the CO_2 concentration can drop above dense vegetation as a result of photosynthesis, but it also clearly points out how rapidly the concentration rises when photosynthesis ceases and respiration plays an unchallenged role.

One more important consideration when speaking of the percentage of CO_2 concentration surrounding a plant is the altitude at which the plant is growing. Although the concentration of atmospheric CO_2 is 300 ppm at sea level and at 15,000 feet, the partial pressure of CO_2 is less at the higher altitude and drops proportionally as the height increases. At 15,000 feet the partial pressure of CO_2 is a little less than half the partial pressure of CO_2 at sea level. The actual significance of this drop in the partial pressure of CO_3 at high altitudes so far as photosynthesis is concerned is interesting since there have been reports of unusually high rates of photosynthesis for some alpine plants (51).

Temperature

As in all life processes, photosynthesis is restricted to a temperature range that corresponds roughly to the range tolerated by protein compounds, which are generally active at temperatures above 0°C and below 60°C. Although the photochemical part of photosynthesis is independent of temperature, the biochemical part, which is controlled by enzyme activity, is strictly temperature dependent. However, plants exhibit a wide variance and adaptability in their ability to tolerate temperature extremes.

Injury at temperature extremes. Cold temperatures retard the rate of photosynthesis both directly and indirectly. Directly, cold temperatures inhibit the rate of photosynthesis by lowering the activity of enzymes involved in the dark reactions of photosynthesis. Indirectly, the process of photosynthesis is affected in an adverse manner by the formation of ice outside and inside the cell. The formation of ice within the outer walls of a plant creates drought conditions by draining water from the living cells. Ice that forms inside a cell not only drains the cell of free water but also causes mechanical injury, which upsets the architecture of the cell and of chloroplasts in the cell. Also, such mechanical injury may destroy the permeability properties of the membranes (including the chloroplast membranes) of the cell. In addition, Rabinowitch (41) pointed out that the colloidal structure of the cytoplasm and the chloroplasts may be modified by mechanical forces.

As is well known, all the vital functions of the cell can be terminated by exposure to high temperatures. Of course, at very high temperatures, thermal death is almost immediate. At temperatures slightly above the temperature range of the organism, death is not immediate but is a slow, steady process, which we can observe by noting the diminishing rate of some vital process (e.g., photosynthesis). The adverse effects of high temperatures may at first be reversible, but when exposure time is lengthened, they become irreversible. Although thermal death usually occurs in most leaves and algae at 55 to 60°C, thermal inhibition of photosynthesis occurs at significantly lower temperatures, thereby suggesting that the effect is on the photosynthetic apparatus itself rather than on the surrounding cytoplasm. With short-time exposure, the stimulation of photosynthesis above the optimum can occur. Therefore, thermal injury appears to be a slow destructive process, which is probably due to the thermal deactivation of enzymes (41). Noddack and Kopp (31) investigated temperature effects on the photosynthesis of *Chlorella* (see Figure 14–17). With short-time

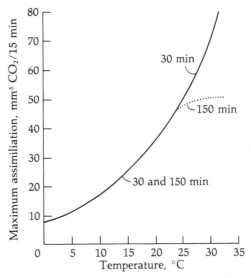

Figure 14–17. Effect of temperature on photosynthesis. Note optimum is reached with short-time exposure. When time of exposure to high temperature is increased, optimum drops.
From W. Noddack and C. Kopp. 1940. Z. Physik. Chem. 187A:79.

exposures, a 30°C temperature optimum can be reached. But when the same organism is exposed longer (150 min) to the same temperature range, an optimum of only about 22°C is reached.

Temperature effects on rate of photosynthesis. In general, increase in temperature results in an acceleration of photosynthesis when other factors are not limiting. This increase is linear at the lower temperatures, starts to drop off as higher temperatures are reached, and finally reaches an optimum above which photosynthesis is inhibited. The optimum response depends on the species tested and the length of time it is exposed (see Figure 14–18).

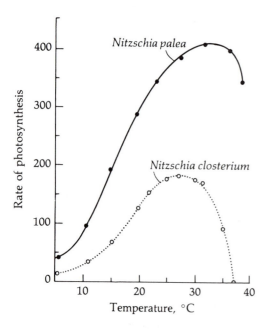

Figure 14–18. Effect of temperature on rate of photosynthesis at high light intensities. Note variations in tolerance for temperature of the two organisms.

From H. Barker. 1935. Archiv. Mikrobiol. 6:141.

In C_3 plants the most likely inhibitory effects of high temperature on photosynthesis are due to a stimulation of photorespiration. CO_2 fixation in C_3 plants is often inhibited at 25°C to 30°C. The C_4 plants, however, show proportional increases and reach an optimum photosynthetic rate above 30°C and sometimes above 35°C because photorespiration is low in C_4 plants.

The effect of temperature on the rate of photosynthesis is roughly comparable to the effect of temperature on enzyme reactions, a fact supporting the theory that the deactivation of enzymes is one cause of the inhibition of photosynthesis at high temperatures. This theory is most likely true. However, we must remember that there are other, maybe undetectable, factors involved. For example, the rate of CO_2 absorption may be limiting at very high rates of photosynthesis, even though the optimum concentration of CO_2 is present. This fact is particularly true for C_3 plants.

Under natural conditions the optimum photosynthetic response is very seldom reached. In most cases, light or CO_2 concentration or both are limiting. The findings of Thomas and Hill (52) clearly illustrate that the influence of temperature on the rate of photosynthesis under field conditions is practically nonexistent in a range from 16°C to 29°C.

Water

It is hard to establish whether or not a deficiency in water supply has a direct inhibitory effect on photosynthesis. The amount of water actually needed for the photosynthetic process is very small in comparison to the amount needed to maintain the living plant. Thus long before a deficiency in the supply of water becomes inhibitory to photosynthesis in a direct manner, the indirect effects of a shortage of water will have had their toll on the entire living system. Water deficiency, of course, would retard photosynthesis along with the other vital processes of the biological mechanism.

Many investigators have noticed reduced rates of photosynthesis in water-deficient soils. For example, Schneider and Childers (46) noted a 50 percent reduction in the photosynthesis of apple trees grown in soil that was allowed to dry gradually. This reduction was observed before evidence of wilting of the leaves could be seen. Similar results were obtained by Loustalot (27) with pecan trees, the greatest reduction in photosynthesis coming when conditions favored high transpiration rates.

Undoubtedly, these inhibitory effects are primarily because of decreased hydration of the protoplasm and stomatal closure. Removal of water from the protoplasm will affect its colloidal structure and its metabolic processes such as respiration and photosynthesis. Enzymatic efficiency is impaired by dehydration of the protoplasm, which, of course, inhibits the rate of vital processes. According to Rabinowitch (39), photosynthesis is more sensitive to dehydration than to some other metabolic processes (e.g., respiration). One reason for this sensitivity might be the physical damage that dehydration may cause to the micromolecular structure of the photosynthetic system.

Many investigators regard stomatal closure as the primary factor in the retarding of photosynthesis by dehydration. When a water deficit occurs in a plant, it causes the stomata of the leaves to close, thereby causing a decrease in the absorption of CO_2. Since CO_2 concentration of the atmosphere is usually low enough to be the limiting factor in photosynthesis under natural conditions, a decrease in its absorption should slow the rate of photosynthesis. However, several workers have seriously challenged this theory. For example, Mitchell (30) found that the rate of photosynthesis remains unchanged until the leaf is wilted. Verduin and Loomis (54) found that CO_2 absorption remains practically unimpaired in visibly wilted *Zea* leaves. Finally, Ting and Loomis (53) concluded that diffusion remains high and approximately uniform until the stomata are closed.

Therefore, stomata that appear to microscopic observance to be closed, are, in fact, opened enough for almost normal CO_2 absorption. Since there is a gradual slowing of the rate of photosynthesis under dehydrating conditions, more than stomatal closure appears to be involved. Stomatal closure is only one of probably many factors involved.

Questions

14–1. Describe briefly the work that led to the identification of the phosphorylated sugar products of photosynthesis.

14–2. What does the term CO_2 fixation mean? Where does it take place in the cell and what is its relationship to the photochemical reactions?

14–3. Compare the CO_2 fixation of C_3 plants with that of C_4 plants. In your answer consider acceptors, intermediates, products, and peculiarities of the plant structures involved.

14–4. Are there any differences in the mechanism of CO_2 fixation between most C_4 plants and those exhibiting Crassulacean acid metabolism?

14–5. From your understanding of the photochemical and CO_2 fixation reactions of photosynthesis, list those factors that dramatically influence photosynthesis.

14–6. Describe light compensation point, shade plants, sun plants, and solarization.

14–7. Explain the Warburg effect. How is the Warburg effect related to photorespiration?

14–8. Correlate the reactions of photorespiration with the structures known to be associated with the process.

14–9. What is the significance of ribulose bisphosphate carboxylase enzyme in the Calvin-Benson pathway and photorespiration? Is the enzyme present in C_4 plants?

14–10. Why do higher intensities of light inhibit CO_2 fixation in C_3 plants?

14–11. Are there any metabolic benefits derived from photorespiration? Explain. What speculations may be made concerning the positive role of photorespiration in plants?

14–12. Describe a situation in which increasing the CO_2 concentration would have no effect on the rate of photosynthesis.

14–13. What effects do high temperatures (30°C to 35°C) have on photosynthesis in C_3 plants? Why?

14–14. Describe some of the physiological reasons why water deficiencies are inhibitory to photosynthesis.

Chollet, R., and W.L. Ogren. 1975. Regulation of photorespiration in C_3 and C_4 species. *Bot. Rev.* 41:137–179.

Galston, A.W., P.J. Davies, and R.L. Satter. 1980. *The Life of the Green Plant*, 3rd ed. Englewood Cliffs, N.J.: Prentice-Hall.

Gifford, R.M., and L.T. Evans. 1981. Photosynthesis, carbon partitioning, and yield. *Ann. Rev. Plant Physiol.* 32:485–509.

Goldsworthy, A. 1970. Photorespiration. *Bot. Rev.* 36:321–340.

Hatch, M.D., and C.R. Slack. 1966. Photosynthesis by sugar cane leaves. A new carboxylation reaction and the pathway of sugar formation. *Biochem. J.* 106:103–111.

Lehninger, A.L. 1982. *Principles of Biochemistry.* New York: Worth.

Lorimer, G.H. 1981. The carboxylation and oxygenation of ribulose 1,5-bisphosphate: the primary events in photosynthesis and photorespiration. *Ann. Rev. Plant Physiol.* 32:349–383.

Osmond, C.B. 1978. Crassulacean acid metabolism: a curiosity in context. *Ann. Rev. Plant Physiol.* 29:379–414.

Raven, P.H., R.F. Evert, and H. Curtis. 1981. *Biology of Plants*, 3rd ed. New York: Worth.

White, A., P. Handler, E.L. Smith, R.L. Hill, and I.R. Lehman. 1978. *Principles of Biochemistry*, 6th ed. New York: McGraw-Hill.

Suggested Readings

Berry, J.A., C.B. Osmond, and G.H. Lorimer. 1978. Fixation of $^{18}O_2$ during photorespiration. *Plant Physiol.* 62:954–967.

Calvin, M., J.A. Bassham, A.A. Benson, V. Lynch, C. Ouellet, L. Schou, W. Stepka, and N.E. Tolbert. 1951. Carbon dioxide assimilation in plants. *Symp. Soc. Exp. Biol.* 5:284–305.

Calvin, M., and A.A. Benson. 1948. The path of carbon in photosynthesis. *Science* 107:476–480.

Canvin, D.T. 1979. Photorespiration: comparisons between C_3 and C_4 plants. In M. Gibbs and E. Latzko, eds. *Encyclopedia of Plant Physiology* 6:368. Berlin: Springer.

Chapter 15

Translocation of Sugars

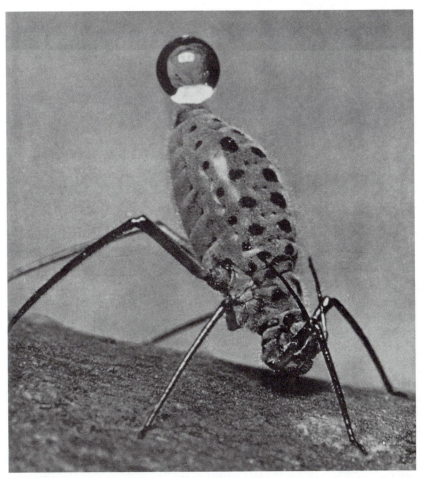

Aphid in twig with its stylet inserted into phloem. Sugary phloem sap passes through aphid's intestines and is released as drop of "honeydew." In early studies, scientists separated insect from stylet, collected exudate directly from stylet, and then analyzed exudate. *From M.H. Zimmermann. 1961. Movement of organic substances in trees.* Science *133:73–79. Copyright 1961 by the American Association for the Advancement of Science. Photo courtesy of M.H. Zimmermann, Harvard Forest.*

We can appreciate the fact that the non-green, living cells of the plant are dependent on the photosynthetic cells for their nutrient supply. However, some of the distances separating photosynthetic cells from other living cells are relatively great. The need for a rapid and efficient translocation system becomes apparent when we consider the distance separating the living cells of the root from those of the leaves. The solution to the problem of nutrient translocation, in the quantity and with the rapidity necessary to carry on normal cell metabolism, is found in specialized cells of the phloem tissue called *sieve tube elements*. These elements, like those of the xylem tissue, form a network of ducts that extends to every part of the plant and brings sugars synthesized in the leaves to all living cells.

Although scientists initiated discussions of the translocation of "elaborate sap" as early as the middle of the seventeenth century, they lacked knowledge of the tissues involved. Indeed, they thought that preformed substances were absorbed from the soil by the roots and translocated through the wood to the leaves, where some changes took place before the now modified substances were retranslocated, also through the wood, in a downward direction. In other words, they believed that both upward and downward translocation occurred in the xylem tissue.

Hartig, in 1837, gave us the first anatomical and physiological description of the tissues involved in the translocation of organic compounds. His discovery of sieve tubes in the bark was the first clue to the plant's elaborate system for nutrient distribution. Hartig demonstrated that nutrients will collect above a stem "girdle" and cause the stem tissues to bulge (see Figure 15–1). In the girdling technique a ring of bark is completely removed from a stem or branch,

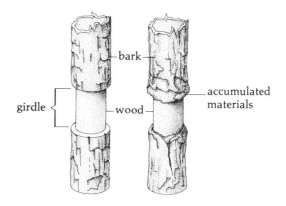

Figure 15–1. Tree trunk immediately after ring of bark has been removed (left) and after longer period of time. Materials translocated from leaves have accumulated in region above girdle and caused it to bulge.

with the wood left intact. Substances being translocated from the leaves will accumulate above the girdle, thus proving that the bark and not the wood is involved in the movement of materials from leaves.

Anatomy of Phloem Tissues

The phloem tissue is composed primarily of *sieve tube elements* and *phloem parenchyma* (11, 17). Companion cells usually accompany the sieve tube elements of angiosperms. An analogous type of cell, called an *albuminous cell*, accompanies the sieve tube elements of conifers. In addition to these cell types, *phloem fibers*, *sclereids*, and *ray cells* are also found. Figure 15–2 shows the position of the phloem tissue in relation to the xylem tissue in a dicot stem.

The large amounts of starch generally found in phloem parenchyma is indicative of the primary function of these cells. However, in addition to storage, phloem parenchyma may play a role in the translocation of sugars in the plant. The parenchyma of

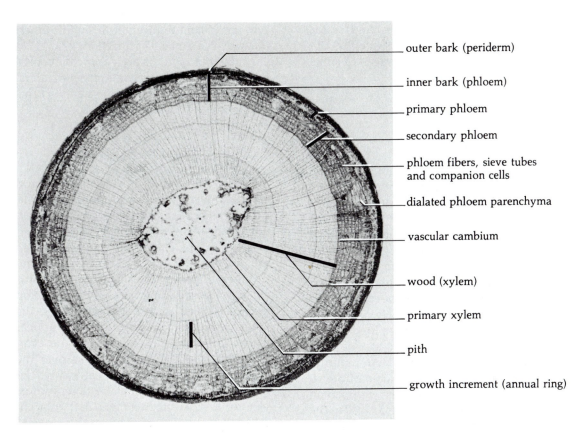

outer bark (periderm)

inner bark (phloem)

primary phloem

secondary phloem

phloem fibers, sieve tubes and companion cells

dialated phloem parenchyma

vascular cambium

wood (xylem)

primary xylem

pith

growth increment (annual ring)

Figure 15–2. Cross section of dicot stem illustrating location of tissues.
Courtesy of C.J. Hillson, The Pennsylvania State University.

phloem tissues of leaves and green stems often contains chloroplasts (11) and functions in polar symplastic movement of sugars to sieve tube elements (54). Crafts (11) pointed out that meristematic tissues and storage areas may obtain nutrients from the sieve tube elements via symplastic movement of these nutrients through nonpigmented parenchyma cells. In an interesting series of experiments with willow stem segments, Weatherley, Peel, and Hill (66) demonstrated that, depending on conditions, a nonpolar exchange of sugars may take place between sieve tube elements and adjacent

parenchyma. Many investigators now realize that parenchyma cells act as metabolic pumps that provide energy for the secretion of nutrients into the sieve tube elements at the source and out of the sieve tube elements at the sink (5, 19, 23). The *sink* is the area in the plant to which nutrients are translocated and either utilized (e.g., meristematic tissue) or stored (e.g., storage organ). In contrast, the storage tissues and photosynthesizing cells serve as important sources of the *assimilates,* or organic compounds produced in the plant by digestion or photosynthesis. Thus the tissues that translocate organic substances and water (including the minerals and water of the transpiration stream) from sources to sinks contain the *assimilate stream*.

Companion Cells

Scientists have focused much attention on companion cells because of the close association of these cells with sieve tube elements. According to Esau (18), the two cell types are not only ontogenetically related but also bear a close physiological relationship. One or more companion cells are cut off from phloem mother cells before their differentiation into mature sieve tube elements. Walls separating the two cells are often very thin or abundantly pitted. Death of the companion cell follows loss of function by the sieve tube element. At maturity, sieve tube elements do not contain nuclei; in contrast, companion cells do. Companion cells, as well as the parenchyma close to the sieve tubes, provide metabolic energy for the movement of assimilates into and out of the sieve tubes. The analogous albuminous cell found in conifers is thought to be similar to the companion cell in its physiological relationship to the sieve tube element.

Some investigators see a strong relationship between the companion cell and the sieve tube element. They suggest that the companion cell and sieve tube element might be viewed as a single functional unit in which the energy produced by the cytoplasm-containing companion cell is utilized by the vacuolar, translocation-adapted sieve tube element (5). This suggestion is supported by electron microscope studies that show a few mitochondria in the sieve tube element and an abundance of mitochondria in the companion cell (15). Also, the two

cells are in communication through numerous plasmodesmatal connections (7).

Phloem ray cells are parenchyma cells that function primarily in storage and lateral transport. Phloem fibers and sclereids function in support of the plant.

Sieve Tube Elements

The sieve tubes, or ducts of the bark, are admirably suited for the rapid and efficient translocation of large quantities of solutes in the plant. Sieve tubes are composed of sieve tube elements, highly specialized cells of the phloem tissue that are strung together to form vertical columns. Transverse walls separating the elements develop into specialized areas called *sieve plates*, the sieve areas being traversed by what appear to be cytoplasmic strands. Thus the cytoplasmic connection is continuous over the entire length of a column of sieve tube elements. Figure 15–3 shows a longitudinal view of a sieve tube element and adjacent cells.

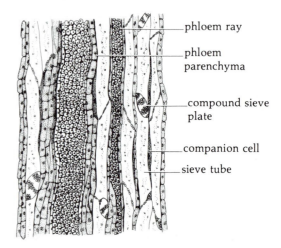

phloem ray

phloem parenchyma

compound sieve plate

companion cell

sieve tube

Figure 15–3. Phloem tissue from stem of *Vitis* species.

The ontogeny of a sieve tube element presents an interesting picture of a cell adapting to a specialized function in the plant. The immature sieve tube element is a rather normal cell, consisting of a nucleus and an actively streaming cytoplasm. The cytoplasm may contain plastids and slime bodies or *P-protein* bodies (18). In the young sieve tube element, cytoplasmic strands may traverse the vacuole, and the nucleus is usually suspended in these strands (11).

As the sieve tube element develops, several changes take place. The nucleus disintegrates and the P-protein bodies disperse. When the plant is wounded the P-protein and a wall material called *callose* produce plugs that seem to block the sieve plates. Normal cells do not develop such plugs. The P-protein remains distributed along the walls, sometimes from one cell to another. Scientists suggest that the major function of P-protein is to seal off the sieve cell by blocking the sieve plates, thereby eliminating leakage of the assimilates when the plant is wounded. Observers have identified spheroid bodies found in the cytoplasm of mature elements as nucleoli released as a result of nuclear disintegration (16). The cytoplasm of a mature element is devoid of an endoplasmic reticulum (1) and appears to be confined to thin layers along the side walls of the cell. Streaming slows and finally stops, and mitochondria appear to be missing. In studies of the ultrastructure of sieve tube elements in *Elodea densa* (13), observers have detected mitochondria, but in relatively low numbers. These features indicate a slowing down of metabolic activity. The cytoplasm at this stage is considered to be highly permeable. Connecting strands of cytoplasm are readily observed traversing the sieve plates in mature sieve elements.

Substances Translocated in Phloem

Carbohydrates

Most of the substances translocated in the phloem are carbohydrates (73). Although this fact has also been shown experimentally, we could have reached this conclusion just from knowing that the bulk of the plant is composed of carbohydrate materials.

Analyses of the phloem exudates of sixteen tree species by Zimmermann (69, 70) have shown that *sucrose* is by far the most abundant carbohydrate translocated. However, in addition to sucrose, some species translocate oligosaccharides such as *raffinose, stachyose,* and *verbascose.* These sugars are similar to each other in that they consist of sucrose with one or more attached D-galactose units. Also, experimenters have detected the sugar alcohols mannitol and sorbitol in the phloem exudates of some species (27, 69, 70). Indeed, sorbitol is known to be translocated in apple trees (27).

Although the hexoses glucose and fructose are commonly present in the phloem tissues of plants, chromatographic analyses of phloem exudates demonstrate the complete absence of these sugars (60, 73). If we consider phloem exudates to be true samples of the substances translocated in the phloem, then we must accept the fact that sucrose is the most prominent sugar translocated and that hexoses are not translocated. The glucose and fructose sugars generally found must occur in the nonconducting cells of the phloem tissues as a result of the hydrolysis of sucrose and related sugars (60).

Swanson and El-Shishiny (62), using a different technique, arrived at the same conclusion. Analyses of grapevine sections

Distance of Translocation (mm)	Counts/min/mg Dry Weight of Bark			Glu/Suc	Fru/Suc
	Suc	*Glu*	*Fru*		
82	8005	661	678	0.083	0.085
202	6268	433	481	0.069	0.077
321	5800	397	402	0.069	0.069
429	4615	220	250	0.048	0.054
652	2942	136	126	0.046	0.043
875	1749	75	69	0.043	0.040
1156	900	34	31	0.037	0.034

Table 15–1. Relative concentrations of ^{14}C-labeled sugars in bark as function of translocation distance.

Source: *From C.A. Swanson and E.D.H. El-Shishiny. 1958. Translocation of sugars in grapes.* Plant Physiol. 33:33.

(*Vitis labruscana* c.v. Concord) at increasing distances from a leaf supplied with $^{14}CO_2$ produced some interesting results. First, the largest amount of radioactivity was found in the sucrose fraction of the bark (see Table 15–1). Table 15–1 also shows that the relative amounts of labeled glucose and labeled fructose are approximately the same at each section of bark analyzed. Now, if we assume that glucose and fructose are equally labeled as a result of $^{14}CO_2$ assimilation in photosynthesis, then the sucrose synthesized from these hexoses should yield, on hydrolysis, equal amounts of labeled glucose and fructose. A reasonable assumption, therefore, would be that the glucose and fructose detected in the bark sections are products of sucrose hydrolysis and not sugars being translocated.

Assuming that the above conclusion is valid, we would expect the ratios of labeled hexoses to labeled sucrose to decrease as the distance from the leaf assimilating $^{14}CO_2$ increases. This line of reasoning is based on the fact that labeled sucrose at a distance from the experimental leaf has had less time to be hydrolyzed than has sucrose in the immediate area of the leaf. Table 15–1 shows these expectations to be fully correct: the ratios decrease from a high of about 0.084 to a low of 0.036. This evidence strongly supports the concept that sucrose is the principal sugar translocated in the phloem and that hexoses are not translocated. The hexoses, which are generally found when phloem tissues are analyzed, are thought to be products of the hydrolysis of sucrose and related sugars.

From this study we can conclude that hexoses appearing in the sieve tube elements were the result of the hydrolysis of sucrose. Burley arrived at essentially the same conclusion in a study of sucrose translocation in soybean and raspberry (9). However, it should be noted that in at least two studies of phloem translocation in sugar cane, the data collected indicate that sucrose remains intact as it moves in the phloem ducts (29, 31).

Nitrogenous Compounds

Amino acids and amides are translocated out of senescent leaves and flowers and relocated in younger areas of the plant. The movement of these nitrogenous compounds takes place primarily in the phloem. Exudate analyses by Mittler (45, 46) for nitrogenous compounds translocated in the sieve tube elements of willow stems have detected the presence of glutamic acid, aspartic acid, threonine, alanine, serine, leucine, valine, phenylalanine, asparagine, glutamine and γ-aminobutyric acid. Although very little work on the detection of these compounds in the phloem has been done, workers will undoubtedly find most of the natural amino acids and amides in sieve tube exudates in the future.

Apparently, the concentration of nitrogenous compounds in phloem exudates is affected by the different developmental stages of the plant. For example, in *Salix* these compounds are present in highest concentration and variety during rapid leaf growth and at the end of the growing season, when leaf senescence is prevalent (60). During the greater portion of the growing season, however, nitrogenous compounds are present in the phloem in very low concentration. Zimmermann (69) found that the concentration of amino acids and amides in the sieve tube exudates of white ash is usually less than 0.001 *M*.

General Aspects of Phloem Translocation

We have discussed the anatomy of the phloem tissue and the organic materials translocated in the phloem ducts. Let us now consider the direction and rate of movement of these substances, as well as other factors that affect translocation, such as temperature, light, metabolic inhibitors, concentration gradients, mineral deficiencies, and hormones.

Direction of Movement

Bidirectional movement. The movement of organic materials in the plant is bidirec-

tional—that is, substances are translocated in opposite directions in the stem simultaneously. Photosynthate moving out of the leaves may be translocated in the direction of the roots, or it may move toward growing points, where flowers, young leaves, or fruits are developing. The mobilization of organic materials in storage organs, such as taproots, tubers, bulbs, for the nourishment of seedling growth is generally in an upward direction. The translocation of materials out of aging leaves and into young actively growing leaves is obviously an upward movement. In a study of phloem transport in bean plants, Biddulph and Cory, using $^{14}CO_2$-feeding and fluorescence techniques, showed that the leaves nearest the root translocate metabolites primarily to the root (2), the leaves nearest the top of the plant transport to the stem apex, and the leaves in an intermediate position translocate metabolites in both directions. Figure 15–4 shows the distribution of labeled metabolites after the primary leaf of a squash

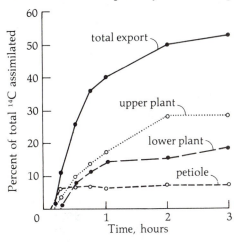

Figure 15–4. Distribution of labeled metabolites after primary leaf of squash plant has been fed $^{14}CO_2$. Radioactivity is found in both upper and lower sections of plant.

From J.A. Webb and P.R. Gorham. 1964. Plant Physiol. 39:663.

plant has been fed $^{14}CO_2$ (67). Note that metabolites moved to both the upper and lower parts of the plant.

Radioactive tagging techniques have shown unequivocally that organic materials move in both directions in the stem simultaneously. What has not been shown is whether materials move in different directions in different phloem ducts or in the same duct simultaneously. This problem can only be solved by the actual demonstration of either unidirectional or bidirectional movement in one sieve duct, a difficult job indeed. However, Biddulph and Cory (2) demonstrated that bidirectional movement in bean plants took place in separate phloem bundles.

Lateral movement in tangential directions. Several studies of translocation patterns show that materials moving in the phloem ducts generally move in a linear fashion— that is, sugars moving out of a leaf into the main translocation stream will move both up and down the stem in line with the supplying leaf. Very little tangential movement takes place. Observers have noted, for example, that annual rings of trees directly under large branches or on the side of a tree receiving less competition from surrounding neighbors are considerably wider than are those on the opposite side. Defoliation of one side of a plant will frequently cause asymmetrical growth, growth on the defoliated side is considerably reduced.

A study of translocation patterns in the sugar beet by Joy (35) produced interesting results. When $^{14}CO_2$ was fed to a leaf for four hours, labeled metabolites were found one week later only in leaves directly above or in the root directly below the supplying leaf. This result agrees with the conclusion that there is an absence of tangential movement. However, Joy found that if he re-

moved all fully expanded leaves from one side of the plant (leaving only young immature leaves) and then fed $^{14}CO_2$ to a mature leaf on the intact side of the plant, he could

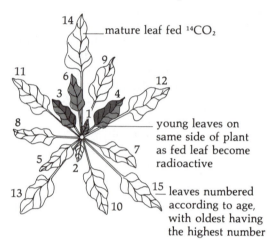

a. ^{14}C translocation from mature leaf to young leaves of intact sugar beet plant.

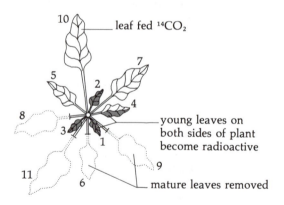

b. ^{14}C translocation from mature leaf to young leaves of sugar beet plant with mature leaves removed from one side of plant.

Figure 15–5. Distribution of ^{14}C in leaves of sugar beet plant one week after single mature leaf was fed $^{14}CO_2$.

From K.W. Joy. 1946. J. Exp. Bot. 15:485.

induce tangential movement. Radioactive metabolites were found not only above and below the supplying leaf but also in the young leaves left intact on the otherwise defoliated side (see Figure 15–5). Apparently, young leaves deprived of photosynthate (by defoliation of the mature leaves) can place a high enough demand for photosynthate from leaves on the opposite side of the plant to cause some tangential movement.

Distribution patterns at different developmental stages in the tobacco plant show both bidirectional and tangential movement (58). The seventh leaf (counting from the bottom) of four different tobacco plants—ages 68, 81, 107, 135 days—was allowed to photosynthesize in $^{14}CO_2$ for 30 minutes, followed by a period of 5 hours and 30 minutes during which it was allowed to photosynthesize under normal conditions. The additional period of 5 hours and 30 minutes allowed for complete distribution of the label without any appreciable redistribution taking place. Then the leaves, stems, and roots were analyzed for ^{14}C (see Figure 15–6 and Table 15–2).

Radioactive carbon was found in the roots of all four plants. However, the bulk of the radioactivity was found in the stems. The pattern of distribution of radioactive carbon, shown in Figure 15–7, demonstrates that areas of high metabolic activity, such as the actively growing stem and young leaves, are particularly good sinks for the deposition of translocated carbohydrates. Note that the radioactive carbon moved both up and down the stem.

Let us consider, now, the lack of ^{14}C in leaves 11 and 19 of plant II. A consideration of Figure 15–7 will show that the *phyllotaxy* (arrangement of leaves on a stem) of plant II was such that leaves 11 and 19 were situated exactly opposite leaf 7 ($^{14}CO_2$-fed leaf). Note also that as the tangential distance from the

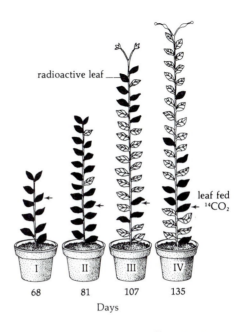

Figure 15–6. Radioactivity (^{14}C) distribution patterns at different developmental stages in tobacco plant showing bidirectional and tangential movement. Shading indicates leaves containing ^{14}C.

Reproduced by permission of the National Research Council from M. Shiroya et al. 1961. Canadian Journal of Botany 39:855.

Table 15–2. Intensity of radioactivity in μc found in treated leaf, other leaves, stem, and root of four tobacco plants shown in Figure 15–7.

	Plant			
Part	*I*	*II*	*III*	*IV*
Treated leaf	131.2	155.9	93.3	136.7
Other leaves	1.3	6.2	trace	trace
Stem	34.4	10.1	10.8	12.7
Root	1.7	0.9	1.8	5.9

Source: *From M. Shiroya, C.D. Nelson, and G. Krotkov. 1961. Translocation of C^{14} in tobacco at different stages of development following assimilation of $C^{14}O_2$ by a single leaf. Can. J. Bot. 39:855.*

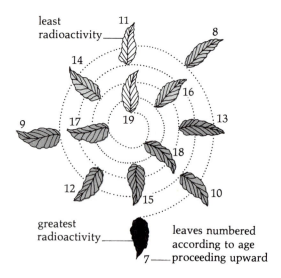

least
radioactivity

leaves numbered
according to age
proceeding upward

greatest
radioactivity

Figure 15–7. Phyllotaxy showing distribution of ^{14}C in plant II of Figure 15–6. Degree of shading indicates intensity of radioactivity. Leaves are numbered from seventh (treated) leaf upward to nineteenth leaf.

Reproduced by permission of the National Research Council from M. Shiroya et al. 1961. Canadian Journal of Botany *39:855.*

fed leaf increases, radioactivity decreases progressively. Leaves 10, 12, 15, 17, and 18 have more ^{14}C than do leaves 9 and 13. Leaves 9 and 13, in turn, have more ^{14}C than do leaves 8, 14, and 16. Leaves 11 and 19, being directly opposite leaf 7, have no ^{14}C. This experiment demonstrates that in the tobacco plant some tangential movement takes place but is definitely secondary to vertical movement.

Lateral movement in radial directions. Experimenters have observed radial transfer from the phloem to the xylem tissues in a wide variety of plants. In fact, loss of labeled metabolites from the phloem to the xylem through radial transport has been shown in the bean plant to reach values of 25 percent or more, as compared to their concentration

in the phloem (3). In another study, ^{14}C-sucrose was detected in the xylem exudate following application of $^{14}CO_2$ to the leafy shoot of a willow stem section (52). Because of their position as continuous connections between the phloem and xylem, vascular rays are thought to facilitate radial movement considerably.

Translocation Rates and Velocities

When we estimate the amount of material needed to maintain the rapid growth of storage organs, we become aware of the importance of translocation rates for the movement of substances in the phloem tissues. Early workers obtained these rates by noting the increase in dry weight of fruits, tubers, storage roots, and other organs that import large amounts of materials from the phloem ducts. However, many difficulties are inherent in this method, and we have to take several other measurements before we can calculate actual rates. For example, we have to make corrections for local synthesis of metabolites if we use photosynthetic tissues. Also, we have to calculate losses due to respiration, condensation, and relocation of metabolites. In many cases we cannot measure these losses directly and have to make certain assumptions. Therefore, translocation rates obtained in this manner may be only indicative of the actual rates occurring.

Tracing techniques. With the advent of tracing techniques, however, fairly accurate translocation rates have been obtained. In this method a leaf is fed $^{14}CO_2$, and this substance is assimilated in the photosynthetic process. The progress of the radioactive metabolites so formed is followed by the detec-

Plants	Rate, cm/hr	Source	
Red kidney bean	107	Biddulph and Cory, 1957	
Sugar beet	85–100	Kursanov et al., 1953	
Concord grape	60	Swanson and El-Shishiny, 1958	
Willow	100	Weatherley, Peel, and Hill, 1959	
Sugarcane	84	Hartt et al., 1963	
Straight-necked squash	290	Webb and Gorham, 1964	
Soybean	100	Vernon and Aronoff, 1952	
Pumpkin	40–60	Pristupa and Kursanov, 1957	

Table 15–3. Translocation rates in different plant species obtained through use of radioactive tracers.

tion of radioactivity at different distances along the stem. Table 15–3 shows some of the translocation rates that have been obtained by the use of this technique.

If we consider what a relatively small amount of area in the complex of sieve tubes is available for translocation, it is remarkable that the high rates shown in Table 15–3 can be obtained. This situation is further complicated by the fact that several thousand sieve plates have to be traversed before a metabolite from the leaf can reach the root of a plant. For example, Weatherley, Peel, and Hill (66) found that for every 16 cm a metabolite traveled in a willow stem, it had to traverse 1,600 to 2,000 sieve plates. Later in this chapter, we will discuss certain mechanisms that may explain how translocation rates can be so high in spite of so much resistance in the phloem ducts.

Different metabolites with different translocation rates. Several investigators have noted that different metabolites are translocated in the phloem ducts at different rates. When a solution containing tritiated water (THO), ^{32}P, and ^{14}C-sucrose is fed to the leaves of a 12-day-old bean plant, different translocation rates are obtained for the individual radioac-

tive substances (3). The ^{14}C-sucrose moves considerably faster (107 cm/hr) than does either THO or ^{32}P, both of which move at the rate of 87 cm/hr. Similar results were obtained by Gage and Aronoff (21) when they introduced a solution of ^{14}C-fructose and THO to a cut petiole of a 3-week-old soybean plant. In this situation, the radioactive sugar moved much faster than THO. According to the authors of these studies, sugars and water may, to a great extent, be independent of each other when they translocated in the phloem. In addition, studies by Nelson and Gorham (51) indicated that different amino acids move at different speeds in the phloem.

Factors Affecting Translocation

Several factors are known to affect translocation rates in plants. The most important of these are temperature, light, metabolic inhibitors, concentration gradients, mineral deficiencies, and hormones. This list of factors is by no means complete but represents those factors that have been studied most frequently.

Temperature. By varying the temperature of a plant and measuring the increase or decrease in the dry weights of different organs, we may obtain indirect measurements of translocation rates. The assumption is that the dry weight of an organ reflects the rate of movement of solutes into that organ. Using the above method, Hewitt and Curtis (33) demonstrated that the optimum temperature for translocation in bean plants is between 20° and 30°C.

When a plant is subjected to a range of temperatures, its entire metabolism is influenced, and we have difficulty obtaining a true picture of the effect of temperature on translocation per se. To circumvent this problem, Swanson and Böhning (61) tried localized temperature treatments. In their experiments, bean plants were grown at a temperature of 20° ± 1°C. The petiole of one leaf of each plant was fitted with a "temperature jacket," and the blade of that leaf was immersed in a sucrose solution. The plants were then placed in dark cabinets where the temperature was kept constant at 20° ± 1°C. Thus with the exception of the treated petiole, the entire plant was subjected to the same constant temperature. After 135 hours of treatment, the increase in length of the stem during this time was taken as a measure of the rate of movement of sucrose through the treated petiole to the stem (see Figure 15–8). The results agree remarkably well with those of the Hewitt and Curtis study in which the whole plant was subjected to fluctuations in temperature. The Swanson and Böhning experiment establishes the important point that the translocation of solutes is influenced by temperature in much the same manner as are other physiological processes—that is, the rate of translocation increases with temperature to a maximum and then decreases due to the detrimental effects of high temperature.

From past experiments we have been able to obtain data on the translocation of radioactive sugars as influenced by different temperatures. Sugarcane plants fed $^{14}CO_2$ have been shown to have translocation rates that increase with a temperature rise. Thus sugarcane plants subjected to air temperatures of 20°, 24.5°, and 33°C have translocation rates of 84.0, 93.6, and 120 cm/hr, respectively (29). The distribution of radioactivity 90 minutes after $^{14}CO_2$ treatment at the above temperatures is shown in Figure 15–9.

The temperature of the root as compared to that of the shoot appears to have an influence on which direction (up or down from a $^{14}CO_2$-supplied leaf) sugars will move in a plant. Hartt (29) found that when root temperature is kept higher than

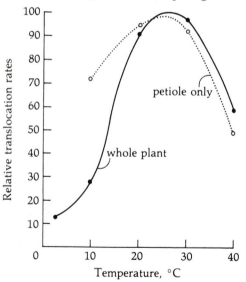

Figure 15–8. Translocation of carbohydrates in bean as affected by temperature in a single leaf petiole (blade immersed in solution of sucrose) and throughout entire plant.

From C.A. Swanson and R.H. Böhning. 1951. Plant Physiol. 26:557 and S.P. Hewitt and O.F. Curtis. 1948. Am. J. Bot. 35:746.

shoot temperature, translocation to the root increases and translocation to the top decreases. When the situation is reversed—shoot temperature higher than root temperature—translocation to the top increases and translocation to the roots decreases. We can assume from Figure 15–10 that the root and top of the sugarcane provide sinks for the utilization of ^{14}C-sugars translocated from the treated leaf. The respiratory activities of these plant parts are enhanced by an increase in temperature. Therefore, an increase in root temperature over shoot temperature will cause an increase in downward translocation. Upward translocation will be increased if shoot temperature is higher than root temperature.

The influence of temperature on source (e.g., leaf blade) and sink (e.g., stor-

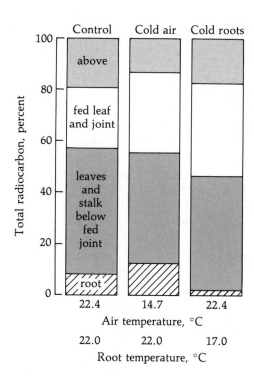

Figure 15–10. Distribution of ^{14}C after 6 days of translocation as affected separately by air and root temperature. When root temperature is kept higher than shoot temperature, translocation increases to root and decreases to top. When root temperature is kept lower than shoot temperature, translocation decreases to root and increases to top.

From C.E. Hartt. 1965. Plant Physiol. 40:74.

age organ) regions primarily reflects the influence of temperature on the rate of translocation. That is, temperature affects the metabolic processes involved in the secretion of sugar into the sieve tubes at the source and out of the sieve tubes at the sink; and, therefore, temperature controls the rate of translocation, as has been demonstrated with experiments on the sugar beet plant (23, 63). When the sink regions of this plant are cooled to about 1°C there is a definite drop in the translocation rate of [14]C-

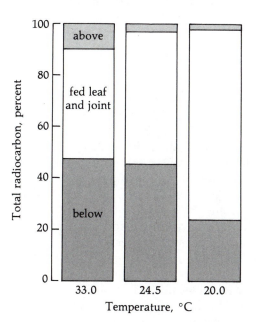

Figure 15–9. Distribution of ^{14}C in sugar cane as affected by air temperature. Data were obtained 90 minutes after [14]CO$_2$ was fed to one leaf.

From C.E. Hartt. 1965. Plant Physiol. 40:74.

labeled photosynthate to a new steady rate approximately 35 to 45 percent of the original rate (23). Rapid recovery to the original rate occurs when cooling is stopped. The fact that translocation continues at a steady but low rate, even though the sink regions are cooled to 1°C, is most likely due to the active secretion of photosynthate into the sieve tubes at the uncooled and therefore uninhibited source region.

Different results are obtained with the same plant when low temperatures (1° to 2°C) are applied to areas other than the source and sink regions (e.g., petiole) (63). When a 2-cm petiole zone is cooled to about 1°C, while the rest of the plant is held at 30°C, a rapid drop in the translocation rate of ^{14}C-photosynthate occurs. However, after a suitable thermal adaptation period, recovery to the original rate occurs. At this time, rewarming of the petiole to 25°C has little effect on the translocation rate (see Figure 15–11).

Because the sugar beet is able to adapt its phloem translocation system to colder conditions, we call it a *chilling-tolerant* plant. Plants, such as the bean, that show a marked inhibition of phloem translocation rates under chilling conditions (1° to 2°C) are called *chilling-sensitive*. There is some evidence that chilling of these latter plants inhibits translocation by causing a physical blockage of sieve plates rather than by inhibiting directly any metabolic process that might drive translocation (24).

Observers have noted a similar situation in cotton plants subjected to high temperatures. They have observed callose buildup in the sieve tube elements in cotton plants exposed to temperatures above 40°C for only 15 minutes (40, 56). The heat-induced callose reduces translocation by causing sieve plate pore constriction. A return to near-normal levels of translocation

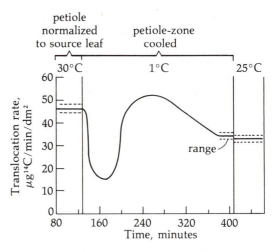

Figure 15–11. Time-course of translocation rate, calculated as carbon accumulating in total sink (all parts distal to cooled region) per minute, normalized to a source leaf area of 1 dm². Time zero equals beginning of labeling with ^{14}CO$_2$. Petiole zone cooled to 1 to 2°C from time 130 to 410 minutes.

From C.A. Swanson and D.R. Geiger. 1967. Plant Physiol. 42:751.

can be achieved within 6 hours after the plants are returned to lower temperatures (41).

Light. As we shall see in a later chapter, the assimilation of CO_2 increases as light intensity increases. The root/shoot dry weight ratio of wheat increases as light intensities increase, thereby indicating that translocation to the root as compared to translocation to the shoot becomes greater with light intensity (see Table 15–4).

A study of the translocation of radioactive metabolites in the light and in the dark with soybean plants by Nelson and Gorham (50) has produced interesting results. They first allowed two 30-day-old soybean plants to photosynthesize in ^{14}CO$_2$ for 15 minutes. Then they allowed one plant to remain in the light an additional 3 hours. They placed

Table 15–4. The root/shoot dry weight ratio of wheat showing an increase with increase in light intensities.

Light Intensity (foot candles)	Root/Shoot Ratio
200	0.14
500	0.17
1000	0.27
1750	0.32
2500	0.32
5000	0.43

Source: *Data of D.J.C. Friend, V.A. Helson, and J.E. Fisher, as reported by C.D. Nelson, 1963.* Environmental Control of Plant Growth. *New York: Academic Press.*

the other plant in the dark, also for a 3-hour period. When they analyzed the plant parts, they found that in 3 hours the light plants translocated about 2 percent of their total radioactivity to the stem tip and 4.4 percent to the roots. In 3 hours, the dark plants translocated only 0.5 percent of their total radioactivity to the stem tip, and the roots received 16.5 percent. We can postulate from this data that in the dark root translocation is favored over shoot translocation.

Studies have shown that translocation rates may be influenced by the quality of light received by the plant. Hartt (30) found that the translocation of ^{14}C-photosynthate in detached blades of sugar cane is accelerated in the presence of red or blue light. Hartt's observations are supported in part by the discovery that red light also enhances the uptake of ^{14}C-sucrose into the plumules of etiolated pea seedlings (25, 26).

Metabolic inhibitors. Metabolic inhibitors have been shown to inhibit carbohydrate translocation (28, 36, 64, 68). Some of the inhibitors used include 2,4-dinitrophenol (DNP), arsenite, azide, iodoacetic acid, fluo-

ride, and hydrogen cyanide. Whether the inhibitor has its effect on the metabolism of the conducting elements per se or on the metabolism of the supplying and receiving cells is difficult to assess. The inhibitor could conceivably be carried to the photosynthesizing mesophyll cells of the leaf, where it would inhibit cell-to-cell transport of photosynthate to the phloem conducting elements. It is just as possible for a metabolic inhibitor to be carried to the receiving cells or sinks, where it might impede the deposition of translocated metabolites. In both cases, the rate of translocation would be inhibited. Indeed, Swanson, in a review of the subject (60), claimed that work with inhibitors indicates that translocation rates are more a function of the metabolism of the supplying and receiving tissues than of the metabolism of the conducting cells themselves. Work with soybean (28) and castor bean seedlings (36) strongly indicates that DNP inhibition of translocation is caused by the effect of DNP on the metabolic processes involved in the movement of photosynthate into and out of the sieve tubes. In these studies, DNP did not appear to affect translocation in the sieve tubes.

Sij and Swanson (59) demonstrated that phloem translocation in squash, following a short adaptation period, progresses in a normal fashion through a zone of petiole tissue, under anaerobic conditions. Again, this suggests that metabolic inhibitors such as cyanide do not inhibit phloem translocation by affecting the metabolism of the conducting elements, but do so by being translocated to the source or sink regions, where they inhibit photosynthesis and loading and unloading processes.

Kursanov (37) emphasized the role of metabolism in phloem transport. Looking at the results of these studies, we can assume that metabolic inhibitors have at least a partial effect on translocation through a direct retarding influence on the metabolism of the conducting elements.

Concentration gradients. Scientists believe that the direction of sugar flow in the sieve tubes is along a gradient of decreasing total sugar concentrations. Early work by Mason and Maskell (42, 43) on the translocation of sugars in the cotton plant demonstrated that the movement of sugars follows a *diffusion pattern*—that is, a correlation is found between rate of transport and the sugar gradient in the bark. They found that the direction of translocation is always from a region of high concentration to a region of low concentration. In addition these workers found that defoliation causes the sugar gradient to disappear. (See also the review by Mason and Phillis [44]).

Zimmermann (69, 70, 71, 72) found concentration gradients in white ash to be about 0.01 mole/m and positive in the downward direction of the trunk. Defoliation experiments by Zimmermann produced some interesting results. As with the work of Mason and Maskell, removal of carbohydrate results. As with the work of Mason and Maskell, removal of carbohydrate supply caused the sugar gradient in the sieve tube system to disappear. However, some of the concentration gradients of individual sugars became negative (see Figure 15–12). We will discuss the significance of sugar concentration gradients in phloem translocation further in the section dealing with translocation mechanisms.

Mineral deficiencies. Significant work concerning the role of minerals in phloem transport has been done with boron. Gauch and Dugger (22) found that the absorption and translocation of sucrose by a leaf of a bean or tomato plant immersed in a solution of ^{14}C-sucrose are greatly facilitated when

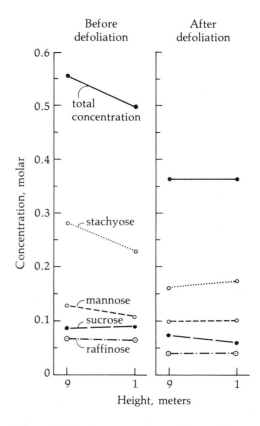

Figure 15–12. Concentration gradients along tree trunk of white ash (*Fraxinus americana*) before and after defoliation. Gradients disappear as result of defoliation, some gradients become slightly negative.
From M.H. Zimmermann. 1958. Plant Physiol. *33:213.*

boron is added to the solution. According to these experimenters, an ionizable complex is formed between boron and sucrose, which moves through cell membranes much more readily than does nonborated sucrose.

Sucrose is not the only compound aided in its translocation by boron. The growth regulators, 2,4-dichlorophenoxyacetic acid, indole acetic acid, 2,4,5-trichlorophenoxyacetic acid, and α-naphthalene acetic acid, when applied with sucrose to the leaves of a bean plant, are translocated much more efficiently in the presence of boron (22).

Aside from the very noticeable effects of boron, we know little about the influence of mineral deficiency on phloem translocation. Whether a mineral deficiency affects phloem translocation per se or has its influence through modifying the metabolism of supplying and receiving tissues is difficult to assess.

Hormones. Plant hormones are closely associated with the actively growing centers of the plant and would, at the very least, have a strong, indirect effect on phloem translocation. Cellular growth and tissue growth, stimulated by hormones, put a heavy demand on translocated metabolites for building materials and energy. Many investigators believe that the metabolism of these growth centers (sinks) has a strong influence on translocation.

Phloem translocation is at least partially under the control of phytohormones, such as the cytokinins, indole-3-acetic acid (IAA), and gibberellic acid (GA). Kinetin, a synthetic cytokinin, appears to affect the translocation of soluble nitrogen compounds (47). If we remove a leaf of *Nicotiana rustica* from the plant, a migration of soluble nitrogen compounds from the blade into the petiole takes place. Because of this migration, a regeneration of protein cannot occur in the blade, and it rapidly turns yellow. However, if we spray the blade with kinetin, it will remain green—that is, the migration of soluble nitrogen compounds out of the blade into the petiole is retarded. Moreover, if only one-half of the blade is sprayed with kinetin, a migration of soluble nitrogen from the unsprayed half to the sprayed half will occur. In other words, kinetin promotes the accumulation of soluble nitrogen. We will discuss this topic further in Chapter 20 on cytokinins.

If we decapitate a bean or pea plant and apply lanolin paste to the cut surface, only a small amount of ^{32}P-phosphate or ^{14}C-sucrose—applied to the lower part of the stem—will accumulate in the decapitated internode. However, the presence of IAA in the lanolin paste causes a marked stimulatory effect on the accumulation of the labeled compounds in the decapitated internode (6, 22). Under similar circumstances, kinetin or GA has little effect. When we consider the lack of activity of kinetin and GA in this respect, we are surprised to find that the stimulatory influence of IAA on phloem translocation is greatly enhanced by the simultaneous application of either one of these compounds. Figure 15–13 shows the time course of accumula-

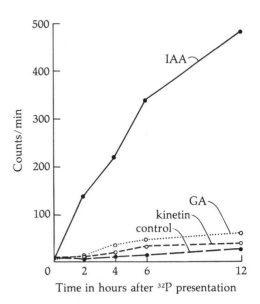

Figure 15–13. Time course of accumulation of ^{32}P in decapitated bean internodes in response to IAA, GA, and kinetin.

From A.K. Seth and P.F. Wareing. 1967. J. Exp. Bot. 18:65.

tion of ^{32}P in decapitated internodes in response to hormone application. Hew, Nelson, and Krotkov (32) obtained somewhat similar results with soybean plants. They removed the apical meristem of the soybean, replaced it with an aqueous solution of IAA or GA, and then exposed one primary leaf to $^{14}CO_2$ for 30 minutes. Determination of the distribution of ^{14}C in various parts of the plant disclosed that both IAA and GA increased the total amount of ^{14}C photosynthate translocated and increased the rate at which it was translocated.

When grapevine roots are treated with benzyladenine (BA), a cytokinin, there is a considerable increase in the amount of ^{14}C-labeled photosynthate transported to the roots from leaves exposed to $^{14}CO_2$ (57). Moreover, the quantity of ^{14}C-labeled amino acids, organic acids, and sugars translocated to the treated roots is also increased, thereby suggesting that BA (or cytokinins in general) has a general enhancing effect on the movement of a number of different compounds in the plant.

Mechanisms of Phloem Transport

Experimenters have presented various theories to explain the transport of assimilates in the phloem. An old and now discarded explanation is that the translocation of materials is a consequence of *protoplasmic streaming*. As formulated by de Vries in 1885, this theory assumed that solute particles are caught up in the circulating cytoplasm of the sieve tube elements and carried from one end of the cell to the other. These particles presumably passed across the sieve plates by diffusing through cytoplasmic strands connecting one element to another. Through the years, however, it has become

clear that the protoplasmic streaming theory does not account for much of our present-day knowledge of phloem transport. The source-to-sink translocation is best explained by a combination of active transport of assimilates into and out of the phloem cells and a pressure flow mechanism.

Münch Pressure Flow Hypothesis

The physiological basis of the Münch pressure flow (also called massflow) hypothesis, described first by Münch in 1930 (48), rests on the assumption that a *turgor pressure gradient* exists between the supplying tissue (source) and the receiving tissue (sink). According to the theory, metabolites are carried passively in the positive direction of the gradient. In other words, in the pressure flow system, we have a unidirec-

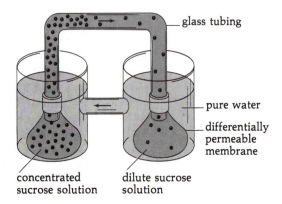

glass tubing

pure water

differentially permeable membrane

concentrated sucrose solution

dilute sucrose solution

Figure 15–14. Physical system demonstrating Münch pressure flow hypothesis. Flow of water into concentrated sugar solution exerts pressure that causes concentrated sugar solution to flow to other compartment. Flow continues until equilibrium or water potentials between compartments and pure water are equal.

tional flow of solutes and water through the sieve ducts, and this flow is driven by a turgor pressure gradient. For example, Figure 15–14 presents a physical system demonstrating the mass, or pressure flow, hypothesis. A and B represent osmometers permeable only to water. Osmometer A contains a higher solute concentration than does B, and both osmometers are submerged in water. Both the osmometers and water containers have open channels that offer little resistance to the flow of solutes and water. Since this is a closed system and the walls of the osmometers are differentially permeable, water will enter A and B and cause a turgor pressure to develop. However, osmometer A will develop a higher turgor pressure since it contains a higher concentration of solute, and this greater pressure will be transmitted throughout the system by virtue of the open channel between the two osmometers. Thus a circulating system is created. Water is induced to flow from A to B, with solutes being carried along passively. Water moves out of B by virtue of the pressure that has developed and is recirculated via the open channel between the water containers. A is the supplying osmometer, and B is the receiving osmometer.

If we apply this system to the plant, A would represent the source and B the sink. The connecting channels between the osmometers and water containers would represent the sieve and xylem ducts, respectively. With this picture in mind, we can see how sugars would be translocated long distances through the sieve tubes.

As Swanson pointed out (60), the movement of sugars out of the leaf chlorenchyma into the sieve tube elements may occur against a concentration gradient. Thus cell-to-cell movement of solutes in the leaf tissues and the final dumping of the solutes into the sieve tube elements could be considered an active process requiring energy. Other investigations suggest that sugar phosphates and an active carrier system may be involved. The findings that sugar beet leaves contain significant quantities of sucrose phosphate (8) and that ATP accelerates the movement of photosynthate from the mesophyll cells to the phloem (38) definitely suggest that the phosphorylation of sugars may be an important factor in their movement across cell membranes. Phosphorylation may facilitate the transport of sucrose across membranes, or it may activate the sucrose molecule, thereby enabling it to unite with a carrier to form a complex that can easily cross cell membranes (37). Figure 15–15 shows the possible path that sucrose may take from the chloroplast to the sieve tube elements. Observers also think that the uptake of sugars by the receiving cells from the phloem ducts proceeds through some active process—a process

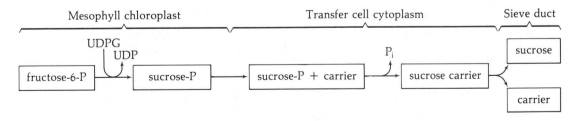

Figure 15–15. Possible path for transport of sucrose from chloroplast to sieve tube elements.

that is perhaps somewhat similar to the transport of sugars into the phloem ducts.

The pressure flow hypothesis accounts only for a unidirectional flow of metabolites. However, it is generally accepted that bidirectional movement takes place in plants. Bidirectional movement cannot occur in the same phloem duct within the physical limits described by the pressure flow hypothesis. Crafts (10, 11) suggested, however, that the leaf may serve two sinks, one toward the apex and one toward the root—that is, metabolites would leave the leaf in separate phloem ducts. Thus we would have bidirectional movement but in separate phloem ducts—a possibility under the pressure flow system.

Studies of phloem translocation have produced considerable evidence in support of the pressure flow theory. Positive concentration gradients have been found in the stems of a number of plants (42, 43, 68, 71, 72). The disappearance of these gradients when we defoliate a plant supports the concept of pressure flow. The common observance of phloem exudate moving out of an incision in the stem, rapidly at first and then at a steadier rate, demonstrates that the sieve tube elements are, in fact, under pressure. Also, the fact that the volume of exuding substance greatly exceeds the volume of any cut sieve tube elements in the immediate area of the incision shows that the exuding substance has been translocated over a considerable distance.

When we consider the evidence for and against the pressure flow concept, we question its workability in the form it was originally conceived of by Münch. The current trend is to confine the pressure flow concept to the sieve tubes only and to accept the fact that energy is required for the absorption of sugars by the sieve tube elements and for the uptake of sugars from these elements by the receiving cells.

Phloem Loading and Unloading

In the movement of assimilates in plants, the assimilates are initially secreted into the phloem at the expense of ATP (active process). This process is termed *phloem loading*. The solutes enter companion cells and then pass into the sieve tubes, possibly via cytoplasmic connections. There is also evidence that parenchyma cells play an important role in the loading process as well. The increased accumulation of sugars in the sieve tubes causes the water potentials to become more negative and facilitates osmosis from surrounding cells and from the transpiration stream. The turgor pressure builds up, and the solutes are moved to a sink, such as the root, meristems, developing leaves, or reproductive structures. At the sink, the sugar is actively moved out of the sieve tubes due to the expenditure of metabolic energy by the companion cells. This active transport process is referred to as *phloem unloading*. Removal of the osmotically active solutes causes the water potentials of the sieve tubes to become less negative and establishes a gradient favorable for the water to diffuse out of the sink and return to the transpiration stream. We should emphasize the pertinent points of the pressure flow hypothesis:

1. The Münch pressure flow hypothesis accounts for the movement of solutes through the sieve tubes along a turgor pressure gradient from source to sink.

2. The loading and unloading of solutes is an active process.

3. The companion cells or parenchyma adjacent to the sieve tubes expend the necessary energy for the loading and unloading processes. The sieve tubes themselves do not seem to be involved in the active transport process.

Questions

15–1. Name and describe the functions of the different cell types found in the phloem tissue of a dicot stem.

15–2. Define the following terms: assimilate stream, sink, metabolic pumps, and assimilate source.

15–3. What is the function of P-protein?

15–4. Discuss the types of substances translocated in the phloem. Are mineral salts translocated in the phloem?

15–5. Phloem tissue is often said to conduct materials in a downward direction. Is this statement true? Explain.

15–6. List and explain the factors that influence translocation in the phloem.

15–7. Explain the role of boron in sucrose translocation.

15–8. List some explanations for the influence of cytokinins in establishing sinks in plants.

15–9. Explain the Münch pressure flow hypothesis.

15–10. Describe the events that take place in the translocation of sucrose from a source to a sink in a plant.

15–11. How are substances translocated in plant parts that do not contain well-developed vascular tissue?

Suggested Readings

Anderson, W.P. 1973. The mechanism of phloem translocation. *Symp. Soc. Exp. Biol.* 28:63–85.

Aronoff, S., J. Dainty, P.R. Gorham, L.M. Srivastava, and C.A. Swanson, eds. 1975. *Phloem Transport*. New York: Plenum Publishing.

Canny, M.J.P. 1973. *Phloem Translocation*. New York: Cambridge University Press.

Crafts, A.S., and C.E. Crisp. 1971. *Phloem Transport in Plants*. San Francisco: Freeman.

Cronshaw, J. 1981. Phloem structure and function. *Ann. Rev. Plant Physiol.* 32:465–484.

Cronshaw, J., and K. Esau. 1968. P-protein in the phloem of *Cucurbita*. 1. The development of P-protein bodies. *J. Cell Biol.* 38:25–39.

Cronshaw, J., J. Gilder, and D. Stone. 1973. Fine structural studies of P-protein in *Cucurbita, Cucumis* and *Nicotiana*. *J. Ultrastruct. Res.* 45:192–205.

Eschrich, W. 1970. Biochemistry and fine structure of phloem in relation to transport. *Ann. Rev. Plant Physiol.* 21:193–214.

Evert, R.F. 1977. Phloem structure and histochemistry. *Ann. Rev. Plant Physiol.* 28:199–222.

Evert, R.F., W. Eschrich, and W. Heyser. 1978. Leaf structure in relation to solute transport and phloem loading in *Zea mays* L. *Planta* 138:279–294.

Lüttge, U., and N. Higinbotham. 1979. *Transport in Plants*. New York: Springer-Verlag.

Moorby, J. 1977. Integration and regulation of translocation within the whole plant. *Symp. Soc. Exp. Bot.* 31:425–454.

Pate, J.S., and B.E.S. Gunning. 1972. Transfer cells. *Ann. Rev. Plant Physiol.* 23:173–196.

Spanner, D.C. 1979. The electroosmotic theory of phloem transport: a final restatement. *Plant Cell Environ.* 2:107–121.

Zimmermann, M.H., and J.A. Milburn, eds. 1975. *Encyclopedia of Plant Physiology*, vol. 1. *Transport in Plants*. 1. *Phloem Transport* Berlin: Springer.

Chapter 16

Respiration and Chemical Interconversions

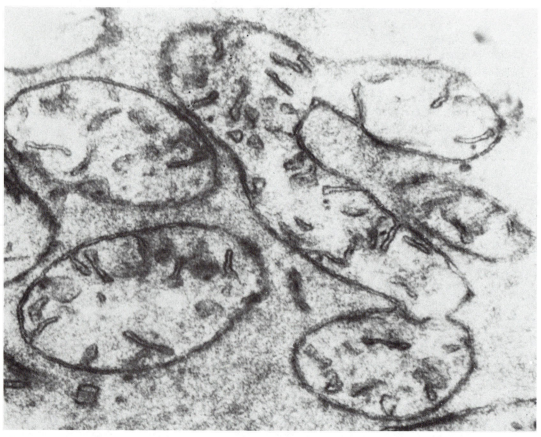

Electron micrograph of dividing mitochondrion in cortical cell of radish (*Raphanus*) root. *Courtesy of M.A. Hayat, Kean College of New Jersey.*

There is considerable diversity in the chemicals that are potential sources of energy for plant cells. Let us consider, for example, the tremendous array of substances classified as carbohydrates, lipids, and proteins. Not only do these substances represent potential energy for driving biological reactions, they are also the building materials for the plant. Indeed, we will observe that, on the basis of elemental composition, there is little distinction between many of the structural components of an organism and its potential fuel. Hence any discussion concerning energy liberation through respiratory pathways must include a consideration of the interconversions (synthesis or breakdown) of organic compounds. It is within these metabolic reactions that the mechanics of energy packaging, into molecules, energy utilization, and architectural construction in plants reside.

During the process of energy utilization, plant structures are maintained because of the preferential utilization and compartmentalization of biochemicals and the control of enzyme synthesis and activity. Carbohydrates provide the readily available and stored energy and are often the preferentially used starting material. The oxidation of glucose and similar components releases considerable amounts of energy. This energy is trapped and utilized as ATP, as other high-energy compounds, and as the reduced coenzymes NADH and NADPH. The amount of bond energy in glucose is not released all at once but in a stepwise series of reactions controlled by enzymes. These reactions may also yield components critical to the cellular structure, depending on a variety of regulatory factors.

In this chapter we consider the various series of reactions, or *metabolic pathways*, that are involved in the relatively precise synthesis and breakdown of biochemicals and in the interconversion of the structural and energy-yielding substrates of respiration in the living plant.

Relationship of Carbohydrate Metabolism to Other Compounds

The carbohydrates produced in photosynthesis are metabolically significant to plants in that they serve as the starting materials in the production of ATP and the reducing power in the form of reduced coenzymes such as NADH and NADPH. The series of oxidation-reduction reactions involved are known collectively as *respiration*. This process may be summarized by the following:

$$C_6H_{12}O_6 + 6\ O_2 + 38\ ADP + 38\ P_i \longrightarrow 6\ CO_2 + 6\ H_2O + 38\ ATP$$

One molecule of glucose ($C_6H_{12}O_6$), six of molecular oxygen, thirty-eight of adenosine diphosphate (ADP), and thirty-eight of inorganic phosphate (P_i) yield six molecules of carbon dioxide, six of water, and thirty-eight of adenosine triphosphate (ATP). We will discuss the energetic considerations inherent in the process later.

Another highly significant aspect of molecular charges during metabolism is that carbohydrates and their derivatives are not always broken down completely but serve as precursors to other substances along the respiratory pathway that lead directly to the synthesis of cell wall materials, nucleic acids, amino acids, proteins, lipids, phytohormones, pigments, and so on (see Figure 16–1). The important point here is that the energy-yielding and the synthetic reactions of metabolism are represented by a dynamic

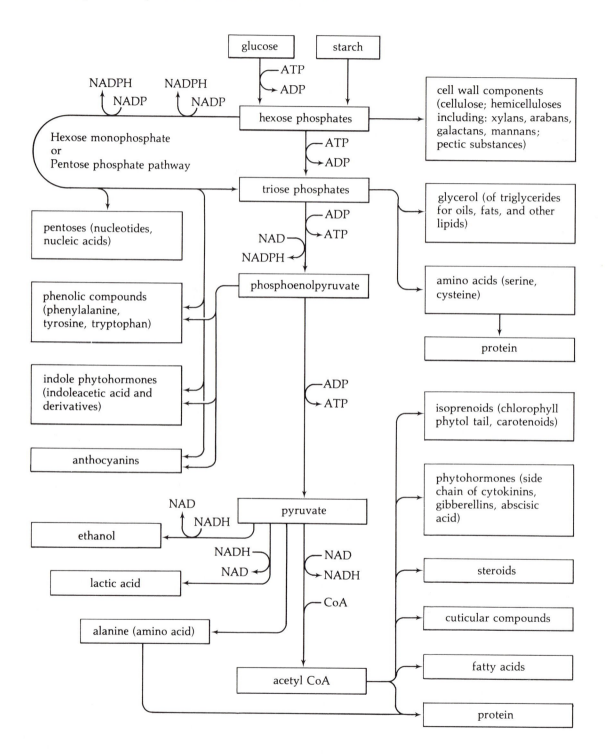

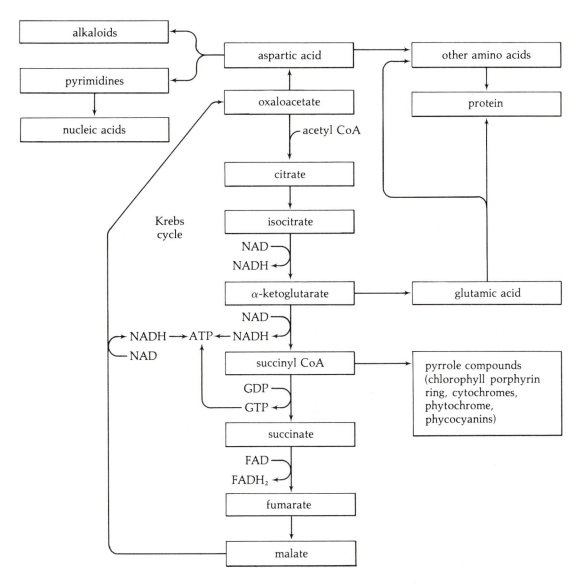

Figure 16–1. Overview of relationship of cellular components and energy-yielding reactions of respiration.

relationship between the interconversion of most biochemicals and the energy liberation and energy utilization by a plant. Figure 16–1 illustrates the general relationship of many plant products.

Energy Liberation and Utilization

Both energy-yielding and energy-consuming reactions occur within the living cell.

The potential, or stored, energy of carbohydrates may be harnessed to drive the synthesis of other compounds (e.g., lipids)—that is, the coupling of energy-yielding and energy-consuming reactions. However, energy-yielding reactions of the cell occur in many instances in the absence of energy-consuming reactions. The energy released in such a situation would be in the form of heat and would be lost to the organism. However, nature has provided the cell with a means of temporary energy storage in the form of adenosine triphosphate (ATP). Thus the energy released in the oxidation of compounds such as carbohydrates, lipids, proteins is immediately utilized in the synthesis of ATP from adenosine diphosphate (ADP) and inorganic phosphate (P_i). The chemical energy transferred to ATP can be used to drive synthetic reactions, ADP and P_i being released in the process.

There is, then, an intermediate compound (ATP) capable of receiving energy from one reaction and transferring this energy to drive another reaction. This energy transfer is of obvious advantage to the living system since ATP can be formed in the oxidation of a variety of compounds and can be used to drive the synthesis of a variety of compounds. In other words, the oxidation of a compound, such as glucose, can provide the energy, through ATP, for the synthesis of a number of cellular materials. In contrast to fuel burned in manufactured engines, which allow a large amount of released energy to be lost in the form of heat, the oxidation of substances in the cell occurs with relatively little energy loss because of the cell's very efficient energy-transfer system mediated by ATP. We must understand that the energy locked in a biological compound may be transferred repeatedly. Thus in a dynamic system, such as the living cell, the stored energy of glucose may be found at one time in ATP and at another time locked in the bonds of a protein molecule. Figure 16–2 shows a schematic representation of the cyclic manner in which ATP is synthesized and broken down as an intermediate between energy-releasing and energy-consuming reactions.

Since the 1940s, our knowledge of the metabolic pathways of respiration has increased immensely. Concepts established through the biochemical investigations of many different organisms leave little doubt that the fundamental aspects of respiration are the same in most forms of life. The glucose molecule that is oxidized in the simple yeast cell travels through the same sequence

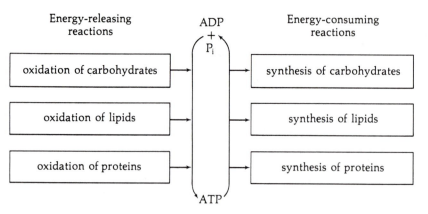

Figure 16–2. Summary of role of ATP as intermediate energy-transfer compound.

of reactions as does a glucose molecule residing in a leaf of the mighty redwood tree. To be sure, there are some differences to be found, but they are minor and may be excluded from the overall picture of respiration as an essential life process.

The most important feature of respiration is the release of usable energy. We will analyze the various metabolic pathways that participate in the release of this energy. In our discussion, we will use the words *oxidation* and *reduction*. *Oxidation* refers to the removal of electrons from a compound, a process usually accompanied in the cell by the removal of hydrogen. *Reduction* refers to the addition of electrons to that compound, usually accompanied in the cell by the addition of hydrogen.

Embden-Myerhof-Parnas Pathway, Glycolysis, Fermentation

The *glycolytic pathway* was the first series of metabolic reactions to be understood. The term describes the sequential breakdown of glucose by a series of reactions operative in a variety of tissues, ending with ethanol and carbon dioxide or with lactic acid. The production of alcohol from hexoses is termed *fermentation*. Since lactic acid and ethanol production is not characteristic of higher plants, the precursor to either compound, *pyruvic acid*, is often considered to be the end product of glycolysis. The pathway from glucose to pyruvic acid is also known as the *Embden-Myerhof-Parnas* (EMP) *pathway*, after the scientists who studied the various intermediates and enzymes.

We should note that the components involved in the various pathways can be referred to by more than one term. For example, pyruvic acid (R—COOH) may be ionized and termed pyruvate (R—COO⁻). The other organic acids may also be referred to as the acid or ionized form.

The EMP pathway results in the conversion of one molecule of glucose to two molecules of the three-carbon compound, pyruvic acid. However, glycolysis is not a one-step reaction but a series of closely integrated reactions that eventually lead to pyruvate. Another point we should stress is that the reactions of glycolysis occur in the cytoplasm and do not require the presence of oxygen.

In order to better understand the general nature of glycolysis, we may subdivide it into two major phases: (1) the production of glucose to fructose-1,6-diphosphate and (2) the splitting of this compound into two three-carbon compounds that lead to the formation of pyruvate (see Figure 16–3).

Three reactions occur in the stepwise conversion of glucose to fructose-1,6-diphosphate. First, the sixth carbon of glucose is phosphorylated in the presence of ATP and the enzyme *hexokinase*. The products of this reaction are *glucose-6-phosphate* and ADP. The next reaction is catalyzed by the enzyme *phosphoglucoisomerase* and results in the conversion of glucose-6-phosphate to *fructose-6-phosphate*. In essence, this reaction entails the conversion of glucose, an aldose sugar, to fructose, a ketose sugar. In the third reaction, the first carbon of fructose-6-phosphate is then phosphorylated in the presence of ATP and the enzyme *phosphofructokinase*. The products of this reaction are *fructose-1,6-diphosphate* and ADP.

The second major phase in glycolysis involves the splitting of fructose-1,6-diphosphate into two three-carbon compounds, *3-phosphoglyceraldehyde* and *dihydroxyacetone phosphate*. *Aldolase* catalyzes this reaction. The two three-carbon compounds are inter-

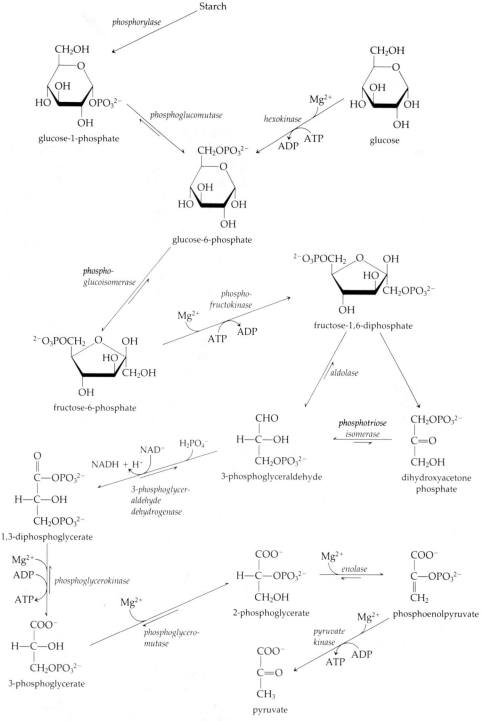

Figure 16–3. Scheme of glycolytic, or EMP, pathway. The *ate* ending of pyruvate, glycerate, and so on indicates ionized form of molecule (e.g. pyruvate equals COO⁻; pyruvic acid equals COOH).

convertible—that is, they are maintained in equilibrium by the action of the enzyme *phosphotriose isomerase*. If for example, 3-phosphoglyceraldehyde is depleted, additional amounts will be formed by the isomerization of dihydroxyacetone phosphate, continuing with glycolysis, 3-phosphoglyceraldehyde is converted to *1,3-diphosphoglycerate*. This reaction involves the incorporation or addition of inorganic phosphate to the first carbon of 3-phosphoglyceraldehyde and the reduction of NAD^+ to NADH and is catalyzed by the enzyme *phosphoglyceraldehyde dehydrogenase*.

Note that the continual conversion of 3-phosphoglyceraldehyde to other intermediates of the glycolytic pathway causes significant decreases in its levels. Thus with continual conversion to other glycolytic intermediates and the fact that there is an equilibrium relationship between the trioses, more and more dihydroxyacetone phosphate is converted to 3-phosphoglyceraldehyde.

The consumption of inorganic phosphate in the oxidation of 3-phosphoglyceraldehyde is important to the plant because this phosphate is involved in the synthesis of ATP in the next reaction of the glycolytic sequence. In the presence of ADP and the enzyme *phosphoglycero kinase*, 1,3-diphosphoglycerate is converted to *3-phosphoglycerate* and ATP. The process of ATP formation by the transfer of phosphate from one of the intermediates of the pathway (in this case; 1,3-diphosphoglycerate is at an appropriate energy level) to ADP is termed *substrate level phosphorylation*. This process represents the major way ATP is generated from bond energy under anaerobic conditions and is particularly important to fermentation.

The 3-phosphoglycerate that is formed in the above reaction is transformed to *2-phosphoglycerate* by the activity of the enzyme *phosphoglyceromutase*. The elimination of the elements of water (dehydration) from 2-phosphoglycerate in the presence of *enolase* results in the formation of *phosphoenolpyruvate*. In the presence of ADP and pyruvate kinase, phosphoenolpyruvate is converted to pyruvate. In this reaction, the phosphoric acid residue of phosphoenolpyruvate is transferred to ADP to form ATP, another example of substrate level phosphorylation.

The EMP pathway, which is also referred to as the *hexose diphosphate pathway*, is the chief pathway in which glucose or intermediates are converted to pyruvate (pyruvic acid). It involves the interconversion of sugars and transfer of phosphate groups and the ultimate conversion of one six-carbon compound to two three-carbon molecules. It is an anaerobic pathway in which some NADH and ATP is generated. The ATP is generated by substrate level phosphorylation. The overall reaction sequence for the EMP pathway is based on the following:

Overall Reaction Sequence

1 glucose + 4 ADP + 2 ATP + 2 P_i + 2 NAD

$$\downarrow$$

2 pyruvate + 2 ADP + 4 ATP + 2 NADH

Balanced Reaction

1 glucose + 2 ADP + 2 P_i + 2 NAD $\longrightarrow$
2 pyruvate + 2 ATP + 2 NADH

In the first phase, the conversion of glucose to fructose-1,6-diphosphate, there is no energy gain. Indeed, two ATP molecules are consumed for every glucose molecule phosphorylated. However, in the second phase, the conversion of fructose-1,6-diphosphate to two molecules of pyruvate, four ATP molecules are formed—two for each triose split off from fructose-1,6-diphosphate. If we consider the complete EMP pathway, the conversion of one molecule of glucose to two molecules of pyruvate results in a net gain of two ATP and two NADH molecules. As we shall see later, the production of NADH from the oxidation-reduction reactions of glycolysis is significant in organisms that respire aerobically.

Fermentation

The overall reaction for fermentation is:

$$C_6H_{12}O_6 \longrightarrow 2\ CH_3\!-\!CH_2OH + 2\ CO_2$$
$$\text{glucose} \qquad\qquad \text{ethanol} \qquad \text{carbon} \atop \text{dioxide}$$

That is, one molecule of glucose is converted to two molecules of ethanol and two molecules of carbon dioxide. Fermentation is a sequential series of reactions that occurs in the absence of oxygen. In fact, there is very little difference between the fermentative process and the glycolytic process; most of the intermediate reactions are found in both pathways.

As in glycolysis, glucose is converted to pyruvate during the process of fermentation. However, in fermentation the process goes one step further and pyruvate is converted to ethanol and CO_2 or to lactic or other organic acids depending on the organism involved (e.g., different types of bacteria). See the equation below. The enzymes catalyzing the two steps of the fermentation reaction are *carboxylase* and *alcohol dehydro-*

genase. Since no ATP is produced in the reaction and the rest of the fermentation process is identical to glycolysis, the net gain of ATP per molecule of glucose fermented is two. We should note, however, that fermentation is not a normal process in plant respiration; it takes place only under appropriate conditions.

Fermentation is the major energy-yielding process of several different microorganisms that are termed *anaerobes*. They are called that since they are capable of existing and breaking down organic compounds in the absence of oxygen. Indeed, some of these organisms, called *obligate anaerobes*, die if they are exposed to an appreciable amount of oxygen. An example of this type of organism is *Clostridium botulinum*, which, in humans, causes the often fatal disease of botulism. The metabolic toxins produced by the bacterium under anaerobic conditions are extremely toxic to humans and other animals. There are other anaerobes, however, that do not rely on fermentation as a source of energy. Rather, they utilize inorganic molecules (NO_3^-, SO_4^-) as hydrogen acceptors in place of O_2.

The best known of the fermenting organisms are the yeasts. The production of alcohol through fermentation by yeast has been known to humans for a very long time. However, real progress in the biochemical analysis of fermentation was not initiated until the beginning of the twentieth century, at which time the Buchner brothers (1897) found that cell-free preparations from yeast could ferment glucose (see Chapter 10 on enzymes). Yeasts are *facultative anaer-*

$$\underset{\text{pyruvate}}{CH_3\!-\!\overset{\displaystyle O}{\overset{\displaystyle \|}{C}}\!-\!COO^-} \xrightarrow{\hspace{1cm}\overset{\textstyle CO_2}{\nearrow}\hspace{1cm}} \underset{\text{acetaldehyde}}{CH_3\!-\!CHO} \underset{\xrightleftharpoons{\hspace{2cm}}}{\overset{\overset{\textstyle H^+}{\overset{\textstyle +}{\overset{\textstyle NADH\quad NAD^+}{\curvearrowright}}}}{}} \underset{\text{ethanol}}{CH_3\!-\!CH_2OH}$$

obes—that is, they can exist in either the presence or absence of oxygen.

Although we have mentioned only ethanol and CO_2 as by-products of fermentation, we should be aware that there are other products that may be produced through this process—for example, lactic acid is a by-product of the fermentation of glucose by lactic acid bacteria. This process is best known for its effect on milk. In lactic acid fermentation, lactic acid, instead of ethanol, is formed from pyruvate. *Lactic acid dehydrogenase*, the enzyme catalyzing this reaction, is also extremely significant in the diagnosis of heart attacks in humans because it is released into the bloodstream from the damaged muscle (the heart).

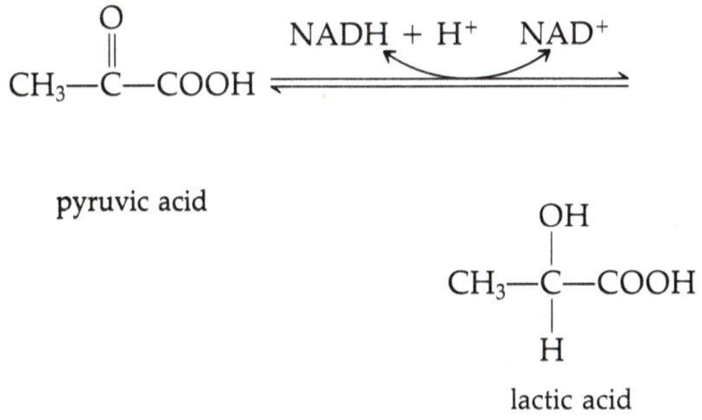

$$CH_3-\overset{\overset{\displaystyle O}{\|}}{C}-COOH \;\rightleftharpoons\; \overset{NADH + H^+ \quad NAD^+}{\xrightarrow{\hspace{2cm}}}$$

pyruvic acid

$$CH_3-\overset{\overset{\displaystyle OH}{|}}{\underset{\underset{\displaystyle H}{|}}{C}}-COOH$$

lactic acid

The products of fermentation—ethanol and lactic acid—still contain a considerable amount of energy. The plant does not benefit from this unreleased energy, an indication that anaerobic respiration is a relatively inefficient process.

Formation of Acetyl Coenzyme A

We have shown that the degradation of carbohydrates under anaerobic conditions proceeds via the EMP pathway to pyruvate. Pyruvate, then, represents the termination of the glycolytic scheme. However, if sufficient oxygen is present, oxidative decarboxylation of pyruvic acid to form *acetyl coenzyme A* takes place. This reaction is very complex and requires the presence of at least five essential cofactors and a complex of enzymes. The five cofactors necessary for the successful formation of acetyl coenzyme A are thiamine pyrophosphate (TPP), Mg ions, NAD^+, coenzyme A (CoA), and lipoic acid. Gunsalus (10) suggested four steps in the formation of acetyl coenzyme A from pyruvate (see Figure 16–4).

The first step involves the formation of a complex between TPP and pyruvate, followed by the decarboxylation of pyruvate. In the second step, the acetaldehyde unit remaining after decarboxylation reacts with the cofactor lipoic acid to form an acetyl-lipoic acid complex. In the reaction, lipoic acid is reduced and the aldehyde is oxidized to an acid. The newly formed acid forms a thioester with lipoic acid. The third step involves the release of the acetyl group from lipoic acid to CoA. The products of this reaction are acetyl CoA and reduced lipoic acid. The final step involves the regeneration of oxidized lipoic acid by the transfer of electrons from reduced lipoic acid to NAD^+. This last reaction is important because it provides for a continuous supply of oxidized lipoic acid necessary for the formation of acetyl CoA from pyruvate. In addition, the two electrons transferred to NAD^+ to form $NADH + H^+$ are eventually passed along to the electron transport system (discussed later), and this transfer results in the formation of three ATP molecules.

The following reaction sequence summarizes these four steps:

$$\text{pyruvate} + \text{CoA} + NAD^+ \longrightarrow$$
$$\text{acetyl CoA} + CO_2 + NADH + H^+$$

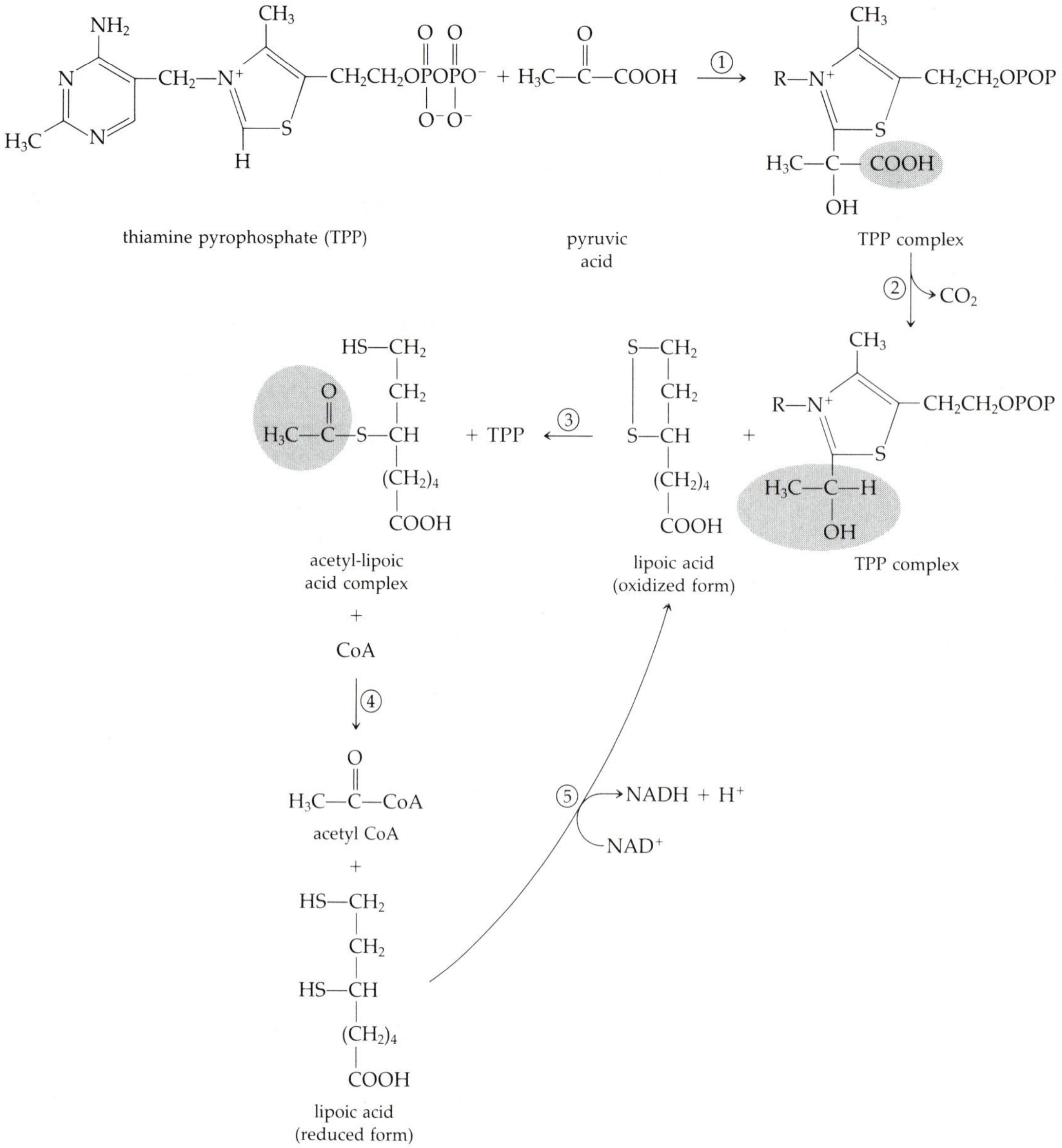

Figure 16–4. Formation of acetyl coenzyme A from pyruvic acid.

Since TPP and lipoic acid are returned to their original state during the reaction sequence, they have been excluded from this summarizing reaction sequence.

Krebs Cycle (Citric Acid Cycle, Tricarboxylic Acid Cycle)

We have noted the relative inefficiency of the glycolysis and fermentation processes in so far as the release of energy is concerned. However, under aerobic conditions, pyruvate, the terminal product of glycolysis, can undergo decarboxylation and, with CoA, form acetyl CoA. Acetyl CoA is the connecting link between glycolysis and the *Krebs cycle* (citric acid cycle or tricarboxylic acid cycle), so-named because of the cylic manner in which the starting compound, *oxaloacetate*, is regenerated. The cycle is named after an English biochemist who played a major role in its discovery. By means of the Krebs cycle and the *electron transport system*, pyruvate is oxidized to CO_2 and H_2O. Thus the complete oxidation of glucose to CO_2 and H_2O may occur through glycolysis, the Krebs cycle, and the electron transport system. Through its association with the electron transport system, the oxidations of the Krebs cycle can account for the formation of twenty-four ATP molecules. Thus the Krebs cycle is far more efficient in the release of energy than are either glycolysis or fermentation. The reactions of the Krebs cycle and the electron transport system require the presence of oxygen and are confined to the mitochondria (see Figure 16–5).

The first reaction of the Krebs cycle is the condensation of acetyl CoA with ox-

aloacetate to form *citric acid* and release CoA. The result of this reaction, catalyzed by *condensing enzyme*, is that a four-carbon dicarboxylic acid is converted to a six-carbon tricarboxylic acid.

Through a series of reactions involving four oxidation steps and three molecules of H_2O (one utilized in the condensation reaction), oxaloacetic acid is regenerated from citric acid. In the process two molecules of CO_2 and eight H atoms are released. The reactions leading to the regeneration of oxaloacetic acid from citric acid are given in Figure 16–5. We should note that these acids and other organic acids of the Krebs cycle are displayed in the ionized form $(R—COO^-)$ and are labeled accordingly (i.e., citrate, oxaloacetate, and so on).

The first reaction involves a dehydration of *citrate* to form *cis-aconitate*. The second reaction calls for the hydration of *cis*-aconitate to yield *isocitrate*.

In the presence of *isocitrate dehydrogenase* and NAD^+, isocitrate is converted to α-*ketoglutarate*. In this first oxidation step of

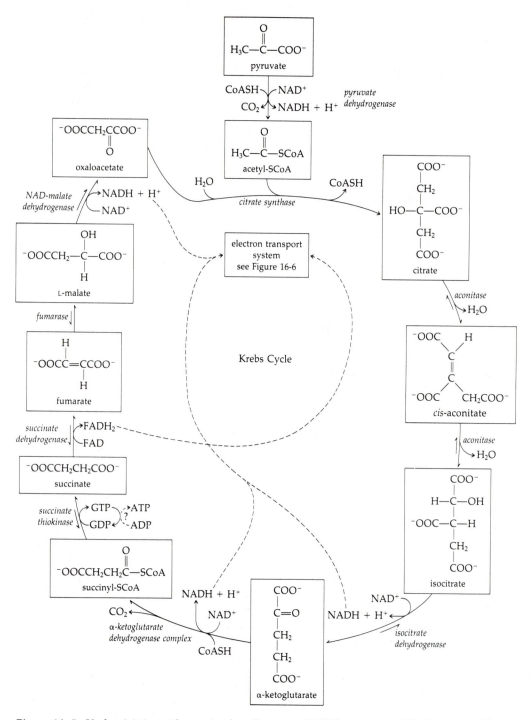

Figure 16–5. Krebs (citric acid or tricarboxylic acid) cycle. Note that NADH and FADH$_2$ are oxidized as part of electron transport system. Each NADH in ETS produces 3 ATP molecules, and FADH$_2$ is responsible for two. All reactions of the Krebs cycle, beginning with formation of citrate from oxaloacetate and acetyl CoA, take place in mitochondria of all eukaryotic cells.

the Krebs cycle, two electrons and two hydrogen ions are removed from isocitrate and are taken up by the coenzyme NAD^+ to form $NADH + H^+$. α-Ketoglutarate is a key compound in the metabolism of the plant. Not only is it involved in carbohydrate and lipid metabolism, but it also plays an important role in the synthesis and degradation of amino acids.

The oxidation of α-ketoglutarate may be considered analogous to that of pyruvate. Thiamine pyrophosphate is required for the initial decarboxylation, and the succinic semialdehyde formed complexes with oxidized lipoic acid. The succinyl moiety of this complex is transferred to CoA, forming succinyl CoA (—SCoA indicates bond to sulfur atom of CoA) and reduced lipoic acid. Reduced lipoic acid is reoxidized by a NAD-containing enzyme, NAD being reduced in the process. The complex of enzymes catalyzing this series of reactions is collectively called α-*ketoglutaric dehydrogenase*. This last reaction also represents the second oxidation step of the cycle. The energy locked within the thioester, succinyl CoA, may be released in the next reaction to form an energy-rich pyrophosphate bond. Thus in the presence of guanosine diphosphate (GDP) and inorganic phosphate, succinyl CoA is converted to *succinate* and guanosine triphosphate (GTP) is formed.

The oxidation of succinic acid to form *fumarate* is interesting, since it is the only Krebs cycle oxidation that does not employ a pyridine nucleotide. Instead, succinate is dehydrogenated (oxidized) by the ferriflavoprotein *succinate dehydrogenase*. Nevertheless, two hydrogen ions and two electrons are removed from succinate and are used to reduce the flavin prosthetic group, flavin adenine dinucleotide (FAD), of the enzyme succinate dehydrogenase. The oxidation of succinate represents the

third oxidation step of the Krebs cycle. The product of this reaction, fumarate, is hydrated in the presence of *fumarase* to yield *malate*.

In the fourth oxidation step of the Krebs cycle, malate is converted to *oxaloacetate* in the presence of NAD^- *malate dehydrogenase*. In the process, NAD^+ is reduced, forming $NADH + H^+$. Thus the regeneration of oxaloacetic acid completes the cycle. In the four oxidation steps four pairs of H ions and four pairs of electrons are removed from intermediates of the cycle. Three of the pairs of H ions and electrons are utilized in the reduction of pyridine nucleotides. The one remaining pair of H ions and electrons is taken up in the reduction of the FAD prosthetic group of succinic dehydrogenase.

The NADH and FADH produced by the reactions of the Krebs cycle represent reducing power that can be harnessed in the presence of oxygen to produce ATP. The production of ATP is accomplished by a series of oxidation-reduction reactions involving chemicals that comprise the *electron transport system* (ETS). The ETS is intimately associated with the mitochondrial membranes and the Krebs cycle apparatus.

Electron Transport System and Phosphorylation

For aerobic organisms it is essential that the enzymes and reduction products of the Krebs cycle be associated with the electron transport system. Through this association, the reduced pyrimidine nucleotide NADH, FADH, and (rarely) NADPH are reoxidized. The energy released from these oxidations is utilized in the synthesis of ATP. The synthesis of ATP via electron flow through the ETS, with oxygen as the terminal electron acceptor, is known as *oxidative phosphorylation* (see Figures 16–4, 16–5) and takes place in mitochondria.

In essence, the ETS is a chain of carriers consisting of nicotinamide adenine dinucleotide (NAD), flavin nucleotides (FAD, sometimes FMN), coenzyme Q (CoQ), and the cytochromes (cyt b, c, a, a_3). Nonheme ion proteins also seem to be involved but their role is not known exactly.

Most important to the living plant is the fact that each step in the system is characterized by a decrease in the energy level from each chemical to the next (Figure 16–6). In other words, the carriers presumably operate in order of an increasing tendency to undergo reduction (reducing potential becomes increasingly positive from NADH through cytochrome a_3). The electron will flow from a higher to a lower energy level. Thus with each step in the system, the energy level of the electron is lowered and the energy difference is transformed into phosphate bond energy by the conversion of ADP to ATP. Note in Figure 16–6 that hydrogen ions are released in the oxidation of reduced coenzyme Q (CoQ). These hydrogen ions may play an important role in ATP production. Only the electrons are passed along the series of cytochromes. Some investigators suggest that the participation of CoQ in the main pathway of the ETS has not been adequately proven. However, the presence of CoQ in the mitochondria of higher plants and its ability to oxidize FADH (actually, FADH + H^+) and to be reoxidized in turn by cytochrome b strongly evidence its participation in the electron transport system (9).

A further study of Figure 16–6 will show that for every pair of electrons passed along this system, three ATP are formed. The synthesis of ATP occurs in the oxidation of NADH, in the oxidation of two cyto-

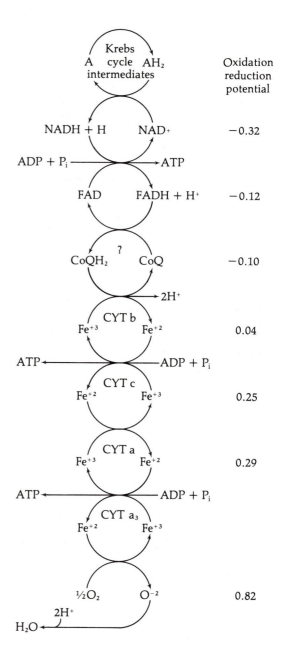

Figure 16–6. Electron transport system. Oxidation reduction potentials are provided to indicate relative reducing power of each component in chain from NADH to molecular oxygen.

chrome b's, and in the oxidation of two cytochrome a's. At their lowest energy level, the electrons are passed to oxygen from reduced cytochrome a_3, thereby activating the oxygen. In this state, oxygen will accept free hydrogen ions to form water.

If we now consider the complete degradation of a molecule of glucose, first to two pyruvic acid molecules via the EMP pathway and then to the two acetyl CoA molecules through the oxygen-requiring Krebs cycle to CO_2 and water, it is possible to have thirty-eight ATP molecules generated.

Inspection of the respiratory pathway (EMP pathway and Krebs cycle) shows that the EMP pathway (Figure 16–3) yields two ATP and two NADH molecules directly from substrate level phosphorylation. These yields are based on the fact that each ATP and NADH is generated for every pyruvic acid molecule used to produce acetyl CoA that enters the Krebs cycle. Also, there are two ATP molecules put into the EMP that are subtracted out in considering total yield. The NADH molecules are generated in the reaction 3-phosphoglyceraldehyde to 1,3-diphosphoglycerate.

Two guanosine triphosphate (GTP) molecules are generated in the Krebs cycle via the reaction succinyl CoA to succinic acid. Phosphate transfer most likely occurs from GTP to ADP, with ATP gained in the reaction.

$$GTP + ADP \rightleftharpoons GDP + ATP$$

Two molecules of NADH are produced in the conversion of two molecules of pyruvate to two molecules of acetyl CoA and six molecules of NADH from the Krebs cycle reactions. Two $FADH_2$ are also produced from the Krebs cycle. These yields are summarized in Table 16–1. Superficially only a very limited amount of ATP appears to be produced directly from the reactions. However, in the presence of molecular oxygen, which acts as the terminal electron acceptor of the electron transport system (ETS), all the NADH and FADH molecules produced may enter the ETS and provide the reducing power necessary to promote electron flow and the production of ATP from ADP and inorganic phosphate. Each NADH generates three ATP molecules in the ETS, and each FADH provides sufficient reducing power for the synthesis of two ATP molecules. These reactions may be summarized as:

$$10 \text{ NAD} + 2 \text{ FAD} + 2 \text{ ADP} + 2 \text{ P}_i \longrightarrow$$
$$10 \text{ NADH} + 2 \text{ FADH}_2 + 2 \text{ ATP}$$

If we used the generally accepted amount of 10,000 calories per mole of ATP, approximately 38,000 calories are generated from one mole of glucose. One mole of glucose contains approximately 673,000 calories, as determined under laboratory condi-

Pathway	NADH (3 ATP)	FADH (2 ATP)	ATP	Total ATP
EMP (yield)	2 (6)	0	2	8
Pyruvic acid to acetyl CoA	2 (6)	0	0	6
Krebs cycle (yield)	6 (18)	2 (4)	2	24
Total ATP	$10 \times 3 = 30$	$2 \times 2 = 4$	4	38 ATP

Table 16–1. Overall energy relationship of EMP pathway and Krebs cycle for the complete oxidation of one molecule of glucose in the presence of oxygen.

tions. However, the actual amount of energy available from each ATP (energy is lost as heat, and so on) is approximately 7,000 calories rather than 10,000. Consequently, the actual energy yield from one mole of glucose is 266,000 calories. This latter amount divided by 673,000 calories present in one mole of glucose indicates an approximate efficiency of 40 percent. Thus organisms that respire aerobically are remarkably efficient in the utilization of chemical bond energy. Conversely, a simple count of the amount of ATP generated in fermentation illustrates the inefficiency of anaerobic respiration.

Oxidative phosphorylation—the chemiosmotic theory. The electron transport system and the coupling of ATP formation from ADP and P_i are established facts. The exact mechanism for the coupling, however, has eluded scientists for some time. The *chemiosmotic theory,* the *chemical hypothesis,* and the *conformational hypothesis* have all been proposed to explain how the electron transport system energizes the mitochondrion and effects the energy transfer mechanisms for ATP formation. Chemiosmotic coupling, proposed by Mitchell (18), is indeed the most widely accepted explanation for the process of ATP formation, both in the chloroplast (photophosphorylation) and in the mitochondrion (oxidative phosphorylation).

The chemiosmotic theory is based on the observation that active mitochondria release hydrogen ions (H^+) to the outside and that the driving force for the proton displacement comes from the flow of electrons through the ETS system. As a result of H^+ accumulation to the outside, a concentration gradient of increasing H^+ concentration from the inside to the outside is established (see photophosphorylation). The hydrogen

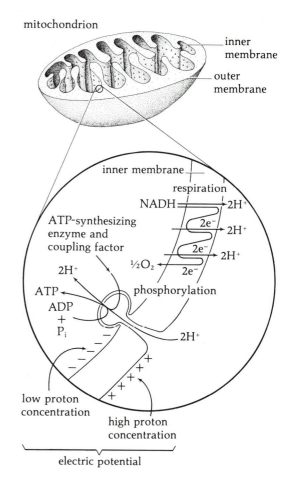

Figure 16–7. Mitochondrion and chemiosmotic explanation for ATP formation.

ion gradient across the mitochondrial membranes represents potential energy for the energy-requiring formation of ATP (18). There are several explanations for how ATP is formed when H^+ reenters the mitochondrion. One explanation is that H^+ moves through channels that terminate as knobs in the membrane (see Figure 16–7) with the flow providing energy to drive the ATPase-catalyzed reaction of ATP formation and removal of water. Another idea is that the H^+, moving through the channel, may initially bind with ATPase to produce an activated phosphate that readily reacts with ADP. The H^+ movement through the channel may also activate the ATPase necessary for the synthesis of ATP. A major modification of Mitchell's original hypothesis is the suggestion that an electrical potential may be established across the membrane, with H^+ or other cation displacement (see Figure 16–7).

Reduced NADH gives up two electrons, which move back and forth through the mitochondrial membrane three times, going from one carrier in the electron transport system to another. The electrons finally reduce oxygen with the formation of water. However, as the electron moves to the outer portion of the membrane each of the three times, the charge differential causes protons to migrate in the same direction and produces the high (from the outside) to low H^+ gradient. The proton buildup initiates the movement of the protons through the diffusion channels into the knobs, thereby providing the energy for the reaction ADP + P_i to ATP. The latter event is catalyzed by ATPase, often referred to as the *coupling factor*. Although further details remain to be elucidated, evidence supports this theory of H^+ flow and the establishment of electrical potential across the mitochondrial membrane during operation of the electron transport system. Also, the action of dinitrophenol, which is a known uncoupler of the ETS and oxidative phosphorylation, is explainable according to the chemiosmotic theory because the ionized phenol will scavenge protons on the outside of the mitochondrion and interfere with the flow necessary for energy transfer for ATP formation.

Cyanide-resistant respiration—the alternative pathway. Cyanide-resistant respiration seems to be widespread in higher plant tissue. Accordingly, the mitochondria in such tissues are also cyanide resistant (12, 20). This resistance results, evidently, from a branch point in the ETS preceding the highly cyanide-sensitive cytochromes. In tissues lacking this branch point, or *alternate pathway*, blockage of the cytochromes by cyanide inhibits the electron flow, the entire electron transport system, the oxidation of NAD-linked substrates, and, hence, the tricarboxylic acid (TCA) or Krebs cycle. Although observers do not know the exact nature of the branch point, they believe it occurs preceding the b cytochromes, presumably near the quinones.

Bendall and Bonner (5) reported the presence of a so-called alternate cyanide-resistant oxidase as part of the alternate pathway. In addition, several chemically uncharacterized flavoproteins, NADH-ubiquinone reductase, and succinic acid dehydrogenase may be associated with the alternate pathway. Further, observers have

Krebs cycle substrates
$\downarrow$
NADH $\longrightarrow$ FADH$_2$ $\longrightarrow$ Q $\rightleftharpoons$ flow $\longrightarrow$ cytochrome
$\downarrow$
flow
$\downarrow$
X $\longrightarrow$ O$_2$ $\rightarrow\rightarrow\rightarrow\rightarrow$ H$_2$O$_2$?
$\qquad\qquad$ Alternate pathway

suggested that ubiquinone is the most likely pivotal molecule at the branch point and that a flavoprotein is the first component of the alternate pathway.

When the alternate pathway is operative, the oxidation of succinate and other NAD-linked substrates seems to be cyanide resistant. Thus the first site of ATP production (see Figure 16–6) may be possible, and the partial conservation of energy is maintained as electrons pass along the alternate pathway to oxygen (20). Controversy exists as to whether the alternate path per se is linked to phosphorylation.

We might ask what the physiological significance of the alternate pathway is. One suggestion is that the alternate pathway may be significant in the respiratory climacteric of ripening fruit and leads to the production of hydrogen peroxide (19) and superoxide, which in turn enhances the oxidation and breakdown of membranes (8)—necessary activities in the ripening process. Solomos (20), in his review of this subject, points out that ethylene may act to implement the alternate pathway in ripening fruit. Peroxides are also necessary for the peroxidase-catalyzed reactions necessary for ethylene biosynthesis.

An attractive explanation for the role of the alternate pathway is that it may provide a means for the continued oxidation of NADH and operation of the tricarboxylic acid cycle, even though ATP may not be sufficiently drained off and may, in high concentrations, inhibit the Krebs cycle reactions via the ETS stoppage of electron flow. In view of the importance of the citric acid cycle intermediates as precursors for cellular components, a suitable mechanism to keep it operative (by oxidation of NADH and regeneration of NAD^+), even though energy demands are low, might be quite significant.

Hexose Monophosphate Shunt

The hexose monophosphate shunt (HMS), also known as *pentose phosphate cycle* or *direct oxidation pathway*, is another pathway that exists in many organisms. This pathway takes place in the cytoplasm and requires oxygen for its entire operation (see Figure 16–8). Reduced NADP occurs in the reactions, thereby forming 6-phosphogluconic acid and ribulose-5-P. If the equivalent of a molecule of glucose is oxidized to CO_2 and H_2O via this cyclic pathway (six turns of the cycle), then twelve molecules of reduced NADP would be formed. In the presence of the enzyme transhydrogenase the hydrogens of NADPH can be transferred to NAD to form NADH. With this in mind we can see where the formation of twelve molecules of reduced NADP via the hexose monophosphate shunt could ultimately lead to the synthesis of thirty-six molecules of ATP. Thus capture of energy released in the oxidation of glucose via this pathway is almost as efficient as that of the glycolytic and Krebs cycle pathway.

In addition to the energy considerations, the in vivo significance of the hexose monophosphate shunt is twofold. First, it is the major means in the cell by which reduced NADP (NADPH + H^+) is produced as the necessary reducing power for synthetic (anabolic) reactions. Second, it is the major pathway by which necessary ribose and deoxyribose are supplied in the biosynthesis of nucleotides and nucleic acids. To be sure, NADPH is produced during the light reactions of photosynthesis in chloroplasts, but it is used in a direct manner for CO_2 reduction (assimilatory) reactions taking place in the cytoplasm.

Inspection of the intermediates of the hexose monophosphate shunt reveals the

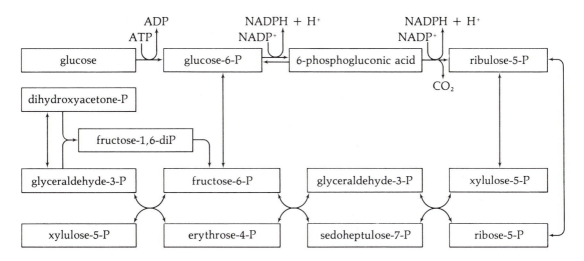

Figure 16–8. Hexose monophosphate shunt.

possibilities available for the direct input of many CO_2 fixation intermediates from photosynthesis and particularly the mechanism for initiation of the pathway from the EMP pathway via the glucose 6-phosphate conversion to 6-phosphogluconic acid. Note the sites of NADPH production. For each turn of the cycle, two NADPH are produced. In addition, erythrose-4-phosphate in plants is a precursor of the aromatic amino acids (phenylalanine, tyrosine, and tryptophan). Tryptophan is the precursor of indole-3-acetic acid, the major phytohormone with auxin activity in plants.

Glyoxylate Cycle

Seeds rich in fat convert storage fats to carbohydrates during germination. The mechanism of this conversion was unknown until the discovery of the glyoxylate cycle in the bacterium *Pseudomonas* by Kornberg and Krebs (16). Later, the reaction (and enzymes) of the β-oxidation of fatty acids and the conversion of acetate units of acetyl CoA to glyoxylate and malate (malic acid) were discovered to take place in microbodies, termed *glyoxysomes* by Beevers, the pioneer in this field. The glyoxysomes are now known to contain all the necessary enzymes for β-oxidation of fatty acids to acetyl CoA and subsequent conversion of the acetate units to malic acid (malate) and succinic acid (succinate). The cycle does not appear to be present in those seeds that store starch rather than fat. In fact, glyoxylate cycle activity in germinating seeds ceases as soon as the fat reserves have been used up. The fact that plants convert fatty acids to carbohydrates is due to the operation of two unique glyoxysome enzymes not known to be present in animals: *isocitrate lyase* and *malate synthetase.* The first major reaction of the glyoxylate pathway shunts isocitrate away from the decarboxylation reactions of the citric acid cycle via the formation of glyoxylate and succinate. The second major reaction results in the condensation of glyoxylate with acetyl CoA to yield malate, which is, in turn, converted to oxaloacetate.

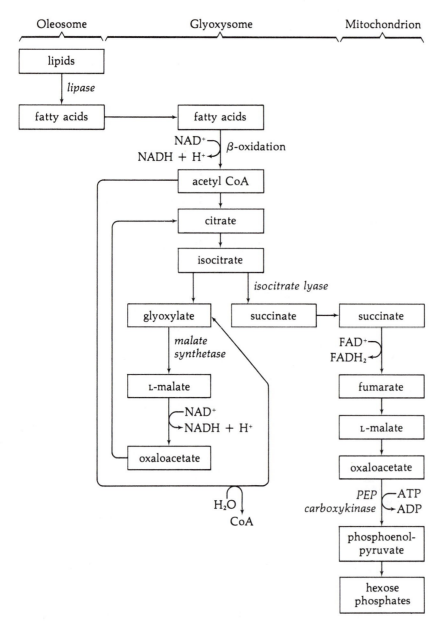

Figure 16–9. Conversion of storage fat to carbohydrates in germinating seed via glyoxylate cycle.

Figure 16–9 illustrates the reactions involved in the conversion of fatty acids to carbohydrates via the glyoxylate cycle. The fatty acids are derived from lipase-mediated enzymolysis of triglycerides occurring in lipid bodies called *oleosomes*. The fatty acids undergo β-oxidation in the glyoxysome with the formation of acetyl CoA. The acetyl CoA reacts with oxaloacetate to form citrate and then isocitrate. The isocitrate is cleaved to succinate and glyoxylate. This reaction is catalyzed by isocitrate lyase. The glyoxylate combines with acetyl CoA to form malate, the reaction being catalyzed by malate synthetase. The malate in the glyoxysome is oxidized to oxaloacetate, which initiates the cycle again by combining with acetyl CoA derived from β-oxidation of fatty acids. The succinate produced moves out of the glyoxysome and into the mitochondrion, where it is converted through the conventional Krebs cycle reactions to oxaloacetate.

The increase of oxaloacetate (OAA) provides ample substrate for amino acid production and carbohydrate formation by reverse glycolysis. Conversion of OAA to phosphoenolpyruvic acid and other glycolytic intermediates takes place in the cytoplasm.

We might ask why the oxaloacetate does not combine with acetyl CoA in the mitochondrion and why it is not oxidized to CO_2 and H_2O via the Krebs cycle. These phenomena do not take place because in seeds containing high lipid reserves there is no acetyl CoA coming into the mitochondrion from pyruvate since carbohydrates are not readily available to produce high amounts of pyruvate. Also, the acetyl CoA produced in the glyoxylate reactions remains in the glyoxysome. Consequently, the OAA built up in the mitochondrion is converted to phosphoenolpyruvate (PEP). This conversion and the reactions of reverse glycolysis take place in the cytoplasm and require energy in the form of ATP and NADH. Reduced pyridine nucleotide is made available in the β-oxidation of fatty acids, a portion of which is used directly in the reaction of reverse glycolysis and another portion to generate ATP (oxidative phosphorylation) by funneling through the reaction transport system of the mitochondrion. Thus the glyoxylate cycle is fundamentally important to the germination and

growth process of seedlings because it provides readily available carbohydrates from lipids.

Measurement of Respiration: Respiratory Quotient

Most methods for measuring the rates of respiration involve quantitative determinations of the CO_2 evolved or the oxygen consumed. One rather simple method involves trapping the CO_2 produced in a barium hydroxide ($Ba(OH)_2$) solution and weighing the barium carbonate ($BaCO_3$) formed. A variation of this method is to have the CO_2 absorbed in NaOH instead of in $Ba(OH)_2$ and the amount of CO_2 absorbed determined by titration. However, most determinations of respiration rates are accomplished by direct measurement of oxygen with an oxygen electrode or probe. The concentration of uncombined oxygen is usually provided by an oxygen analyzer as a direct meter readout that compensates automatically for the effect of temperature on oxygen solubility and membrane permeability. Due to the variations in the instrument designs of oxygen analyzers and oxygen probes, we will not describe them here.

A major method of measuring and analyzing CO_2 is by means of infrared spectrophotometry. We measure respiration by noting the gas changes that are indicative of respiratory activity. In the past, experimenters measured changes in gas pressure in living material by means of a manometer attached to a flask in which gas exchanges were taking place due to the presence of respiring seeds and tissues. However, experimenters no longer use this kind of apparatus extensively.

When respiration is being measured, it is usually desirable to measure both the oxygen consumed and the CO_2 evolved. The ratio of CO_2 produced to oxygen consumed is called the *respiratory quotient* (RQ).

When a carbohydrate is respired, this ratio is equal to 1.0. However, the RQ of different substrates (proteins, fats, carbohydrates) may vary considerably. For example, substrates that are highly oxidized, such as the acids of the Krebs cycle, will give RQ values greater than 1.0; substrates that are relatively reduced, such as fats, will yield RQ values less than 1.0.

Generally, when a carbohydrate is respired in the cell, one molecule of oxygen is consumed for every molecule of CO_2 given off. Krebs cycle intermediates are more highly oxidized than are carbohydrates and, consequently, need less oxygen for their oxidation to CO_2 and water. For example, the oxidation of malic acid to CO_2 and water gives an RQ value of 1.33. Fats are more reduced than carbohydrates, and thus more oxygen is needed for their respiration. The respiration of a fat may have an RQ value as low as 0.7.

The RQ value of a respiring tissue may provide valuable information to an investigator. From an RQ value, we can obtain a rough indication of the nature of the substrate being oxidized. However, we must realize that a precise identification of the type of substrate being respired by a tissue through RQ values is impossible. If different substrates are being respired simultaneously, the RQ value obtained is only an average of the RQ values of each individual substrate.

As we might expect, the organs of most mature plants that are well supplied with carbohydrates show little variation in their RQ values, which range from 0.97 to 1.17 (14). This fact suggests that the predominant substrate of respiration under normal conditions is carbohydrate. However, plants under starvation conditions will exhibit RQ values consistently below unity.

James (14) cites examples such as aging green leaves, leaves in the dark, or detached embryos. The drop in RQ value is a result of more reduced substrates (such as fatty acids and proteins) being respired. For example, Yemm (24, 25) has observed RQ values of 0.85 and less for green leaves in the dark.

Germinating seeds offer a good study of the agreement between the RQ value and the substrate being respired. In the seed, fatty oils are usually stored in addition to carbohydrates; and in many cases fatty oils may be the predominant stored reserve. During germination, proteins are degraded in the storage organs and synthesized in the embryo. In seeds containing large amounts of fat relative to carbohydrate, the RQ values during germination will be considerably lower than 1.0. Seeds with carbohydrate as the principal food reserve will have RQ values near 1.0 during germination.

Factors Affecting Rate of Respiration

Temperature

As with all chemical reactions, the chemical reactions of respiration are sensitive to temperature changes. Since the reactions of respiration are under the control of enzymes, the temperature range in which they may occur is quite narrow. At temperatures approaching 0°C, the rate of respiration becomes very low. As the temperature rises, so also does the rate of respiration, until temperatures destructive to enzyme activity are reached. A maximum rate may be attained somewhere between 35° and 45°C.

However, when studying the effect of temperature on respiration, we must consider the length of time an organ or plant is exposed to any one temperature level. For example, a four-day-old pea seedling (*Pisum sativum*) will exhibit an initial increase in respiration rate when the temperature is raised from 25° to 45°C. When the seedling is left for any length of time at this high temperature, however, its respiration rate will decrease. In other words, we must consider a time factor when we study the effect of temperature on respiration. Apparently, at temperatures above 30°C, factors leading to denaturation of enzymes involved in respiration begin to have an adverse effect. Since denaturation at these temperatures is not immediate, there will be an initial in-

crease in the respiration rate. However, in time, this adverse effect will show, and the rate of respiration will drop off. In general, the higher the temperature, the shorter is the time period before the rate of respiration drops off.

Work with pea seedlings by Fernandes (7) illustrates the importance of the time factor in studies of the effect of temperature on respiration. Figure 16–10 indicates that 30°C is the optimal temperature for four-day-old pea seedlings, since there is no dropping off of respiration rate over a long period of time.

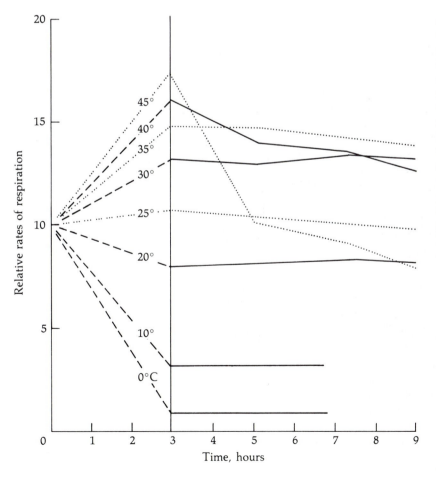

Figure 16–10. Effect of temperature on respiration rate in four-day-old pea seedlings (*Pisum sativum*). Note relationship between temperature, time, and rate of respiration. Broken lines indicate time interval between changes of temperature from 25°C to temperatures indicated in figure.

From D.S. Fernandes. 1923. Rec. Trav. Bot. Néerl. 20:107.

Oxygen

Oxygen is necessary for Krebs cycle reactions to occur, and oxygen is the terminal acceptor of electrons in the electron transport system. Considering these facts, we naturally assume that the rate of respiration is sensitive to changes in oxygen concentration. In general, at low oxygen concentrations, both aerobic and anaerobic respiration can be expected to occur in the plant. Under these circumstances, RQ values would be greater than 1.0 and, in fact, would approach infinity as the concentration of oxygen approached zero. That is, under complete anaerobic conditions, the CO_2 produced would be a product of anaerobic respiration (fermentation) exclusively. As the oxygen concentration is increased, anaerobic production of CO_2 falls off rapidly, aerobic respiration increases, and RQ values approach unity. The point at which an RQ value reaches unity at a certain oxygen concentration is called the *extinction point* (21). At this point, anaerobic respiration ceases. A typical example of these relationships may be seen in Watson's work (14) with Bramley seedling apples (see Figure 16–11).

Over a complete range of oxygen concentrations, it is desirable to measure both CO_2 production and oxygen consumption. Oxygen consumption gives a measure of aerobic respiration, as does CO_2 production after the extinction point has been passed. However, below the extinction point, CO_2 production results from both aerobic and anaerobic respiration; oxygen consumption below this point is still strictly a measure of aerobic respiration. Determining the participation of both gases over a complete range of oxygen concentrations allows us to measure both aerobic and anaerobic respiration.

Numerous studies of the respiration rates of a variety of plants point to a general conclusion: as the oxygen concentration increases from zero, the rate of aerobic respiration increases. With most plants this increase is hyperbolic—that is, the rate of increase falls off with increase in oxygen concentration. In some plant material, the increase in rate of aerobic respiration is linear over a range of oxygen concentration. For example, Taylor (23) found this phenomenon to be true for germinating rice grains. A possible explanation is that oxygen consumption is limited by an oxygen diffusion barrier, such as might be found in the outer covering of the rice grain. James (14) has indicated that oxygen consumption in this case is proportional to the amount of oxygen diffusing across the barrier, not to oxygen being consumed in respiration.

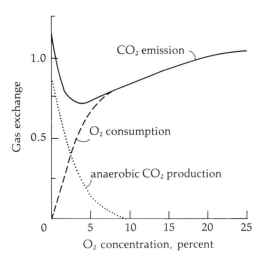

Figure 16–11. Production of CO_2 by Bramley seedling apples at different oxygen concentrations. Production rate in air equals 1.0.

From W.O. James. 1953. Plant Respiration. *Oxford: Clarendon Press.*

Carbon Dioxide

Increasing the concentration of CO_2 has a definite repressing effect on respira-

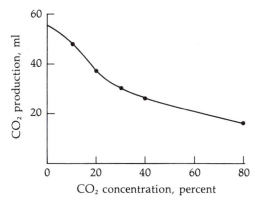

Figure 16–12. Retardation of respiration rate in germinating white mustard seeds as result of increase in CO_2 concentration.

From W. Stiles and W. Leach. 1960. Respiration in Plants. New York: Wiley.

tion, as evidenced by Kidd's studies (15) on the respiration of germinating white mustard seeds (see Figure 16–12). Although many studies of leaf respiration have indicated a depressing effect of CO_2, there is evidence that this effect may be partially indirect. Heath (11) demonstrated that CO_2 can cause stomatal closure, thus limiting gaseous exchange. This closure may have the effect of raising the internal concentration of CO_2 considerably, thereby limiting respiration.

Inorganic Salts

Lundegårdh and Burström (17) observed that the rate of respiration increases when a plant or tissue is transferred from water to a salt solution. The amount by which respiration is increased over normal has been labeled *salt respiration*. This aspect of respiration is discussed in detail in Chapter 7.

Mechanical Stimulation

In a series of studies, Audus (1, 2, 3, 4) demonstrated that leaf respiration could be increased by handling, stroking, or bending leaves. In the cherry laurel leaf, this increase caused by handling may reach as high as 18.3 percent. Response to handling decreases if the handling is repeated over a period of time.

Wounding

For many years plant scientists have known that the wounding of plant organs stimulates respiration in that organ. Generally, wounding initiates meristematic activity in the area of the wound and results in the development of wound callus. We can only speculate what connection this development might have with the stimulatory effect of wounding on respiration. An interesting study on potatoes by Hopkins (13) showed a considerable increase in sugar content after a potato had been cut. Perhaps increase in respiration because of wounding is caused by increased availability of respiratory substrate.

Questions

16–1. Does the general equation for respiration give an accurate account of the process? Explain.

16–2. Would it be theoretically possible for carbon in a cell to be transferred from glucose to starch to glucose to pyruvic acid to alanine to protein? Explain.

16–3. What is significant about coupled reactions in biological systems?

16–4. Suggest a reason why low night temperatures seem to favor the movement of nutrients and growth in some plants.

16–5. Explain the following terms: fermentation, glycolysis, hexose diphosphate pathway, and Embden-Myerhof-Parnas pathway.

16–6. Name some of the compounds that are synthesized in plants when pyruvate is one of the starting materials.

16–7. Indicate some of the major features of the EMP pathway in terms of reactants, products, and types of reactions.

16–8. Explain the term *substrate-level phosphorylation.*

16–9. What is the role of CoA and how is it formed?

16–10. From what respiratory intermediates are fatty acids, gibberellins, and the phytol tail of chlorophyll derived?

16–11. In the presence of oxygen, why is the Krebs cycle more efficient in terms of ATP production than the EMP pathway?

16–12. What is the function of the electron transport system? How does it work and from what source does it derive the reducing power for operation?

16–13. Why is oxygen necessary for the operation of the electron transport process?

16–14. How is the mitochondrion structured for the operation of the ETS?

16–15. Explain the chemiosmotic theory as it pertains to oxidative phosphorylation.

16–16. How might the alternative pathway play an important role with respect to the functioning of the Krebs cycle and the production of intermediates?

16–17. Name two major products of the hexose monophosphate pathway.

16–18. Many seeds are high in lipids and low in carbohydrates. How is the energy for the growing embryo derived from the lipid? What is the major pathway involved and how does it work?

16–19. Explain the terms *oleosomes* and *RQ.*

16–20. Indicate the similarities and differences of photophosphorylation, oxidative phosphorylation, and substrate level phosphorylation.

16–21. Explain some of the major factors that influence respiration.

16–22. What are some of the regulatory mechanisms in plant cells responsible for storage, growth, respiration, or the formation of respiratory intermediates? Why might one process (synthesis) take place over that of another (respiration)?

Suggested Readings

Bonner, W.D., Jr. 1973. Mitochrondria and plant respiration. In L.P. Miller, ed., *Phytochemistry.* New York: Van Nostrand Reinhold.

Ikuma, H. 1972. Electron transport in plant respiration. *Ann. Rev. Plant Physiol.* 23:419–436.

Laties, G.G. 1982. The cyanide-resistant alternative path in higher plant respiration. *Ann. Rev. Plant Physiol.* 33:519–555.

Lehninger, A.L. 1982. *Principles of Biochemistry.* New York: Worth.

Meeuse, B.J.D. 1975. Thermogenic respiration in aroids. *Ann. Rev. Plant Physiol.* 26:117–126.

Solomos, T. 1977. Cyanide-resistant respiration in higher plants. *Ann. Rev. Plant Physiol.* 28:279–297.

Solomos, T., and G.G. Laties. 1976. Induction by ethylene of cyanide-resistant respiration. *Biochem. Biophys. Res. Commun.* 70:663–671.

Stryer, L. 1981. *Biochemistry,* 2nd ed. San Francisco: Freeman.

Theologis, A. 1979. The genesis development and participation of cyanide-resistant respiration in plant tissue. Ph.D. Thesis, University of California, Los Angeles.

White, A., P. Handler, E.L. Smith, R.L. Hill, and I.R. Lehman. 1978. *Principles of Biochemistry,* 6th ed. New York: McGraw-Hill.

Chapter 17

Phytohormones: The Auxins

Root initiation and growth in auxin-treated (left) and untreated (right) mung beans. *Courtesy of C.W. Heuser, The Pennsylvania State University.*

The presence of growth-regulating chemicals in plants was first suggested by Sachs (50) in the latter half of the nineteenth century. He proposed that "organ-forming substances" were produced in the leaves of plants and translocated downward. This astute theory foreshadowed the intensive study of plant growth regulation that has been going on during the twentieth century.

History

While Sachs was forming his theories on growth regulation, another famous scientist was studying plant tropisms. Darwin, although better known for his theory on evolution, studied the effect of gravity and unilateral light on the movement of plants (11). He demonstrated that the effects of light and gravity on the bending of both roots and shoots are mediated by the tip, and that this influence can be transmitted to other parts of the plant. He concluded that when seedlings are freely exposed to a lateral light, some influence is transmitted from the upper to the lower part, causing the latter to bend. With respect to geotropism of roots, he concluded that only the tip is acted on and that this part transmits some influence to the adjoining parts, thus causing them to curve downwards (11).

Darwin was primarily interested in the *coleoptile*, which is a specialized leaf in the form of a hollow cylinder that encloses the epicotyl and is attached to the first node. It affords protection to the delicate, growing tip of a grass seedling until, eventually, the more rapidly growing first leaf emerges above ground.

Darwin found that if he exposed the tips of intact coleoptiles to a unilateral source of light, the coleoptiles would bend toward the light. As we know today, this bending response is termed *phototropism*, and the light stimulus results from hormone activity. Darwin further observed that if the tip of the coleoptile was covered or removed, the coleoptile did not bend. These results led Darwin to report that the tip of the coleoptile is involved in the phototropic response.

About the same time Darwin was making his observations, Salkowski and Salkowski detected indole-3-acetic acid in fermentation media. This chemical was implicated with the phototropism of coleoptiles many years later. Also, in the early 1900s Bayliss and Starling (2), working on the digestive system of dogs, introduced the concept of hormones and suggested the following characteristics: (1) They are specific chemical substances. (2) They are produced in specific areas of the organism. (3) They travel to other areas (the targets). (4) At the targets and in small quantities, they regulate physiological responses (growth, movement, reproduction). It was not until much later that plant scientists adopted the term *hormone*.

In 1907 Fitting (15) demonstrated that lateral incisions either on one or both sides of *Avena* coleoptiles did not prevent the phototropic response as long as the cut surfaces were not separated. Observations suggested that complete cellular integrity was not necessary for passage of the internal stimulus. Early in the twentieth century, Boysen-Jensen (8) provided further proof of the material nature of the stimulus for coleoptile phototropism. He first decapitated a coleoptile a few millimeters from the tip and placed a block of gelatin on the stump. When he put the tip back and illuminated the coleoptile unilaterally above the incision, curvature toward the light was the same as in the intact coleoptile. Boysen-Jensen also demonstrated that the normal phototropic bending of the coleoptile could be prevented by inserting a piece of mica in a transverse slit halfway through the coleoptile below the tip on the dark side of a unilaterally illuminated grass seedling. Inserting a piece of mica on the illuminated side of the seedling produced no such interference, thereby providing evidence that the stimulus for bending passes down the dark side of the coleoptile. In 1918 Paàl (44) decapitated the coleoptile, replaced the tip asymmetrically, and discovered that the coleoptile, even in darkness, bends away from the side with the tip. Paàl's experiments strongly suggested that a substance emanating from the tip was responsible for the curvature. Finally, Söding (16) demonstrated that a substance coming from the tip must be responsible for elongation of the coleoptile. Stark (1) in his studies found that the sap collected from *Avena* coleoptiles could be dispersed in agar blocks. When he applied the blocks asymmetrically to the stumps of detipped coleoptiles placed in darkness, curvature resulted. Figure 17–1 illustrates the results of these experiments.

The next logical steps were to isolate this substance from the plant and to demonstrate that it could stimulate growth when introduced into the plant. This task was accomplished by the Dutch botanist Went. He placed freshly cut coleoptile tips on small blocks of agar for a measured period of time, and then he placed the agar blocks asymmetrically on decapitated coleoptiles for 2 hours in the dark. The coleoptiles exhibited a curvature similar to that obtained when coleoptile tips were placed asymmetrically on coleoptile stumps. He then developed a method for determining the amount of active substance in coleoptile tips—that is, he developed a bioassay for auxin. Went found that the degree of curvature of the

coleoptile is proportional, within limits, to the amount of active substance in the agar blocks. Because of the use of the *Avena* plant for this bioassay, it subsequently became known as the *Avena coleoptile curvature test.*

Application of the *Avena* test to a great variety of substances led to the finding that human urine is rich in growth substance. Starting with 33 gallons of human urine, Kögl and Haagen-Smit (35) performed a series of purification steps. The activity of the products of each purification step was determined by the *Avena* coleoptile curvature test. After distillation in high vacuum, the final step yielded 40 mg of crystals that had a specific activity 50,000 times that of the original urine. The final product was given the name *auxin-A* (auxentriolic acid).

Using approximately the same purification methods, Kögl, Erxleben, and Haagen-Smit, (32) isolated another active substance from corn germ oil. They found this substance to be very similar in structure and activity to auxin-A and gave it the name *auxin-B* (auxenolonic acid). In the same year, still another substance was isolated from human urine. Repeating the isolation from urine on a larger scale and with the use of a charcoal absorption method for removing the active substance, Kögl, Haagen-Smit, and Erxleben (34) isolated the compound *heteroauxin* (other auxin), or, as it is known today, *indole-3-acetic acid* (IAA). IAA was not a new compound but had been discovered and isolated from fermentations in 1885 by Salkowski and Salkowski.

indole-3-acetic acid (IAA)

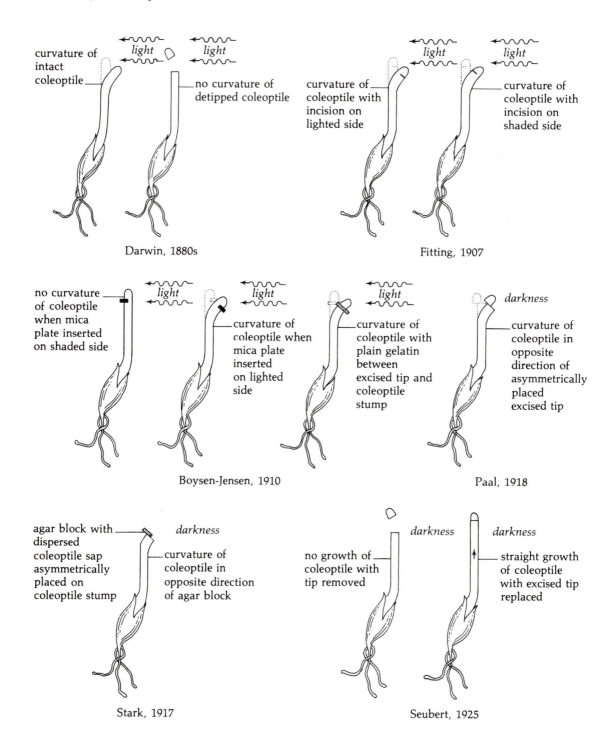

curvature of intact coleoptile

light   *light*

no curvature of detipped coleoptile

Darwin, 1880s

light   *light*

curvature of coleoptile with incision on lighted side

curvature of coleoptile with incision on shaded side

Fitting, 1907

no curvature of coleoptile when mica plate inserted on shaded side

light   *light*   *light*

curvature of coleoptile when mica plate inserted on lighted side

curvature of coleoptile with plain gelatin between excised tip and coleoptile stump

darkness

curvature of coleoptile in opposite direction of asymmetrically placed excised tip

Boysen-Jensen, 1910

Paal, 1918

agar block with dispersed coleoptile sap asymmetrically placed on coleoptile stump

darkness

curvature of coleoptile in opposite direction of agar block

Stark, 1917

darkness   *darkness*

no growth of coleoptile with tip removed

straight growth of coleoptile with excised tip replaced

Seubert, 1925

Today, there is considerable doubt as to the existence of auxin-A and auxin-B. Since their first isolation by Kögl and his colleagues, auxin-A and auxin-B have never been isolated again. In contrast, a number of different investigators have isolated IAA in the crystalline form many times from different sources.

Kögl and Kostermans (35) isolated IAA from yeast plasmolysates in 1934. Thimann (60) followed shortly with the isolation of IAA from cultures of *Rizopus suinus*. In 1946 Haagen-Smit and colleagues reported the occurrence of IAA in a higher plant (26). To date IAA has been identified in many higher plants and is recognized as the principal auxin in higher plants. It is interesting to note that experimenters have not isolated IAA from coleoptile tips. One calculation is that 20,000 tons of coleoptile tips would have to be processed for a yield of one gram of IAA. Auxins are obviously very active in minute amounts. Went's auxin is probably IAA, but plant parts seem to contain other compounds, possibly IAA derivatives, that are active in the phototropic response.

It is worthy of note here that Kögl, Haagen-Smit, and Went used the term *auxin* (Greek, *auxein:* to grow) in their studies involving the growth and phototropism of *Avena* coleoptiles. Initially, Thimann (59) provided the following definition for auxin: "an organic substance which at low concentration promotes growth along the longitudinal axis when applied to shoots of plants freed as much as possible from their growth-promoting substances." To distinguish auxins from gibberellins, another group of phytohormones, the definition of-

ten includes the idea that auxins, unlike gibberellins, inhibit root elongation at certain concentrations. There are also other differences which we will discuss later.

Bioassays

One of the most important principles employed in the early investigations on auxins and subsequent hormonal characterizations was the establishment of the *Avena* coleoptile curvature test specifically and the bioassay concept in general. The term *bioassay* is used to describe the use of living material to test the effect of known and putative biologically active substances. It is now evident that when we deal with biologically active substances such as phytohormones, we must have a means of measuring their biological activity. In most cases, the plant material we use to measure the activity of a growth regulator must respond specifically to that compound or to a group of similar compounds. Also, there must be a correlation between the extent of the response of the biological material and the concentration of the chemical.

The bioassays described are based primarily on cellular enlargement. However, there are numerous responses that might be used in a bioassay for auxins. Just a mention of the physiological responses attributed to compounds with auxin activity will reinforce this point. Later we will consider in detail some of the following physiological effects influenced by auxins:

Cell enlargement of stems, leaves, and roots

Cell and organ differentiation

Flower initiation and development, fruit set, fruit growth, and embryo growth

Figure 17–1. Summary of experiments leading to isolation and discovery of auxin (IAA), based on its activity in coleoptile phototropism.

Abscission of leaves, flowers, and fruits

Direction of growth (tropism of stems or roots)

Parthenocarpy in some plants

Apical dominance

Enlargement and cell division of callus tissue cultures

Even though many of the bioassays for chemicals eliciting different phytohormonal activities are based on varied physiological responses (auxin and cell enlargement, cytokinins and cell division, and so on), all bioassays must have certain requirements and similarities in order to be effective. The characteristics of a useful bioassay must include (1) specificity, (2) sensitivity, (3) ease in measuring a detectable and relatively quick response, (4) relative ease in setting up and controlling, and (5) absence, in the plant material, of the substance or related chemicals being tested. Any researcher interested in phytohormone isolation will consider these features. As we discuss in this and the following chapters, some of these common features will be evident.

Although numerous bioassays for auxin activity have been devised since the discovery of auxins in plants, only a few have found general use. We will briefly consider the following four bioassays that are applicable to the study of auxins: *Avena* coleoptile curvature test, *Avena* coleoptile section test, split pea stem curvature test, and cress root inhibition test.

Avena Coleoptile Curvature Test

The *Avena* coleoptile curvature test, developed by Went (65), was the first and best bioassay to lead to the isolation and

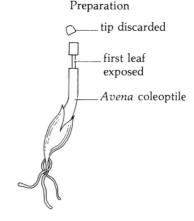

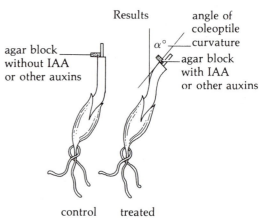

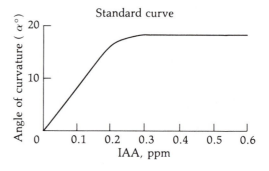

Figure 17–2. Avena coleoptile curvature test. *Redrawn from L.J. Audus. 1959.* Plant Growth Substances. *New York: Interscience Publishers.*

eventual characterization of auxin (IAA) and its derivatives (see Figure 17–2). Because of its sensitivity and reliability experimenters use this bioassay extensively even today, more than fifty years after its development.

The measurement of auxin activity by the *Avena* coleoptile curvature test depends on the strict, rapid polar transport (from the morphological tip toward the morphological base of the plant axis) of auxin in the *Avena* coleoptile. Because of this property, auxin applied to one side of the coleoptile will diffuse down that side rapidly. However, it will not diffuse laterally to any significant extent. The differential growth caused by the transport of auxin down only one side of the coleoptile produces a curvature that is proportional, within limits, to the amount of auxin applied.

The procedure for the *Avena* curvature test is as follows:

1. *Avena* seedlings are germinated and grown in the dark. There is a reduction in sensitivity of the coleoptile to auxin when it is exposed to blue light. Inconvenient elongation of the first internode may be reduced by exposing the seedling two days after germination to 2 to 4 hours of red light.

2. After the seedling reaches a height of 15 to 30 mm, 1 mm of the apical tip of the coleoptile is removed, thus removing the natural source of auxin.

3. A second decapitation of 2 to 4 mm is necessary after a 3-hour period to remove tissue that has regenerated and produces auxin.

4. The primary leaf, which has been exposed by the second decapitation, is then gently pulled. The primary leaf should extend a few mm out of the coleoptile. Note that we now have vertical support (tip of the primary leaf) for the agar block that will be placed on the coleoptile.

5. An agar block containing auxin is now placed on one side of the severed end of the coleoptile. Auxin, will be transported in a polar fashion down the side of the coleoptile to which the auxin-agar block has been applied.

6. After 90 min, the shadows of the seedlings are projected onto a strip of bromide paper and photographed, thereby giving the investigator a permanent record.

7. Curvature is measured by recording the angle made by a vertical line and a line drawn parallel to the curved portion of the stem.

A linear relationship between the concentration and the amount of curvature obtained exists within a certain range of concentrations of IAA. As shown in Figure 17–3, this range for IAA reaches an optimum peak at around 0.2 mg/l.

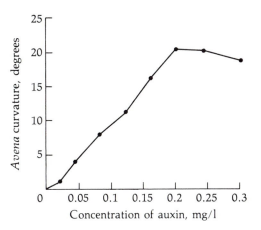

Figure 17–3. Response of *Avena* coleoptile to increasing concentration of IAA.

From F.W. Went and K.V. Thimann. 1937. Phytohormones. New York: Macmillan.

Avena Coleoptile Section Test

The *Avena* coleoptile section test (see Figure 17–4) is based only on the ability of auxin to stimulate cell elongation. Transport of auxin or differential growth caused by auxin are not involved here.

This test, utilizing sections of the oat coleoptile, was first used by Bonner in 1933 (7). Since then, this bioassay has found wide use because of its simplicity and its applicability. The *Avena* coleoptile section test, unlike the *Avena* coleoptile curvature test, measures the effect of growth regulators over a wide range of concentrations. In addition, unlike the *Avena* coleoptile curvature test, the section test is not hindered by problems of transport of the growth regulator. Some growth regulators are not transported as readily as IAA and therefore could not be used in the *Avena* coleoptile curvature test. However, the *Avena* coleoptile curvature test is much more sensitive to low concentrations of auxin than is the section test, thus the former has a major advantage in this respect, particularly in plant extrac-

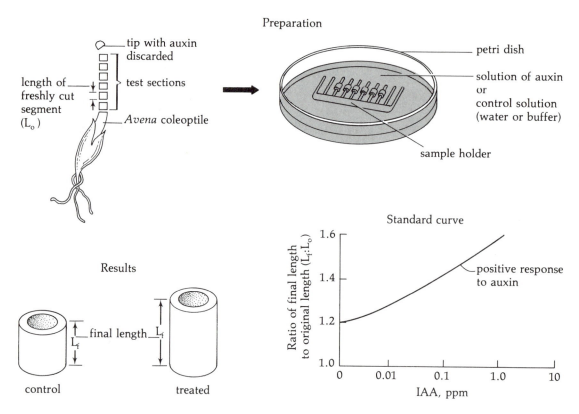

Figure 17–4. Avena coleoptile section test. L_o equals length of freshly cut segment; L_x equals length of untreated segment after floating in water for length of test period; L equals length of treated segment after floating in test solution for length of test period.

Redrawn from L.J. Audus. 1959. Plant Growth Substances. *New York: Interscience Publishers.*

tion procedures in which only very small quantities of auxin are present. In order to detect the presence of auxin in these cases, we would have to use the *Avena* coleoptile curvature test.

The procedure for the *Avena* coleoptile section test is as follows:

1. *Avena* seeds (caryopsis) of a pure strain (e.g., victory) are germinated and grown in the dark at 25°C and a relative humidity of about 85 percent. Only weak red light may be used in the growth room.

2. When the coleoptiles are about 25 to 30 mm in length, they are harvested; the apical 4 mm are removed, and a section 3 to 5 mm in length is cut from each coleoptile cylinder.

3. All sections are soaked in distilled water for a minimum of 1 hour and then distributed at random to petri dishes containing 20 ml of test solution.

4. After a 12, 24, or 48-hour incubation period at 25°C, the sections are measured with the aid of a dissecting microscope equipped with an ocular micrometer. If determination of growth rate is desired, the sections are measured after a 12-hour incubation period. If growth is to be determined, 24- or 48-hour incubation periods are usually used.

In the *Avena* coleoptile section test, the growth response of the sections is found to be directly proportional to the logarithm of the concentration of growth regulator used (see dose response curve in Figure 17–4). This correlation contrasts with the *Avena* coleoptile curvature test in which the growth response is directly proportional to the amount of auxin used. The curvature test, therefore, is a much more sensitive test but is confined to a short concentration range.

Split Pea Stem Curvature Test

The split pea stem curvature test (see Figure 17–5), first described by Went (67) in 1934, depends, as does the *Avena* coleoptile curvature test, on a differential growth response. A stem section of a pea seedling of a pure strain (e.g., Alaska) is slit longitudinally and floated on the test solution. At first a negative curvature (curvature outward) occurs because of the uptake of water by the inner cortical cells. The epidermal cells respond to auxin with considerable growth in length and a negligible growth in width, the cortical cells respond to auxin with more growth in width than in length. Consequently, after an incubation period with a physiological concentration of auxin, a positive curvature results. Within a certain range, the response of the slit halves of the stem is roughly proportional to the logarithm of the concentration of auxin used.

The procedure for the split pea stem curvature test is as follows:

1. Pea seeds are germinated and grown in the dark for 8 days. The seedlings are exposed to 3 hours of red light per day to increase sensitivity to auxin.

2. The stems are harvested and decapitated and a section about 1 cm long between the second and third internode is removed.

3. The sections are soaked in distilled water for 1 hour to remove any endogenous auxin that might be present in the stem sections.

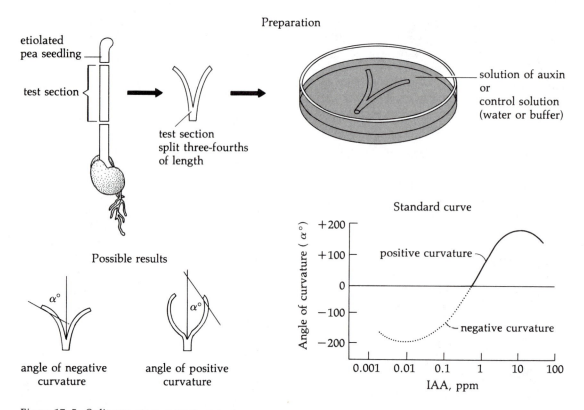

Figure 17–5. Split pea stem curvature test.
Redrawn from L.J. Audus. 1959. Plant Growth Substances. *New York: Interscience Publishers.*

4. The stem sections are then slit longitudinally a standard 3 cm and placed in a petri dish containing 25 ml of auxin solution. Five or six sections to a petri dish is the usual procedure.

5. After an incubation period of 6 hours, curvature of the slit stem tips is recorded.

As in the *Avena* coleoptile section test, transport of auxin is not involved in the split pea stem curvature test. Therefore, the effect of growth regulators that are not easily transported in plant tissues can be measured by the split pea stem curvature test.

Cress Root Inhibition Test

Roots are much more sensitive to auxin than are stems, and roots are, in fact, inhibited by concentrations of auxin that normally stimulate stem growth. However, at very low concentrations of auxin, root growth may be stimulated. The value, then, of the cress root inhibition test (see Figure 17–6) is that the effect of extremely low concentrations of auxin, such as found in plant extracts, may be measured.

The procedure for the cress root inhibition test is as follows:

1. Seeds are sterilized and then germinated on moist filter paper.

2. When the roots of the seedling have reached a desired length, they are placed in petri dishes containing 15 ml of test solution.

3. Growth of the root is measured after 48 hours.

Many other bioassays for auxin activity have been devised, some for specific use and others for more general application. However, the bioassays mentioned here are the ones most generally used. Of the four assays described, the *Avena* coleoptile curvature test is the best for quantitative determinations, but it is restricted to compounds that are transported rapidly in a polar manner. The *Avena* coleoptile section test and split pea stem curvature test are applicable to a wide range of concentrations, but they cannot be used for quantitative determinations of low concentrations of auxin, such as those found in plant extracts. The cress root inhibition test is even more sensitive than the *Avena* coleoptile curvature test since it can detect very low concentrations of IAA. However, small differences in auxin concentration cannot be detected by the root test, its response being roughly proportional to the logarithm of the auxin concentration.

Definitions

Since the discovery and chemical characterization of auxin, there has been an immense amount of research in the field of plant growth regulation. This prodigious amount of work brought forth a number of synthetic as well as natural compounds that were similar to IAA in their physiological activity.

Preparation

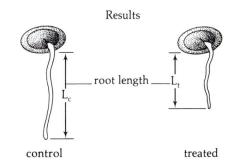

Results

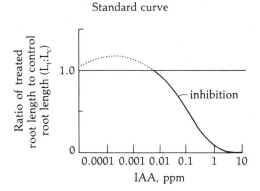

Figure 17–6. Cress root inhibition test. L_c equals length of control seedling root at termination of test period, L_t equals length of treated seedling root at termination of test period.

Redrawn from L.J. Audus. 1959. Plant Growth Substances. New York: Interscience Publishers.

Let us now consider some of the terms relating to growth regulation in plants.

In most cases the synthetic compounds were chemical analogues of the natural auxin. Also, many compounds were discovered that counteracted the effect of growth regulators. Because of the number of biologically active compounds being introduced and the confusion in terminology that arose, a committee formed by the American Society of Plant Physiologists suggested the following definitions (63):

Plant regulators are organic compounds other than nutrients that in small amounts promote, inhibit, or otherwise modify a physiological process in plants.

Plant hormones, or *phytohormones*, are regulators produced by plants, which in low concentrations regulate plant physiological processes. Hormones usually move within the plant from a site of production to a site of action.

Growth regulators, or *growth substances*, are regulators that affect growth.

Growth hormones are hormones that regulate growth.

Flowering regulators are regulators that affect flowering.

Flowering hormones are hormones that initiate the formation of floral primordia or promote their development.

Auxin is a generic term for compounds characterized by their capacity to induce elongation in shoot cells. Auxins resemble indole-3-acetic acid in physiological action. They may, and generally do, affect processes besides elongation, but elongation is considered critical. They are generally acids with an unsaturated cyclic nucleus or derivatives of such acids.

Auxin precursors are compounds that in the plant can be converted into auxins.

Antiauxins are compounds that inhibit the action of auxins.

Synthetic Auxins

A natural consequence of the discovery of auxin activity was the isolation and characterization of IAA. As soon as this isolation was accomplished, scientists began an intensive search for compounds chemically similar to IAA and having auxin activity. The results of this search brought forth other indole derivatives, such as indole-3-butyric acid (73) and indole-3-propionic acid, which demonstrate physiological activity similar to that of IAA. Experimenters synthesized other compounds similar in activity (and therefore called auxins) but not in chemical structure to IAA. Of these, the more notable active ones are α- and β-naphthalene acetic acids, α-naphthoxyacetic acid (30), phenoxyacetic acid (72), and derivatives (i.e., chlorophenoxy acids) benzoic acids and picolinic acid (see Figure 17–7). Many of these compounds are herbicides that have been used effectively in modern agriculture. In most instances, compounds with auxin activity at low concentrations become very phytotoxic at relatively higher concentrations. The first selective herbicides to be discovered and used widely were 2,4-dichlorophenoxyacetic acid (2,4-D) and its derivatives. These compounds are very potent auxins.

Not until Zimmerman and Hitchcock (72) discovered the auxin activity of the phenoxyacetic acid series was the profound effect of the substitution of various groups onto the ring or side chain in auxin and herbicidal function truly appreciated. The na-

Indoles

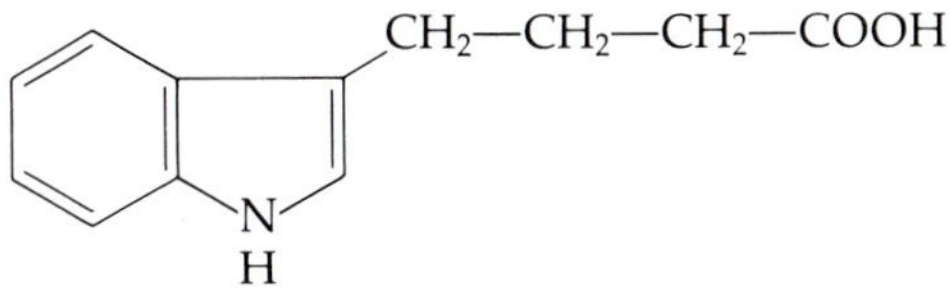

indole-3-butyric acid

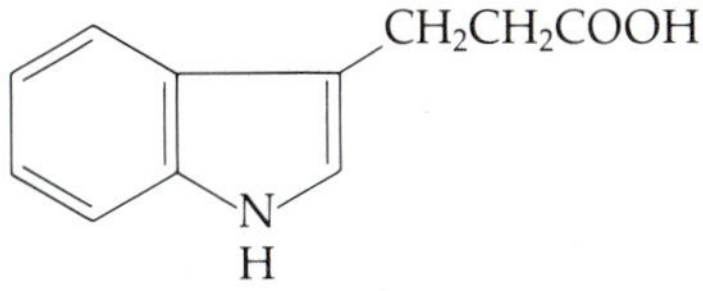

indole-3-propionic acid

Benzoic acids

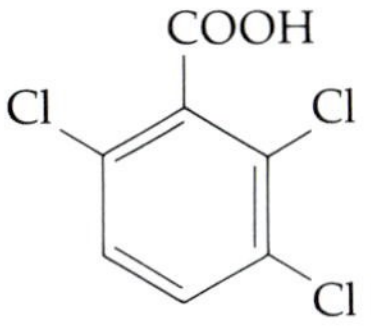

2,3,6-trichlorobenzoic acid

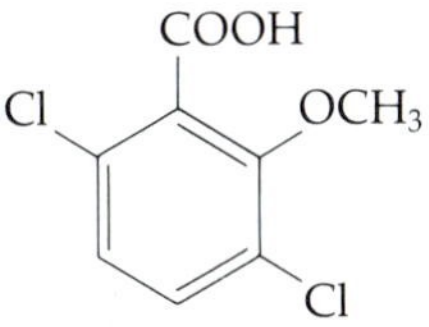

2-methoxy-3,6-dichlorobenzoic acid
(dicamba)

Naphthalene acids

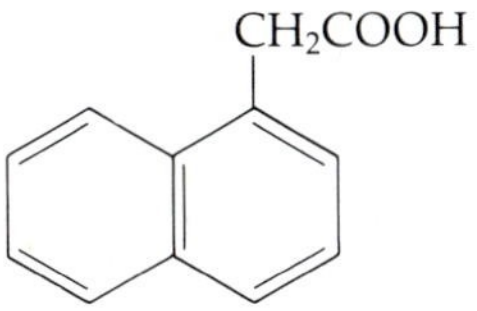

α-naphthalene acetic acid

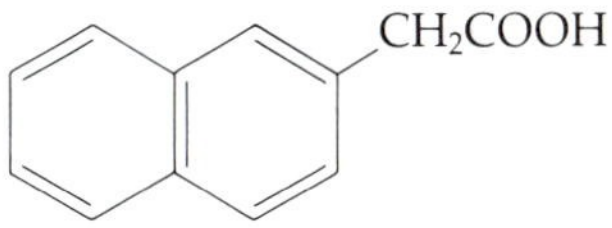

β-naphthalene acetic acid

Chlorophenoxy acids

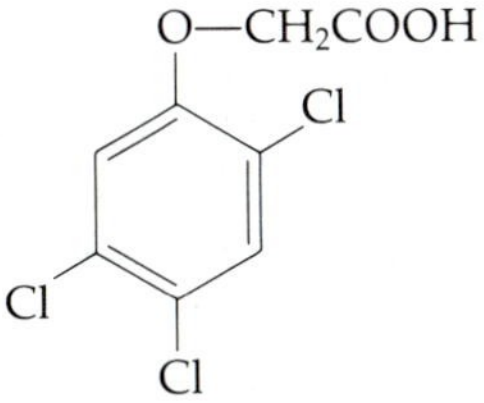

2,4,5-trichlorophenoxyacetic acid
(2,4,5-T)

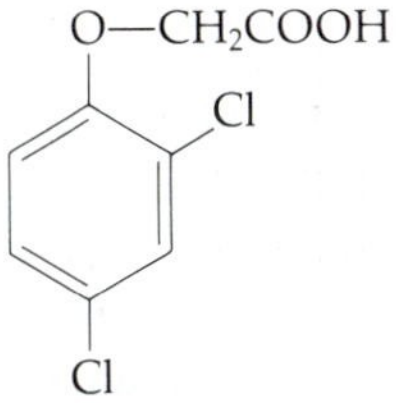

2,4-dichlorophenoxyacetic
acid (2,4-D)

Naphthoxy acid

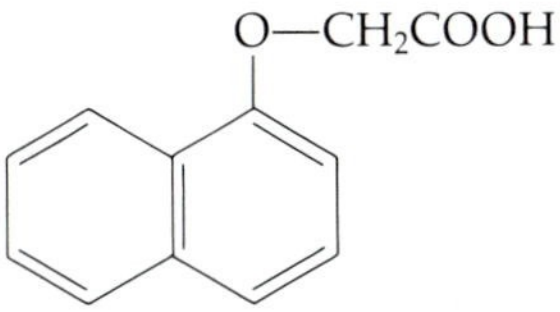

Picolinic
acid

α-naphthoxyacetic acid

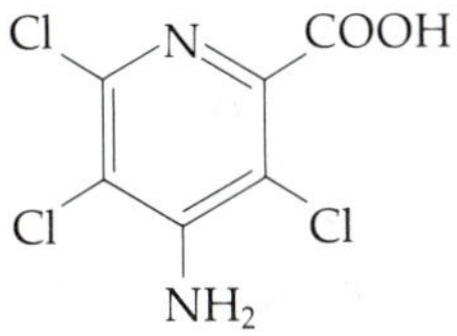

4-amino-3,5,6-trichloropicolinic
acid (tordon or pichloram)

Figure 17–7. Types of synthetic auxins.

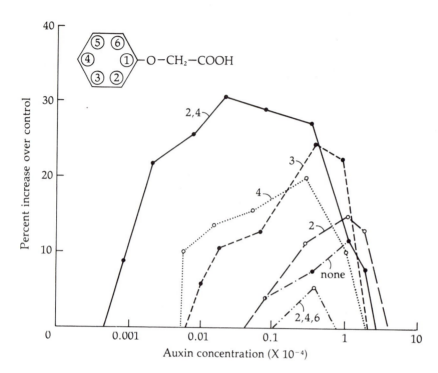

Figure 17–8. Effect of different concentrations of chlorinated phenoxyacetic acids on *Avena* section test. Numbers on curves represent position of chlorine substitution on phenyl ring.

From R.M. Muir et al. 1949. Plant Physiol. 24:359.

ture of the group substituted and the location of the substitution were found to influence the activity of the compound. We can see a striking example of the substitution of the chlorine atom at various positions of the phenyl ring of phenoxyacetic acid (see Figure 17–8).

Because of these selective herbicidal properties, the phenoxyacetic acids particularly 2,4-D and 2,4,5-trichlorophenoxyacetic acid (2,4,5-T), have been widely used commercially for the past thirty years. They were developed because of their potential usefulness in chemical warfare. In fact they were used during the early sixties as defoliants. They are very stable and not subject to destruction in plants by the IAA-oxidase enzyme system, which is normally involved in IAA breakdown. Consequently, the phen-

oxyacetic acids are selectively effective on the broad-leaved dicots at relatively low concentrations. Although different formulations, consisting of free acids, salts, and amine salts, are the most common, effective preparations (of 2,4,5-T) include a variety of esters. For example, *agent orange,* which was used in the war in Vietnam as a defoliant, is an effective mixture of free 2,4-D and the n-butyl ester of 2,4,5-T. However, the reactions employed in the synthesis of 2,4,5-T and other chlorinated phenols are recognized as potential sources of numerous side products, such as the chlorodioxins, which are harmful to humans and other animals. One side product in particular is 2,3,7,8-tetrachlorodibenzo-para-dioxin (TCDD), the most toxic synthetic chemical known. It is the possible presence of such compounds in preparation of phenoxyacetic acid that is of major concern. Moore (41) summarizes the evaluation and regulation of 2,4-D and 2,4,5-T.

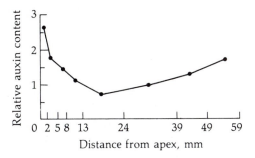

TCDD

Figure 17–9. Auxin distribution in etiolated *Avena* seedling.

From K.V. Thimann. 1934. J. Gen. Physiol. 18:23. Redrawn from A.C. Leopold. 1955. Auxins and Plant Growth. Los Angeles: University of California Press.

Distribution of Auxin in the Plant

The highest concentrations of auxin are found in the growing tips of the plant—that is, in the tip of the coleoptile, in buds, and in the growing tips of leaves and roots. However, auxin is also found widely distributed throughout the plant, undoubtedly transported from the meristematic regions, as clearly illustrated by Thimann (59) in his determination of the auxin content in different areas of the *Avena* seedling (see Figure 17–9). The concentration of auxin drops as we progress from the tip to the base of the coleoptile; the highest content is found at the tip, and the lowest at the base. Continuing from the base of the coleoptile along the root, we find a steady increase in auxin content until a high point is reached at the tip of the root. The concentration of auxin found at the tip of the root is significantly lower than the concentration found at the tip of the coleoptile. Since Thimann's early work, several studies on auxin distribution have been made (61, 64), confirming the widespread occurrence of auxin in the plant.

Free versus Bound Auxin

There are two general categories of auxins in plants: free and bound. *Free auxins* include diffusible auxins, which move out of the tissue quite readily (e.g., auxins diffusing out of coleoptile tips into agar), and auxins that are readily extractable in various solvents (e.g., diethyl ether at 0° to 5°C). Conversely, bound auxins are auxins that are released from plant tissues only after they are subjected to hydrolysis, autolysis, or enzymolysis. For example, spinach leaves heated in a weak alkaline solution or treated with enzymes that hydrolyze protein (where auxins may be bound) give up a much higher content of auxin than is found when only direct extraction procedures are performed.

Free Indole Compounds Other Than IAA

The most abundant free indole compounds other than IAA that have been found in various plants are indole-3-acetaldehyde, indole-3-pyruvic acid, indole-3-acetonitrile and indole-3-ethanol. The chemical structures of these compounds are illustrated below.

Although experimenters have isolated all of these compounds from plants, most studies support the idea that they are converted to IAA and are not in themselves significant for activity. For example, aldehyde dehydrogenase, the enzyme that catalyzes the conversion of IAALD to IAA is active in tissues in which experimenters have detected IAALD. Similarly, IAN, found in the Cruciferae and Gramineae families, is also accompanied by the enzyme nitrilase, which is involved in the conversion of IAN to IAA. Thus these similar circumstances in a range of plants indicate that IAA is the major active free auxin in plants. Furthermore, free auxin forms are probably the most immediately utilizable by the plant in the growth process. Some synthetic auxins (chemicals not occurring naturally) that are active seemingly remain at least partly free when they are taken up by plants. They may, however, become bound or detoxified.

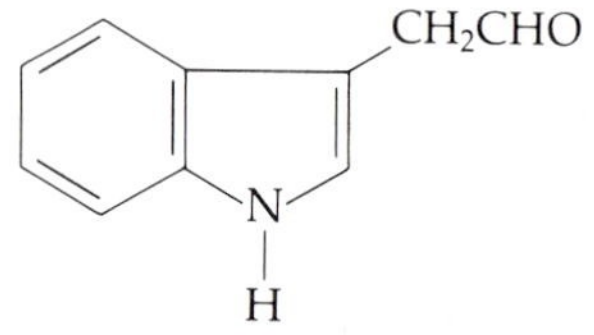

Bound Auxins

Some auxins are combined with substances in the cell that do not allow for easy extraction of the auxin. Bound auxins include *reserve*, or *storage*, forms and *detoxification* forms. Detoxification products are not active per se. They are often formed from excess IAA or from high levels of synthetic auxins that may be introduced into plant tissues. Auxin glucosyl esters, abundant in seeds, are prime examples of bound auxins that are inactive until IAA is released by enzymolysis.

Auxin–amino acid or protein complexes, ascorbigen, and glucobrassicin, which are found in representatives of the Cruciferae and Brassicaceae families, may be detoxification products. Similarly, some of the synthetic auxins may be conjugated as amino acid complexes (aspartic and glutamic acid are very common) and glycosyl esters. The most common sugar conjugates include glucose and arabinose. Also, inositol and other sugar alcohols may complex with various auxins (see Figure 17–10).

indole-3-acetaldehyde (IAALD)

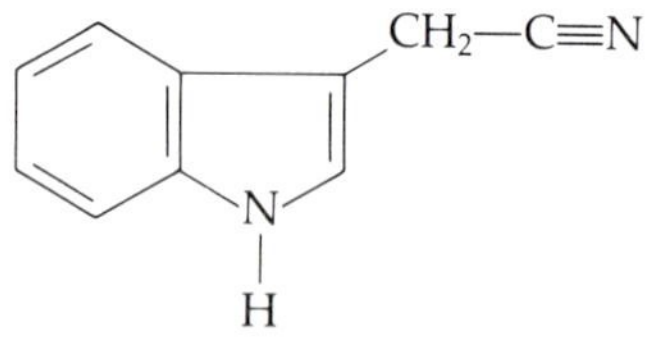

indole-3-acetonitrile (IAN)

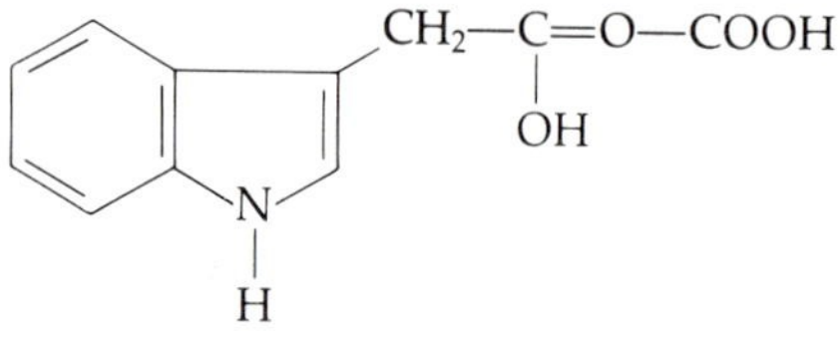

indole-3-pyruvic acid

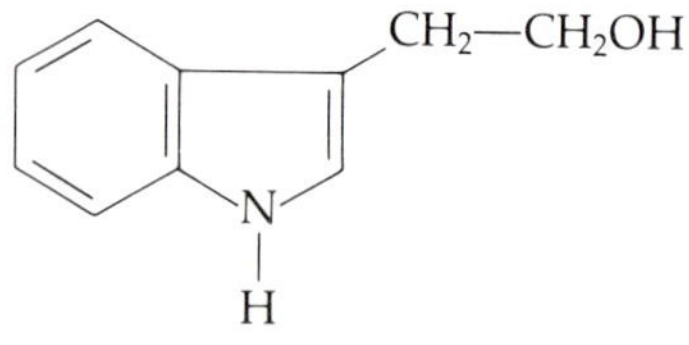

indole-3-ethanol

Figure 17–10. Forms of bound auxin.

Thioglucoside

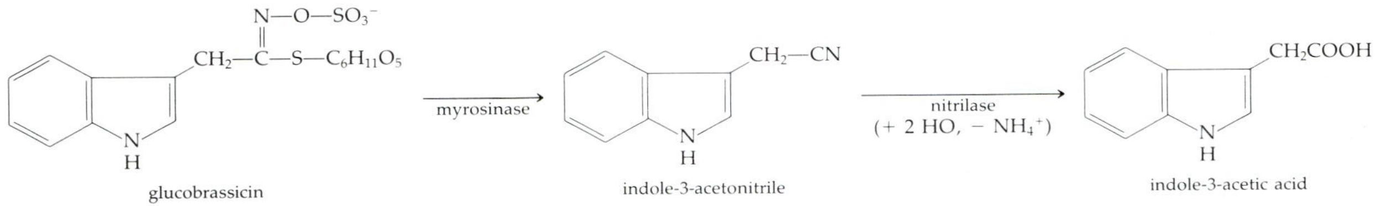

$N-O-SO_3^-$

$CH_2-C-S-C_6H_{11}O_5$

glucobrassicin

myrosinase →

CH_2-CN

indole-3-acetonitrile

nitrilase
(+ 2 HO, − NH$_4^+$) →

CH_2COOH

indole-3-acetic acid

Glycosyl esters

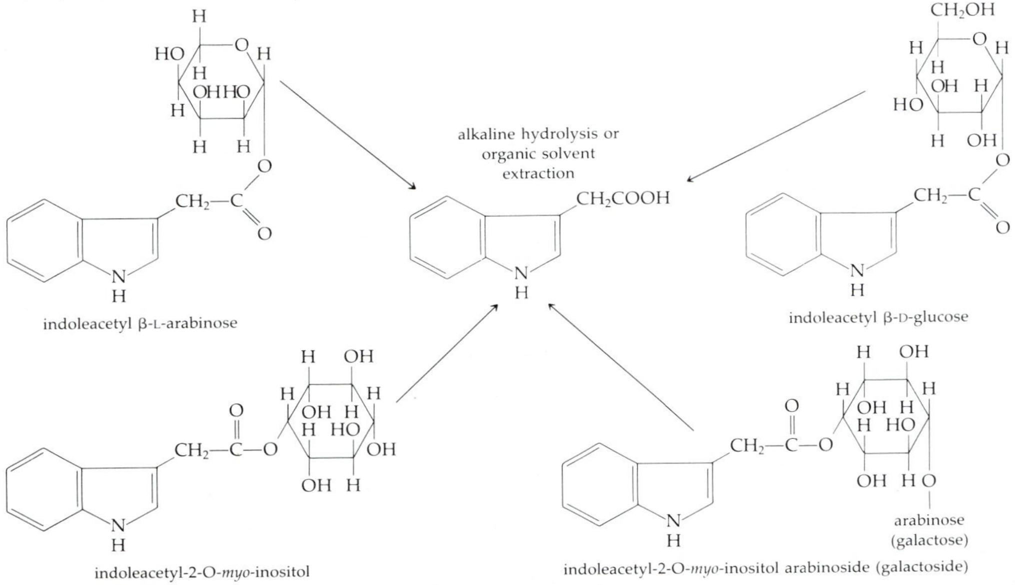

alkaline hydrolysis or organic solvent extraction

CH_2OH

indoleacetyl β-D-glucose

CH_2COOH

indoleacetyl β-L-arabinose

arabinose (galactose)

indoleacetyl-2-O-myo-inositol arabinoside (galactoside)

indoleacetyl-2-O-myo-inositol

IAA Peptides

COOH
$-C-H$
CH_2
CH_2
COOH
$O=C-N-H$
CH_2

indoleacetylglutamate

COOH
$-C-H$
CH_2
COOH
$O=C-N-H$
CH_2

indoleacetylaspartate

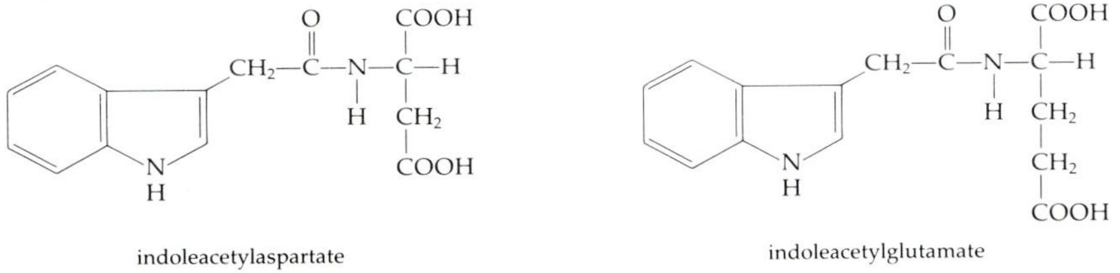

Indole-3-Acetic Acid Biosynthesis

In the earlier years of auxin study, Bonner (6) found that the mold *Rhizopus suinus*, which at that time was one of the best sources of natural auxin, increased its output of natural auxin if grown in a medium containing peptone. This increase in auxin supply undoubtedly occurred through the oxidation of the amino acids of peptone. Three years later, Thimann (60) demonstrated that this mold could convert the amino acid tryptophan to IAA. To this day, tryptophan is considered a primary precursor of IAA of plants.

The synthesis of auxin during lengthy extraction procedures was a source of error in early work with IAA. Experimenters soon discovered that boiling the plant material (25) or extracting at low temperatures (70) effectively limited the synthesis of IAA. These discoveries gave support to the suggestion by Skoog and Thimann (57) that the production of auxin is an enzymatic process. Finally, an enzymatic system capable of converting tryptophan to IAA was isolated by Wildman, Ferri, and Bonner (69) from spinach leaves. The enzymes involved in the conversion of tryptophan to IAA in *Avena* coleoptiles are similar in distribution to IAA. They are in greatest amount at the tip and progressively less concentrated toward the base.

Figure 17–11 illustrates the biosynthetic pathways by which tryptophan might be converted to IAA. Gordon and Nieva (23) found that if leaf disks or crude extracts of pineapple leaves are incubated with tryptophan, tryptamine, or indolepyruvic acid, IAA is formed. They proposed that IAA could be formed from tryptophan via two different pathways: by the deamination of tryptophan to form indolepyruvic acid, followed by decarboxylation to form indole-acetaldehyde; or by the decarboxylation of tryptophan to form tryptamine, followed by deamination to form indoleacetaldehyde. By either pathway, indoleacetaldehyde is formed and thus must be considered the immediate precursor of IAA in plants. One or both pathways have been detected in a variety of plant material (36, 42, 45). Sherwin (53) detected, in cucumber seedlings, the presence of tryptophan decarboxylase, an enzyme that enables these plants to convert tryptophan to tryptamine. Also, tryptophan transaminase activity has been detected in numerous plant species by Truelsen (62). Indolepyruvic acid is thought to be derived from tryptophan by transamination. Indole-acetaldehyde is readily oxidized to form IAA. Experimenters, using crude enzyme preparations from different plants have demonstrated its conversion to auxin on several occasions.

There have been suggestions over the years that tryptophan is not the precursor of the indole ring of IAA. In addition, the pathway (Figure 17–11) was challenged for higher plants because of the possible contamination by IAA-producing bacteria associated with the plants under study. But the possibility that bacterial contamination may provide erroneous results has essentially been eliminated by modern experiments in which plants were treated with antibiotics or grown aseptically and were still able to convert tryptophan into IAA. Also, the enzymes indicated in Figure 17–11 can be extracted from aseptically grown plants and used in vitro to convert tryptophan to IAA.

The natural occurrence in some plants of indole-3-acetonitrile (IAN) suggests another pathway for auxin biosynthesis. In some species, IAN, which has no auxin activity, can be readily converted to IAA in the

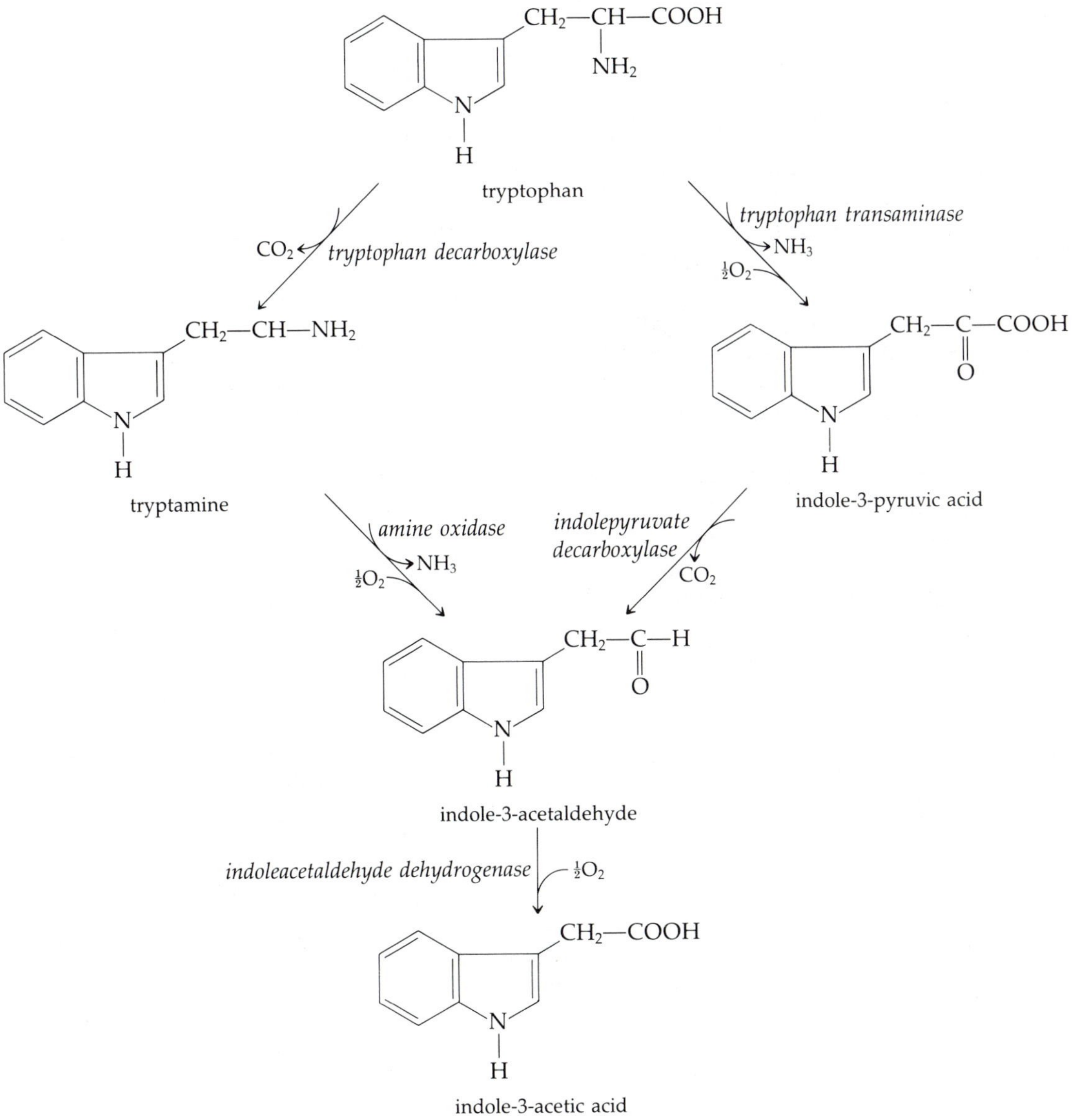

Figure 17–11. Pathways of auxin synthesis from tryptophan.

presence of the enzyme nitrilase. Further, the biochemistry of auxin formation in germinating seeds may be different from that in leaves, coleoptile tips, and other growing areas of the plant. Nevertheless, until additional experiments prove differently, the pathway illustrated in Figure 17–11 is based on the best evidence available today.

Auxin Transport

Experiments by Darwin and by Boysen-Jensen, which demonstrated the movement of an active stimulus from the tip to the base of the coleoptile, led other investigators to assume that the movement of this stimulus was polar. Experiments by Went (67) and Beyer (4) early in the study of auxin supported this concept, and for many years auxin movement in the plant was believed to be strictly polar. Investigators thought this movement occurred in a *basipetal* fashion—that is from the morphological tip to the morphological base (see Figure 17–12). The early work on plant movements (tropisms) also indicated lateral movement.

Although basipetal movement seems to predominate in coleoptiles and some shoots, Jacobs (31) found that in *Coleus* stem sections the ratio of basipetal to *acropetal* (from the morphological base to the morphological tip) transport of auxin is 3:1, respectively. Although acropetal movement is only one-third that of basipetal movement, it is real and significant.

The movement of auxin in the root system is also polar. Movement in roots, however, unlike that in shoots, is primarily acropetal. Work by Scott (52) provides appreciable evidence for the predominantly acropetal movement of auxin in roots, a phenomenon that undoubtedly is significant to the mechanism of auxin action in the geotropism of roots. Also, some of the auxin produced by leaves is transported in the phloem tissues to other parts of the plant (1), a type of transport that is definitely not polar. Finally, in a number of studies, Goldsmith (20, 21) clearly showed that auxin movement occurs acropetally as well as basipetally, although basipetal movement is probably the predominant type.

The translocation of auxin in plant tissues occurs at such high rates as to exclude diffusion as a major method of auxin transport. Also, another reason for eliminating diffusion is the fact that auxin in the plant can move against a concentration gradient. Velocities for auxin transport vary with the type of plant being studied and the conditions under which experiments were per-

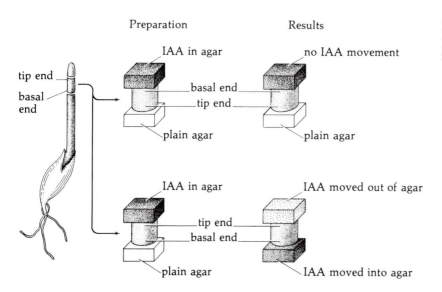

Preparation   Results

tip end

basal end

IAA in agar   no IAA movement

basal end

tip end

plain agar   plain agar

IAA in agar   IAA moved out of agar

tip end

basal end

plain agar   IAA moved into agar

Figure 17–12. Basipetal polar transport of IAA in coleoptile sections.

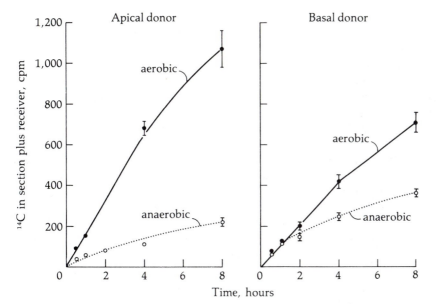

Figure 17–13. Comparison of effect of aerobic conditions or anaerobic conditions on total uptake by *Avena* coleoptile sections from either apical or basal donors containing ^{14}C carboxyl-labeled IAA ($10^{-5}M$).

From M.H.M. Goldsmith. 1966. Plant Physiol. 41:15.

formed. Investigators (46, 47, 49) have observed velocities from 6.4 mm/hr to 26 mm/hr. There is also nonpolar movement of exogenously (from outside) applied regulators in the phloem and xylem of plants. In these instances the velocities may be as high as 40 to 60 mm/hr.

The polar transport of auxin seems to require metabolic energy. Anaerobic conditions (43) or metabolic inhibitors (43) often inhibit auxin transport. As might be expected of a metabolically driven phenomenon, polar transport requires oxygen, is temperature sensitive, and may take place against a concentration gradient. Most auxin present in *Avena* coleoptile sections seems to be there through one of two ways: one way is dependent on metabolic energy and the other is through diffusion (20, 21).

Basipetal movement in *Avena* sections occurs as a result of both diffusion and metabolic transport; acropetal movement relies only on diffusion. We can demonstrate this phenomenon by comparing auxin movement in the sections under aerobic and anaerobic conditions. If we place a cylindrical section of an *Avena* coleoptile between two agar blocks, the top block acting as a donor (contains auxin) and the bottom block as a receiver (contains no auxin); we can clearly see polar basipetal transport. However, if we perform the experiment under anaerobic conditions, polar transport no longer exists, and all auxin movement proceeds by passive diffusion (20). Figure 17–13 shows a comparison of basipetal and acropetal movement under aerobic and anaerobic conditions. Note that under anaerobic conditions, basipetal movement is not much different from acropetal movement. Also, acropetal auxin translocation in coleoptiles and shoots seems to be due to diffusion and is therefore nonmetabolic.

The actual mechanism involved in the transport of auxin is still unknown. In the past, several investigators have proposed that a difference in electrical potential between the tip and the base of the coleoptile controls auxin transport (40, 51). Went was the first to suggest that tropistic responses might be caused by a difference in electrical potential. According to this theory, the base of the *Avena* coleoptile is more electropositive than the tip, the dark side of a unilaterally illuminated coleoptile is more electropositive than the lighted side, and, in a horizontally placed coleoptile, the lower side is more electropositive than the upper side. In each of these situations, the translocation of auxin is toward the highest positive charge. A serious objection to this theory, however, is that when the coleoptile is exposed to an external electrical field, initial curvature is toward the positive pole of the applied charge (51). This movement is the opposite of the direction natural tropism takes, which is toward the negatively charged side. Also recent evidence indicates that electrical potential gradients in coleoptile tissues after an appropriate phototropic or geotropic stimulus seem to arise as a result of auxin migration into an area of tissue rather than before this migration. Thus the auxin itself seems to promote the charge differential. However, one intriguing idea proposed by Scott (52) is that a combination of membrane permeability changes and an electric field induced by auxin acts to "push" the translocating auxin down the coleoptile.

Leopold and Hall (38) suggested that polar movement of auxin in coleoptiles is due to secretion of auxin from the cell's basal end. They estimated that if the net amount of auxin moving basipetally through one corn coleoptile cell exceeded that of acropetal movement by 3 percent, then, after passing through 4 mm (about 30 cells) of tissue, the auxin found in the basal end, or the receiver (e.g., agar block), would be 54 times as much as that found in the apical end. Under these same condi-

tions, if 52.5 percent of the total auxin were secreted out of the basal end of each cell in a file of 100 cells, then over 10,000 times more auxin would accumulate at the basal end than at the acropetal (toward the tip) end of the file. Speculation exists that IAA may be transported across the membrane in the basal end of the cell by complexing with a specific carrier located in the membrane. After moving toward the outside, IAA is released and moves freely into the next cell. Thus the direction of auxin transport may be determined by the location of the carrier, especially if the upper cellular membrane does not contain carriers and if its permeability features favor unimpeded passage of IAA only from the direction of the apex. Further consideration of auxin transport requires direct evidence and not more speculation.

Destruction of Auxin

Just as the production, compartmentalization (free or bound), transport, and utilization of auxin are important to plant growth, the inactivation of auxin is equally significant to the regulation of plant morphogenesis. Auxin is important, by its presence or absence, to the events of vegetative and reproductive growth and aging of plant tissues.

Considerable work has been performed on the mechanisms of auxin inactivation. Two means of destruction of IAA in plants appear to dominate: (1) enzymatic oxidation and (2) photooxidation. In 1947 Tang and Bonner (58) isolated an enzyme involved in the oxidation of IAA. This enzyme system is called IAA oxidase and is the major cause for the disappearance of the phytohormone in vivo. IAA oxidase seems to be ubiquitous, it has been isolated from numerous plant sources (9, 18, 19). However, observers usually study one system in greater detail than they do others; in this case the focus is on the IAA oxidase from etiolated pea epicotyls.

The oxidation of IAA in pea epicotyls appears to be catalyzed by a peroxidase in which one mole of O_2 is consumed (hence the term oxidase) and CO_2 is released for every mole of IAA inactivated. A flavin protein closely associated with the peroxidase appears to generate the hydrogen peroxide required for the reaction. The idea for in vivo hydrogen peroxide generation was originally proposed by Galston, Bonner, and Baker (17) in 1953 and seems to be supported by more current evidence. Peroxidase activity is typified by the oxidation of phenols with H_2O_2 as the electron acceptor according to the following reaction:

$$H_2O_2 + \text{phenol (reduced)} \xrightarrow{\text{peroxidase}} \text{phenol (oxidized)} + 2H_2O$$

The basic distinction between peroxidase and oxidase activity is that a peroxidase reaction does not require the addition of oxygen. However, in the breakdown of IAA, the enzyme exhibiting peroxidase activity also operates as an oxidase, and oxygen is consumed in the reaction. The reaction sequence for the IAA oxidase–mediated destruction of IAA may be summarized as shown below.

The major end products are 3-methyleneoxindole or indolealdehyde, the relative production of either one may vary from one system to the next. The reaction of some systems requires Mn^{+2} ion and a phenolic factor such as 2,4-dichlorophenol.

Early studies showing different and broad pH optima for enzymes from different sources suggested the existence of multiple enzyme forms, a fact now known to be true from electrophoresis studies on IAA oxidases. Certain natural products and chemicals are known to inhibit IAA oxidase reactions. They include chlorogenic acid, caffeic acid, scopoletin, and ferulic acid (71). Chlorogenic acid is preferentially broken down by peroxidase and in this manner IAA is spared. Chlorogenic acid and caffeic acid inhibition can be reversed by the addition of H_2O_2, thereby suggesting the involvement of these inhibitors with the hydrogen peroxide–generating mechanism.

Significance of IAA Oxidase to Growth

In 1954 Galston and Dalberg (18) measured IAA oxidase activity and the growth responses of 7- to 8-day-old etiolated pea seedlings. The IAA oxidase content of the various plant parts was measured according to in vivo and in vitro methods. For the in vivo method, sections of seedlings from the tip down were incubated in a standard IAA

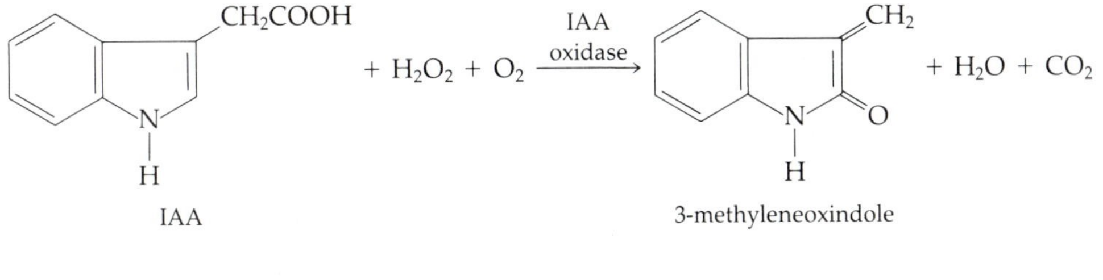

IAA + H_2O_2 + O_2 $\xrightarrow[\text{oxidase}]{\text{IAA}}$ 3-methyleneoxindole + H_2O + CO_2

oxidase reaction mixture. The in vitro method entailed the extraction of IAA oxidase and incubation of the extract in a standard reaction mixture. Residual IAA was measured to assess IAA oxidase activity. Galston and Dalberg found that the ability of stem sections to grow decreases markedly from apical to basal sections, whereas both IAA oxidase assays revealed just the opposite situation for the occurrence of the enzyme. The enzyme increases in activity from the tip down. Thus IAA oxidase activity seems to be low in regions of high auxin content (high growth) and high in regions with low IAA content (low growth). The data suggest that the levels of IAA oxidase in certain areas of the plant regulate auxin levels and, therefore, plant growth. These same workers showed that experimentally aged subapical tissues lose their sensitivity to applied auxin. The tissues also exhibit an increase in IAA oxidase activity. These results illustrate the aging process as another area of IAA and IAA oxidase involvement.

Photooxidation

It has long been known that IAA can be inactivated by ionizing radiation. Skoog (55, 56) demonstrated that rapid inactivation of pure IAA takes place when it is subjected to X- and gamma-radiation. He also noted that little inactivation takes place in a nitrogen atmosphere, thereby suggesting that inactivation is due to oxidation by peroxides formed during irradiation (19). There is some evidence that only a small amount of IAA is inactivated or oxidized in this manner, most of the detrimental effect of this type of irradiation to IAA is of an indirect nature. For example, Gordon (22) claimed that the major effect of ionizing ra-

diation on auxin metabolism may be found in the destructive effect of the radiation on the enzyme system converting tryptophan to IAA.

Ultraviolet light also inactivates IAA. This phenomenon might have been predicted because of the ring structure of the IAA molecule, which absorbs in ultraviolet (maximum absorption at about 280 nm). Here, there is a direct effect on the IAA molecule due to the absorption of ultraviolet light. Determinations of auxin content before and after irradiation with ultraviolet light showed that this type of irradiation reduces auxin levels in plants (10, 48).

Questions

17–1. Discuss the contributions of Darwin, Fitting, Boysen-Hensen, Paal, Stark, Kögl, Went, and Haagen-Smit to the discovery of auxin and its role in plants.

17–2. Provide the chemical formula for indole-3-acetic acid.

17–3. What is a bioassay and what characteristics of a bioassay are important?

17–4. Name some of the major bioassays employed to study auxins. What are the biological responses influenced by auxins that serve as the basis of many bioassays?

17–5. What is the current theory of the role of auxins in phototropism?

17–6. Define the following terms: growth regulator, phytohormone, auxin, antiauxin, and plant growth regulator.

17–7. Name the categories of synthetic auxins and provide the chemical formula for each category.

17–8. What is the significance to plants of bound versus free auxin in plants?

17–9. Explain the process of polar transport with respect to auxin. How might the process occur?

17–10. Describe the mechanisms in a plant that are involved in auxin destruction? How are the levels of auxin maintained in certain plant tissues?

17–11. What role might you ascribe to IAA oxidases in different plant tissues?

Suggested Readings

Brenner, M.L. 1981. Modern methods for plant growth substance analysis. *Ann. Rev. Plant. Physiol.* 32:511–538.

Cohen, J.D., and R.S. Bandurski. 1982. Chemistry and physiology of bound auxins. *Ann. Rev. Plant Physiol.* 33:403–430.

Galston, A.W., P.J. Davies, and R.L. Satter. 1980. *The Life of the Green Plant,* 3rd ed. Englewood Cliffs, N.J.: Prentice-Hall.

Goldsmith, M.H.M. 1977. The polar transport of auxin. *Ann. Rev. Plant Physiol.* 28:439–478.

Leopold, A.C., and P.E. Kriedemann. 1975. *Plant Growth and Development,* 2nd ed. New York: McGraw-Hill.

Moore, T.C. 1979. *Biochemistry and Physiology of Plant Hormones.* New York: Springer-Verlag.

Skoog, F., ed. 1980. *Plant Growth Substances.* pp. 37–105. Proc. 10th Int. Conf. 1979. *Plant Growth Substances.* New York: Springer-Verlag.

Torrey, J.G. 1976. Root hormones and plant growth. *Ann. Rev. Plant Physiol.* 27:435–459.

Varner, J.E., and D.T.H. Ho. 1976. Hormones. In J. Bonner and J.E. Varner, eds., *Plant Biochemistry,* 3rd ed. New York: Academic Press.

Wareing, P.F., and I.D.J. Phillips. 1978. *The Control of Growth and Differentiation in Plants,* 2nd ed. New York: Pergamon Press.

Went, F.W. 1974. Reflections and speculations. *Ann. Rev. Plant Physiol.* 25:1–26.

Chapter 18

Physiological Effects and Mechanisms of Auxin Action

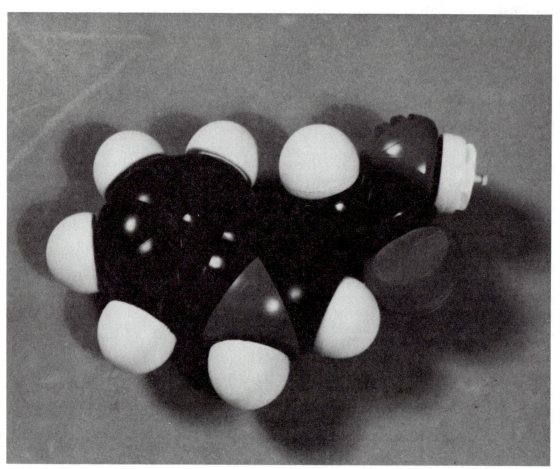

Cori, Pauling, Koltun (CPK) space-filling model of indole-3-acetic acid. Note wedge-shaped ring nitrogen at first position and carboxyl group of acetic acid substituent at third carbon position. *Photo by F.H. Witham.*

Since the discovery of IAA and its characterization as a phytohormone, an enormous amount of information describing the physiological effects of auxin in plants has been published. Many of the auxin effects are both of basic and applied interest. Chemicals exhibiting auxin activity, such as herbicides, have contributed significantly to advances in agriculture. Current knowledge of auxin action is based on studies of the relationship of auxin chemical structure and activity, the effects of auxin on cell walls and cellular enlargement, and the nature of the auxin receptor site in the cell. As with all chemically induced biological responses, the receptor is especially important in the initial cellular translation of auxin chemistry into a physiological response.

Certain working facts have developed from experimentation on auxin action. These phytohormones, IAA for the most part, operate in minute amounts and must be continuously available at the site of action for continued growth (enlargement) to take place. Many of the biochemical and some easily observable physiological events attributed to auxin action occur soon after exposure to auxin. Such responses are referred to as *rapid responses.* Explicit examples of rapid responses include coleoptile and stem segment cell wall deformation and elongation, which occur within 10 minutes after auxin application. In addition, auxins stimulate and sustain mRNA and protein synthesis to form enzymes that catalyze the production of cell wall materials, sugars, and other cellular components. Many auxin-evoked reactions are involved in the so-called *long-term responses.* Both rapid and long-term responses (to be discussed later) provide the plant with mechanisms to adapt to changes in the environment during the course of its morphogenesis.

Let us start our discussion of the action of auxins by considering auxin-mediated responses. Obviously a detailed treatment of the physiological processes influenced by auxins is beyond the scope of this text. However, we can consider some well-known responses in plants that seem to be regulated very precisely by auxins. The following are the most notable responses: cellular elongation, phototropism, geotropism, apical dominance, root initiation, parthenocarpy, abscission, respiration, and callus formation.

Due to the characteristic nature of phytohormonal action, the extent of the induced response depends on several factors. Hormonal concentration and the physiological state of the cells receiving the hormone, the chronological and physiological age of the cells, and other sometimes unknown factors are all important. In a tissue that is sensitive to auxin and other phytohormones, we can observe a characteristic, usually predictable, response curve depending on the level of hormonal concentration (see Figure 18–1).

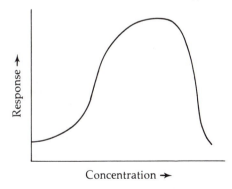

Figure 18–1. Hypothetical dose-response curve illustrating general influence of phytohormone at different concentrations.

This *dose-response* curve is in fact often a means of diagnosing the presence of the hormone. Relatively low concentrations of phytohormone stimulate a given response. As hormonal concentration increases, the response increases and reaches an optimum or peak. At higher levels of hormonal concentration, the extent of the response eventually levels off and declines. The declining side of the curve does not always imply death of the cells but is usually the result of hormonal inhibition. This inhibition seems to be due to the very same chemical characteristics and activities that promote the response at lower levels of hormonal concentration. Thus one mode of regulation of plant response by phytohormones is *stimulation* (turning on) or *inhibition* (turning off) of a response.

Cellular Elongation

Cellular elongation is the basic component of many of the responses influenced by auxins. Most of the studies of the effects of auxins on cell elongation are performed on excised plant material (e.g., *Avena* coleoptile sections or excised root sections) that have little or no endogenous auxin supply. This plant material is ideal for measuring the influence of auxin on cellular elongation because the effect of exogenously applied auxin may be measured without interference from endogenous auxin. The response of coleoptile sections to optimal concentrations of IAA is ten times greater than their response in the absence of IAA (see Figure 18–2).

Over the years, several theories were proposed to account for the action of auxin in cell elongation. Scientists suggested that auxins somehow decrease the osmotic potential of the cell, increase the permeability

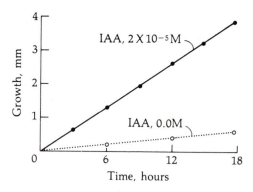

Figure 18–2. Growth of *Avena* coleoptile sections in auxin growth medium and in absence of auxin. Initial length of section was 5.0 mm.

From R.M. Klein, ed. 1961. Plant Growth Regulation. Ames: Iowa State University Press.

of the cell to water, induce the synthesis of RNA and protein (enzymes) for wall components, and cause a reduction in wall pressure. Recently accumulated evidence indicates that auxins may act at the gene level but that, more directly, reduction in wall pressure, which takes place as a result of auxin-induced wall changes (i.e., loosening or deformation), is the primary means by which auxins stimulate cell elongation.

Reduction in wall pressure and cell wall loosening (deformation). As pointed out in Chapter 1, "slippage" or "sliding" of the wall components is necessary for wall extension. Most importantly, breakage of the cross-links between cell wall components probably occurs, with ultimate reestablishment of cross-links at the termination of elongation. Thus the noncovalent cross-links between the xyloglucan polymers and the cellulose microfibrils are most likely broken as a result of some unknown action of auxin, which may be either enzymatic or nonenzymatic. This process tends to favor increased plasticity or loosening of the wall, with decreased elasticity.

As a result of the continued breakage and reformation of the hydrogen bonds, the xyloglucans are thought to "creep" along the cellulose, thus resulting in irreversible stretching of the wall. Observations have shown that low pH favors this process. In fact, as we will discuss later, cell wall loosening may be accomplished without auxin under conditions of acid pH.

Auxin-induced cellular enlargement and water relation changes. For some time researchers have directed their attention toward the suspected changes in the pressure potential, osmotic potential, and water potential of enlarging cells. Ample evidence shows an increase in the amount of solutes in the cell sap of an auxin-treated cell. The concentration of osmotically active solutes does not increase nor does the osmotic potential change. However, even though the osmotic potential does not become more negative in auxin-treated cells, the water potential still becomes negative. If we consider the relationship of these parameters, $\psi_w = \psi_\pi + \psi_p$, we can see that if the osmotic potential does not change, the turgor pressure, or pressure potential (ψ_p), must change. Thus with auxin-induced cell wall loosening and an accompanying decrease in resistance to stretching and internal pressure, the cell membrane pushes outward, with a decrease in turgor. When the internal pressure becomes less positive, the water potential of the cell sap becomes more negative than that of the surrounding cells. Consequently, water diffuses inward along the newly established gradient, thereby causing stretching and an increase in cellular volume. The addition of new cell wall material and the reestablishment of noncovalent linkages between the cellulose and polysaccharides

(xyloglucans) leave an enlarged cell, with increased volume and irreversibly stretched cell walls.

"Acid Growth" and Auxin Action

Implicit in the action of auxin is the idea that auxin induces a decrease in pH in the vicinity of the wall, presumably by activating a membrane-bound H^+ ion pump. Some observers believe the pump is in the plasmalemma. In 1934 Bonner (8) reported that lowered pH of the incubation medium increased the extensibility and growth of coleoptile sections. In 1956 Thimann (65) reported that acidification of the incubation medium accompanied auxin-induced elongation of coleoptile sections. It was not until 1970, however, that Rayle and Cleland (53, 54) proposed that auxin-induced acidification was the mechanism by which cell wall deformation takes place. According to this theory as the pH at a site in the cell wall becomes acidic, loosening enzymes become activated. Another possibility is that H^+ ions act directly on the wall cross-linkages and cause breaks between the noncovalent bonds (bonds between the cellulose micelles and xyloglucans).

Auxins in themselves do not contribute to the acidic pH within the wall matrix. Auxins, however, may somehow interact with membranes, possibly the plasmalemma. One speculation (32) is that auxin action at the plasmalemma causes a release of an unknown substance that is transmitted to the nucleus. This substance then induces alterations in DNA transcription and brings about the production of new mRNA, thus promoting the production of cell wall–loosening enzymes and enzymes for increased respiration necessary for auxin-induced growth. We might further speculate that a release of H⁺ ions from the plasmalemma or H⁺ ion pump activation occurs. Auxins may also influence other membranes, such as the endoplasmic reticulum. However, little evidence clearly supports any of these ideas. The mechanism of auxin action is an area of intensive research that should yield answers in time.

Auxin Action, Specific RNA, and Protein Synthesis

In addition to the plasmalemma or the cell wall being considered likely sites for auxin reception, auxins may also interact at the gene level. We do not know whether an auxin-induced factor is released from some other location in the cell or whether auxin works directly on DNA. Interactions between auxins and DNA are chemically feasible (34, 68).

The relationship between the effects of auxins on nucleic acids and growth was first suggested by Skoog in 1954. Since that time there have been numerous studies supporting Skoog's suggestion that the action of auxins in regulating growth is associated with nucleic acid metabolism (16, 39, 47, 51).

That exogenously applied IAA can induce the synthesis of new RNA and protein has been demonstrated in a variety of plant tissues. For example, applied IAA has induced RNA and protein synthesis in *Rhoeo* leaves (57), yeast cells (60), green pea stem sections (17), bean endocarp (56), and *Avena* coleoptile sections (47). With the use of specific metabolic inhibitors, this activity of IAA has been proven to be associated with auxin-induced cell wall plasticity and extension. Four inhibitors frequently used in this type of study are *actinomycin D, chloramphen-*

icol, 8-azaguanine, and *puromycin*. All four inhibit RNA and protein biosynthesis. Let us examine a study in which metabolic inhibitors were used to clarify the role of IAA in cell expansion.

Artichoke tuber disks that were aged for 24 hours in water responded to exogenously applied IAA with a considerable amount of growth. This increase was accompanied by a substantial amount of new RNA and protein synthesis. However, when actinomycin D (50 mg/l) or 8-azaguanine (0.8 mM) were added simultaneously with IAA, the effect of the auxin was almost completely negated (51) (see Figure 18–3). The fact that metabolic inhibitors of RNA and protein biosynthesis completely offset the effect of IAA on artichoke tuber disks suggests that auxins in cell wall extension are associated with nucleic acid metabolism. Researchers experimenting with these inhibitors in a variety of plant tissues have observed the same results.

These findings place the primary influence of auxins close to the gene level. All of the cells of a plant contain the complete complement of DNA that is characteristic for that plant. All of the genes are present but not all of them are active at any one time—that is, each individual cell contains a number of active genes and a number of repressed genes. Thus we find differences among cells containing the same complement of genes in that some genes may be repressed (63).

One attractive theory is that auxins in some way derepress a repressed gene, thereby releasing DNA template for RNA synthesis. The new RNA—presumably mRNA—would then induce the formation of one or more new enzymes thus increasing wall plasticity and extension. This hypothesis is supported by the finding that the growth of *Avena* coleoptile sections is in-

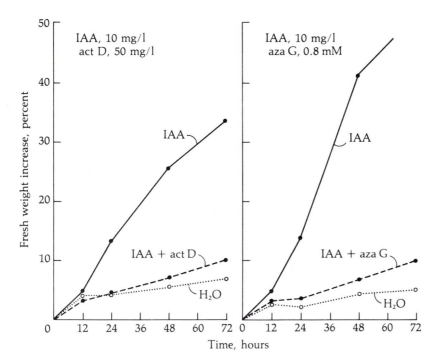

Figure 18–3. Effect of actinomycin D and 8-azaguanine of IAA-induced growth in aged artichoke tuber disks.

From L.D. Nooden. 1968. Plant Physiol. 43:140.

creased when the sections are treated with the enzyme β-1, 3-gluconase, which hydrolyzes β-1,3-glucose links in the coleoptile cell wall. Also, the enzymes hemicellulase, invertase, pectin methylesterase, and ascorbic acid oxidase have been found to be important components of the cell wall protein. Finally, Fan and Maclachlan (23) demonstrated that production of the enzyme cellulase could be induced by the application of IAA to pea epicotyl tissue.

One serious deficiency in the hypothesis that auxin causes cell wall extension by inducing the synthesis of cell wall enzymes is the slow rate at which the cell wall enzymes are produced. Observers note an increase in growth rate in response to IAA treatment in 10 minutes or less. In contrast, changes in protein levels following IAA treatment take longer than 10 minutes. As

we indicated earlier, the initial increased rate of growth is obviously part of the rapid response system, and the synthesis of enzymes is part of the long-term response system.

The two types or sites of auxin action are not necessarily mutually exclusive. The rapid auxin response may be due to preformed proteins with the influence of auxins on protein synthesis being necessary to replace the protein for long-term effects. The rapid response system may work with auxin activation of the hydrogen pump. According to this theory, the cell wall material is initially provided from cellular pools, or reservoirs. The long-term response system provides additional cell wall materials, enzymatic protein, and ATP for sustained growth. Figure 18–4 presents these ideas diagrammatically.

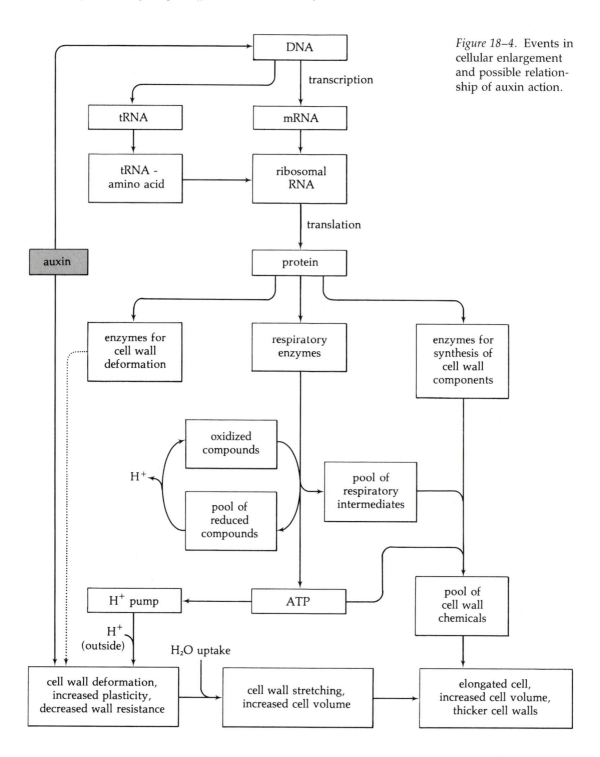

Figure 18–4. Events in cellular enlargement and possible relationship of auxin action.

Plant Growth Movements (Terminology)

The basis for most plant movements resides in cellular growth. These movements are classified according to the nature of the stimulus, the response of the plant organ being stimulated with respect to the direction of the stimulus, endogenous biological-timing mechanisms, and phytohormonal cellular levels.

Tropisms. Movements of a plant organ in response to directional fluxes or gradients in environmental stimuli are called tropisms. Usually the response is directionally influenced by the stimulus. The direction of the tropism is based on the physiological state of the cells and the spatial relationship between the stimulus and the responding plant part.

Nastic movements. Movements whose direction is determined by the morphology of the plant. This type of movement is not necessarily directed toward or away from the stimulus. "Touch" responses in *Mimosa* are an example of movements that do not necessarily occur toward or away from the stimulus.

Epinasty (or hyponasty). Epinasty is a differential growth response, with greater growth on the upper surface (lower surface in hyponasty) of an organ, resulting in downward bending (leaves of many species). Epinasty may be due to differential localization of phytohormones, including stimulants or inhibitors.

Nutations. Nutations represent growth movements with fluxes in their rate due to varying rates of growth on different sides of the organ. This spiraling type of growth, which is often recorded by time-lapse photography, may be superimposed upon (or negated) by those stimuli that favor tropisms.

Biological clock (time-measuring) growth regulation. Often movements of leaves and other plant parts may occur within a predictable time span even though the plants are exposed to nonvarying conditions of gravity, light, and so on. The movements may be cyclic—for example diurnal rhythms that are regulated by a biological clock mechanism. The biological clock may be set by entrainment (conditioning) under appropriate conditions (e.g., red light).

Table 18–1 gives some examples of tropisms and nastic movements. In hydrotropism and hydronasty, roots do not seek out water but respond to applied water. They will exhibit growth in well-watered soils or

Table 18–1. Addition of prefix to denote tropism or nastic movement in accordance with contributing structure.

Stimulus	Tropism	Nastic Movement
gravity	geotropism*	
light	phototropism*	photonasty
darkness		nyctinasty*
temperature	thermotropism	thermonasty
touch	thigmatropism	thigmonasty*
chemical	chemotropism*	chemonasty
water	hydrotropism	hydronasty

* Most widely observed movements. Other movements (no asterisk) illustrate prefix and tropism of movement that might exist.

increasing water gradients and areas of least resistance (e.g., water drainage pipes).

Phototropism

When a growing plant is illuminated by a unilateral light, it responds by bending toward the light. The bending of the plant is caused by cells on the shaded side elongating at a much greater rate than cells on the illuminated side. This differential growth response of the plant to light, called *phototropism*, is caused by an unequal distribution of auxin, the higher concentration of the growth hormone being on the shaded side.

Study of the plant phototropic system is complicated by the fact that the response varies with the intensity of light. DuBuy and Nuerenbergk (21) showed that the phototropic response of the *Avena* coleoptile to unilateral light over a wide range of intensities amounted to one negative and three positive curvatures. If the proper intensity of light is used, the coleoptile can actually be made to bend away from the source of light (negative curvature). In our discussion we will concern ourselves only with the first positive curvature, since it is in this area that most of the work on phototropism has been accomplished.

The Cholodny-Went theory states that there is a higher concentration of auxin on the shaded side than on the light side of a unilaterally illuminated coleoptile. This unequal distribution of auxin could be the result of light-induced inactivation of auxin on the light side, light-induced lateral transport of auxin, or inhibition of basipetal transport of auxin. The current evidence does not favor the explanation of light-induced inactivation of auxin. Either light-induced lateral transport of auxin or inhibition of basipetal transport is a more likely mechanism for auxin distribution in stems and coleoptiles.

Geotropism

If we place an intact seedling in a horizontal position, it will respond to the earth's gravitational field with a particular pattern of growth. The stem under these circumstances will curve upward until it is vertical again, and the root system will curve downward until it too is vertical again. Accordingly, we refer to the stem as an organ that exhibits *negative geotropism* and to the root as an organ that exhibits *positive geotropism.* Thus the perception of gravity by a plant part may result in different orientation in response to a gravitational field. Primary roots and primary stems are positively and negatively geotropic, respectively; secondary roots and shoots are *plagiogeotropic*— that is, they grow to a position at an oblique angle to the gravitational force. Rhizomes are referred to as *diageotropic* because they grow horizontally.

Cholodny-Went theory and geotropism. The Cholodny-Went theory is a reasonably sound explanation for geotropism and phototropism. Cholodny (13, 14) and Went (66) proposed that the differential growth exhibited by a horizontally placed stem or root is due to the accumulation of auxin on the lower side. This accumulation of auxin on the lower side of the horizontally placed stem causes accelerated growth on that lower side and results in stem curvature upward (negative geotropism). This theory as applied to stems still appears to be essentially correct. Conversely, a horizontally placed root will exhibit a positive geotropism when auxin concentrates on the lower side.

According to the Cholodny-Went theory, roots are much more sensitive to IAA than are stems, and the concentration of IAA that stimulates cell elongation in stems is actually inhibitory to cell elongation in roots. The accumulation of auxin on the lower side of a horizontally placed root, therefore, retards cell elongation on that side. The concentration of IAA on the upper side may be reduced to the stimulatory level for root cell elongation.

The inhibition of cell elongation of roots by auxin may be due to the auxin-induced stimulation of ethylene (71). When auxin builds up to a relatively high, or threshold, level, ethylene synthesis is initiated and its presence influences geotropism. Stems, however, do not seem to be as sensitive to ethylene, with respect to geotropism. The combined effect of retardation of cell elongation of the lower side and slightly stimulated cell elongation of the upper side results in the root's curving downward. We will later present a different view of the role of auxins and growth inhibitors in the positive geotropism of roots. It is entirely possible that gravitational pull influences lateral transport of other growth regulating factors as well as of auxins.

Perception of gravity. The simplest explanation concerning the "perception" of gravity by plant parts is based on the naturally induced change in distribution of a cellular component in response to gravitational pull. In addition, several studies indicate that geotropically induced lateral movement represents an active transport process (33, 67). If active transport is involved, we should not be able to detect a geotropic response by plants under anaerobic conditions. Some studies have shown a lack of geotropic response by plants under anaerobic conditions, other studies (33, 67) have

shown the opposite. Some investigators believe that bodies termed *statoliths*, which move under the influence of gravity, are responsible for the lateral movement of auxin in geotropism (35, 41, 42). How the movement of statoliths influences lateral transport of phytohormones is not yet clear.

Statolith theory. A *statolith* is a body that changes position in the cell or plant organ or part as a direct consequence of the change in direction of the axis of an organ with reference to the direction of the force of gravity. The existence of statoliths was originally suggested by Barthold in 1886. Later, Haberlandt (31) suggested that statolith-containing cells, termed *statocysts* or *statocytes*, are found in such sensitive areas of the plant as the root cap cells, the tips of coleoptiles, the root endodermis, the sheath of vascular bundles of hypocotyls, epicotyls, and young foliage. As research continued, it became obvious that the particles to be sedimented would have a density considerably greater than the cell cytoplasm. Audus (4) in 1962 calculated that only starch grains (or amyloplasts) were large enough for their rate of displacement to be correlated with the gravity response. He further calculated that statoliths were not ribosomes or smaller particles because their rate of sedimentation under gravitational influence is much too slow. With few exceptions, even plants that do not normally synthesize storage starch still possess amyloplast statoliths in root caps and bundle sheaths.

How the sedimentation of statoliths relates to the distribution of growth regulators is open to speculation. Also, other ideas relating to hormone distribution will have to be tested. For example, one idea is that gravity may cause a polarization of lateral membranes and produce a one-way flow of hormones from cell to cell. Another suggestion is that in the reorientation of cells to gravity, the vacuole may "float" in the cytoplasm. The thicker layer of cytoplasm toward the bottom may explain the increased auxin concentration on the lower side of a horizontal organ.

Geotropism, auxin, and inhibitors. Current evidence shows that the Cholodny-Went theory, which once explained geotropism of roots simply on the basis of differential auxin concentration, is no longer valid. Let us look at some of the most recent observations. Although IAA is present in root tips (58), its transport in roots is highly polarized in the acropetal direction (58). The evidence is equally clear that geotropism of roots is regulated by the root cap (38). When the root cap is removed, most geotropism is inhibited (38). However, when the cap is regenerated, geotropism of the root is restored. When half of the root cap of maize root tips is removed, the roots (placed horizontally or vertically) will develop curvature toward the side of the remaining half cap (see Figure 18–5).

Further, the growth rate of maize roots increases after the cap is removed. Also, when maize caps are placed on the tip of intact lentil roots, a decrease in root elongation results (see 58). These observations and others not mentioned here strongly suggest that a growth inhibitor is produced by the cap cells of maize and that it is not species-specific. This inhibitor, possibly abscisic acid (ABA), is transported basipetally into the elongation zone and, through the action of gravity (possibly via statoliths or other gravity perception mechanisms), may accumulate and inhibit cellular elongation on the lower side of roots placed in a horizontal position.

Thus an increasing number of proponents of the inhibitor theory support the

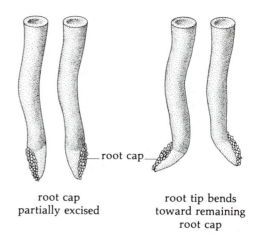

Figure 18–5. Direction of root growth after microdissection of root cap is toward remaining portion of root cap.

idea that root growth and geotropism are controlled by the basipetal transport of an inhibitor (possibly ABA) produced in the root cap and displaced by gravity and the growth-promoting auxin that flows uniformly from the basal end of the root. The key departure from the Cholodny-Went theory is the idea that auxin is not the inhibitor of growth, that its effect as a growth promoter is dependent on accumulations in the root tip via acropetal transport, and that the inhibitor, ABA, is transported basipetally and is distributed by some gravity-sensing mechanism.

Apical Dominance

Long before the discovery of hormonal regulation in plant growth, botanists noted the peculiar dominance of apical over lateral growth in a great many species of plants. They observed that the apical, or terminal, bud of many vascular plants is very active in growth and lateral buds remain inactive. They observed the same phenomenon in the new shoot growth of many tree species. In fact, the characteristic growth patterns of many plant species reflect the influence of apical dominance. Plants that grow tall and unbranched reflect a strong influence; plants that are short and shrubby give evidence of a weak influence of apical dominance.

The strong influence of the apical bud on the growth of lateral buds is easily demonstrated by removing it from the plant. In the absence of the apical bud, active growth begins in the lateral bud. However, in a short time, the lateral bud nearest the apex will establish dominance over the remaining buds and cause them to become inactive again.

The first hint that apical dominance might be because of auxin produced at the terminal bud and transported downward through the stem was given in studies by Skoog and Thimann (62). Removal of the terminal bud of the broad bean and its replacement with a block of agar resulted, as might be expected, in lateral bud growth. Replacement of the terminal bud with agar blocks containing IAA suppressed lateral bud growth in much the same manner as did presence of the terminal bud (see Figure 18–6).

Prior to the Skoog and Thimann experiments, scientists had noted that the apical bud contained a much higher auxin content than did the lateral buds. This fact undoubtedly led to the experiments with the broad bean. However, physiologists to this day

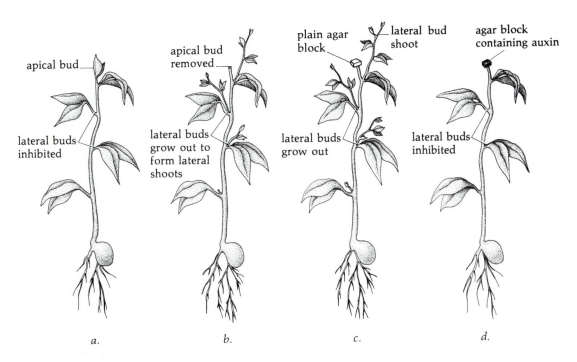

Figure 18–6. Effect of apical bud removal and auxin on lateral bud growth.

have been unable to explain why lateral bud growth should be inhibited by a much smaller amount of auxins than is found in the apical bud, which, to make the problem more complex, grows vigorously in the presence of this relatively high concentration of auxins.

Although the problem of apical dominance did not lend itself to easy solution, it did cause a great deal of speculation in the botanical world. Many theories were proposed with varying degrees of acceptance until Thimann, in 1937, suggested that lateral buds respond to auxins in much the same manner as do roots and shoots—that is, to minimum, optimum, and maximum concentrations (64). Concentrations of auxins above that which will give the maximum concentration will cause inhibition (see Figure 18-7). Thimann claimed that lateral buds are more sensitive than stems to auxins and that the concentration of auxins that stimulates stem growth is inhibitory to lateral bud growth. This theory received general acceptance, although it still failed to explain why the apical bud should be less sensitive to auxins merely because of its location on the stem.

The apical bud is not the only source of auxins. Young developing leaves also produce auxins, and it has been shown that auxins from this source may inhibit lateral bud growth (52).

This explanation of apical bud dominance has been receiving an increasing amount of criticism. For example, studies done on the lilac (*Syringa vulgaris*) demonstrated that the auxin-poor mature leaves of this plant have a much greater influence on lateral bud inhibition than does the auxin-rich terminal bud (10). In addition, lateral bud inhibition not only occurs below the mature leaves on the stem but above them also. Because of the upward movement of the auxin influence on the stem, Champagnat (10) claimed that auxins may not be involved in apical dominance. But, as we discussed earlier, nonpolar movement of auxins has been demonstrated in several cases, thereby making it possible for an influence of auxins to be felt in an upward direction from its origin as well as in a downward direction.

The most provocative criticism of Thimann's theory on apical dominance was given by Gregory and Veale (28). They investigated the nutritional aspects of apical dominance, with surprising results. They found that the influence of auxins on lateral bud growth is controlled by the nutritional status of the plant. If the nitrogen needs of a flax plant are completely supplied during its growth, then, at the optimal period of growth, lateral bud inhibition by applied auxins cannot be demonstrated. However, in flax plants grown under conditions of inadequate nitrogen supply, the influence of applied auxins on lateral bud growth is easily demonstrated.

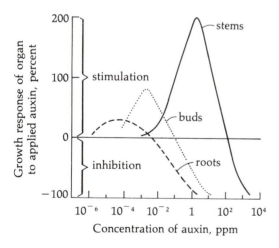

Figure 18-7. Dose-response curves showing effect of different concentrations on growth of three plant organs.

From L.J. Audus. 1959. Plant Growth Substances. *New York: Interscience Publishers.*

Root Initiation

Removal of the tip of a shoot greatly reduces the shoot's growth rate. In contrast, the removal of the tip of a root does not appreciably affect the growth rate (66). In fact, removal of less than 1 mm of the root tip results in a very small but significant stimulation of growth rate (13). Replacement of the tip will retard root growth (13, 14).

The question arises whether the action of auxins is fundamentally different in roots as compared to stems. The action of auxins in roots is similar to that in stems, but the concentrations of auxins that are stimulatory to stem growth are inhibitory to root growth. In other words, roots are much more sensitive to auxins than are stems (see Figure 18–7); and real stimulation of root elongation may be achieved if low enough concentrations are used (20, 36).

The application of relatively high concentrations of IAA to roots not only retards root elongation but causes a noticeable increase in the number of branch roots. Application of IAA in lanolin paste to the severed end of a young stem stimulates the rate of root formation and the number of roots initiated. This discovery is not only of scientific interest but has also opened the door to commercial application of IAA to promote root formation in cuttings of economically useful plants. Figure 18–8 illustrates the effect of IAA and two synthetic auxins on root formation in bean seedlings.

Parthenocarpy

With pollination and the subsequent fertilization of the ovule of a flower, the complex growth patterns leading to fruit set begin. Growth of the ovary wall and, in some cases, the tissues associated with the

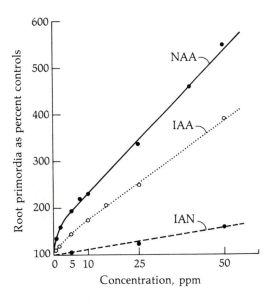

Figure 18–8. Dose-response curves illustrating effect of three auxins on promotion of root primordia formation in bean seedlings.

From L.C. Luckwill. 1956. J. Hort. Sci. 31:89. Redrawn from L.J. Audus. 1959. Plant Growth Substances. New York: Interscience Publishers.

receptacle is greatly accelerated. Most of this acceleration of growth is due to cell enlargement, a phenomenon we have now learned to associate with auxins.

Pollination and fertilization are in some way connected with development of the fruit—perhaps with the release of a stimulus of some kind. Fruit development in the absence of pollination does, however, occur and is, in fact, relatively common in the plant world. The development of fruit in this manner is called *parthenocarpic development*, and the fruit that is formed is called *parthenocarpic fruit.*

In the great majority of cases, however, fruit development does not occur if fertilization does not take place. In what manner, then, does fertilization of the ovule trigger off responses leading to fruit set? In

1902 Massart (46) demonstrated that swelling of the ovary wall of orchids could be stimulated by dead pollen grains. Following Massart's work, Fitting (24) observed that water extracts of pollen are capable of inhibiting floral abscission and stimulating ovary wall swelling in orchids. Due to either a lack of interest in or the complexity of the investigation, the problem of parthenocarpic fruit development lay dormant at this level for over 20 years. In 1934 Yasuda (69) succeeded in causing the development of parthenocarpic fruit with the application of pollen extracts to cucumber flowers. An analysis of the materials in such extracts showed the presence of auxins. Finally, Gustafson (29) demonstrated that parthenocarpic development of fruit could be induced by application of IAA in lanolin paste to the stigma of the flower.

Muir (48) later found that immediately after pollination, a sharp rise in the auxin content of tobacco ovaries occurs. In the absence of pollination, no increase in auxin content is observed (see Figure 18–9). He also observed that growth of the pollen tube considerably increases the amount of extractable auxins in the style of tobacco plants, a phenomenon leading him to suggest that an enzyme may be released by the pollen tubes that catalyzes the production of auxin. This suggestion was later given support by Lund (45), who found that pollen tubes secrete an enzyme capable of converting tryptophan to auxins.

It is obvious that auxins play an important role in the development of fruit. Pollination, growth of the pollen tube, and fertilization all contribute to the "gush" of auxins responsible for fruit development. Although significant, the amount of auxins found in pollen grains is not sufficient to account for the high concentration of auxins found in the ovary after fertilization. How-

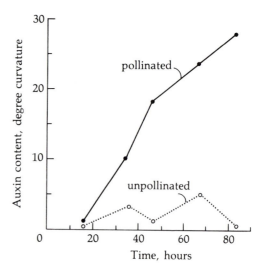

Figure 18–9. Increase in diffusible auxin content in tobacco ovary due to pollination.
From R.M. Muir. 1942. Am. J. Bot. 29:716. Redrawn from A.C. Leopold. 1955. Auxins and Plant Growth. *Los Angeles: University of California Press.*

ever, by the growth of the pollen tube, an enzyme may be released that is involved in the synthesis of auxin, perhaps from a precursor such as tryptophan.

Natural parthenocarpic fruit development is common in the plant world and leads some to suggest that auxins are not involved after all in the development of fruit. However, in the ovaries of species capable of natural parthenocarpy, the auxin content is much higher than that found in the ovaries of species needing fertilization to produce fruit (30).

Abscission

The controlling influence of natural auxins on the abscission of leaves was first suspected in 1933, when Laibach (40) showed that a substance contained in the extract of orchid pollinia is capable of preventing abscission. This observation was supported by LaRue (44), who demonstrated the delaying effects of various synthetic auxins on the abscission of *Coleus* leaves. Since then a great deal of confirmatory work has been performed that clearly establishes indole-3-acetic acid (IAA) as an important controlling factor in the abscission of plant organs (3).

Before the abscission of a plant organ, a layer of tissue is usually formed at the base of the organ; this tissue is easily distinguished from the surrounding tissues. This layer of tissue is referred to as the *abscission zone*. Cells in the abscission zone appear to be thin-walled and are almost completely lacking in lignin and suberin (59). In most cases, a series of cell divisions precedes separation, although separation in the absence of cell division has been found in several species (3).

Three types of dissolution may cause abscission. In some cases, the middle lamella dissolves between two layers of cells and the primary walls remain intact. The middle lamella and the primary wall may both dissolve. And in a few examples, whole cells have dissolved.

What are the factors leading up to the abscission of a plant organ? That removal of a leaf blade will cause, in a short period of time, the abscission of the petiole is well known. As discussed before, one of the sites of auxin production is the leaf blade from which auxins are transported through the petiole into the stem. Auxins, therefore, are suspected as a controlling factor in abscission, as illustrated by Shoji, Addicott, and Swets (61). These investigators found that auxin content is high in the immature bean leaf blade compared to the petiole, but that as the leaf ages, the auxin content of the blade falls to a point comparable to that

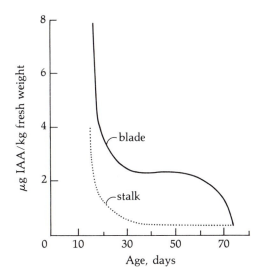

Figure 18–10. Decrease in diffusible auxin content in bean leaf blades and petioles with age. *From K. Shoji et al. 1951. Plant Physiol. 26:189.*

found in the petiole (see Figure 18–10). At this point, the leaves are yellow and ready to abscise.

Senescence of the leaf, therefore, is one of the prerequisites for abscission. In addition to hormonal changes, the internal processes of senescence appear to be directly influenced by changes in the photoperiod (decreased light period), which are characteristic of the northern temperate region of the earth. The photoperiodic induction of abscission and the relationship between the accompanying changes in phytohormones are not clear. It is clear, however, that auxin and ethylene regulate leaf abscission. Their presence and influence are definitely related to the physiological condition and age of the leaf. Although another phytohormone, abscisic acid (ABA), accelerates abscission in cotton plants, a number of studies with other plants since then have revealed that ABA is not particularly active

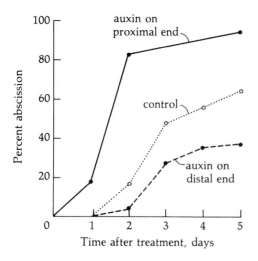

Figure 18–11. Effect of proximal and distal applications of auxin (105 mg/liter) on rate of abscission of debladed leaf petioles.

From F.T. Addicott and R.S. Lynch. 1951. Science *114:688.*

as an abscission agent. Hence many plant physiologists do not consider ABA an important regulator of leaf abscission.

To understand the work related to auxins and leaf abscission, we must consider the earlier experiments of Addicott and Lynch (2). These investigators suggested that the most important factor controlling abscission is the condition of the *auxin gradient* across the abscission zone. Application of IAA in lanolin paste to either the proximal or distal (away from the stem) end of debladed bean leaf petioles has a profound effect on the rate of abscission of those petioles. Proximal application accelerates the rate of abscission and distal application retards it (see Figure 18–11). A critical auxin concentration gradient across the abscission zone rather than the concentration itself may be necessary to prevent abscission. Accordingly, abscission does not occur when the gradient is steep—that is, when the endogenous auxin concentration is high

on the distal side and low on the proximal side of the abscission zone. Abscission occurs when the gradient becomes slight or neutral and is accelerated when the gradient is reversed. Figure 18–12 shows these relationships diagrammatically. It is interesting to note that Rossetter and Jacobs (52) found that the intact leaves of *Coleus* speed abscis-

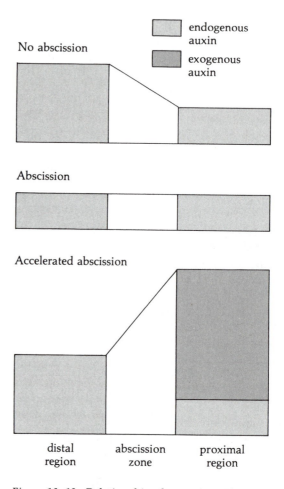

Figure 18–12. Relationships between auxin gradient across abscission zone and abscission.

Reproduced with permission, from F.T. Addicott and R.S. Lynch, The Annual Review of Plant Physiology, *Volume 6. © 1955 by Annual Reviews Inc.*

sion of nearby debladed petioles, thus suggesting that the intact leaves act as sources of proximal auxin for nearby petioles. Also in the bean, application of IAA to the distal tip of one of a pair of opposite debladed petioles accelerates the abscission of the untreated petiole (19, 20).

The ideas of Addicott and Lynch were later refined by Chatterjee and Leopold (11, 12, 55), who showed that the simple gradient theory was not entirely adequate as a universal explanation of auxin action in leaf abscission. These workers showed that the abscission inhibitory action of auxin, or the auxin gradient, across the abscission zone is a function of leaf age. After the leaf has aged, distal applications only promote abscission. The latter effect is probably due to auxin-induced ethylene synthesis. Leopold and his colleagues suggested that a young leaf experiences a long period of auxin-inhibiting potential (stage I leaf); but as the leaf ages, it gradually loses its auxin-inhibiting capacity, and abscission occurs. (stage II leaf).

The most important natural abscission-accelerating factor in senescing leaves appears to be ethylene. Exposing plants to air containing ethylene gas in concentrations as low as one part per million causes the rapid abscission of older leaves (see Figure 18–13). Young leaves, because of their ability to produce higher levels of auxin, can resist abscission in the presence of ethylene. Young, actively growing leaves also produce relatively large amounts of ethylene, which, perhaps, accounts for the fact that the presence of young leaves tends to accelerate the abscission of older leaves. Ethylene produced by the young leaves may diffuse to the older leaves, which also produce ethylene, and induce the abscission of the older leaves. Removal of the young leaves of a plant will delay the abscission of the

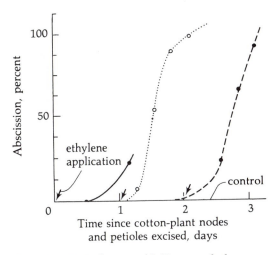

Figure 18–13. Influence of 0.25 ppm ethylene applied at different times (see arrows) on cotton petiole abscission.
From S.P. Burg. 1968. Plant Physiol. 43:1503.

older leaves. This delay may be due to the lowering of ethylene concentration around the older leaves, but removing the young leaves may also reduce competition for nutrients. In fact, the preferential flow of nutrients to the younger leaves at the expense of the more mature leaves may be a more important factor than the effect of ethylene in the abscission of senescent leaves.

Auxin and ethylene appear to be the main hormones that control leaf abscission. Ethylene is certainly the main promoter of abscission. During the early and developing stages of the leaf, auxin is continuously translocated from the lamina to the abscission zone and maintains the attached leaf. As the leaf ages, auxin production tends to decline. It has also been suggested that cytokinins, which act as antisenescent hormones in leaves, decline, and senescence factors build up. When cytokinins are added directly to the abscission layer, senescence of the zone is retarded; but if cytokinins are injected just outside the abscission zone, abscission is accelerated. This effect may be due to the sink, or nutrient accumulation effect, of cytokinins. Ethylene will promote the abscission process in aged petiole tissue (stage II). Added auxin at this time results in accelerated abscission because of the auxin-induced synthesis of more ethylene.

It is now known that ethylene promotes abscission because of its direct effects on the stimulation of cellulase (cellulose-degrading enzyme) synthesis and its release from the cells of the abscission zone. Ethylene in the petiole acts primarily on abscission zone cells and only when those cells have aged or reached a specific physiological condition.

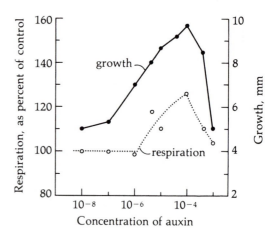

Figure 18–14. Effect of different concentrations of auxin on rate of growth and respiration of corn coleoptile sections.

From R.C. French and H. Beevers. 1953. Am. J. Bot. 40:660.

Respiration

James Bonner recognized in 1933 that auxins have a stimulatory effect on respiration (7). His work led him to suggest that auxin activity only takes place in the presence of oxidative metabolism. Since Bonner's pioneer work, many studies have confirmed that auxins stimulate respiration and that there is a correlation between an increase in growth due to auxin treatment and increase in respiration. Figure 18–14 shows a striking similarity between the response of growth and the response of respiration to different concentrations of IAA. The optimal response for both curves occurs at almost the same concentration of IAA.

Physiologists are still faced with the problem of explaining how auxins induce a stimulation of respiration. An attractive approach to the problem was made by French and Beevers (25). They demonstrated that respiration may be increased by substances that have no effect or an inhibitory effect on growth. Dinitrophenol (DNP), a substance that inhibits oxidative phosphorylation increases the rate of respiration while inhibiting growth. Since the rate of respiration is normally limited by the supply of ADP, treatment of living tissues with DNP should cause an increased supply of ADP and thus stimulate respiration. Auxins may also increase the supply of ADP by causing ATP to be rapidly used up in the expanding cell, thus increasing the supply of ADP. This occurrence would appear to give auxins an indirect role in the stimulation of respiration rather than the direct role that was earlier postulated.

We have already discussed the stimulatory influence of IAA on the synthesis of new RNA and protein. Both synthetic reactions require an expenditure of energy and as such would cause an increase in respiration. Also, in all probability, the activity of enzymes synthesized as a result of IAA stimulation would bring about an increase in respiration.

Callus Formation

Although we have stressed that auxin activity manifests itself in the plant primarily through its stimulant effect on cell elongation, it also is active in cell division. For example, application of 1 percent IAA in lanolin paste to a debladed petiole of a bean plant will cause a yellow swelling where the auxin is applied. This swelling is caused by the development of callus tissue made up of rapidly dividing parenchyma cells. If a succulent stem is cut a few millimeters below a mature leaf and the wound is treated with IAA in lanolin paste, this same proliferation of parenchyma cells is observed. After a period of time, young adventitious roots will develop. Thus IAA not only causes a proliferation of cells but also, under some conditions, may cause a dedifferentiation of these cells—that is, cause the formation of adventitious roots.

Also, in many tissue cultures where callus growth is quite normal, the addition of auxins is necessary for the continued growth of such callus. The amount of callus tissue formed is related to the concentration of IAA applied, higher concentrations causing greater development of the callus tissue.

Questions

18–1. Explain the benefits derived from the occurrence of rapid and long-term responses in plants.

18–2. Describe the details of a dose-response curve as it applies to phytohormones. How is it important in determining whether a compound works as a phytohormone or general nutrient?

18–3. Explain the action of auxin on cellular enlargement through increase in cell wall plasticity.

18–4. Observation of coleoptiles incubated in solutions at pH 4.5 reveal that these acid conditions promote elongation. How has this observation been explained?

18–5. Define nutation, epinasty, nastic movement, tropism, and biological clock.

18–6. Explain the current theories that account for phototropism and geotropism in plants.

18–7. What is a possible explanation for apical dominance? Why do auxins inhibit the growth of lateral buds but not that of the apical bud?

18–8. How can you account for the fact that auxins in relatively low concentration may promote a given response but at relatively higher concentrations inhibit the same process? Is the molecular activity of auxins different at high concentrations?

18–9. Discuss the current thoughts concerning the role of phytohormones in abscission.

18–10. What is callus tissue and its role in intact plants?

Suggested Readings

Audus, L.J. 1972. *Plant Growth Substances*, vol. 1. *Chemistry and Physiology*. London: Leonard Hill Books.

Cleland, R. 1971. Cell wall extension. *Ann. Rev. Plant Physiol.* 22:197–222.

Evans, M.L. 1974. Rapid responses to plant hormones. *Ann. Rev. Plant Physiol.* 25:195–223.

Firn, R.D., and J. Digby. 1980. The establishment of tropic curvatures in plants. *Ann. Rev. Plant Physiol.* 31:131–148.

Marré, E., P. Lado, F. Rasi-Caldogno, R. Colombo, M. Cocucci, and M.I. de Michelis. 1975. Regulation of proton extrusion of plant hormones and cell elongation. *Physiol. Vég.* 13:797–811.

Moore, T.C. 1979. *Biochemistry and Physiology of Plant Hormones*. New York: Springer-Verlag.

Morré, D.J., and J.H. Cherry. 1977. Auxin hormone–plasma membrane interactions. In P.E. Pilet, ed., *Plant Growth Regulation*. New York: Springer-Verlag.

Rayle, D.L., and R. Cleland. 1977. Control of plant cell enlargement by hydrogen ions. In A.A. Moscona and A. Monroy, eds., *Current Topics in Developmental Biology*. vol. 11. *Pattern Development*. New York: Academic Press.

Rubery, P.H. 1981. Auxin receptors. *Ann. Rev. Plant Physiol.* 32:569–596.

Sexton, R., and J.A. Roberts. 1982. Cell biology of abscission. *Ann. Rev. Plant Physiol.* 33:133–162.

Thimann, K.V. 1977. *Hormone Action in the Whole Life of Plants*. Amherst: University of Massachusetts Press.

Chapter 19

Gibberellins

Effects of varying concentrations of gibberellic acid in nutrient solutions on corn (*Zea mays*) plants. Concentrations of GA₃ from left to right are: (1) control (no phytohormone), (2) 0.005 ppm, (3) 0.05 ppm, (4) 0.5 ppm, and (5) 5.0 ppm. *Courtesy of R.N. Arteca, The Pennsylvania State University.*

The *bakanae* ("foolish seedling") disease, which had devastating effects on the rice economy of Japan during the nineteenth and early twentieth centuries, led to the discovery and characterization of gibberellins. During the late 1800s, Japanese farmers reported that rice plants affected with the disease were taller and paler (chlorotic) than their normal counterparts. Most importantly from an agricultural standpoint, the diseased plants were sterile and devoid of fruit (59). Crop losses as high as 40 percent were reported. Needless to say, Japanese scientists were interested in the cause of the disease and its control.

In the early part of the twentieth century, an extensive program of research into the cause of *bakanae* disease was initiated. Japanese pathologists first demonstrated the connection between the *bakanae* disease and the fungus *Gibberella fujikuroi*. Sawada (63) then postulated that the disease was caused by something secreted by the fungus. This postulation was given experimental support by Kurosawa (40), who demonstrated that sterile filtrates of the fungus caused symptoms of *bakanae* disease in otherwise normal rice seedlings. We now know that the ascomycete *Gibberella fujikuroi* is the *perfect* (sexual) stage and *Fusarium moniliforme* is the *imperfect* (asexual) stage of the fungus. Finally, in 1935, Yabuta and Hayashi (79) were able to isolate two crystalline forms of active material in the culture filtrate of *Gibberella fujikuroi*, which they called gibberellin A and B. In 1954 the chemical nature of gibberellin (GA) was determined. Almost simultaneously, workers in England—Brian, Bonow, Elson, Cross, and others—isolated and identified a gibberellin (see 58); and workers in the United States, headed by Stodola and colleagues, isolated gibberellic acid (GA₃) from filtrates of *Gibberella fujikuroi*.

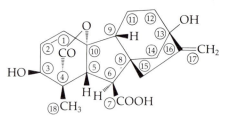

gibberellic acid (GA_3)

Since the initial discovery of GA_3 in fungus filtrates, scientists have shown the widespread distribution of gibberellins in higher plants.

Chemistry of Gibberellins

At the present time fifty-two gibberellins have been isolated, and in some cases as

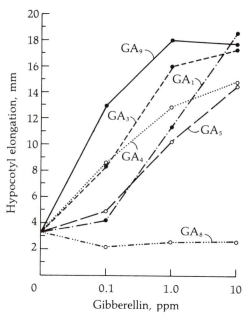

Figure 19–1. Stimulation of hypocotyl elongation of *Lactuca sativa*. Hypocotyl growth was measured 3 days after treatment; each point on graph represents a mean of 30 seedlings.

Reproduced from data of V.K. Rai and M.M. Laloraya. 1967. Physiol. Plant 20:879.

many as seven of these compounds have been found in one plant. For example, seven gibberellins—GA_{17}, GA_{38}, GA_{44}, GA_9, GA_{20}, GA_{29}, and GA_{51}—have been isolated from *Pisum sativum*. All the gibberellins are able to promote either stem elongation or cell division or both in plants, but their relative effectiveness varies greatly (see Figure 19–1).

Figure 19–2 illustrates the basic chemical structures of twelve, free, naturally occurring gibberellins. The structures of the

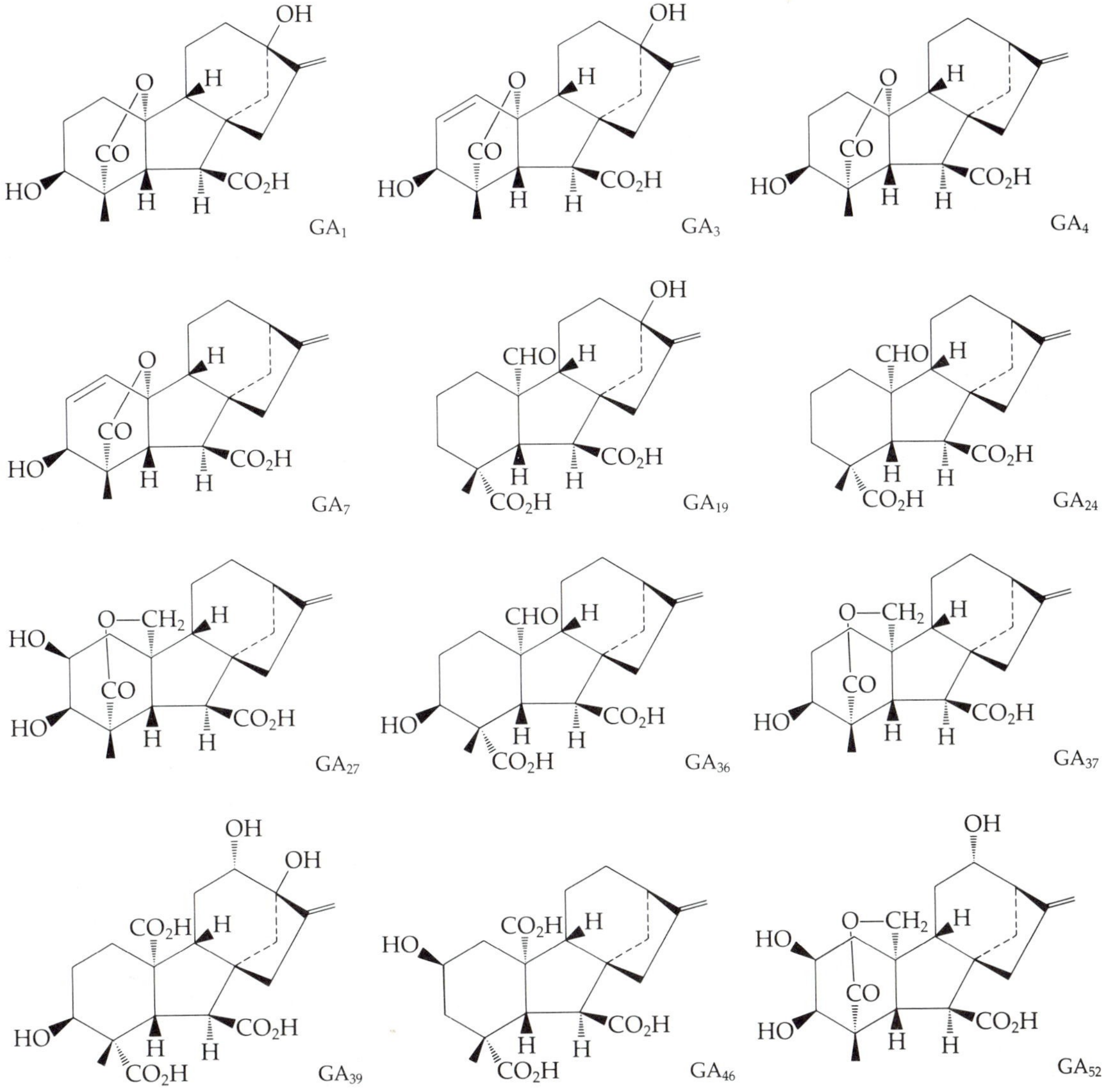

Figure 19–2. Chemical structures of twelve, free, naturally occurring gibberellins. They differ mainly in position and number of substituents.

fifty-two known and identified gibberellins appear in a work by Hedden, MacMillan, and Phinney (24). We can see at a glance that the gibberellins are closely related to one another chemically; all possess the same general carbon skeleton and are similar structurally. They are chemically related to a large group of naturally occurring compounds called *terpenoids*, a great number of which (e.g., sterols, carotenoids) occur in plants. Terpenoids are built from five-carbon *isoprene units;* two units form a *monoterpene* (C-10), three units form a *sesquiterpene* (C-15), and four units form a *diterpene* (C-20). The immediate precursor of gibberellin is the diterpene *kaurene.* The gibberellins are compounds consisting of the ent-gibberellane skeleton (20 carbons) or ent-20 norgibberellane (19 carbons). They are distinguished also by the presence or absence of the lactone configuration (internal ester) in the A ring and the substituents, particularly hydroxyl (OH) groups, about the entire ring structure. The gibberellins are easily interconvertible in the organism by changes in the hydroxyl substitutions. This process can be important in the production of active over inactive forms and vice versa, depending on the presence of hydroxylating enzymes during the different developmental stages of the plant.

Gibberellin Biosynthesis

Most of the information relating to gibberellin biosynthesis in plants comes from studies of immature seeds. West and his associates (74), experimenting with liquid endosperm of immature seeds, accomplished much of this initial work.

Radioactive isotope experiments leave no question as to the participation of acetate as a primary precursor of gibberellin synthesis (see Figure 19–3). Research also indicates that, as in the case of many other biosynthetic pathways, the transfer of active acetyl groups involves the acetyl transfer coenzyme, coenzyme A (CoA). The first few steps in gibberellin biosynthesis involve the formation of three acetyl CoA molecules and their eventual condensation to form mevalonic acid. In the presence of two ATP molecules and a kinase enzyme, mevalonic acid is phosphorylated in a two-step reaction to form mevalonic acid pyrophosphate. Decarboxylation of this latter compound in the presence of ATP and a decarboxylating enzyme yields isopentenyl pyrophosphate (IPP), a five-carbon isoprenoid unit from which all carotenoids, gibberellins, ABA, and a portion of cytokinins are derived.

Isomerization of IPP to form dimethylallyl pyrophosphate constitutes the first step towards the synthesis of higher terpenoids. The reaction is catalyzed by the enzyme IPP isomerase. Dimethylallyl pyrophosphate then acts as an acceptor of one molecule of IPP, the condensation reaction yielding the ten-carbon compound geraniol pyrophosphate. Two successive additions of IPP units to geraniol pyrophosphate leads first to the formation of farnesol pyrophosphate (C-15) and then to the diterpene geranylgeraniol pyrophosphate (C-20). Geranylgeraniol pyrophosphate is first converted to the diterpene alcohol copalyl pyrophosphate and then to kaurene. Kaurene can readily be converted to gibberellin in the plant. Finally, we should note that Milborrow (49) showed that abscisic acid, a sesquiterpenoid, is synthesized from mevalonate and follows the same initial pathway as the gibberellins. Both regulators are mutually antagonistic in certain plant growth systems. Appreciable quantities of gibberellins and abscisic acid are produced in chloroplasts (49).

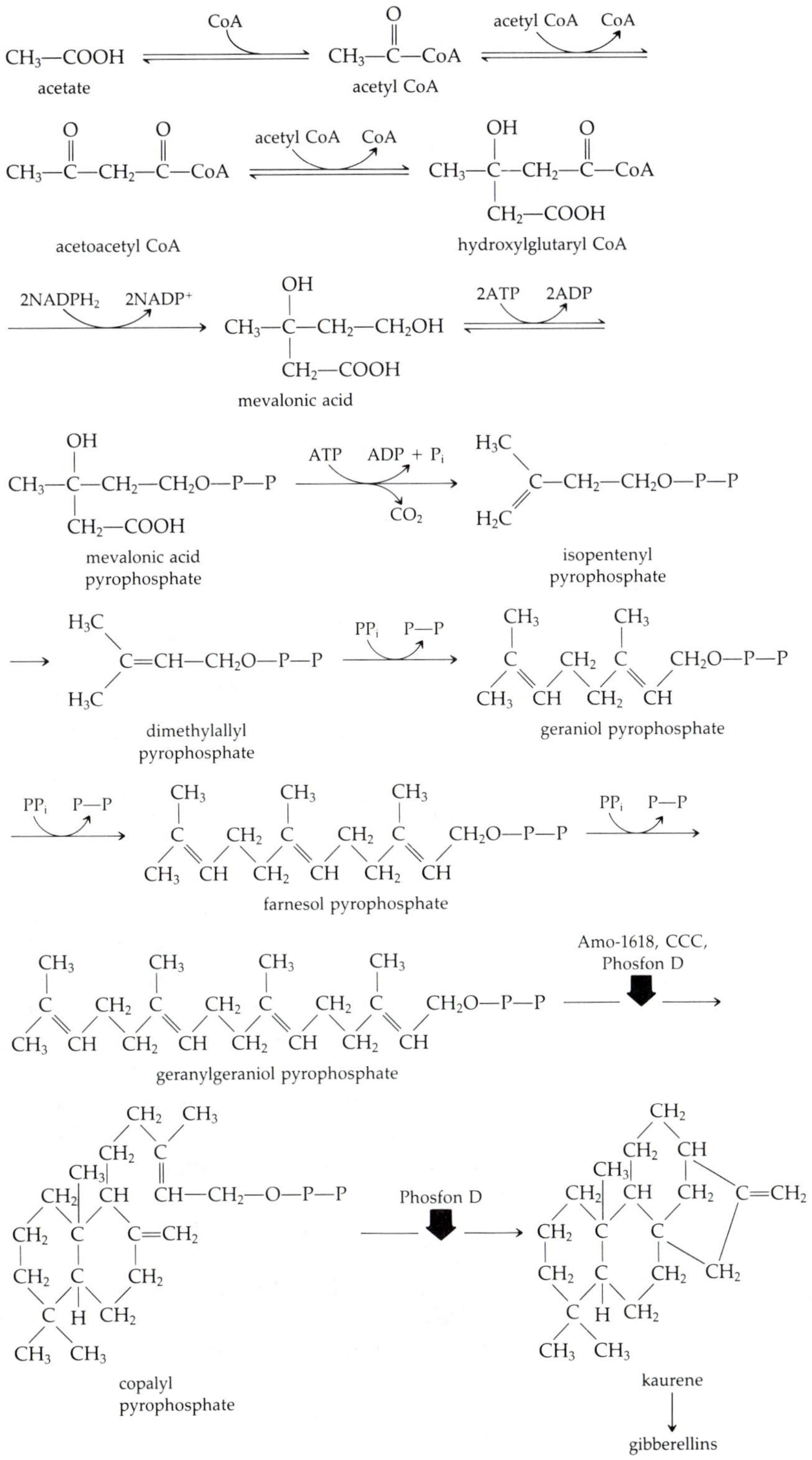

Figure 19–3. Biosynthetic steps leading to formation of gibberellins from acetate. Note sites of inhibitory action for Amo-1618, CCC, and Phosfon D.

Bound Gibberellins

Interconversion of gibberellins in plant tissues is quite common. There is evidence that bound gibberellins exist in plant tissues as gibberellin glycosides (i.e., conjugates with sugar). Scientists do not know whether this phenomenon is a form of inactivation or a storage mechanism. Conjugated gibberellins are believed to exist in the bleeding sap of maple and elm trees, developing bean seeds, developing and germinating pea (*Pisum sativum*) seeds, and Japanese morning glory (see 52).

Antigibberellins or Growth Retardants

Within the last two decades, scientists have synthesized a number of compounds that have an antigibberellin effect on growth. Because of their dwarfing effect on plant growth, we refer to the antigibberellin compounds as growth retardants. The most notable of these compounds are 2'-isopropyl-4'-(trimethylammonium chloride)-5'-methylphenyl piperidine carboxylate (AMO-1618), β-chloroethyltrimethylammonium chloride (cycocel, CCC), tributyl-2,4-dichlorobenzylphosphonium chloride (Phosfon D), and N-dimethylamminosuccinamic acid (B-995, B-Nine, alar). Figure 19-4 gives the chemical structures of the three retardants that are known to inhibit the biosynthesis of gibberellins.

Studies have shown that the inhibitory influence of retardants on plant growth can usually be counteracted by application of gibberellic acid (GA). For example, Lockhart (46) demonstrated that the inhibitory influence of CCC and Phosfon D on stem elongation in beans could be entirely counteracted by GA_3. In another study, Kende, Nunne-

Figure 19-4. Chemical structures of growth retardants, Amo-1618, CCC, and Phosfon D.

mann, and Lang (34) found that Amo-1618 and CCC inhibited GA_3 production by cultures of *Gibberella* but did not affect the growth of the fungus in any other way. From these and other studies we can conclude that Amo-1618, CCC, and Phosfon D retard plant growth by inhibiting the biosynthesis of GA (see Figure 19-3). Some might argue that these growth retardants produce their effect by interfering with the action of GA rather than by somehow blocking its biosynthesis. However, we should note that in plant tissues where the response to GA depends entirely on an exogenous supply of the hormone, even massive doses of the retardants have little effect (43).

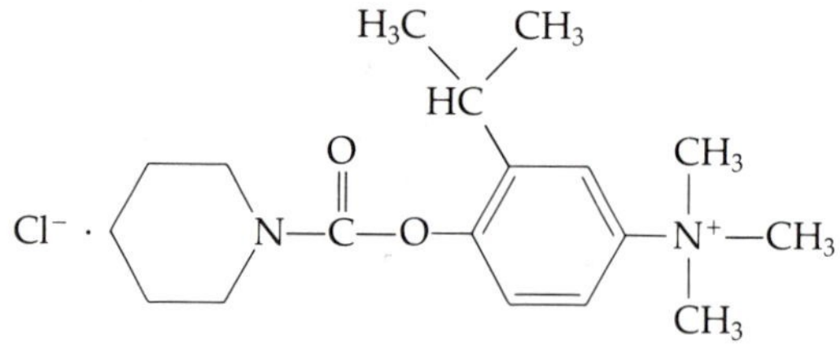

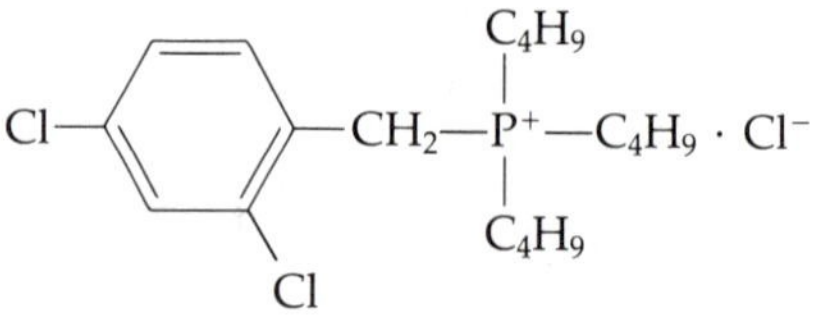

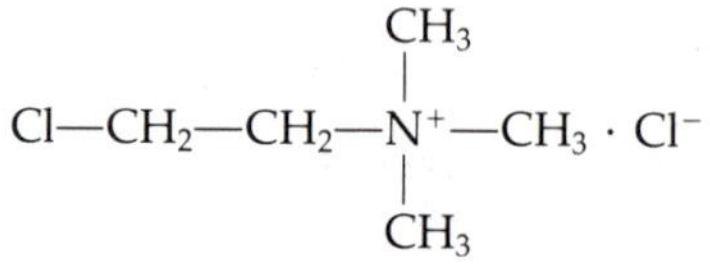

Through good chemical detective work West and his colleagues pinpointed the actual site of the inhibitory action by Amo-1618, CCC, and Phosfon D (14, 15, 65). All three retardants block the conversion of geranylgeraniol pyrophosphate to copalyl pyrophosphate; in so doing they inhibit the synthesis of kaurene and other such compounds (e.g., the gibberellins) that are derived from this intermediate compound. Phosfon D is less specific in its inhibitory action than are Amo-1618 and CCC in that it will also block the conversion of copalyl pyrophosphate to kaurene (see Figure 19–3).

In addition to abscisic acid, two naturally occurring diterpenoids, epiallogibberellic acid and atractyligenin, inhibit gibberellic acid activity. However, we know little about the mechanism of the inhibition.

Gibberellin Transport

Much of the work concerning GA transport in plants is based on studies of the movement of externally applied, or exogenous, radioactive GA to excised stem, petiole, or coleoptile sections. Transport, for the most part, is nonpolar (although some experimenters claim to have observed polar transport on occasion). GA transport occurs in the phloem according to flow patterns similar to those of carbohydrates and other organic substances. Experimenters have isolated gibberellins from sieve elements. GA is translocated in the xylem as well, due to the lateral movement between the two vascular tissues. We do not know the actual mechanism involved in the distribution of gibberellins from their biosynthetic source to the site of action (i.e., growth center, or sink). No special mechanisms seem to operate, aside from those regulating metabolite movement in the vascular system.

Bioassays

Although experimenters have improved isolation techniques and have made significant analytical advances during the last decade, various bioassays played an important role in the early isolation and chemical characterization of gibberellins. A summary of some of these historically important bioassays follows:

Dwarf corn. GA solution is applied to the ligule of the first leaf of dwarf corn seedlings, thereby causing a marked stimulation of the next node or leaf sheath. This test requires an assay period of 10 days and will detect approximately 10 nanograms of GA_3.

Dwarf pea. A solution of GA is applied to the seedling and results in the stimulation of stem elongation. Epicotyl length is measured after 5 days of assay time. The test is sensitive to as little as 1 nanogram.

Lettuce hypocotyl. The entire seedling is placed in a GA solution for 2 to 3 days, and the elongation of the hypocotyl is taken as a measure of the GA response. The minimum detectable amount of GA_3 in the system is 0.1 nanogram.

Avena leaf. Leaf sections are incubated for 3 days and then measured for increased length. The minimum detectable amount of GA_3 is 1 nanogram.

Barley endosperm. This bioassay is probably the most widely used. Half seeds of barley (the half not containing the embryo) are incubated in a solution of GA_3 for 1 day. If gibberellins are present, they will stimulate amylase activity, disappearance of starch, and buildup of reducing sugars. This test will detect a minimum of 0.2 nanogram of GA_3.

Rumex *leaf.* Disks or leaf sections incubated in GA will retain chlorophyll significantly longer than controls. The assay period takes approximately 5 days and will detect a minimum of 0.2 nanogram of GA_3.

Workers have modified these bioassays with different plant materials. The basic responses mentioned for each bioassay, however, are generally used to detect gibberellins in plant extracts. Although experimenters assess the presence of gibberellin activity in plant extracts with bioassays, they analyze and structurally characterize the chemicals present by subjecting the extracts to gas chromatography, a combination of gas chromatography and mass spectroscopy, high-pressure-liquid chromatography, and other techniques.

Physiological Effects

Gibberellins have often been compared in biological activity to auxins. Indeed, in some instances the two classes of phytohormones act similarly. For example, gibberellins and IAA promote cell elongation, induce parthenocarpy, promote cambial activity, and stimulate nucleic acid and protein synthesis. In fact, as we shall see later, gibberellins are also similar in activity to the cytokinins. It is almost as though the major phytohormones operate via similar receptors or similar metabolic pathways and cellular structures. This aspect of phytohormone activity is one of the interesting areas of current research. We will cover, in the following discussion, the influence of gibberellins on dwarfism, bolting and flowering, light-inhibited stem growth, parthenocarpy, and mobilization of storage carbohydrates during germination.

Genetic Dwarfism

One of the most striking properties of gibberellins is that they overcome the phenotypic expression of dwarfism in certain plants. Genetic dwarfism in some instances may be due to a gene mutation. The best-known example is a mutant of corn, called dwarf-5 (d_5), which, due to a single gene mutation, appears to be expressed phenotypically because of a deficiency of gibberellins. The mutation causes a block in a portion of the gibberellin synthetic pathway between copalyl pyrophosphate (see Figure 19–3) and kaurene. Without kaurene, gibberellins are not produced.

Generally, plants with this type of dwarfism exhibit a shortening of internodes and attain approximately one-fifth the size of normal plants. For this reason when gibberellins are applied to d_5 or other single-gene dwarf mutants, such as *Pisum sativum,* *Vicia faba,* and *Phaseolus multiflorus,* the plants elongate to become indistinguishable from their normal counterparts (4). When a plant of one genotype is altered phenotypically by chemical or physical treatment to mimic the phenotype of another plant of different genotype, we refer to it as a *phenocopy.* Gibberellins, however, have little effect when they are applied to the normal plant. Figure 19–5 illustrates the influence of gibberellins on dwarf, intermediate, and normal pea plants. Note the lack of response of the normal plants and the excellent response of the dwarf pea plants. Note, also, that the response increases with an increase in the gibberellin concentration applied.

Although certain dwarfism appears to be due to deficiencies in gibberellins, some dwarf plants do not appear to respond to exogenously applied gibberellins nor do they exhibit differences in GA content as

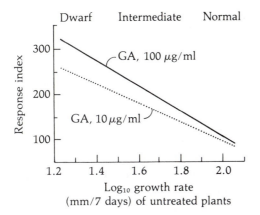

Figure 19–5. Relationship between growth rate and response of pea varieties to gibberellic acid. Response index equals mean growth increment of treated plants per mean growth increment of untreated plants multiplied by 100.

From P.W. Brian and H.G. Hemming. 1955. Physiol. Plant. 8:669.

compared with their normal counterparts. Some investigators suggest that there may be ineffective receptors, metabolic blockages, or possibly an excess of natural inhibitors in certain dwarf plants. These speculations are interesting but of no significance without sufficient evidence.

Bolting and Flowering

In addition to their role in internode elongation, gibberellins function in many plants as a controlling factor in a balance between internode growth—a form of growth called a *rosette*—and leaf development. For example, in many plants, leaf development is profuse and internode growth is retarded. Just before the reproductive stage, there is a striking stimulation of internode elongation, the stem sometimes elongating from five to six times the original height of the plant.

Usually this type of plant is a rosetted *long-day* plant, which requires a certain minimum number of hours of daylight to bolt and flower, or a rosetted *cold-requiring* plant, which needs a cold treatment to bolt and flower. If the long-day plant is kept under short-day conditions and the cold-requiring plant is not given a cold treatment, the rosette form of growth is maintained.

Treatment of these plants with gibberellin during conditions that would normally maintain the rosette form will cause the plant to bolt and flower (41, 43, 70). Experimenters can even separate bolting from flowering by controlling the amount of gibberellin applied—that is, a plant will bolt but not flower if a smaller dosage of gibberellin is applied (59).

The separation of bolting and flowering in gibberellin treatment of rosetted plants has led some investigators to suggest that flowering is only an indirect result of gibberellin treatment. The stimulated elongation of the stem necessitates the production of the many compounds needed to maintain such internodal growth. Some of these compounds, by either their concentration or presence, may ultimately lead to the differentiation of floral primordia. In addition, gibberellin treatment of short-day plants under photoperiods unsuitable for flowering does not promote flowering (66). In fact, in at least one case, gibberellin treatment of short-day plants under conditions favorable to flowering actually reduced flowering (23).

The reason a plant either remains in the rosette form or bolts and flowers appears to be related to the amount of native gibberellin present in the plant. For example, there is some evidence that native gibberellin-like substances are found in greater amounts in the bolted plant than in the nonbolted form. In addition, higher concentrations of gibberellin-like substances have been found in the bolted cold-requiring plant *Chrysanthemum morifolium* Ram. cv. Shuokan and in the long-day plant *Rudbeckia speciosa* Wenderoth than in the nonbolted forms (22, 53).

The influence of gibberellin on bolting includes a stimulation of cell division as well as cell elongation. Responsive plants treated with gibberellin will show a pronounced increase in cell division in the subapical meristem, as evidenced by work with growth retardants that are mutually antagonistic with gibberellin. For example, the growth retardants Amo, CCC, and Phosfon D, compounds that act by blocking gibberellin biosynthesis, inhibit subapical cell division and induce lateral expansion of the apex. However, if gibberellin is applied with one of

these compounds, the retardant effect is neutralized (62) (see Figure 19–6).

Light-Inhibited Stem Growth

If we compare the stem growth of an etiolated plant with that of a light-grown plant we will immediately conclude that light has an inhibitory effect on stem elongation. Application of gibberellins to certain plants growing in the light will greatly increase their stem growth.

Is there a relationship or interaction between endogenous gibberellin and light absorbed by the plant? The reversal of light-induced inhibition of stem elongation by application of gibberellin suggests that endogenous gibberellin is the limiting factor in

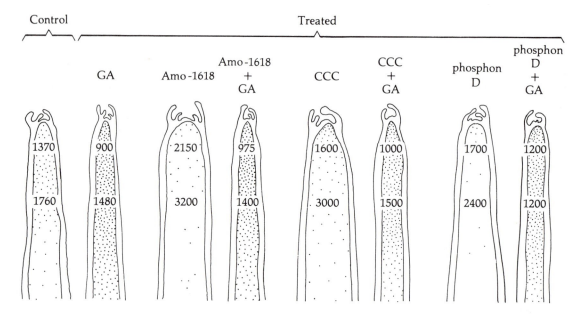

Figure 19–6. Density and distribution of mitotic activity in pith tissue of chrysanthemum stems treated with Amo-1618, CCC, or Phosfon D in presence or absence of added GA. Each dot represents one mitotic figure. Growth retardants in-

hibit subapical cell division and induce lateral expansion of apex.

From R.M. Sachs and A.M. Kofranek. 1963. Am. J. Bot. 50:772.

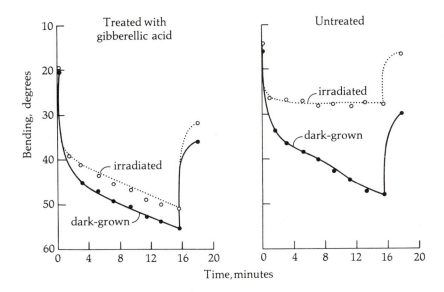

Figure 19–7. Plasticity of cell wall of elongating cell in dark-grown and irradiated Alaska pea stems. Irradiation was 3-hr exposure to red light. Gibberellic acid was applied 3 hr prior to irradiation. Plasticity is measured as amount of residual bending after weight is removed. Plasticity was not decreased by irradiation when gibberellic acid was applied.

From R.M. Klein, ed. 1961. Plant Growth Regulation. Ames: Iowa State University Press.

stem growth. The most obvious conclusion is that light causes inhibition of stem growth by lowering the level of available gibberellin in the plant. This inhibition is overcome by applying exogenous gibberellin to the plant. However, investigation has cast doubt on this simple solution.

Lockhart (44) showed that an increase in the level of available gibberellin causes an increase in the plasticity of young cell walls. Earlier we mentioned the importance of cell wall plasticity in cell elongation. Lockhart also demonstrated that plasticity of the cell wall is decreased in the light-grown cells (see Figure 19–7). Application of exogenous gibberellin counteracts light-induced decrease in plasticity. Evidence that red-light irradiation retards the conversion of a gibberellin precursor to gibberellin was found in a study of stem elongation in the bean *Phaseolus vulgaris* cv. Pinto (46). Obviously, this results in a red-light inhibition of stem elongation that can be overcome by exogenously applied GA. Thus, in this plant at

least, we have evidence of a light-mediated decrease in the level of gibberellin in the plant.

In work by Mohr and Mohr and Appuhn (50, 51), an argument may be found against the theory that light inhibition of stem elongation is caused by light-induced lowering of the gibberellin level in plants. Stem elongation of mustard seedlings grown in the dark may also be stimulated by application of gibberellin. In fact, the concentration of gibberellin needed for maximum response is the same for both dark- and light-grown mustard seedlings. This could not be so if light lowered the endogenous level of available gibberellin in the plant. The possibility exists that light stimulates the production of inhibitors that interfere with the activity of GA in stem elongation. Evidence supporting this possibility has been found in studies of GA activity in pea stem elongation (33, 37).

We do not as yet know whether gibberellin-induced elongation and light-induced inhibition of stem growth act independently of each other. We can provide arguments for either conclusion.

Parthenocarpy

In a previous discussion, we described how application of auxins could cause fruit to develop parthenocarpically. In the early years after this discovery, scientists thought that auxin activity after fertilization was the primary mechanism of fruit development. In fact, substitution of exogenous auxin for fertilization became a highly valuable venture into the economics of fruit growing.

However, auxins are not the only natural growth hormones capable of inducing parthenocarpy. Gibberellins have been found very reliable in producing parthenocarpic fruit-set and, in many cases, show higher activity than do the native auxins in this respect. In fact, several examples exist in which auxins proved to be ineffective and gibberellins proved to be active (16). For example, pome and stone fruits have been generally unresponsive to auxin treatment (77). Yet gibberellins have induced parthenocarpy in both pome fruits (13, 47) and stone fruits (12, 61). Native gibberellins and gibberellin-like substances undoubtedly play a major role in the development of fruit under natural conditions. Whether this is a direct action by gibberellins or an interaction with the native auxin of the plant has not been conclusively shown. We do know, however, that young developing seeds contain relatively high amounts of gibberellin. As the seed matures and growth slows, workers have observed a simultaneous drop in gibberellin content. Gibberellin that is produced during seed development is probably translocated out into the fruit tissues, where it exerts some control over fruit development.

Mobilization of Storage Compounds during Germination

A longitudinal section through a mature cereal grain will reveal that the greater part of its bulk is composed of the embryo and the endosperm. The endosperm consists of a mass of starch-laden cells surrounded by a layer of cells called the *aleurone*. The embryo, of course, represents the future adult plant. Growth of the embryo during germination depends on the mobilization of stored starch in the endosperm. By mobilization, we mean the enzymatic breakdown of the stored starch to simple sugars and the translocation of these sugars to the embryo, where they will provide an energy source for growth.

Up until 1958 scientists thought that the endosperm played only a passive role in germination and that the embryo provided the enzymes for the breakdown and mobilization of the endosperm starch reserves. However, Yomo, a Japanese scientist, demonstrated that under aerobic conditions barley endosperm, separated from the embryo but incubated with it in the same culture flask, could exhibit amylase activity (80). No amylase activity was observed in culture flasks containing either the embryo or the endosperm alone. From his experiments, Yomo concluded that amylase activity in the endosperm is controlled by an unknown factor produced in the embryo. Yomo (81, 82) and Paleg (55, 56), working independently, concluded that the unknown factor was gibberellin (see Figure 19–8). Both investigators demonstrated that exogenously applied GA could stimulate amylase activity in *isolated* barley endosperm. Paleg (55, 56, 58) was also able to show that the enzymes α- and β-amylase were present in the cul-

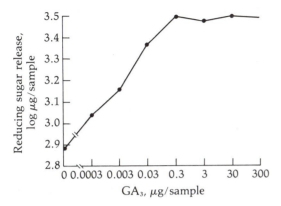

Figure 19–8. GA-induced breakdown of starch in endosperm tissue incubated for 24 hours.

Reproduced from data of L.G. Paleg and B.G. Coombe. 1967. Plant Physiol. 42:445.

ture medium containing GA-treated endosperm. Figure 19–9 shows breakdown of endosperm starch under the influence of GA in barley grains from which the embryos have been removed.

Other studies soon showed that it is the aleurone layer of the endosperm that is sensitive to GA. As Figure 19–10 depicts, removing the aleurone renders the endosperm almost totally insensitive to GA treatment (48). Subsequent studies showed that GA treatment of isolated aleurone (see Figure 19–11) can cause the release of the hydrolytic enzymes required for the digestion of the endosperm starch (6, 57, 72, 73). Finally, electron microscope studies revealed that GA treatment of the aleurone has a profound affect on the cellular ultrastructure of that tissue (27). Changes are particularly apparent in the aleurone grains and their membranes.

Results obtained after isolated aleurone layers of barley are incubated in gibberellins clearly show that the aleurone cells produce and secrete various hydrolytic enzymes. The enzymes released from aleurone cells are α-amylase, ribonuclease, β-1,3-glucanase, protease, β-amylase, and

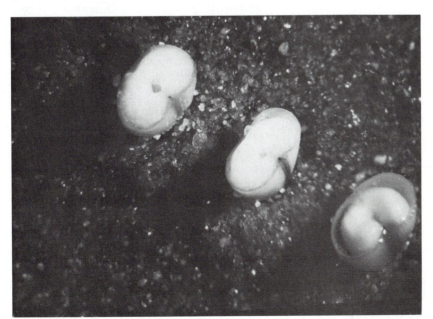

Figure 19–9. Sterile halves of barley grains treated on surface with 0.5 ml H_2O (left), 1 part per billion GA (center), and 100 parts per billion GA (right). Photograph was taken 48 hours after grains were treated. Digestion of starch has taken place in two grains treated with GA.

Courtesy of J.E. Varner, Washington University, St. Louis.

possibly other unidentified hydrolases. As a result of the work of Varner and his associates (72, 73) and of Bennett and Chrispeels (2, 8, 9, 10), we can now clearly see that, with the exception of β-amylase, the above-mentioned enzymes are synthesized *de novo* by the GA-stimulated aleurone cells. β-amylase is preformed but is released only when GA is present. *De novo* synthesis of the hydrolases refers to the fact that they are newly formed after GA stimulation of the aleurone cells. Conversely, the β-amylase is present in the form of previously synthesized inactive protein that becomes activated and released in the presence of GA.

By using the heavy isotope of oxygen in water ($H_2^{18}O$), Varner showed the *de novo* synthesis of α-amylase and protease. The initial events in the stimulation of aleurone cells by GA include the hydrolysis of stored proteins and the release of amino acids. During peptide hydrolysis, the formation of

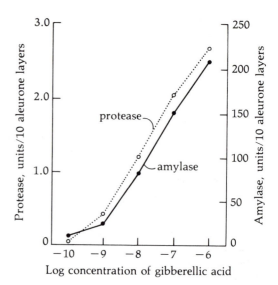

Figure 19–11. Release of amylase and protease by aleurone layers in response to various concentrations of GA₃.

From J.V. Jacobsen and J.E. Varner. 1967. Plant Physiol. 42:1596.

the carboxyl end of the free amino acids requires the addition of H_2O. If aleurone cells are incubated in GA₃ and $H_2^{18}O$, the resulting amino acids formed will be labeled with the heavy isotope of oxygen. Thus in the next steps, when new enzymes (i.e., α-amylase, protease, and so on) are formed (*de novo*), with a primary structure consisting of labeled amino acids, the enzymes will be heavier than those produced by aleurone cells incubated in GA and regular water ($H_2^{16}O$). The process by which newly formed enzymes (proteins) are labeled with a heavy isotope is known as *density labeling*. Once the enzymes are released and recovered from the aleurone cells, extracts of the control and density-labeled protein extracts are subjected to isopycnic equilibrium centrifugation (at 45,000 RPM for 70 hours in a cesium chloride gradient). The results are illustrated in Figure 19–12. As we can

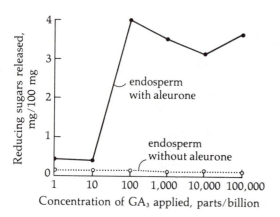

Figure 19–10. Effect of GA₃ on breakdown of starch in endosperm tissue with aleurone layer and endosperm tissue without aleurone layer.

From data of A.M. MacLeod and A.S. Millar. 1962. J. Inst. Brewing 68:322. Redrawn from J. van Overbeek. 1962. Science 152:721. Copyright 1962 by the American Association for the Advancement of Science.

see, the heavier protein fraction indicates the presence of ^{18}O. In those aleurone cells provided with $H_2{}^{18}O$, the heavy isotope was incorporated into the carboxyl groups of the amino acids resulting from the hydrolysis of existing proteins. The amino acids were in turn incorporated into the newly synthesized (*de novo*) α-amylase. The evidence showed that all the amylase formed under the influence of GA_3 was newly synthesized from the amino acids of preexisting protein.

Mechanism of Action of Gibberellins

The induction of enzyme activity in the endosperm by GA_3 suggests that the primary action of this plant growth regulator may be—as is the case with auxins—close to the gene level. Indeed, studies have shown that at least four of the enzymes (α-amylase,

protease, ribonuclease, and β-1,3-glucanase), induced by GA_3 treatment, arise through *de novo* synthesis (29, 73). This evidence certainly indicates the participation of newly synthesized RNA, possibly as a result of DNA being activated through the derepression or activation of one or more genes. The fact that compounds that inhibit RNA synthesis (8-azaguanine and actinomycin D) and compounds that inhibit protein synthesis (cycloheximide and puromycin) inhibit GA-induced enzyme synthesis in barley aleurone layers lends support to this hypothesis (72, 73).

The simplest explanation of the events involved is that the genes for α-amylase or protease synthesis, for example, are repressed prior to seed germination. In the early stages of germination, an effector, gibberellic acid, is released by the embryo and is translocated to the aleurone cells. Once there, it causes the derepression of the genes controlling enzyme (α-amylase, protease, ribonuclease, β-1,3-glucanase) syn-

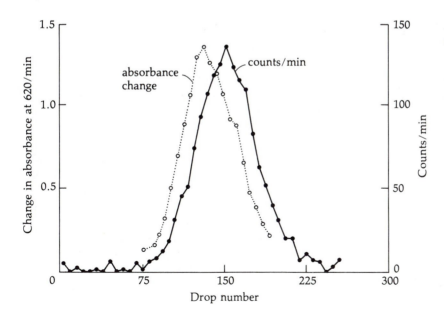

Figure 19–12. Relative distribution of radioactively labeled (with ^3H) α-amylase and ^{18}O-labeled α-amylase from aleurone cells treated with GA_3 by isopycnic equilibrium centrifugation on a CsCl gradient. Distribution of ^{18}O-labeled α-amylase synthesized in aleurone cells treated with GA_3 and $H_2{}^{18}O$ showed higher density than radioactively labeled enzyme.

From P. Filner and J.E. Varner. 1967. Proc. Natl. Acad. Sci., U.S. 58:1520.

thesis. DNA, activated by the derepression, produces new RNA, which in turn produces new protein. In support of this explanation, Evins and Varner (17) found an increase in polyribosome formation and an increase in synthesis of ribosomes and endoplasmic reticular membranes 2 to 4 hours after GA_3 treatment of the isolated aleurone system. In addition, studies showed that the newly formed polyribosomes are responsible for at least some of the new enzymes formed (e.g., α-amylase) in this system (17). It is interesting to note that in addition to inhibitors of RNA and protein synthesis, the naturally occurring inhibitor ABA also inhibits GA-induced enzyme synthesis in barley aleurone layers (7, 8, 9, 10, 17). The effect of ABA, in this respect, is similar to 8-azaguanine, an inhibitor of RNA synthesis.

Gibberellins appear to be capable of retarding leaf senescence in certain species of plants, and this effect may have something to do with the ability of gibberellins to induce new RNA and protein synthesis. We can observe the influence of gibberellins on leaf senescence in a simple laboratory exercise in which we compare the chlorophyll retention of leaf disks of *Rumex* floating on a gibberellin solution with that of leaf disks floating on water. The gibberellin-treated disks retain their chlorophyll for a much longer period of time. The yellowing of leaves is the first visible sign of senescence, and it is accompanied by a reduced capacity for RNA and protein synthesis.

Gibberellin Interaction with DNA

If, in fact, phytohormones such as the gibberellins stimulate RNA, protein synthesis, and specific enzyme induction, it is very likely that interactions between DNA and/or RNA and active gibberellins take place. Existing information strongly suggests that at least some auxins, cytokinins, and gibberellins influence the physical properties of DNA (28, 35). For example, IAA, kinetin, and GA_3, when incubated appropriately with DNA, cause changes in DNA melting points and hence physical properties because of their likely hydrogen-bonding to DNA. In addition, they seem to influence the *in vitro* coiling of the DNA helix and reduce chromatin-protein binding (28). GA_3 has also been shown to bind specifically to adenine-thymine-rich DNAs (35, 36) and, with DNA ligase, induces the formation of circlelike loops in nuclear DNA. Further, Snir and Kessler (68) observed synergistic and antagonistic interactions between ethidium bromide and GA_3 in promoting cucumber hypocotyl growth. Ethidium bromide is a known *intercalating agent*—that is, it bonds to DNA between successive base pairs—thus suggesting that the gibberellins and possibly other hormones operate similarly. The work of Witham, Hendry, and Chapman with space-filling models shows that a likely interaction between DNA and phytohormones is chemically feasible (25, 76). Whether these interactions are necessary for the broad spectrum of biochemical and physiological responses is hypothetical at this point. Nevertheless, we can use a space-filling model of GA_3 to show the feasible intercalation of the phytohormone into DNA (Figure 19–13). Therefore, the chemical information inherent in the structure of a given phytohormone intercalated into DNA may possibly produce template modifications that control numerous growth responses and, indeed, enzyme induction at the very least.

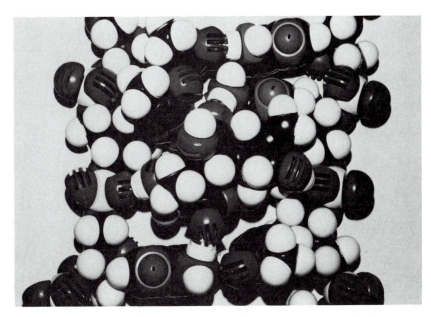

Figure 19–13. Space-filling models showing highly precise interaction of gibberellin with DNA.

Photo by F.H. Witham.

hydrogen

oxygen

methyl group

ring carbons of adenine

phosphate

ring nitrogen

adenine amino group

gibberellic acid

thymine nucleotide

Gibberellin and Auxin Interactions

Gibberellins have often been compared to auxins in their biological activity. Table 19–1 presents a summary of some of the major effects of these two classes of hormones.

We have learned that gibberellins are active in many of the same plant growth systems in which auxins are active (e.g.,

cell elongation, fruit-set, flowering). The question arises whether gibberellins act through an auxin-mediated system—that is, whether gibberellins promote the synthesis, transport, action, or inactivation of auxins in the plant. An answer to this question may possibly be found in a study of the effects of gibberellin on Progress No. 9, a dwarf pea plant (54). Application of GA to the intact green plant will cause the inter-

Response Activity	Auxins	Gibberellins
Apical dominance	promote	no effect
Avena coleoptile elongation	promote	no effect
Bolting and flowering in non-vernalized biennials and long-day plants	no effect	promote
Callus formation of tobacco pith	promote	no effect
Chlorophyll retention of detached leaves	no effect	promote
Growth of intact cucumber hypocotyls	promote	promote
Growth of dwarf pea stem sections	promote	no effect
Growth of intact dwarf pea stem	no effect	promote
Epinastic responses	promote	no effect
Leaf abscission	promote or inhibit	no effect
Pathenocarpic fruit growth (tomato)	promote	promote
Polar transport	yes (stems) no effect (roots)	no (sometimes) no effect (usually)
Root initiation	promote	no effect
Root growth	promote	no effect
Seed germination and breaking of dormancy	no effect	promote

Table 19–1. Summary of major effects of auxins and gibberellins.

nodes to elongate considerably. In contrast, application of IAA has no effect. When these same internodes are removed from the plant (excised) and placed in a buffered medium, their response to either GA or IAA alone is only slight. However, a pronounced synergistic (more than additive) effect on elongation is observed when both IAA and GA are added together to the medium supporting the stem sections. The suggestion in this study is that GA is dependent on IAA for its action.

The excised internode is separated from the apical meristem from which it received its supply of auxin, but the addition of exogenous IAA to the buffered solution relieves this shortage. In addition, decapi-

tated plants do not respond to applications of gibberellic acid (1).

Several studies have demonstrated that gibberellins and auxins are quite different and that they act independently of each other. For example, etiolated pea stem sections respond to both gibberellin and IAA when these growth regulators are applied separately. When they are applied simultaneously, their effect is merely additive (32, 60), thus indicating independent action. In fact, Hillman and Purves (26) found that gibberellic acid could promote pea stem section elongation in the presence of inhibitory levels of IAA, thereby again indicating independent action. Finally, GA-induced mobilization of carbohydrate reserves in barley endosperm does not require the presence of endogenous auxin (11).

Experimenters found that competitive inhibitors of IAA activity (antiauxins) are noncompetitive with gibberellins in pea stem section elongation (32)—that is, increasing the concentration of gibberellin could not overcome the inhibitory effect of the antiauxin.

Many investigators believe that gibberellins may have an influence on IAA oxidase that results in an auxin-saving mechanism—that is, the auxin level in the plant is raised because of an influence of gibberellins on IAA oxidase. Studies have shown that gibberellic acid partially inhibits IAA oxidase activity in an enzyme extract from crown gall tissue cultures of *Parthenocissus tricuspidata* (75).

Galston and McCune (20) found that gibberellin treatment of dwarf pea and corn plants lowers peroxidase activity in both plants. This phenomenon would have the effect of protecting endogenous IAA from oxidation. Earlier we discussed the essential role of peroxidase as a constituent of the IAA oxidase system. However, the findings by Hillman and Purves (26) that gibberellins will promote etiolated pea stem section elongation in the presence of inhibitory concentrations of IAA seem to contradict any suggestion that gibberellins act through an auxin-saving or -protecting mechanism in the plant. We should also consider several studies indicating that gibberellins actually promote the synthesis of IAA. Auxin levels in pea and sunflower seedlings increase following GA treatment (38) and, of even greater significance, GA has been shown to enhance the conversion of tryptophan to IAA (39). Valdovinos and his co-workers demonstrated that $^{14}CO_2$ release from tryptophan-1-^{14}C in cell-free preparations of the apical regions of *Coleus* and sunflower

stems is enhanced if the apical regions are pretreated with GA. Decarboxylation of tryptophan is an initial step in the conversion of the amino acid to IAA.

Gibberellins and auxins appear to act both independently and together, depending on the species of plant, the conditions under which the plant is growing, and the type of response being measured. The study of whether or not auxins and gibberellins interact is still far from conclusive. A lot of work remains to be done on this aspect of plant growth regulation.

Commercial Uses of Gibberellins

With the exception of herbicides, the exploitation of phytohormones for commercial use has not been extensive in past years. However, the emphasis on food production, successful food storage, and food distribution has led plant scientists to investigate the possibilities of utilizing many of the known physiological effects of the gibberellins and other phytohormones for commercial purposes.

Currently, GA_3 and sometimes GA_4 and GA_7 are used to increase the number of grapes in the cluster. In fact, most growers today spray their table grapes with gibberellins for increased grape and cluster size.

As we might expect, gibberellins are also used to increase the amount of α-amylase in germinating barley, which is used in the production of malt for the beer industry. Other commercial uses include the stimulation of flower bud formation and of fruit-set in apples and pears and the improvement in size, color, and quality of the fruit of many plants. Because of their antisenescence activity, the gibberellins, like the cytokinins, may prove to be very valuable in moderating the post-harvest physiology of crops. Plant scientists are studying citrus fruits in particular in this regard.

One of the most notable uses of gibberellins is their application to sugarcane. They promote the elongation of the stalk and, most important, do so with no diminution of sugar concentration (amount of sugar per mass of tissue).

An interesting application of phytohormones is their activity as mixtures. For example, a preparation called *promalin*, which is a mixture of the cytokinin 6-benzyladenine and GA$_4$ and GA$_7$, seems to be very active in stimulating increases in apple size, particularly in red Delicious apples.

We must also keep in mind the growth-regulating activities of gibberellin inhibitors. For instance, cycocel (CCC), a known inhibitor of gibberellin biosynthesis (see Figure 19–3), is used extensively by the greenhouse floriculture industry to retard the growth of such plants as chrysanthemums and poinsettias (see Figure 19–14). Not only do these inhibitors add to the aesthetic qualities of these plants by retarding growth but also add to the greenhouse growers economical utilization of bench space. The gibberellins and their inhibitors will play an increasingly significant role in enhancing the quality of life through their action as plant growth regulators.

Figure 19–14. Poinsettia plants, treated with SADH (B-9) or CCC. In left photo, treatments were (from left to right): (1) control (drenched with water); (2) drenched once with CCC (3,000 ppm) 3 weeks after potting; (3) drenched twice with CCC (3,000 ppm) 3 weeks and 6 weeks after potting. Plants are approximately 6 months old. Control plant is 5 feet high. In right photo, treatments were (from left to right): (1) control (sprayed with water); (2) sprayed once with SADH (B-9); (3) sprayed once with cycocel (3,000 ppm); (4) drenched once with cycocel (3,000 ppm). All treatments were given 2 weeks after potting. Plants were grown $3\frac{1}{2}$ months as they would be for commercial purposes.

Courtesy of J.W. Mastalerz, The Pennsylvania State University.

Questions

19–1. What is *bakanae* disease and what is its relationship to the discovery of gibberellins?

19–2. Chemically, the gibberellins are diterpenes. Describe the meaning of the term. Provide the names of other natural products that fall into this category of compounds.

19–3. Indicate some of the reasons that account for the occurrence of so many gibberellins throughout the plant world. Do they all play a role in growth or were they formed merely as side products of other reactions?

19–4. How are mevalonic acid, geranylgeraniol pyrophosphate, and copalyl pyrophosphate related?

19–5. Is gibberellic acid transported in a polar fashion? Explain.

19–6. List six of the bioassays used to detect gibberellins. Indicate the basic response used to detect gibberellins with each bioassay. Should bioassays still be used? Why or why not?

19–7. List the plant responses influenced by gibberellins.

19–8. Are all forms of dwarfism in plants due to deficiencies of gibberellic acid?

19–9. From the information available, might gibberellic acid be called a "flowering" hormone? Explain.

19–10. Describe the events that take place during the germination of a barley grain from the time gibberellic acid is released from the embryo. Indicate the site of action of the phytohormone (cellular components and tissue).

19–11. Name some of the inhibitors that seem to block the action of GA_3 in barley grains. Where in the target cells do they most likely act?

19–12. What evidence suggests that gibberellins interact with nucleic acids?

19–13. What are some of the differences and similarities exhibited by auxins and gibberellins with respect to individual plant responses?

19–14. Describe some of the commercial uses of gibberellins. Can you suggest any other uses that might be employed in the future?

19–15. Exogenously applied gibberellins may not be as active as those occurring naturally in the plant being tested. What might be some of the reasons for the differences?

Suggested Readings

Barendse, G.W.M. 1975. Biosynthesis, metabolism, transport and distribution of gibberellins. In H.N. Krishnamoorthy, ed., *Gibberellins and Plant Growth*. New Delhi: Wiley Eastern Limited.

Chrispeels, M.J., and J.E. Varner. 1967. Hormonal control of enzyme synthesis: on the mode of action of gibberellic acid and abscisin in aleurone layers of barley. *Plant Physiol.* 42:1008–1016.

Hedden, P., J. MacMillan, and B.O. Phinney. 1978. The metabolism of the gibberellins. *Ann. Rev. Plant Physiol* 29:149–192.

Jacobsen, J.V. 1977. Regulation of ribonucleic acid metabolism by plant hormones. *Ann. Rev. Plant Physiol.* 28:537–564.

Leopold, A.C., and P.E. Kriedemann. 1975. *Plant Growth and Development*, 2nd ed. New York: McGraw-Hill.

MacMillan, J. 1977. Some aspects of gibberellin metabolism in higher plants. In P.E. Pilet, ed., *Plant Growth Regulation*. New York: Springer-Verlag.

Moore, T.C. 1979. *Biochemistry and Physiology of Plant Hormones*. New York: Springer-Verlag.

Skoog, F. 1980. *Plant Growth Substances*. 1979. Proc. 10th Int. Conf. *Plant Growth Substances*. New York: Springer-Verlag.

Thimann, K.V. 1974. Fifty years of plant hormone research. *Plant Physiol.* 54:450–453.

Varner, J.E., and D.T. Ho. 1976. Hormones. In J. Bonner and J.E. Varner, eds., *Plant Biochemistry*, 3rd ed. New York: Academic Press.

Wareing, P.F., and I.D.J. Philips. 1978. *The Control of Growth and Differentiation in Plants*, 2nd ed. New York: Pergamon Press.

Chapter 20

Cytokinins, Ethylene, and Abscisic Acid

Mature leaf of *Begonia rex* sprayed with cytokinin. Note extensive shoot formation along veins. Individual shoots can be propagated through tissue culture. *Courtesy of Chiko Haramaki, The Pennsylvania State University.*

Up to this point we have emphasized the role of phytohormones primarily in cellular elongation. Although auxins and gibberellins influence a broad spectrum of responses, including increases in cell numbers, plant scientists believe that the cytokinins are the major promoters of cellular division. Perhaps the most exciting discovery in the search for specific cell division–inducing compounds was that of kinetin (6-furfurylamino purine). Kinetin and closely related chemicals, generically termed cytokinins, represent the addition of new information about the hormonal control of plant morphogenesis.

History

The formation of wound callus on plants that have been damaged in one manner or another (e.g., in pruning) is a common observation. In the latter half of the nineteenth century scientists postulated that damaged tissue may cause the production of a substance, which, upon diffusion into nearby undamaged cells, would stimulate meristematic activity. In 1913 Haberlandt (32) demonstrated that phloem diffusates stimulate cell proliferation of potato tuber tissue and that extracts of damaged cells are capable of inducing meristematic activity when applied to undamaged cells.

Subsequent investigations, notably those of Wehnelt (124) and Bonner and English (8), led to the isolation of a compound that is very active in inducing meristematic activity in undamaged green bean pods. This hormone, a straight chain dicarboxylic acid, was given the name *traumatic acid*. Its structure is $HOOCCH{=}CH(CH_2)_8COOH$.

The effect of traumatic acid, which is to induce cells to divide, does not appear to be general. In fact, most plant tissues do not respond to traumatic acid, thus suggesting that it is a specific wound hormone for bean-pod tissue (8).

The existence of other unknown naturally occuring compounds that promote cellular proliferation was initially demonstrated by van Overbeek and his colleagues (119) in studies of young *Datura* embryos grown on tissue culture medium. Van Overbeek demonstrated that substances present in unautoclaved coconut milk, the liquid endosperm of coconut (milk), stimulated cell division and differentiation of very young *Datura stramonium* embryos, which did not synthesize these growth factors. A few years later Caplin and Steward (15) showed that factors present in liquid endosperm of coconut plus IAA stimulated mature cells to divide and grow rapidly in culture.

In 1944 van Overbeek (120) reported that unpurified extracts of *Datura* ovules, yeast, wheat germ, and almond meal promoted growth of *Datura* embryo cultures. Because the responsible substance(s) seemed to be of widespread occurrence, scientists investigated other plant sources for the presence of endogenous growth-promoting substances. McLane and Murneek (60) detected one such substance in maize kernels at the milky stage. They termed it *syngamin* because it occurred in high amounts in developing corn kernels only after fertilization, or syngamy. In many respects syngamin appeared to be chemically and biologically similar to zeatin, which was later isolated and characterized from milk-stage kernels. Syngamin was neither purified nor identified. However, much later, Netien and Beauchesne (86, 87) demonstrated that extracts of milk-stage corn kernels promoted cell division due to the presence of several active factors that were distinct from auxin.

Haberlandt demonstrated in the early 1900s that phloem diffusates stimulate cell proliferation of potato tuber tissue. During the early 1950s, Jablonski and Skoog (42) observed that vascular tissue cells contain materials that stimulate cell division of tobacco (*Nicotiana tabacum* cv. Wisconsin No. 38) plants. As an extension of this work, Carlos O. Miller, then a postdoctoral student at Wisconsin, investigated several natural sources of cell division substances. To detect activity, the Wisconsin workers perfected techniques for using tobacco stem pith segments, grown in culture, as a specific and sensitive bioassay for cell division factors. They initially observed that yeast extract in combination with IAA is as active as coconut milk in promoting continuous cell division of cultured tobacco stem pith segments. Since yeast extract is readily available in large quantities, they used it as an initial source of active material. Subsequent work by Miller and his colleagues revealed that the active material was possibly a purine. Previously, Skoog and his associates showed that the purine adenine elicited slight cell division activity when tested in the tobacco stem pith assay. Later, rich sources of purines, particularly herring sperm DNA that had aged on the shelf for some time, were found to contain a very active stimulator of cell division in the tobacco stem pith assay.

Although relatively fresh (nonaged) DNA was not active, aged or autoclaved DNA yielded active material. As a result, Miller and his colleagues (76, 77) isolated and purified, in crystalline form, a purine from herring sperm DNA, which they identified as 6-furfurylaminopurine. They named this compound *kinetin* because it induced cytokinesis of cultured tobacco cells. It seems surprising and paradoxical that compounds (such as IAA and kinetin) important in plant growth were isolated initially from nonplant sources.

Hall and deRopp (34) synthesized kinetin. They found that a mixture of autoclaved furfuryl alcohol and adenine yielded kinetin. In view of this fact and the knowledge that kinetin is a degradation product from deoxyadenosine of aged or autoclaved DNA, kinetin seemed to be an artifact of isolation. As a result of kinetin's discovery, a new era of research on cell division stimulants in plants was initiated. To this day, kinetin is not known to have been detected in plants, although there is reason to suspect that it might occur naturally, possibly as a breakdown product as part of DNA turnover.

When kinetin was discovered, scientists originally proposed the term *kinin* as the generic name for kinetin-like substances. However, since kinin was an already existing term used to classify certain polypeptides (isolated from animals) that stimulate the constriction of blood vessels and smooth muscles, Skoog and others (107) adopted the term *cytokinin* to distinguish the phytohormones from the animal kinins. Hence *cytokinin* is the most widely accepted term for those compounds that exhibit kinetin-like plant growth–regulating properties.

The discovery of kinetin stimulated the synthesis of literally hundreds of analogues, studies on the effects of kinetin on many diverse biological responses, and the search for naturally occurring cytokinins. Although Miller, Skoog, and others have synthesized and tested several hundred analogues of kinetin in numerous bioassays, kinetin and 6-benzyladenine are the most widely used in studies on the physiological effects of cytokinins. Figure 20–1 illustrates the structures of kinetin and several analogues of kinetin, all active in cell division.

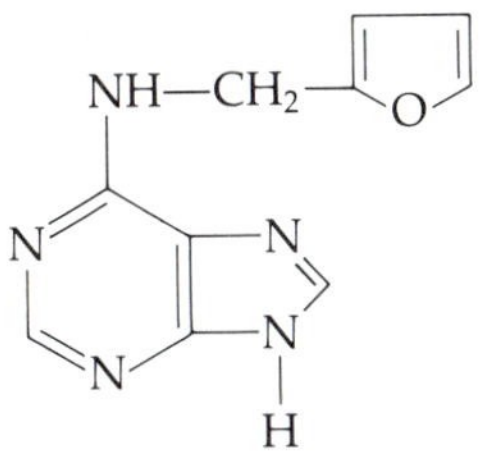

Figure 20–1. Structural formulas for kinetin and three analogues. All four compounds are active in promotion of cell division.

kinetin (6-furfurylaminopurine)

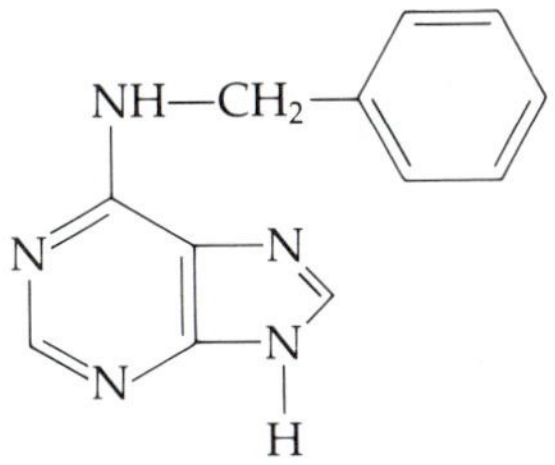

6-benzylaminopurine

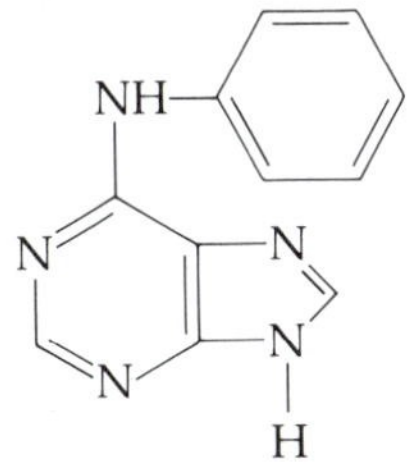

6-phenylaminopurine

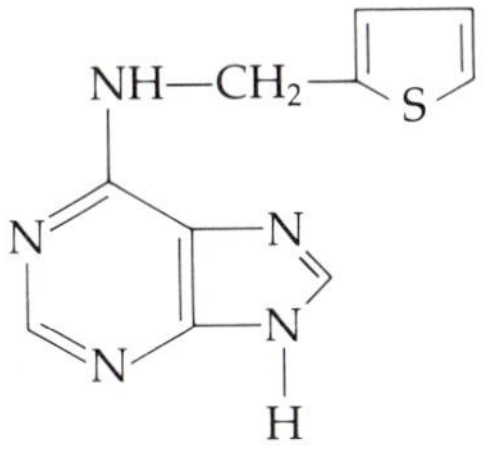

6-(2-thenylamino) purine

From well over one hundred active N^6-substituted adenine chemicals that have been synthesized, most evidence supports the idea that adenine derivatives are the major naturally occurring cytokinins. We should keep in mind, however, that other compounds, such as the phenylureas, auxins, and gibberellins, exhibit cell division activity under certain circumstances.

Detection and Isolation of Zeatin and Derivatives

Early workers considered the milk-stage kernels of corn (*Zea mays*) to be somewhat analogous to the milky endosperm of the coconut. As we have mentioned, McClane and Murneek reported, in 1952, the detection of syngamin, a growth factor associated with embryo development in corn kernels. Five years after the appearance of the publi-

cations on the discovery of kinetin, Miller reported the isolation of a purine with cytokinin properties from milk-stage corn kernels. In 1961 Miller characterized the substance as a 6-substituted amino purine, consisting of a five-carbon substituent containing a hydroxyl group, a methyl group, and a double bond. The publication of the information, however, was delayed until 1964 (78). In the meantime Letham crystallized a compound that he termed *zeatin* and that had the same components as Miller's substance. Following the mass spectral characterization of zeatin as 6-(4-hydroxy-3-methyl-trans-2-butenylamino) purine by Letham (50, 53), zeatin was synthesized by Shaw and Wilson (104) and proven to occur naturally and not as an artifact of isolation. Letham and Miller (55) reported in 1965 that the compound they had worked on independently was in fact zeatin. Table 20–1 illustrates the structure of zeatin and some other naturally occurring cytokinins.

Source: *From F. Skoog, personal communication.*

Table 20–1. Chemical names, abbreviations, structures, and sources of eighteen naturally occurring cytokinins.

| Substance | Structure | | | Source | | | |

Chemical Name and Abbreviation	R_1	R_2	R_3	Bacteria	Fungi	Higher Plants	Animals
6-(3-methyl-2-butenylamino) purine; i^6Ade	(isopentenylamino structure)	H	H	+	+	+	?
6-(3-methyl-2-butenylamino)-9-β-D-ribofuranosylpurine; i^6A	"	H	rib*	+	+	+	+
6-(3-methyl-2-butenylamino)-2-methyl-thiopurine; ms^2i^6 Ade	"	H₃CS	H	?		?	
6-(3-methyl-2-butenylamino)-2-methyl-thio-9-β-D-ribopuranosylpurine; msi^6 A	"	H₃CS	rib	+		+	
6-(4-hydroxy-3-methyl-*trans*-2-butenyl-amino) purine; zeatin (io^6Ade)	(hydroxymethyl butenylamino structure)	H	H		+	+	
6-(4-hydroxy-3-methyl-*trans*-2-butenyl-amino)-9-β-D-ribofuranosylpurine; ribosylzeatin (io^6A)	"	H	rib		+	+	
6-(4-hydroxy-3-methyl-*trans*-2-butenyl-amino)-2-methylthiopurine; mszeatin (tms^2io^6Ade)	"	H₃CS	H			?	
6-(4-hydroxy-3-methyl-*trans*-2-butenyl-amino)-2-methylthio-9-β-D-ribofurano-sylpurine; msribosylzeatin (t-ms^2io^6A)	"	H₃CS	rib			+	
6-(4-hydroxy-3-methyl-*cis*-2-butenyl-amino) purine; *cis*- zeatin (c-io^6Ade)	(cis hydroxymethyl butenylamino structure)	H	H	+		?	
6-(4-hydroxy-3-methyl-*cis*-2-butenyl-amino)-2-β-D-ribofuranosylpurine; ribosyl-*cis*-zeatin	"	H	rib		+?	+?	

Table 20–1. (continued)

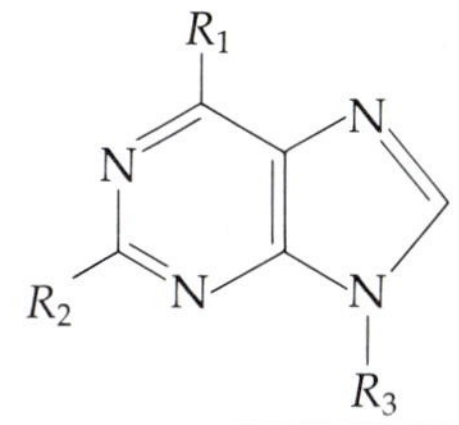

		Structure			Source			

R_1

R_2

R_3

Chemical Name and Abbreviation	R_1	R_2	R_3	Bac-teria	Fungi	Higher Plants	Ani-mals
6-(4-hydroxy-3-methyl-*cis*-2-butenyl-amino)-2-methylthio-9-β-D-ribofurano-sylpurine; msribosyl-*cis*-zeatin (cms²io⁶ A)	"	H_3CS	rib		+?	+	
6-(3-methylbutylamino) purine; hi⁶Ade	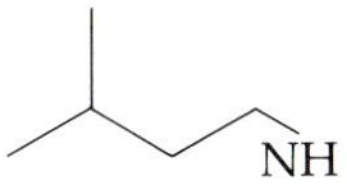	H	H			+?	
6-(4-hydroxy-3-methylbutylamino) purine, dihydrozeatin (hio⁶Ade)	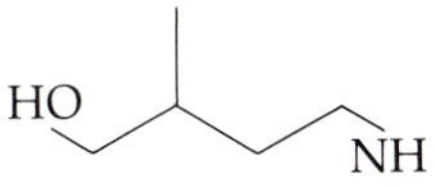	H	H			+	
6-(4-hydroxy-3-methylbutylamino)-9-β-D-ribofuranosylpurine; ribosyldihydro-zeatin (hio⁶A)	"	H	rib			+	
6-(3-hydroxy-3-methylbutylamino) purine; 30HiP†	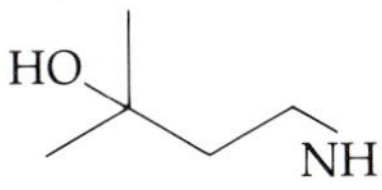	H	H	?	?	+	?
6-(3-hydroxy-3-methylbutylamino-9-β-D-ribofuranosylpurine; 30HiPA†	"	H	rib	?	?	?	?
6-furfurylaminopurine; kinetin	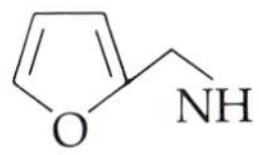	H	H	?	?	?	?
6-(2-hydroxylbenzylamino) purine	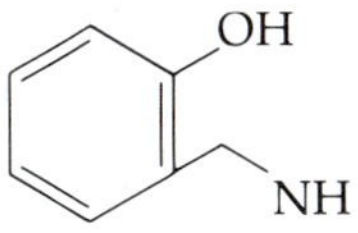	H	H			+	

* *rib* = ribosyl
† Not generally accepted as naturally occurring.

415

Studies on the distribution of zeatin in maize plants by Miller and Witham (78) revealed it to be present in the roots, stems, and leaves, but with the greatest amount present in milk-stage kernels. The biological properties of zeatin are similar to those of kinetin, although in some instances zeatin is more active (127).

Other Naturally Occurring Cytokinins and Their Distribution

Zeatin was identified initially as a free base, but other derivatives of zeatin exist. In 1965 Miller (73) provided evidence for the presence of zeatin ribonucleoside and zeatin ribonucleotide in corn kernels. Letham (51, 52, 53) later substantiated this finding. By 1972 cytokinins had been isolated from numerous sources, some of which are shown with the common naturally occurring cytokinins and several synthetic compounds in Table 20–1. All the cytokinins found to occur naturally to date are isopentenyl adenine derivatives. For examples, researchers have isolated zeatin from culture filtrates of *Rhizopogon roseolus* (74), pumpkin seed (31), and bean (*Phaseolus vulgaris*) fruit (48). They have also isolated the riboxyl form of zeatin from the culture filtrates of *Rhizopogon* (74), chickory root extracts (10), a variant strain of soybean callus tissues (79), crown gall tumor tissue of *Vinca rosea*, and bean fruit (48).

The *cis* form of the riboside was purified from tRNA extracts of corn, peas, and spinach (33). Also, methylthio-zeatin riboside was isolated from wheat germ tRNA fractions (14, 36). The only definitive reports on the presence of the zeatin ribonucleotide

were its isolation from corn by Letham (51, 52) and detection of representative activity in a variant strain of soybean callus tissue by Miura and Miller (79).

In 1966 Matsubara and Koshimizu detected cytokinin activity in lupine (*Lupinus luteus*) seeds (64). This cytokinin was chemically characterized as 6-(4-hydroxy-3-methylbutylamino) purine and was given the trivial name of dihydrozeatin (47). Krasnuk, Witham, and Tegley (48) later isolated dihydrozeatin and the nucleoside form of dihydrozeatin from bean fruit. Dihydrozeatin differs from zeatin in that the alkyl group of the former compound is saturated (i.e., unlike zeatin, it does not contain a double bond between the second and third carbons).

Another species of cytokinin (referred to in the early literature as 2ip, or 6-γ,γ-dimethylallylamino purine), now commonly called *isopentenyl adenine* (i^6 Ade), differs from zeatin in that a methyl group is present at the third carbon position in place of the hydroxyl.

A particularly interesting situation with respect to cytokinin distribution is the presence of a cytokinin as an odd base in certain tRNA molecules. In fact, the discovery of i^6 Ade was due to RNA base sequence studies by Zachau, Dutting, and Feldmann (132) when it was noticed as an odd base adjacent to the anticodon for yeast serine tRNA. In addition, i^6 Ade has been identified in tRNA fractions of yeast, *Escherichia coli*, liver (3, 6, 101, 118), *Corynebacterium fascians* (63), and filtrates of *Corynebacterium* (38).

It is now well established by such studies that cytokinins are indeed present as the odd base adjacent to the anticodon of all tRNA molecules that bind to those condons of mRNA in which the first code letter

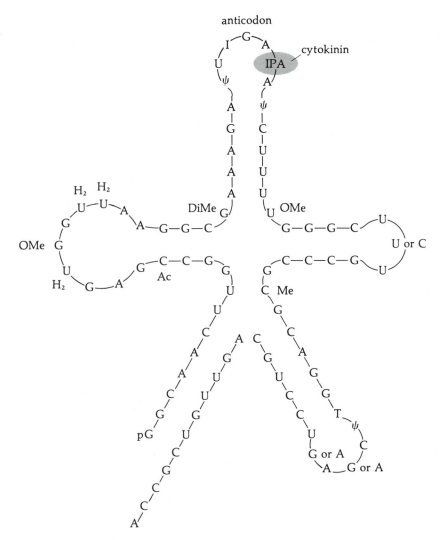

Figure 20–2. Structure of serine tRNA showing position of N[6]-delta-2-isopentenyl-adenosine (IPA) adjacent to anticodon.

From A.W. Galston and P.J. Davies. Control Mechanisms in Plant Development. *© 1970. Reprinted by permission of Prentice-Hall, Inc., Englewood Cliffs, New Jersey.*

(or *base*) is "u" (see Figure 20–2). We do not know the significance of the exact positioning of certain cytokinins in specific tRNA molecules. One possibility is that the cytokinin at a specific location close to the anticodon maintains the stability of the struc-ture and may enhance tRNA and amino acid binding during the translation process (24, 27). Thus, cytokinins appear to be involved in the regulation of protein synthesis during the translation process. Experimenters reported isolation of the ribonucleoside of i⁶Ade (previously referred to as 2iPA but now called i⁶A) from tRNA hydrolysates of yeast, peas, spinach, wheat germ, and tRNA extracts of human, chick, and calf liver (see 35, 37, 105). Another derivative of i⁶A, 2-methylthioisopentenyl adenine (abbreviated as mSi⁶A) has been purified from wheat germ tRNA and *Escherichia coli* tRNA fractions (see 80). Scientists have not detected the nucleotide form of these latter compounds.

Conjugated Cytokinins

Direct evidence for the conjugation of the free base cytokinins with amino acids (as for the auxins), peptides, or proteins does not exist. Scientists believe that specific cellular receptors of cytokinins exist, possibly in membranes of organelles or in the cytoplasm where they function as phytohormonal carriers. These possible interactions, however, are not known as cytokinin storage complexes.

One mechanism that might be important to the storage or inactivation of cytokinins is their glucosylation, or the formation of other carbohydrate derivatives. For example, glucose has been found to be conjugated to the seventh carbon position (nitrogen atom) of zeatin forming a *glucosylzeatin* named *raphanatin*. Letham and his associates (54) found this substance initially in great abundance in radish (*Raphanus sativus*).

Another glucoside, 9-glucosylzeatin, is characterized by the attachment of glucose at the ninth carbon position, which in zeatin riboside or nucleotide is occupied by ribose (ribosyl moiety). In other instances, less common glucosides may be formed by the addition of glucose to the hydroxyl group on the N-6-substituted side chain.

When the glucosides and ribosides of the cytokinin free bases are tested in appropriate bioassay systems, they all seem to exhibit some biological activity, but with different effectiveness. We do not know if the complexes are active per se or if, as storage forms upon metabolism, they yield the active free base. The same situation may hold true for the methylthio derivative (CH_3S^- at the second carbon position of the purine ring). Further research with radioactively labeled compounds should yield information concerning the role of cytokinin conjugates in plant growth regulation.

Distribution of Cytokinins in the Plant

We do not know a great deal concerning the details of in vivo cytokinin synthesis. We do know, however, that the cytokinins are produced in meristematic regions and areas of continued growth potential. During the vegetative portion of the life cycle, they are apparently synthesized in the roots, particularly during the seedling stage (78, 127), and translocated to the upper plant parts. Cytokinins are frequently detected in the xylem sap that exudes from cut surfaces or in stem and root extracts. They are probably transported through the xylem. From the early isolation and distribution studies (78), we can conclude that cytokinins seem to be most abundant in roots, young leaves, and developing fruit. Scientists have, therefore, suggested that the establishment of sinks in these areas is primarily due to the relatively higher concentration levels of cytokinins and other phytohormones as compared with these levels in other areas of the plant. As we will discuss later, the cytokinins are probably very important in the establishment of sinks at areas of high metabolic activity. We do not definitely know, however, if the cytokinins are exclusively translocated to or manufactured in sink areas.

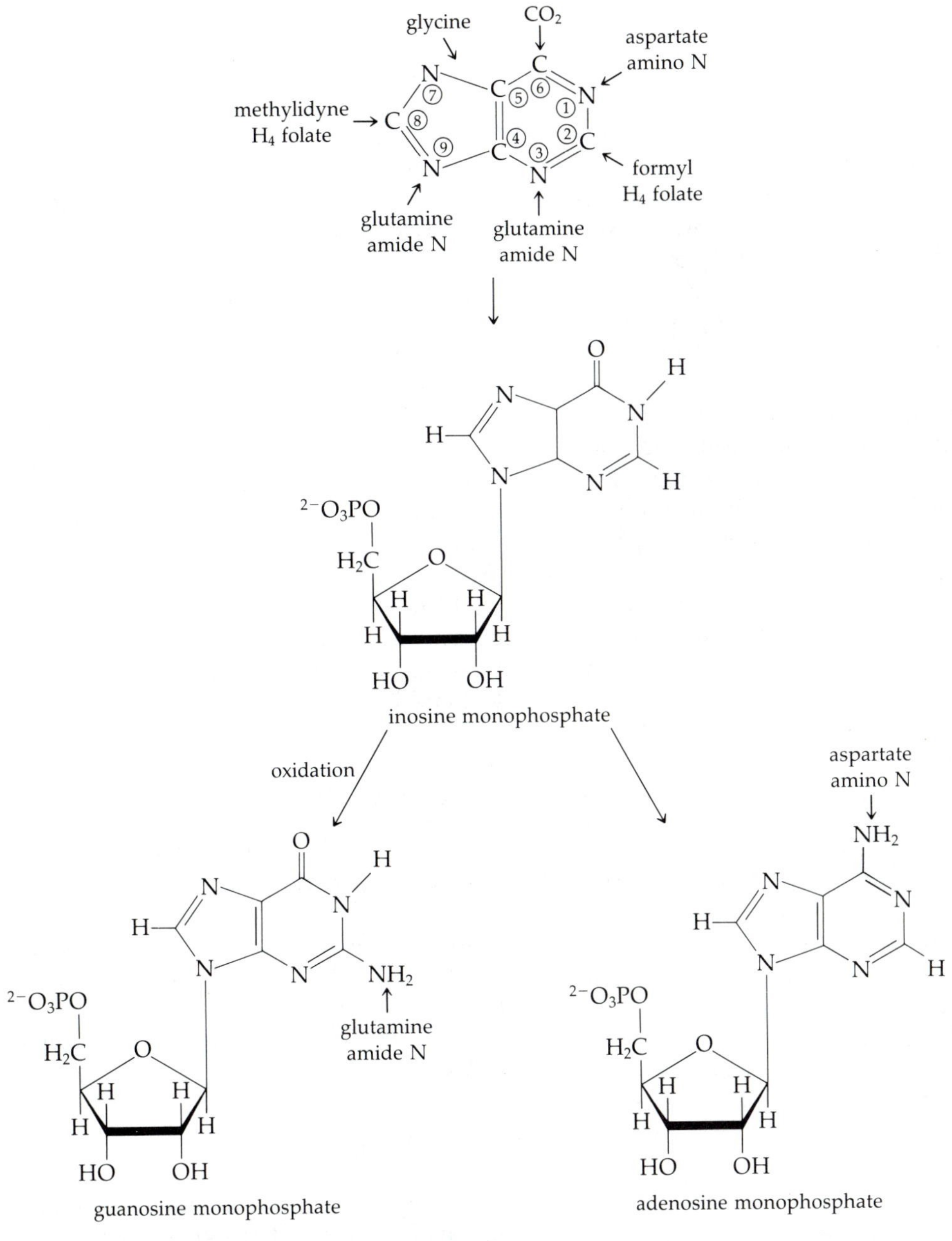

Figure 20–3. Purine atom sources. Rings are assembled on ribose-5-phosphate residue obtained from 5-phosphoribosylpyrophosphate (not shown).

Biosynthesis

The most definitive information about cytokinin synthesis concerns the formation of adenine. The ring skeleton (see Figure 20–3) is derived from several small molecules. The nitrogen at the first carbon position is provided from amino N of aspartate, the second carbon position from formyl H_4 folate, the nitrogen at the third position from a glutamine amide N, carbons at the fourth and fifth positions and the nitrogen at the seventh position are formed from an intact glycine molecule. The sixth carbon position is derived from CO_2. The eighth carbon position comes from methylidyne H_4 folate, and the ninth nitrogen position comes from glutamine amide N. The amino group, at the sixth carbon position is derived from the amino N of aspartate.

The in vivo synthesis of the purine ring takes place on ribose phosphate. Biochemistry texts can provide the details of this assembly. In view of the biosynthetic sequence, with adenosine monophosphate as the end product, it is probable that free adenine or even the substituted adenine cytokinins are derived from a six-substituted adenosine monophosphate or riboside. The cytokinin ring structure in tRNA is most likely produced according to the sequence illustrated in Figure 20–3, followed by the addition of the five carbon substituent from isopentenyl pyrophosphate. The cytokinins zeatin and dihydrozeatin probably arise from modifications in the size chain. The methyl and sulfur groups of the methylthiocytokinins also seem to be added either to cytokinins present in tRNA molecules or to adenosine monophosphate or adenosine (the riboside). Speculation also exists as to whether the free cytokinins are released from nucleic acids or are synthesized independently.

Cytokinin Bioassays

Although experimenters have used numerous bioassay systems in the past to detect cytokinins (see Table 20–2), the most sensitive and specific are those of plant callus tissue cultures. Of these we have already mentioned the historical significance of the tobacco stem pith. The excised cotyledon callus cultures of soybean developed by Miller have been used since the early 1960s (71, 72). Letham developed the excised radish cotyledon enlargement test later, which, as is true of many of the bioassays, has been used to study the action of cytokinins.

Inspection of the list of bioassay systems indicates the broad spectrum of responses attributed to the cytokinins. The cytokinins influence lettuce seed germination, root growth, cell division, cellular enlargement and differentiation, lateral bud development and shoot formation, leaf expansion, chlorophyll retention in detached leaves (or delay of senescence), and the establishment of sinks. There are also many biochemical effects in plants due to cytokinin action. Some are mentioned in this text and in the Suggested Readings.

Physiological Effects

Shortly after the discovery of kinetin, a number of papers appeared describing its effect on different plant growth systems. Most of these were related to kinetin's promotion of cell division and cell enlargement. However, cytokinins regulate many responses that may or may not be a direct result of cytokinesis.

Cell Division

Jablonski and Skoog (42) observed the stimulation of cell division in plant tissue culture of tobacco stem pith callus. They noted that, in addition to kinetin, an auxin (indole-3-acetic acid) had to be incorporated into the growth medium for continuous growth and survival of the pith tissue in culture. Although either growth regulator produces a slight response when used alone,

Table 20–2. Cytokinin bioassays.

Source: *From D.S. Letham. 1967. Chemistry and physiology of kinetin-like compounds. Ann. Rev. Plant Physiol. 18:349. © 1967 by Annual Review Inc.*

Bioassay	Key References	Normal Assay Time (days)	Lowest Detectable Kinetin Concentration (μg/l)	Range of Kinetin Concentrations over Which Linear Relationship Exists with Response (μg/l)	Substances Other Than Cytokinins Known to Be Active in Assay
Radish leaf disk	58	0.8	10	10–1,000	adenine
Etiolated bean leaf disk	88, 92	2	200	unknown	gibberellins, cobaltous ions
Lemma minor (growth in darkness)	43	2	60	60–600	adenosine, cobaltous ions
Lettuce seed germination	89, 133	2	unknown but <1,000	unknown	gibberellins, thiourea urea, certain urea derivatives
Etiolated pea stem section	136	1	10	10–10,000	sucrose, benzimidazole
Xanthium leaf senescence	105	2	100	100–10,000	benzimidazole, sugars, adenine,* adenosine,* guanosine*
Barley leaf senescence	52, 53	2	3	3–3,000	inorganic salts (high conc. only)
Tobacco stem pith callus	75, 102, 148, 119	35	1	1–15	gibberellic acid*
Tobacco stem pith	13	21	40	unknown	none†
Soybean callus	92	21	1–4	4–10,000	none†
Carrot root tissue	16, 68, 71	21	1	1–100	gibberellins‡

* The activity of these substances is slight compared with the activity of cytokinins.

† The activity of gibberellins in this assay does not appear to have been determined.

‡ Gibberellic acid does not appear to promote growth in this assay, but certain other gibberellins cause very slight growth increments.

the effect does not last. Possibly the response evoked by kinetin or IAA alone in tobacco pith culture is due to small amounts of endogenous kinetin-like compounds and IAA already present. When the two substances are present in the right amounts (for example, kinetin at 0.5 mg/l and IAA at 5.0 mg/l), along with the proper vitamins and minerals, the cells divide and enlarge and produce a mass of loosely arranged, mostly triploid and undifferentiated cells. This mass of undifferentiated cells is referred to as *callus tissue*. The stimulation of cell division and subsequent callus growth with auxin present appear to be characteristic of all cytokinins. Figure 20–4 illustrates the promotion of cell division by kinetin and IAA. When the ratio is changed to favor a high cytokinin to auxin ratio, either by addition of more kinetin or the use of less IAA, the callus may be stimulated to differentiate into cultured plantlets with stem and leaves.

In order for cell division to occur, an ordered sequence (DNA synthesis, mitosis, and cytokinesis) must take place. Does IAA or cytokinin alone have a specific influence on any step in this sequence? The answer, apparently, is yes. Das, Patau, and Skoog (20) found that both IAA and kinetin, when used alone, stimulate DNA synthesis in tobacco pith cultures. They also found that both growth regulators are needed for mitosis, although IAA appears to dominate in this step. In addition, they suggested that when either kinetin or IAA is present in high concentration, the other may become limiting for at least one of the three steps needed to complete cell division. Later studies, which appeared to support this line of reasoning, concluded that cytokinins act as a specific trigger for cytokinesis. Here again, as in our discussion of gibberellins and IAA, we are presented with the importance of balance between growth hormones in plant growth and development.

Figure 20–4. Tissue cultures of tobacco (*Nicotiana tabacum*) callus. By altering cytokinin-to-auxin ratio, tobacco stem pith tissue may be maintained in culture as undifferentiated callus (left) or induced to differentiate and bud into plantlets (right).

From work of F. Skoog and C.O. Miller. Photo by F.H. Witham.

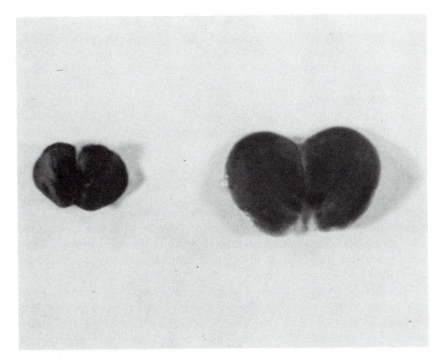

Figure 20–5. Kinetin-induced enlargement of radish (*Raphanus sativus*) cotyledons. Cotyledons were incubated on filter paper wetted with 2 mM potassium phosphate buffer (right) for 72 hours at 26°C and light intensity of 450 lux.

Courtesy of I.S. Bewli and F.H. Witham.

How a cytokinin induces cell division is still an unanswered question. The adenine moiety of the cytokinin molecule appears essential for this process, many different substituted side chains being applicable.

Cell and Organ Enlargement

Cytokinins also induce cell enlargement, an effect usually associated with IAA and gibberellins. Treatment of leaf disks cut from etiolated leaves of *Phaseolus vulgaris* (bean) with kinetin causes significant cell enlargement (70, 95). This effect of kinetin can occur in the absence of IAA. Experimenters have also observed cell enlargement after kinetin treatment in tobacco pith cultures (29), in tabocco roots (4), and in excised artichoke tissue (1). Stimulation of cell enlargment by cytokinins other than

kinetin has also been observed (53). Since cytokinin-induced cell enlargement has been clearly shown, cytokinins should not be considered solely as cell division factors.

It is a curious fact that cytokinins promote enlargement in tobacco tissue and elongation of the upper parts of dark-grown bean seedlings but inhibit the elongation process in stem sections of various plants. Apparent contradictory effects such as these may result from different physiological conditions in the plant material rather than from different molecular activities of the active material.

Of special interest is the cytokinin enlargement of excised cotyledons of radish (54), pumpkin (5), cocklebur (23), flax (109), and fenugreek (100). Figure 20–5 shows the enlargement of radish cotyledons as stimulated by kinetin. Letham (49) indicated that cytokinin-induced expansion of detached radish cotyledons is due to cell enlargement and not cell division. The enlargement is, at least in part, due to stimulated water uptake. The water uptake is in response to the production of reducing sugars in the cotyledonary cells (7, 40). In the presence of cytokinin, the reducing sugars appear to build up due to the conversion of lipids. Although the cytokinins do not appear to affect the lipid conversion enzymes, kinetin does increase invertase activity in the cotyledons, thus suggesting that sucrose, which initially builds up from the lipid conversions, is hydrolyzed rapidly to the osmotically active sugars, glucose and fructose. Ross and his colleagues have shown that zeatin promotes cell wall changes by some unknown mechanism in which cell wall modification (increased plasticity) takes place. Hence, cytokinins stimulate cotyledon enlargement by their action on at least two physiological processes and other initiating and supporting biochemical processes.

Another point of interest is that, similar to cotyledons of other dicots that rely on the digestion of lipid-stored minerals, radish cotyledons are stimulated by light to enlarge. The light effect is mediated by phytochrome, although the ultimate response-inducing chemicals are most likely cytokinins (40). We do not know, however, the exact relationship between phytohormones and the phytochrome system. How the phytochrome system, activated by light, induces phytohormonal activity continues to be a challenging enigma.

Lettuce Seed Germination

The germination of lettuce seeds (*Lactuca sativa* c.v. Grand Rapids) is stimulated by red light and inhibited by infrared (far-red) light. If lettuce seeds grown in darkness imbibe a solution of kinetin or other cytokinin, germination is significantly higher than in the dark-grown controls. The percentage of germination of the cytokinin-treated seeds is similar to the percentage of those that receive a red-light treatment. Furthermore, the inhibition of seed germination by far-red light is reversed at least partially when the seeds imbibe a cytokinin solution for 12 to 18 hours prior to the light treatment. Tables 20–3 and 20–4 illustrate these results and those taken from work with bean leaf disks.

Concentration of Kinetin (M)	Light Treatment*	Increase in Diameter (mm)
0	none	1.05 ± 0.04†
5×10^{-5}	none	2.48 ± 0.03
0	5 min red	2.58 ± 0.08
0	5 min far red	1.01 ± 0.06
0	5 min red and then 5 min far red	1.17 ± 0.07
5×10^{-5}	5 min far red	2.49 ± 0.08

* *Light treatments given at beginning of experiment.*
† *Standard error: ten disks per treatment.*

Table 20–3. Effect of kinetin and red and far-red irradiation on growth of bean leaf disks during 48-hour growth period.

Source: *From C.O. Miller. 1956. Plant Physiol. 31:318.*

Concentration of Kinetin (M)	Light Treatment*	Germination (%†)	
		Expt. 1	Expt. 2
0	none	8	7
5×10^{-5}	none	84	86
0	8 min red	96	96
0	5 min red and then 8 min far red	5	7
5×10^{-5}	8 min far red	86	83

* Light treatments given 16 hours after start of experiment.
† Percent given as nearest whole number, 95–105 seeds per treatment.

Table 20–4. Effect of kinetin and red and far-red irradiation on germination of Grand Rapids lettuce seeds during 72-hour period. Source: *From C.O. Miller. 1956. Plant Physiol. 31:318.*

Root Initiation and Growth

Although relatively few studies have been made of cytokinin effects on the root system, cytokinins appear to be able both to stimulate and to inhibit root initiation and development. Kinetin, in the presence of casein hydrolysate and IAA, stimulates root initiation and development in tobacco stem callus cultures (106). Fries (26) found an increase in dry weight and elongation of the roots of lupin seedlings to be promoted by kinetin. He found that all concentrations of kinetin increase the dry weight of the root system even though root elongation is inhibited at the higher concentration.

In excised pea root segments, lateral root development is slightly stimulated by low concentrations of kinetin (5×10^{-8} M). At higher concentrations, however, kinetin is inhibitory (114). There is some evidence that interaction between cytokinins and auxin may influence the site of lateral root initiation. Bonnett and Torrey (9), for example, demonstrated that by applying auxin and cytokinin at different concentrations to the opposite ends of excised root segments of the common bindweed (*Convolvulus*), they could alter the site of lateral root formation.

Bud Development and Shoot Growth

The original work with tobacco callus cultures and kinetin indicated the potentially important role of cytokinins in regulating shoot initiation and apical dominance. Earlier, we discussed apical dominance, the inhibition of lateral bud growth by auxin emanating from the apical bud. The controlling features of this phenomenon are not clearly understood and may involve not only IAA but other factors that interact with the auxin, as was suggested by Wickson and Thimann (125) in a study on the interaction of IAA and kinetin in apical dominance. They found that the growth of the lateral buds of pea stem sections in culture solutions containing IAA is inhibited, as might be expected. The growth of lateral buds on stem sections in nutrient solutions not containing IAA, of course, is uninhibited. However, the addition of kinetin along with IAA stimulates the growth of these buds (see Figure 20–6).

Wickson and Thimann also demonstrated that the effect of kinetin on apical dominance can be observed in entire shoots—that is, with the apical bud present. They found, as in the classical

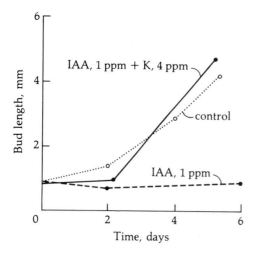

Figure 20–6. Effect of kinetin-IAA interaction on bud growth of stem sections of *Pisum sativum.* IAA inhibition is overcome by kinetin. Concentrations used: 1 ppm IAA and 4 ppm kinetin.

From M. Wickson and K.V. Thimann. 1958. Physiol. Plant. 11:62.

studies of apical dominance, that removal of the apical bud stimulates the growth of the lateral buds. If the apical bud is retained, the lateral buds are completely inhibited. However, if the intact shoot is soaked in a kinetin solution, inhibition of the lateral buds by the apical bud is overcome to a large extent (see Figure 20–7).

Several other investigations demonstrated the stimulating influence of cytokinins on lateral bud growth (84, 113). For example, Torrey (113) observed that kinetin initiated bud primordia on root segments of *Convolvulus arvensis* (bindweed). This effect was more marked when the segments were grown in darkness.

Five-day-old bean seedlings soaked in kinetin solutions and then allowed to grow for an additional 46 hours responded with an increase in fresh weight of the epicotyl, increase in leaf expansion, and increase in the elongation of stems and petioles. Further, zeatin stimulated secondary shooting

in pea seedlings (127) and is believed to be the chemical responsible for *fasciation* (excessive secondary shoot development), induced by *Corynebacterium* (46).

Apical dominance appears to be controlled by a balance between endogenous cytokinin and auxin concentrations (125). Some investigators suggest the possible direct inhibitory influence of cytokinin on IAA oxidase production. The cytokinin application to the lateral buds may inhibit the synthesis of certain multiple forms of IAA oxidase normally induced by IAA translocated from the terminal bud. With the repression of the IAA oxidase, the spared auxin may stimulate lateral bud growth and shoot development. In addition to the possible inhibition of IAA degradation, the cytokinins may initiate a sink mechanism at the lateral

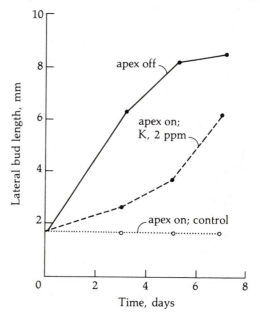

Figure 20–7. Effect of kinetin on apical bud dominance in *Pisum sativum.* Two ppm of kinetin partially counteracted inhibitory effect of apical bud on lateral bud growth.

From M. Wickson and K.V. Thimann. 1958. Physiol. Plant. 11:62.

bud sites that enhances translocation of nutrients, including other growth substances, vitamins, and minerals necessary for bud initiation and shoot growth. The actual correlation of direct effects of cytokinins on IAA oxidase synthesis or establishment of sinks are good hypotheses for further experiments with intact plants.

Retention of Chlorophyll and Delayed Senescence in Leaves

When mature functional leaves are detached from the plant, there is a rapid breakdown of protein in the blades. This breakdown is accompanied by an outflow of nonprotein nitrogen, lipid, and nucleic acid components through organelle membranes and into the petioles. Soon thereafter, chlo-

roplast degradation takes place, with the accelerated disappearance of chlorophyll. Chibnall (18) first demonstrated, in 1954, that the formation of roots induced by IAA on detached leaves retarded the onset of these senescent symptoms so that detached leaves with initiated and growing roots remained in healthy condition for weeks. Chibnall suggested that the roots or the petioles of detached leaves produce a hormone that is translocated to the lamina and there acts to retard senescence. The initial observations of Richmond and Lang (99) and others (91) indicate that cytokinin treatment extends the life span of detached leaves by delaying protein degradation and loss of chlorophyll. Hence the disappearance of chlorophyll in detached leaves (loss of green color) serves as a useful index for studying the effect of various compounds on senescence (see Figure 20–8).

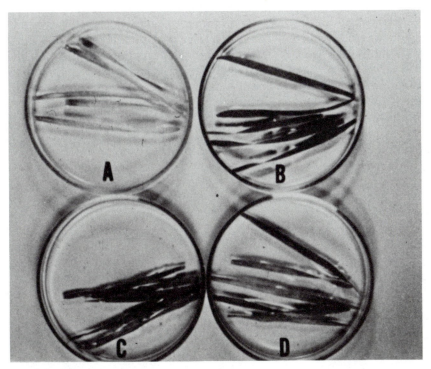

Figure 20–8. Effect of kinetin and zeatin on chlorophyll retention in detached wheat leaves. Leaves were floated on (A) double-distilled water; (B) kinetin, 5.0 mg/l; (C) zeatin, 5.0 mg/l; and (D) zeatin, 0.5 mg/l. *Photo by F.H. Witham.*

During the early studies on the chlorophyll retention properties of cytokinins, observers noted that the conservation of protein and chlorophyll was not restricted to the action of cytokinins. Person, Samborski, and Forsyth (93) originally observed that benzimidazole delayed senescence in detached wheat leaves. Others have shown that auxins, including IAA (90, 92), chlorinated phenoxyacetic acids, and gibberellins have the same effect on detached leaves from various plants. Chlorophyll preservation in detached leaves is used to detect cytokinins and gibberellins in several bioassay systems.

With respect to chlorophyll retention in detached leaves, Osborne (91) in the early 1960s proposed that the cytokinin effects are mediated by a stimulation of the RNA-protein system. Sugiura, Umemura, and Oota (108) reported that kinetin stimulated net synthesis of RNA in both microsomal and soluble cytoplasmic fraction of tobacco leaf disks. Other studies since then have demonstrated the effects of cytokinins on RNA and protein synthesis. We do not know the exact mechanism.

The so-called green islands resulting from clusters of chlorophyll (green areas) and starch-rich cells in otherwise chlorotic and necrotic leaf tissue are characteristic of plant disease induced by certain fungi (particularly rust organisms) and viruses. These green islands are maintained by cytokinins that are either synthesized by the host cells or by invading organisms. This aspect of the host-parasite relationship is not clear. The localization of cytokinins in these areas seems to cause the chlorophyll retention and possibly acts to produce sinks and the accumulation of nutrients to support the proliferating parasite.

Exogenously applied cytokinins are effective on fresh flowers, vegetables, and fruits as post-harvest preservatives. The use of cytokinins for the preservation of plant materials commercially, however, is not practiced widely in the United States because of governmental restrictions on exposing foodstuffs to various chemicals.

Cytokinins and Virus Infectivity

The synthetic cytokinins 6-BAP and kinetin influence the production of viruses in certain host systems (45, 85, 110). Király and Szirmai (45) initially observed that tobacco mosaic virus (TMV) production was inhibited in leaf disks of tobacco (*Nicotiana glutinosa*) when the disks were incubated in kinetin (50 mg/l) immediately after inoculation of the virus. For maximum inhibition, however, treatment of the whole leaf with kinetin prior to leaf disk preparation seemed to be necessary.

Observers have noted fewer and smaller virus lesions on leaf strips floated on kinetin prior to inoculation of virus than on nonkinetin-treated plants. As we would expect, 6-benzyladenine is as active or more active than kinetin in inhibiting virus increases and local lesions in various plant materials (2).

The levels of cytokinins and the interactions in the host and the virus are important for constant effects. For example, Tavantzis, Smith, and Witham (110) observed that intact tobacco leaves sprayed daily with kinetin at a relatively low concentration (0.1 mg/l) for several days prior to inoculation with tobacco ringspot virus (TRSV) exhibited high levels of virus infectivity; higher cytokinin concentration of kinetin (1.0 mg/l and 10.0 mg/l) inhibited virus infectivity (see Figure 20–9). Extracts from tobacco plants infected with TRSV

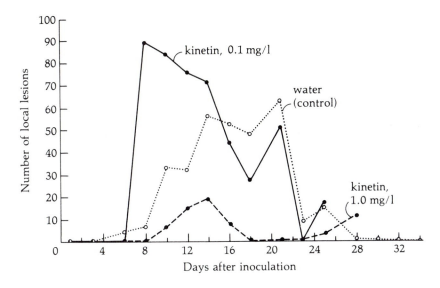

Figure 20–9. Local lesion bioassay. Viral infectivity expressed as number of local lesions per half leaf of each cowpea test plant inoculated with extracts, taken at indicated times from intact tobacco leaves. Leaves were sprayed with water or kinetin (0.1 mg/l or 1.0 mg/l) daily, beginning 9 days prior to inoculation and continuing afterwards. *From S.M. Tavantzis, S.H. Smith, and F.H. Witham. 1979. Physiol. Plant Path. 14:227–233.*

contain significantly less cytokinin activity than do leaf extracts from noninfected plants. Also, Tavantizis, Smith, and Witham showed that root exudates of TRSV-infected cowpea plants contain less cytokinin activity than do root exudates of noninfected plants. These results indicate important interactions for the cytokinins in host-parasite relationships. More information will provide exciting opportunities to develop control methods of virus infectivity via cytokinins in a broad range of plants.

Translocation of Nutrients and Organic Substances

During the late 1950s and early 1960s, Mothes and his colleagues, Englebrecht and Shutte (82, 83) demonstrated the kinetin-induced transport of soluble nitrogen from intact leaves of *Nicotina rustica* to localized areas of other leaves on the same plant. They also showed that labeled glycine applied to the lower left quadrants of a leaf blade was translocated to another quadrant previously sprayed with kinetin. Figure 20–10 illustrates the action of cytokinins on translocation. In fact even α-aminoisobutyric acid, which is not incorporated into protein, is accumulated at the kinetin locus (area sprayed with kinetin). The results of the induced translocation of substances suggest that one effect of kinetin in metabolite accumulation is not necessarily due to the direct stimulation of RNA and protein synthesis. Regardless of the exact mechanism, the evidence strongly indicates that cytokinins in plants influence the establishment of sinks, or areas that preferentially attract and concentrate nutrients.

Some workers have speculated (127) that equivalent levels of active cytokinins in physiologically active leaves and stems regulate nutrient flow so that the nutrients are drawn to certain areas (growing tips, expanding leaves, and so on) while the plant is in the vegetative state. The leaves at maturity may loose the capacity to produce or accumulate cytokinins, thereby causing

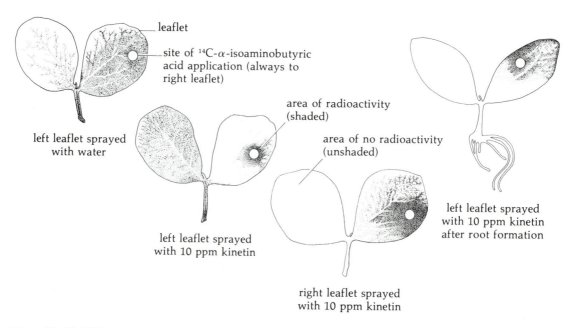

leaflet

site of ^{14}C-α-isoaminobutyric acid application (always to right leaflet)

area of radioactivity (shaded)

left leaflet sprayed with water

area of no radioactivity (unshaded)

left leaflet sprayed with 10 ppm kinetin

left leaflet sprayed with 10 ppm kinetin after root formation

right leaflet sprayed with 10 ppm kinetin

Figure 20–10. Effect of kinetin on translocation of ^{14}C-labeled α-aminoisobutyric acid. *Data from work of K. Mothes.*

subsequent changes in RNA, protein, and chlorophyll synthesis. Presumably the senescence and chlorophyll loss and decreased cytokinins of mature leaves are related to the development of newly unfolded leaves, which tend to accumulate cytokinins and the necessary nutrients.

Witham and Miller (127) have observed in developing maize kernels that there is a marked increase in cytokinins after fertilization and subsequent kernel development. The cytokinins appear to be at their highest level in milk-stage kernels, and this level is much higher than in stems and leaves. If cytokinins do in fact promote sink formation, they might cause the observed preferential transport of nutrients from the vegetative regions of the plants to the developing reproductive structures. It is interesting to observe in many annual plants the concomitant mobilization and movement of

nutrients to the reproductive structures, while the vegetative parts exhibit chlorophyll loss and sensescence. There is much to be learned about the hormonal regulation of nutrient and photosynthate translocation in plants. Even the casual observer cannot help but see the agricultural importance of cytokinin-induced nutrient translocation.

Action of Cytokinins

Preparations of tRNA from both plant and animal sources contain cytokinins, with the purine moiety of the cytokinin molecule being a component of the tRNA chain. Located adjacent to the anticodon, cytokinins most likely influence protein synthesis through their involvement in the attachment of tRNA to the ribosome-mRNA complex during protein synthesis. Control of amino acid binding in this manner could offer an explanation for the participation of cytokinins in many physiological effects. The major argument against this idea as being the primary mechanism of action of cytokinins is that when cytokinins are supplied exogenously, they are not incorporated as an intact molecule into the specific tRNA during synthesis. At least investigators have not demonstrated this incorporation experimentally.

Interaction of Cytokinins and Nucleic Acids

Some scientists suggest that the nucleic acids act as the cellular source of soluble cytokinins and that, aside from their presence in tRNA, other sites of action for cytoplasmic or exogenously supplied cytokinins exist. In view of the production of kinetin from degraded deoxyadenosine, this idea has merit, although direct production from nucleic acids is difficult to envisage because there seem to be inconsistencies between the stereochemistry of cytokinins coming directly from nucleic acids (*cis* forms) and the free derivatives (*trans* forms). In this respect we must obtain a great deal more information about the levels of different forms in vivo.

Several major factors indicate that cytokinins interact directly with nucleic acids. These factors are: (1) cytokinin composition, particularly the adenine ring and, therefore, reactivity; (2) existence in cells of active cytokinin ribonucleosides and nucleotides; (3) cytokinin-stimulated RNA and protein synthesis; (4) cytokinin stimulation of certain enzyme activities and product formation; (5) presumed in vivo presence of cytokinins in RNA; and (6) observed binding of kinetin to oligonucleotides.

According to some early ideas presented by animal physiologists, the interaction of hormones with genomic or organelle genetic components is accomplished by transport of hormones via receptors. In plants scientists have not unequivocally demonstrated the existence of a specific receptor for any phytohormone. Nevertheless, the cytokinins, as well as the auxins and gibberellins, clearly affect the physical properties of DNA (43). Hendry, Witham, and Chapman (39, 126) have shown that biologically active molecules, which operate as regulators of many physiological processes in plants and animals, can, at least theoretically, interact molecularly with double-stranded DNA by the process of *intercalation*. As indicated in the section on gibberellins, this process involves the proper alignment of a given molecule between base pairs of double-stranded DNA (see Figure 20–11). A cytokinin intercalated into DNA might cause template modifications (frameshifts, misreading, gene repression, derepression, and so on) that are important to the mechanics of transcription and translocation and hence important to numerous physiological and morphogenetic processes. The same kind of binding of cytokinins to the double-stranded RNA (e.g., complexes of anticodon and codon sequences of tRNA and mRNA) are also highly probable. To date, however, no definite experimentation supports these speculations.

There is evidence that cytokinins bind to ribosomal proteins, thus suggesting at least one site of action at the ribosome and an influence on protein synthesis. Also, in consideration of recent observed effects of zeatin on cell wall modifications, it seems possible that cytokinins have multiple sites of activity in plant cells. Active sites in a cell must reflect the information of DNA.

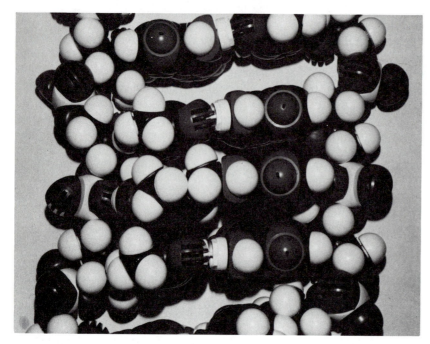

Figure 20–11. CPK space-filling models of proposed DNA-cytokinin interactions. Zeatin interaction with DNA between A-T and A-T base pairs.
Photo by F.H. Witham.

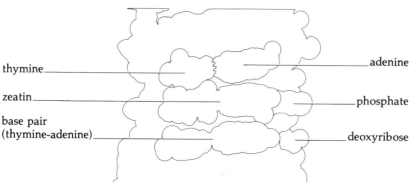

Cytokinins and Alternative Pathway of Respiration

Of recent interest from the studies of C.O. Miller is the finding that the cytokinins 6-BAP, kinetin, and 6-isopentenylaminopurine influence the respiration in mitochondria isolated from six species of plants: bush bean, mung bean, soybean, maize, peas, and wheat (75). The cytokinins 6-BAP and kinetin inhibit oxygen uptake by the mitochondria when malate is provided as the substrate. Zeatin and adenine, however, do not inhibit malate oxidation in the mitochondria. Also, under certain conditions, 6-BAP inhibits NADH and succinate oxidation.

The cytokinin inhibition of succinate

oxidation by the mitochondria in the presence of antimycin A is similar to that caused by salicylhydroxamic acid. The antimycin A blocks electron transport through the conventional cytochrome system. But salicylhydroxamic acid is a known inhibitor of the alternative respiration pathway. Therefore, the portion of the respiration process inhibited by certain cytokinins in the presence of antimycin A must belong to the alternative pathway. Figure 20–12 presents a diagrammatic scheme for the effect of cytokinin on the alternative pathway.

At point CK, cytokinins block the flow of electrons from malate by way of internal NADH (NAHD$_i$). Flow of electrons from externally supplied NADH (NADH$_{ex}$) and succinate to the cytochrome system takes place through routes b and d, respectively. Cytokinins (CK) would not block here. However, the flow of electrons to the alternative oxidase (alternative pathway) from NADH$_{ex}$ and succinate would be via routes a and c, respectively, and would be blocked by cytokinins or salicylhydroxamide.

Although the evidence presented by Miller is intriguing with respect to cytokinin effects on the alternative pathway, there is doubt as to whether these cytokinin effects are physiologically important. As pointed out by Miller, to inhibit mitochondrial respiration the active cytokinins have to be used

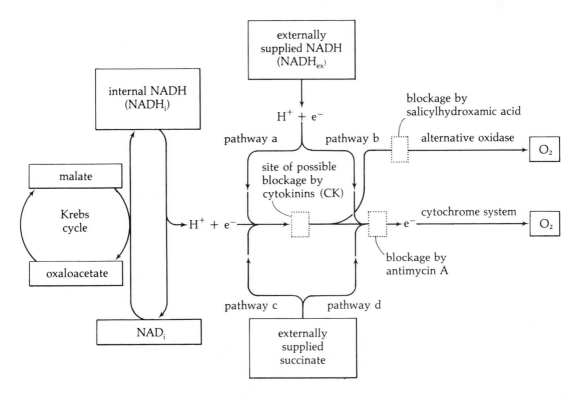

Figure 20–12. Proposed effect of 6-benzyladenine and other cytokinins on alternative pathway of mitochondrial respiration in bush bean, mung bean, soybean, maize, pea, and wheat.

at concentrations higher than necessary for hormonal action. The concentrations are more in the range necessary for pharmacological effects. Also zeatin, the widespread naturally occurring cytokinin, is not an effective inhibitor of substrate oxidation and electron flow via the alternative pathway. This lack of effectiveness of the natural hormone may reflect the cell's ability to control its accumulation. The influence of some cytokinins on respiration represents the basis for further in-depth studies that may provide answers to the effects of cytokinins on various aspects of cellular metabolism.

Physiological Effects of Ethylene

Scientists have known for some time that ethylene influences varying plant processes, from germination to fruit ripening. Even the ancient agriculturalists realized that this gas would promote the ripening of various tree fruit. The overripe apple, the so-called bad apple in the barrel, promotes overripening of other apples in the vicinity through the production of ethylene. The ethylene in turn stimulates degradation enzymes, cell loosening, and other physiological reactions of ripening.

Initially, studies on the physiology of fruit ripening and the advent of the gas chromatograph were primarily responsible for the discovery and identification of ethylene as an important phytohormone. Some of the physiological processes influenced by ethylene are release of seed, bud dormancy, seedling etiolation, seedling growth, stem growth, flower and fruit initiation, fruit growth and ripening, and promotion of leaf, flower, and fruit abscission.

Obviously, ethylene is physically quite different from the other plant hormones. At normal physiological temperatures, ethylene is a gas and is quite simple with respect to its molecular structure ($CH_2\!=\!CH_2$). However, it is similar to other plant hormones in that minute quantities of ethylene are produced in healthy plant tissues and cause dramatic changes in plant processes. One difference is that it diffuses out of plant tissues quite rapidly. It also seems probable that many of the effects once attributed to auxin alone are in fact caused by ethylene, acting alone or in concert with auxins. In addition, ethylene production is caused in vivo by wounding, rubbing, radiation, and some chemicals, including the auxins.

Fruit Ripening

In most fruits the rate of respiration will undergo a sharp rise and then fall near the end of ripening. Kidd and West (44) termed this phenomenon *climacteric rise* in 1930, when they published their studies on the pattern of respiration in stored apples. The term, shortened to *climacteric,* has been adopted almost universally. The climacteric acts as a trigger that sets in progress those changes that rapidly transform the fruit from an unripe to a ripe (and edible) condition.

Before ethylene was identified as a natural product of plants, it was noted with some surprise that ripening fruit gave off some volatile substance, with accelerated ripening of other fruits stored nearby. This substance was identified as ethylene, and its presence was detected in very small amounts in almost all fruits examined. Measurements made continuously of the amount of ethylene present in fruit tissue while it ripens will show that, although a very small amount of ethylene is present at all times, there is about a hundredfold increase just before or during the climacteric. Conditions that slow or prevent ripening, such as low-temperature storage, also retard ethylene production. Finally, application of ethylene to unripe fruit will bring on a premature climacteric and accelerate ripening. Thus ethylene has been firmly established as a true fruit-ripening hormone.

In some fruits ethylene production parallels the increase in respiration during the climacteric; in other fruits it occurs at the onset of the climacteric and actually declines as the rate of respiration reaches its peak (see Figure 20–13). The "gush" of ethylene that occurs in fruit tissue appears not to be simply a product of the climacteric; it seems, rather, to act as a trigger for other factors that initiate the ripening process. We must understand that fruit ripening is a dynamic active process that includes (1) hydrolysis of stored materials, (2) softening through enzymatic changes of pectic substances, (3) changes in pigmentation, (4) changes in flavor components, (5) dramatic changes in respiration, and (6) other biochemical reac-

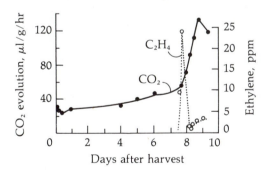

Figure 20–13. Relationship between ethylene production and respiration during climacteric rise in banana.

Reprinted from Botanical Gazette *126:200 by S.P. Burg and E.A. Burg by permission of The University of Chicago Press. Copyright 1965 The University of Chicago Press.*

tions. We do not yet know how ethylene induces ripening. However, two current theories explain the metabolic changes that occur during ripening, and ethylene could play a prominent role in both.

Early workers attempted to explain the climacteric in terms of a change in organization resistance, or permeability, of the tissue—that is, a change in the permeability properties of membranes that separate certain enzymes and substrates occurs during the climacteric and this change, in turn, affects respiration and other metabolic events. More recent studies of membrane permeability changes have led to a revival of this theory. For example, Sacher (102) found that an increase in the leakage of solutes in banana tissue preceded the onset of the climacteric by 44 hours, and that maximum membrane permeability occurred at the respiratory peak. Also, Young and Biale (131) concluded, from their study of ^{32}P uptake by disks of the avocado pear, that the climacteric was initiated as cellular membranes changed their permeability properties. We should note, therefore, that ethylene has been shown to cause an increase in tissue permeability (58, 121). However, the influence of ethylene on membrane permeability may be indirect. Mayak and Halevy (65) showed that exogenous applications of ethylene to rose petals induced an increase in ABA activity. Glinka (30) demonstrated that ABA altered the permeability characteristics of sunflower root cell membranes.

The other theory, induced enzyme formation, draws its support from several studies showing that an increase in protein content accompanies the climacteric (25, 41). New "ripening" enzymes may possibly be formed during the climacteric, and the activity of these enzymes may in turn cause the various metabolic changes that occur during and after the climacteric. Frenkel, Klein, and Dilley (25) demonstrated that fruit ripening can be prevented by blocking protein synthesis with cycloheximide at the early climacteric stage. The induction of protein synthesis by ethylene has been demonstrated in several different plant species (17, 98, 123). In addition, ethylene synthesis requires protein synthesis, especially during early climacteric. We do not know, however, whether ethylene, when it is in action, induces new protein (enzyme) formation. Further work is required on the nature of ethylene reception and action in plant cells.

Seedling Growth and Emergence

During the germination process both the radicle and the shoot tip may be protected by specialized tissues. In monocots the coleoptile and coleorhiza protect the epicotyl and root tips, respectively. However, particularly in dicots, the mode of seedling growth in the soil and during emergence is particularly important. One manner of seedling growth, characterized in the garden bean (*Phaseolus vulgaris*), for example, is *epigean* germination, whereby the cotyledons emerge above ground with the growing tip due to the elongation of the hypocotyl and the formation of the hypocotyl hook (see Figure 20–14). As the hypocotyl elongates, the shoot tip and cotyledons are protected and "pulled" upward through the soil. When the hypocotyl hook emerges and is exposed to light, symmetrical growth of the stem takes place and causes straight growth and a straightening of the hook.

Certain other dicots exhibit *hypogean* germination, in which the cotyledons remain below the soil surface and the hypo-

cotyl does not elongate. The plumule, however, is arched. The epicotyl elongates and the sensitive tip is protected as the arch is pushed upward. When the arch breaks through the surface, light initiates symmetrical growth and causes the epicotyl arch to straighten out.

During seedling development of dicots, ethylene is produced either in the plumule and plumular arch (hypogean germination) or hypocotyl areas (epigean germination). The localized production of ethylene is responsible for the initiation, formation, and maintenance of either the plumular arch or hypocotyl hook, depending on the mode of germination. As the etiolated seedling develops, ethylene inhibits growth in the hook or arch region. When the arch or hook emerges from the soil, light (white light or red light of 660 nm) causes a marked decrease of ethylene synthesis and allows the hypocotyl arch or plumular area to expand through symmetrical growth. Also, in experimentally grown seedlings, red light (660 nm) promotes straightening of the hypocotyl hook or epicotyl arch and far-red light (730 nm) reverses the effect of red light. The morphogenetic expression (plumular or hypocotyl expansion of seedlings) is regulated by localized levels of ethylene. Green tissues of seedlings are not as sensitive to ethylene as their etiolated counterparts.

Leaf Abscission

Leaf abscission is a dynamic process, important in the replacement of leaves during the plant vegetative cycle and as part of the overwintering (hardening) mechanism in deciduous trees. Leaf abscission is accomplished through the formation of parenchyma cell layers, usually at the base of the

Figure 20–14. Epigean germination. Seedling growth of garden bean (*Phaseolus vulgaris*) illustrating hypocotyl hook and symmetrical growth above ground.

Courtesy of Nickerson-Zwaan B.V., Barendrecht, the Netherlands.

petiole. The smaller-sized cells and the presence of vascular elements and fibers contribute to the fact that the abscission zone is weaker than the surrounding area.

During development and maturation of a leaf, auxin, presumably produced in the leaf blade, flows through the abscission area and inhibits abscission layer formation. At some point, however, formation of the abscission layer is initiated. The dynamic nature of abscission layer formation is charac-

terized by an increased synthesis of cell wall–modifying enzmyes and other proteins, with increases in respiration. In addition, there is a decrease in sensitivity of the aging cells to the inhibitory influence of auxin. Conversely, the cells become responsive to ethylene, which accelerates senescence and abscission layer formation. After the abscission process has begun, exogenously applied auxins also accelerate abscission. This effect is due to the auxin-induced stimulation of ethylene biosynthesis. Once concentrated in the abscission zone cells, ethylene stimulates cellulase production, which results in cellulose hydrolysis, cell wall disruption, and separation of the cells. Mechanical forces, such as the wind, bring leaf abscississon to its culmination.

Other Responses

Some of the additional effects of ethylene on plant growth include the inhibition of elongation in roots, stems, and leaves; the stimulation of adventitious root formation on stems; the inhibition of geotropism in peas; the inhibition of flowering; and epinasty.

Ethylene is a powerful inhibitor of root and stem growth. The inhibitory effect of auxins at high concentrations is known to be entirely due to auxin-stimulated ethylene synthesis. In fact, root sections incubated in auxin synthesize ethylene. Although ethylene inhibits root growth, scientists do not believe it to be the only growth-inhibiting substance involved in root geotropism. As pointed out previously (see auxin and geotropism), considerable evidence indicates that abscisic acid is one inhibitor produced in the root cap that migrates laterally and inhibits growth.

Ethylene is a potent inhibitor of bud growth, and in this respect may have a controlling influence on apical dominance. Ethylene seems to be more prevalent in meristematic tissues where auxin is produced. In the mature light-grown plant, lateral bud growth seems to be retarded by IAA-induced ethylene formation in the nodal regions as a result of auxin translocated there from the apical bud and leaf blades.

We learned earlier that cytokinins can overcome the inhibitory effect of IAA on lat-

eral bud growth. Studies by Burg and Burg (12) revealed that inhibition of lateral bud growth by ethylene or auxin is completely overcome by kinetin. Other studies demonstrated that lateral bud growth is partially released in intact pea plants placed in a 5 percent CO_2 atmosphere (112). Carbon dioxide is a competitive inhibitor of ethylene.

The inhibition of elongation of roots and stems is characterized by lateral swelling, particularly in the usual areas of elongation. Somewhat consistent with these effects is the fact that etiolated pea stems in ethylene do not appear to be affected by gravity. As a result they grow in an *ageotropic* (not sensitive or responsive to gravity) manner. We can explain this influence of ethylene as a blocking of the normal movement of auxins in response to gravity. Investigators have observed that pea stem sections growing in dilute auxin solutions often exhibit a pronounced curvature of 40° or more. As we might expect, the curvature of the sections is due to an asymmetrical distribution of auxin. Pea stem sections incubated on their side in a solution of IAA-^{14}C exhibited a lower- to upper-side ^{14}C ratio of 72:28. However, the ratio was 55:45 if ethylene was included (11). Thus, in pea stem sections at least, the normal lateral movement of auxin in response to gravity is almost entirely blocked by ethylene. Observers have noted no immediate effect of ethylene on the longitudinal transport of auxin, but prolonged exposure to ethylene does inhibit longitudinal transport.

A number of investigators found that low concentrations of IAA and other auxins induce ethylene formation in the roots, stems, leaves, flowers, and fruits of all plants they examined. Thus most inhibitory effects of high auxin concentrations are due to excessive amounts of ethylene being formed.

Exposure of a leaf to ethylene stimulates differential growth, with faster growth on the upper side. We can observe this growth as lateral swelling of the cells at the base of the petiole and midvein. This phenomenon is referred to as *epinasty* and will cause downward bending of the leaf. The auxins produce the same effect due to auxin-stimulated ethylene synthesis in vivo. With respect to the auxin effect, the auxin herbicides such as the phenoxyacetic acids often induce epinasty on plants unintentionally sprayed or from "drift" in greenhouses and in the field. Under "normal" circumstances (physiological concentrations), ethylene is important in regulating the usual angles of leaves through the control and balance of epinastic and *hyponastic* (lower-side growth) responses.

Ethylene inhibits flowering in most plants, although pineapple is a notable exception. In fact, exogenously applied auxins have been used to stimulate ethylene production and promote flowering in pineapple. The commercial substance ethrel has also been used to study the effect of ethylene on flowering. The interesting feature of ethrel is that ethylene is released from the chemical. Scientists anticipate that such phytohormone-releasing chemicals or precursors to ethylene biosynthesis in plants will have widespread applications in future commercial uses for plant growth regulators.

Ethylene Biosynthesis

Lieberman, Mapson, Kupnishi, and Wardale (56) were the first to suggest that the sulfur-containing amino acid methionine is the primary natural precursor of ethylene in plants. In 1966 Yang, Ku, and Pratt (130) first showed, in a model system, that ethylene can be formed from methionine. Shortly thereafter studies showed that treatments of both fruit and vegetative tissue with methionine accelerated ethylene production (13, 56). In addition, Yang (128), using labeled methionine, demonstrated that the third and fourth carbons of methionine provided the carbon for ethylene. Today considerable evidence exists that methionine is the primary precursor of ethylene in many higher plants (129). Figure 20–15 shows the biosynthetic pathway for the formation of ethylene from methionine.

The important characteristics of ethylene biosynthesis as outlined in Figure 20–15, are that the first carbon of methionine gives rise to CO_2, liberated with NH_3. The second carbon is converted to formic acid, and the third and fourth carbon give rise to ethylene, as demonstrated by Burg and Clagett (13) in 1967 and later by Yang (128). The sulfur remains and is recycled into the production of methionine.

The important features of the pathway shown in Figure 20–15 may be clarified further as follows:

Step 1. Methionine (MET) conversion to S-adenosylmethionine (SAM) requires ATP-yielding pyrophosphate and inorganic phosphate.

Step 2. SAM is then converted to 1-aminocyclopropane-1-carboxylic acid (ACC). This reaction, at least in tomato fruit tissue, is catalyzed by the enzyme ACC synthase (129). This enzyme controls the rate of formation of ethylene. Its ACC synthase activity or levels are regulated by certain chemicals, including IAA, by wounding, by the presence or absence of oxygen (decreased under anaerobic conditions), and by factors (possibly phytohormones) of the ripening process. More specifically, by some unknown mechanism, all of the above factors act directly on enzyme induction and in

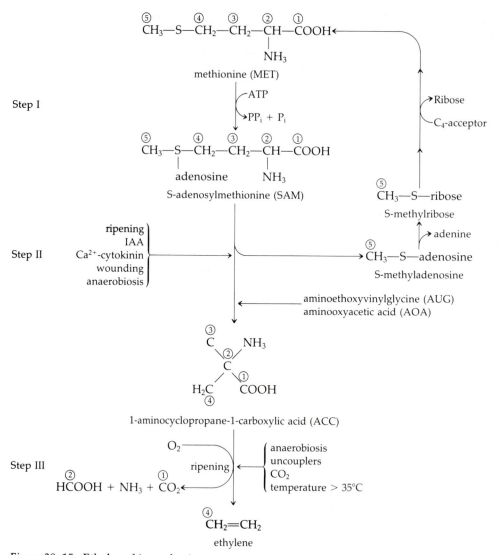

Figure 20–15. Ethylene biosynthesis.
Data courtesy S.F. Yang.

this way control the rate of formation of ACC synthase and ultimately ethylene production. The important point to be emphasized is that IAA stimulates ethylene production through its primary effort on ACC synthase induction (129), a fact that is significant to the understanding of phytohormonal action of genetic information. Further, phytohormones that induce enzymes must interact chemically with nucleic acids. As shown in Figure 20–15, Step 2 can also be inhibited by substances that inhibit the enzyme. Aminoethoxyvinylglycine (AVG) and aminooxyacetic acid (AOA) are two such inhibitors. Another significant feature of Step 2 is the recycling of sulfur and its reutilization in the biosynthesis of additional methionine (129). Although not shown here, in this pathway an S-methyl-adenosine is found, which leads to S-methylribose and ultimately methionine.

Step 3. The conversion of ACC to ethylene results in the production of carbon dioxide, ammonia, and formic acid. The other carbon products arise respectively from the first and second carbons; and ethylene comes from the third and fourth carbons of methionine. Factors that influence this reaction are factors that promote ripening and high CO_2 levels that are inhibitory. In addition, the reaction is inhibited by high CO_2 levels, temperatures greater than 35°C, and uncouplers of oxidative phosphorylation such as dinitrophenol (DNP).

Yang (129) points out that an understanding of this pathway and the mechanism of enzyme formation in plant tissues should lead to the successful control of adverse post-harvest physiological processes and the control of morphogenetic events influenced by the phytohormone.

Abscisic Acid

In 1961 Liu and Carns (57) isolated a substance in crystalline form from mature cotton fruit. This substance stimulated the abscission of debladed cotton petioles. The structure of the isolated product, which they called *abscisin I*, was not determined. Nevertheless, its detection led to the discovery of another similar substance, which Ohkuma and colleagues (89) isolated from young cotton fruit and termed *abscisin II*. Partial chemical characterization of abscisin II at the time revealed that it was a fifteen-carbon compound.

Almost simultaneously with the re-

ports on abscisin II, Eagles and Wareing (22) published a study on the extraction of an inhibitor that accumulated in birch leaves held under short-day conditions. When the inhibitor was reapplied to leaves of birch seedlings, apical growth was completely arrested. This result led Eagles and Wareing to suggest that the compound was a dormancy inducer. They termed this uncharacterized chemical as *dormin*.

In 1965 Ohkuma and colleagues (88) proposed the chemical structure of abscisin II. Conforth and his colleagues (19) isolated dormin in pure form from methanolic extracts of sycamore leaves. Most importantly, they showed that abscisin II and dormin were the same compound, with the chemical structure illustrated in Figure 20–16.

To eliminate the confusion that arose from the different names for the same substance, the principal scientists involved in the early work decided on a standardized system of nomenclature and termed the newly discovered compound *abscisic acid*, or *ABA*. The terms abscisin I, abscisin II, and dormin have been dropped and appear only in the early pioneering studies.

Chemistry of ABA

Abscisic acid is a sesquiterpene consisting of fifteen carbons and characterized by a six-membered ring, with a chiral (assymetrical) center and an unsaturated six-carbon substituent (see Figure 20–16). The chiral center is

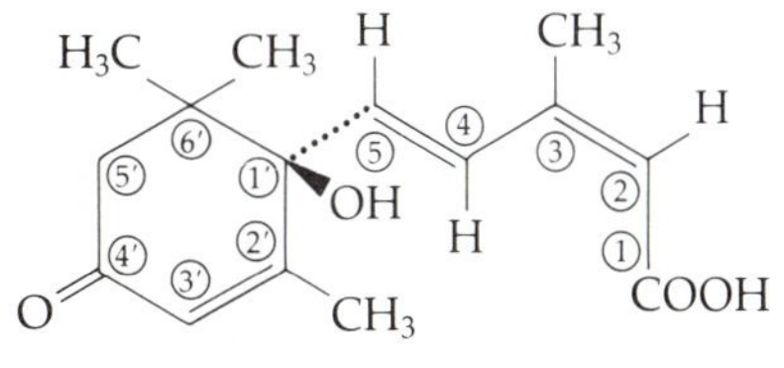

Figure 20–16. S-Abscisic acid.

responsible for the occurrence of two enantiomorphic forms, (R)-abscisic acid and (S)-abscisic acid. However, (S)-abscisic acid is the naturally occurring active form, and for this reason it is the only form designated simply as ABA, or abscisic acid, without clarification. In addition, the enantiomers are optically active. For example, ABA, which is (S)-(+)-abscisic acid, is dextrorotatory, or shifts a plane of monochromatic light strongly to the right. In fact, ABA is often detected by optical rotatory dispersion techniques because it is probably one of the most optically active natural products known to humans.

Methods of Detection

The major historically significant methods used to detect and identify ABA are *spectropolarimetric analysis, gas chromatographic analysis,* and *bioassays.*

Spectropolarimetric Analysis

ABA is structurally unique in that the one prime carbon of the ring (a chiral center) confers easily detectable *optical rotatory dispersion* (ORD) properties to ABA. As a result we can use this property to make qualitative and quantitative determinations about ABA in a reasonably purified extract (67).

A refinement of the method employing ORD entails the use of both ORD and ultraviolet absorption of ABA. It is particularly useful in determining the amount of ABA remaining after initial isolation and final purification. Readers interested in the details of this method are referred to some of the articles on this subject (67) although it is not now the most commonly used since it has, essentially, been replaced by gas chromatography.

Gas Chromatographic Analysis

Gas liquid chromatography (GLC) is based on the preparation and volatilization of the various ABA trimethylsilyl derivatives. A sample, usually purified with activated cabon, is mixed with a material such as bis-(trimethylsilyl) acetamide to produce the trimethylsilyl derivatives of ABA. The different derivatives are then measured by GLC against a known amount of similarly treated ABA in solution. This method is very sensitive. As little as .030 μg ABA can be detected and separated from remarkably similar substances. Care must be taken, however, in the preparation of the trimethylsilyl derivatives, the use of the GLC columns, and the calibration of the instrument.

The most powerful current analytical means available for studying ABA in plant extracts is the combination of gas chromatography–mass spectrometry (GC–MS). The principle behind this method is the separation of sample components into chromatographic peaks that can be identified directly from their mass spectra. The mass spectrum of a compound, based on charged molecular fragments, is diagnostic of a compound's structure.

Scientists are using high performance liquid chromatography and UV detectors increasingly to purify and identify ABA and other phytohormones. Thus the number of analytical techniques and devices has increased considerably since the time of Went and his pioneering studies on auxins.

Bioassays

The bioassays that have been used to detect ABA are numerous and include the following examples of biological responses: inhibition of seed germination, inhibition of coleoptile curvature (straight growth), inhibition of α-amylase synthesis in aleurone cells, and acceleration of abscission in excised abscision zones.

Although many bioassays have been developed for detecting ABA, we must not depend on a single bioassay or combination of bioassays for the identification of a given biologically active substance. Bioassays are useful in the tracking of "new" phytohormones present in extracts. They are effective in indicating the presence of a category of compounds, through auxin-, gibberellin-, or cytokinin-type activity. However, they are subject to variability, nonspecificity, and other growth factors.

An important feature of some of the widely used bioassay systems developed early in studies on new substances is their use as tools to study physiological processes and the mechanism of action of new phytohormones. A selectively specific bioassay that has been analyzed statistically by different experimenters is extremely useful in evaluating the action of a phytohormone.

Biosynthesis of ABA

We might simply state that ABA, as a sesquiterpene, is a compound of three isoprene units. The indications are good that ABA, like other isoprenoids, is derived from mevalonic acid (67). We should note that the mevalonic acid pathway is of great importance in the synthesis of other phytohormones. As we have mentioned previously, this pathway is necessary for the synthesis of gibberellins, the five-carbon substituent of zeatin, and other cytokinins, as well as ABA.

According to Milborrow (66), who was involved in much of the early structural work on ABA, synthesis of ABA in the leaves of bean and avocado plants may occur primarily in chloroplasts. Some investigators suggest that ABA is a product of the photoconversion of xanthophylls; there is certainly a similarity between the two types of compounds, particularly between the immediate precursor xanthoxin and ABA. However, no evidence elevates these ideas beyond the level of speculation at this time.

Translocation of ABA

The synthesis of ABA seems to take place predominantly in the mature leaf, and from there ABA is readily transported to various regions via the petioles and stem tissue. The rate of movement, at least in cotton, is about 20 to 30 mm per hour. Most likely, the translocation of ABA takes place in the phloem and xylem. Further, ABA may be translocated from the root cap, where it influences the geotropic response (see section on auxins and geotropism). As we will discuss later, the levels of ABA in the plant seem to be controlled by stress, which is clearly related to some of the physiological activities of ABA as a phytohormone. The following are some of the physiological responses in which ABA is implicated: dormancy, inhibition of bud growth and shoot formation, seed development and germination, geotropism, and stomatal closing.

Dormancy

Since the early work on the discovery of ABA as an inducer of bud dormancy in trees, many studies have been directed to provide evidence in support of ABA as a dormancy phytohormone. A majority of studies, however, show that applied ABA does not induce dormancy in various woody species (see 80). Also, no definitive correlation has been shown between the photoperiodic induction of dormancy and increased ABA levels in plants—that is, ABA levels are not higher in plants exposed to (photoperiodically induced) short days compared to those exposed to long days.

The evidence supporting the induction and maintenance of seed and bud dormancy, therefore, is not conclusive. As pointed out by Walton (122) in a review of this subject, an understanding of the role of ABA in dormancy is difficult because we do not know the biochemical events leading to dormancy induction and reversal. Thus the role of ABA in dormancy is controversial at this time.

Inhibition of Bud Growth and Shoot Formation

Tucker (115, 116, 117) suggested that abscisic acid inhibits lateral bud growth in tomato plants. Using cultivars of tomato that exhibit varying degrees of apical dominance, he showed that cultivars with strong apical dominance contained high ABA levels. Plants in which bud growth is suppressed by far-red light treatments also show high ABA levels. Conversely, cultivars with weak apical dominance have substantially lower levels of endogenous ABA.

In addition Tucker showed that plants with high ABA levels also have high levels of auxin (IAA) activity. Based on these results, Tucker (115, 116, 117) suggested that IAA is responsible for maintaining the levels of ABA sufficiently high for strong apical dominance. The reverse seems true in other tomato mutations. However, the cor-

relation between high IAA and ABA levels in tomatoes is not a general phenomenon in that ABA levels do not always change when IAA levels drop. We cannot say at this time that ABA plays a general and fundamental role in apical dominance in all plants.

Seed Development and Germination

Evidence indicates that ABA builds up in embryos of developing seeds. Workers believe that the increased ABA level comes primarily from *de novo* synthesis. However, there may also be translocation of ABA or its precursors from the leaves since the radioactivity from exogenously applied ^{14}C-ABA to leaves can be recovered in developing seeds. According to Dure (21), in his review on seed formation, ABA in developing ovules inhibits the formation of germination enzymes in the embryo. Thus ABA appears to play a fundamental role in developing seeds by inhibiting *vivipary*, the phenomenon of precocious germination before maturity or the release of the seeds from the fruit.

Exogenous ABA inhibits seed germination even in the presence of gibberellins and cytokinins, known promoters of germination. Some plant scientists have suggested that exogenously applied ABA to mature, otherwise nondormant, seeds may inhibit germination by suppressing nucleic acid–directed synthesis of specific germination enzymes. In fact, Dure suggested that ABA inhibits the translation of specific mRNA and in this way blocks protein synthesis. This hypothesis requires considerable experimentation, even though similar evidence implicates the action of most other phytohormones through their nucleic acid interactions.

Geotropism

The root cap responds to light and gravity by synthesizing or accumulating growth inhibitors (16, 28, 94, 103). Scientists currently believe that inhibitors are produced in the lower portion of the root cap in response to gravity and are translocated basipetally into the zone of cell enlargement. There they may inhibit cell enlargement of the cells on the lower side of the root. Differential growth of the root (due to concentrated inhibitors on the lower side) results in geotropism. Scientists have recently shown that ABA is present in corn (*Zea mays*) root caps. The buildup of ABA in the cap appears to require light and gravity. ABA produced in maize root caps seems to translocate basipetally and to stimulate a positive geotropic response (103).

Unfortunately, certain discrepancies and contradictions exist in the results of past experiments. We cannot, therefore emphatically state that ABA is responsible for geotropism in roots. Other unidentified inhibitors are present in plant root caps. They may also translocate basipetally. Further, in some plants ABA has only a small inhibitory effect on the root systems. The idea that geotropism is controlled by auxins alone is not a suitable explanation for the phenomenon. There must be some involvement of ABA or other inhibitors.

Stomatal Closing

The most significant and best-known role of ABA is its control of stomatal closing. The primary action of ABA on guard cells seems to be its inhibition of potassium uptake by the guard cells (96). Certain substances, such as the fungal toxin fusicoccin,

are known to overcome the effect of ABA by stimulating potassium uptake and proton release. The flow of other solutes also seem to be influenced by ABA during the stimulated closing of stomata. For example, malic acid is released with potassium from the guard cells more quickly in the presence of ABA. One idea that accounts for the effect of ABA on the loss of guard cell turgidity is that ABA inhibits the H^+/K^+ exchange and promotes the leakage of malic acid. The reduction of osmotically active solutes would of course render the guard cells flaccid and keep the stomatal pore closed.

Another significance of malic acid is that it may be a primary source of protons necessary for the H^+/K^+ exchange process during stomatal opening. Obviously, if ABA promotes leakage of malic acid, the loss of the source of protons would promote closing. Although CO_2 seems to interact with ABA to promote stomatal closing, we do not understand the nature of the interaction at this time.

Water Stress and ABA

ABA seems to play an important role in plants during water stress and during drought conditions. We know that exogenously applied ABA promotes stomatal closing in several experimental systems (96, 97). We also know that ABA accumulates in water-stressed mesophytic plants and that the ABA level decreases when the plant is no longer under stress. A fact that indicates the importance of ABA in water-stressed plants is the existence of the wilty mutant of tomato. These mutant plants seem to contain very low levels of ABA; and these levels do not change when water is withheld. The tendency of these plants to wilt, however, is reduced by exogenous applications of ABA; the ABA acts directly on the stomatal apparatus.

Evidence explaining what factor or factors stimulate increased ABA levels during water stress is meager. There are some observations, however, that, in a parallel fashion, ABA levels in the plant rise as the water potentials become more negative. The exact stimulus (osmotic potential, pressure potential) and the mechanism of ABA accumulation in the leaf during water stress, however, is not definitely known.

We might ask how the levels of ABA are regulated in stressed and nonstressed plants. Mansfield, Wellburn, and Moreira (61, 62) and Milborrow (68, 69) suggest that ABA is produced in chloroplasts and remains there in unstressed leaves. When the plant is water stressed, the permeability of the chloroplast membranes to ABA increases and allows movement of the hormone to the epidermal cells, including the guard cells. ABA then acts on the guard cells to promote stomatal closing. The loss of ABA in the plastids promotes additional hormone biosynthesis, and this promotion continues until the water stress is relieved. When the water stress is relieved, the chloroplast membranes become impermeable to ABA and ABA becomes restricted to the chloroplasts. Presumably the buildup of ABA is eventually turned off by end product inhibition.

Although these considerations represent a good working hypothesis, each individual step has not been conclusively shown to be true. A considerable amount of work on the metabolism of ABA within and outside chloroplasts remains to be accomplished. Nevertheless, in many instances ABA plays a significant role in water-stressed plants, and in some cases ABA may be involved in part of the mechanism of drought resistance. As pointed out by Walton (122), the establishment of plant lines that produce high levels of ABA may be highly significant to the growing of crops in arid parts of the world. We must realize that ABA activity may not be the means by which all plants exhibit drought resistance. Other means of drought resistance are undoubtedly operative in plants.

Questions

20–1.   Discuss the circumstances concerning the discovery and isolation of 6-furfurylaminopurine. What is the common name of this substance? Is it related to zeatin or to syngamin?

20–2.   Why does coconut milk serve as a supplement for certain plant tissue cultures?

20–3.   Describe the techniques you might use to determine the presence of auxins, cytokinins, and gibberellins in an extract of plant tissue.

20–4.   Are the cytokinins restricted to certain plant species? Explain.

20–5.   Where in the plant might you find the highest levels of cytokinins?

20–6.   Why are the cytokinin-free bases often more active than the ribosides or glucosides?

20–7.   List several bioassays for cytokinins. What primary activity of cytokinins does each one exhibit?

20–8.   Describe eight plant responses that are influenced by cytokinins.

20–9.   Provide several possible commercial applications of cytokinins.

20–10. How might cytokinins influence the formation of sinks in plants?

20–11. Provide at least five reasons supporting the idea that cytokinins operate through their action on the nucleic acids.

20–12. Do cytokinins influence the plants reducing potential and respiration?

20–13. What is the meaning of the following terms: climacteric, ethylene, epigean and hypogean germination, and hypocotyl hook and epicotyl arch.

20–14. What is the role of ethylene in seedling growth and emergence?

20–15. Does abscisic acid play a major role in abscission of leaves from most plants? Explain.

20–16. What is the relationship of ethylene to the inhibitory effects of auxin on root and stem growth?

20–17. Trace the biosynthetic pathway for the formation of ethylene from methionine in plants. Where in the pathway might cytokinins, wounding, ripening, and anaerobiosis act?

20–18. Describe the early work leading to the discovery and final identification of abscisic acid.

20–19. Name some of the responses employed in bioassays that are used to detect ABA in extracts from plant tissues.

20–20. What do the following substances have in common: the 6-substituent of zeatin, gibberellic acid, abscisic acid, and carotenoids?

20–21. List four major involvements of abscisic acid in plant growth.

20–22. Why may abscisic acid be important in geotropism?

20–23. Discuss the role and importance of ABA in water stress.

20–24. Why does one rotten apple spoil the rest of the apples in the barrel?

20–25. Suggest a possible mechanism that might lead to the synthesis of abscisic acid in plants subjected to water stress.

Suggested Readings

Abeles, F.B. 1973. *Ethylene in Plant Biology.* New York: Academic Press.

Adams, D.O., and S.F. Yang. 1979. Ethylene biosynthesis: identification of 1-aminocyclopropane-1-carboxylic acid as an intermediate in the conversion of methionine to ethylene. *Proc. Natl. Acad. Sci., U.S.* 76:170–174.

Audus, L.J. 1975. Geotropism in roots. In J.G. Torrey and D.T. Clarkson, eds., *The Development and Function of Roots.* New York: Academic Press.

Burrows, W.J. 1975. Mechanism of action of cytokinins. *Current Adv. Plant Sci.* 7:837–847.

Harrison, M.A., and D.C. Walton. 1975. Abscisic acid metabolism in water-stressed bean leaves. *Plant Physiol.* 56:250–254.

Juniper, B.E. 1976. Geotropism. *Ann. Rev. Plant Physiol.* 27:385–406.

Leonard, N.J. 1974. Chemistry of the cytokinins. In V.C. Runeckles, E. Sondheimer, and D.C. Walton, eds., *The Chemistry and Biochemistry of Plant Hormones,* vol. 7. *Recent Advances in Phytochemistry.* New York: Academic Press.

Lieberman, M. 1979. Biosynthesis and action of ethylene. *Ann. Rev. Plant Physiol.* 30:533–591.

Pilet, P.E. 1977. Growth inhibitors in growing and geostimulated maize roots. In P.E. Pilet, ed., *Plant Growth Regulation.* New York: Springer-Verlag.

Thomas, H., and J.L. Stoddart. 1980. Leaf senescence. *Ann. Rev. Plant Physiol.* 31:83–111.

Walton, D.C. 1980. Biochemistry and physiology of abscisic acid. *Ann. Rev. Plant Physiol.* 31:453–489.

Chapter 21

Photoperiodism and Phytochrome

Flower of poinsettia (*Euphorbia pulcherrima*). Courtesy of W.R. Fortney, The Pennsylvania State University.

Even to the casual observer it must be evident that light controls the growth and development of plants. In fact, we can easily demonstrate that a plant cannot grow in darkness, thus indicating that light is essential for development, or *morphogenesis*. Surprisingly enough, the fact that light is necessary for morphogenesis was not even recorded until 1779, when Ingenhousz recognized the importance of light in photosynthesis. Since that time, there has been slow but steady progress toward the recognition that many light-controlled processes regulate plant morphogenesis. Today, *photomorphogenesis* is a term encountered regularly in the scientific literature.

A plant must absorb light before it can begin to respond to that light. A receptor of some sort (usually a pigment) must, therefore, be present to absorb the wavelength(s) of light responsible for the response. In many cases the absorption of light by the receptor causes the receptor to become activated, thus initiating a sequence of chemical reactions that leads ultimately to a general plant response. The absorption of light, with subsequent activation of the absorbing molecule, followed by a series of chemical reactions leading to a general plant response, may be termed a *photobiological process*.

Scientists have studied extensively many of the photobiological processes occurring in plants; and, in several cases, they have isolated and characterized individual constituents of these processes. Some of the photobiological processes they have studied in detail are photosynthesis, chlorophyll synthesis, phototropism, photooxidation, leaf expansion, inhibition of stem elongation, flowering, and photoperiodism.

Photoperiodism is a term that escapes precise definition. We define it, generally, as the response of a plant to the relative lengths of light and dark periods. However, we can modify this definition in several ways. For example, the duration of the dark period is much more important than the duration of the light period. Intensity and quality of light can be modifying features in the magnitude of the response. The total quantity of light received can have an influencing effect. That the duration and order of sequence is most important in the initiation of a photoperiodic response is, however, generally accepted. Any response, then, by a plant to the duration and order of sequence of light and dark periods may be called a *photoperiodic response*.

Plants respond to alterations of light and dark periods in a variety of ways. Flowering, vegetative growth, internode elongation, seed germination, and leaf abscission are examples of photoperiodic responses exhibited by plants. Since flowering was the first photoperiodic response to be discovered and the one most extensively studied, our discussion of photoperiodism will, largely, be an analysis of this phenomenon.

Flowering

Although controlling effects of photoperiod on flowering were observed before the twentieth century, the first good experimental evidence supporting this concept was presented during the early years of this century. Tournois (48), in 1912, attempted to explain why hemp flowers vigorously if planted early in the spring but remains in a vegetative stage if planted in late spring or summer. Tournois showed that if hemp is provided with short photoperiods (6 hours), it will flower; but if it is provided with relatively longer photoperiods, it will remain in a vegetative state.

A study of the flowering habits of *Sempervivum* by Klebs (28) in 1913 showed that flowering can be induced by artificial illumination in midwinter in a greenhouse, although the normal time of flowering for this plant is June. Klebs concluded that flowering of *Sempervivum* is controlled by the length of the photoperiod and that light serves as a catalytic factor.

The first clearly stated hypothesis on photoperiodism was given in 1920 by Garner and Allard (15). Initially they observed that Biloxi soybean (*Glycine max*) plants flowered in September and October even if they were germinated over a 3-month span in May, June, or August (see Table 21–1). They observed that a variation of 59 days in germination date during May and June caused a difference of only 11 days to the first open flowers (15). The number of days from germination to blossoming and the final height of plants decreased as the season continued. Thus the data indicate that in soybean plants there is a seasonal timing mechanism.

For another experimental plant, they used a large-leaf tobacco plant mutant noted for its vigorous vegetative growth and a flowering habit that differs radically from normal tobacco plants. The mutant, Maryland Mammoth, did not flower in the field during the summer months of relatively long photoperiods (days) at Beltsville, Maryland. However, when the plant was grown in the greenhouse under the relatively shorter photoperiods of winter, it flowered profusely in mid-December. The following year, the experimenters sowed the seeds from this plant along with "normal" tobacco plants and the pattern was repeated. The mutant remained vegetative in the field during the summer months, but when it was brought into the greenhouse for the winter months, it flowered under the relatively shorter days of December.

Next, the experimenters subjected the Maryland Mammoth tobacco plant to short day lengths during the summer by placing the plant in darkness after exposing it to a day length that would be equivalent to a winter day. The mutant then flowered in the summer. After ruling out other factors, such as temperature, nutrition, and light intensity, Garner and Allard concluded that the length of day controlled flowering. This conclusion was supported by the fact that the mutant could be kept in the vegetative state during the winter months by merely lengthening the days with artificial light. Garner and Allard termed the response of Maryland Mammoth to day length *photoperiodism* (see Figure 21–1).

Germination Date	Date of First Open Blossoms	Maximum Height (inches)	Days to Blossoming
May 2	September 4	52	125
June 2	September 4	52	94
June 16	September 11	48	92
June 30	September 15	48	77
July 15	September 22	44	69
August 2	September 29	28	58
August 16	October 16	20	61

Table 21–1. Growth and flowering of Biloxi soybeans as function of germination date.

Source: *Data from W.W. Garner and H.A. Allard. 1920. Effect of length of day on plant growth.* J. Agr. Res. *18:553.*

tobacco plant grown
in unlighted greenhouse
(short days)

tobacco plant grown
in electrically lighted
greenhouse (long days)

Figure 21–1. Maryland Mammoth tobacco plants in which photoperiodism was first observed. Plant on left grew in unlighted greenhouse (short days). Plant on right grew in electrically lighted greenhouse (long days).

Terminology

As a result of their studies, Garner and Allard developed terms necessary to describe the photoperiodic response of flowering. The Maryland Mammoth mutant was called a *short-day plant* because of its habit of flowering only under short-day conditions. Plants vary considerably in their response to day length. In some plants, long-day photoperiods induce flowering. Other plants appear unresponsive and flower under both long- and short-day conditions. Still others respond to photoperiods somewhere between short- and long-day lengths. The definitions given here are based on a 24-hour cycle of light and darkness.

Short-day flowering plants. These plants flower when the day length is less than a certain critical length. Day lengths in excess of this critical point will keep the short-day flowering plant vegetative. The so-called critical day length differs with different spe-

cies. Some examples of short-day flowering plants are *Nicotiana tabacum* (Maryland Mammoth), *Xanthium pennsylvanicum* (cocklebur), and *Glycine max* (Biloxi soybean).

Long-day flowering plants. These plants flower after a critical day length is exceeded. Again this critical day length differs from species to species. Some examples of long-day plants are *Spinacea oleracea* (spinach), *Beta vulgaris* (sugar beet), and *Hyoscyamus niger* (black henbane).

Day-neutral flowering plants. These plants flower after a period of vegetative growth, regardless of the photoperiod. Some examples of day-neutral plants are *Lycopersicum esculentum* (tomato), *Mirabilis* (four-o'clock), and certain varieties of peas (*Pisum sativum*).

Although relatively rare, some plants require a long photoperiod succeeded by a short photoperiod in order to flower. Also, relatively few plants are induced to flower when short photoperiods are followed by long photoperiods. These plants, which require a long-short-day sequence or a short-long-day sequence to flower, will not flower if kept under continuous short or long photoperiods.

We should note that the above classification is based on whether or not a plant will flower when it is subjected to a photoperiod that exceeds or is less than a critical length. The classification does not mean to imply that all short-day plants flower under photoperiods that are shorter than photoperiods inducing flowering in long-day plants. An example to amplify this point would be to compare the short-day plant *Xanthium* with the long-day plant *Hyoscyamus*. *Xanthium* has a critical day length of 15½ hours and flowers if this critical value is not exceeded. *Hyoscyamus* has a critical day length of 11 hours and flowers when this critical value is exceeded. The significant point here is that both *Xanthium*, a short-day plant, and *Hyoscyamus*, a long-day plant, will flower if subjected to a photoperiod of 13 hours. The delimiting factor, then, is not the number of hours of light received but when a plant will flower—before or after a critical period (see Figure 21–2).

Importance of Dark Period

Plants under normal conditions are subjected to a 24-hour cycle of light and darkness. Early workers on photoperiodism also used a 24-hour cycle, thus simulating the natural condition. Workers soon found that they could obtain a more-sophisticated analysis of photoperiodism by changing the

normal cycle, for example, by following an 8-hour light period with an 8-hour dark period or following a 16-hour light period with a 16-hour dark period. Subjecting long- and short-day flowering plants to cycles other than 24 hours convincingly demonstrated that flowering in plants is more of a response to the dark period than to the light period. In other words, short-day plants flower when a certain critical dark period is exceeded, and long-day plants flower when the duration of the dark period is less than a critical value. For example, observers noted, after Garner and Allard's original work, that a plant can be kept from flowering, even though on the correct photoinductive cycle, by the interruption of its dark period with a brief light period, or light break (see Figure 21–3). Breaking up the light period with a brief period of darkness had very little effect

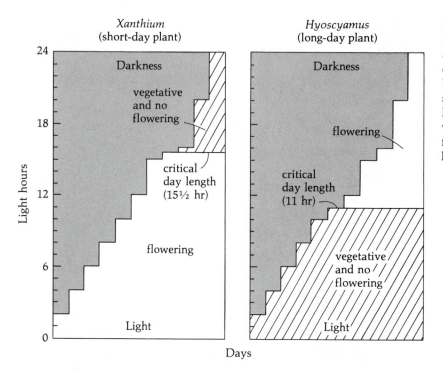

Figure 21–2. Relationship of short-day plant and long-day plant to critical day length. Both plants growing side by side at 13 hours of light per day will flower. Limiting factor is critical day length.

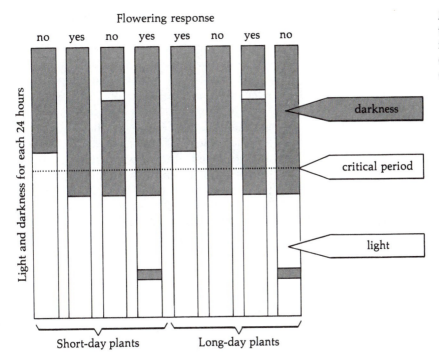

Flowering response

no   yes   no   yes   yes   no   yes   no

Light and darkness for each 24 hours

darkness

critical period

light

Short-day plants   Long-day plants

Figure 21–3. Importance of dark period for flowering in short-day plants and long-day plants.

(17). These findings showed that flowering is more of a response to the dark period than to the photoperiod.

The findings of Hamner (16) gave considerable support to the concept that the dark period is the critical part of the photoperiodic cycle. Working with Biloxi soybeans, short-day plants, he found that flowering could not be induced unless the plants received dark periods in excess of 10 hours. We now know that the length of the dark period is important for flower induction, but that the photoperiod has a quantitative effect on flowering.

Importance of Photoperiod

While the length of the dark period determines actual initiation of floral primor-

dia, the length of the light period influences the number of floral primordia initiated (16). The optimal response for the Biloxi soybean is found in a photocycle consisting of 16 hours of darkness and 11 hours of light (photoperiod) (see Figure 21–4). Photoperiods of more than 11 hours result in the differentiation of a smaller number of floral primordia.

As indicated in Figure 21–5, there is a quantitative response to the length of the photoperiod. With this in mind, we should ask whether the intensity of light has an influence on the number of floral primordia differentiated. The answer to this question is a complex one. Intensity of light could have an indirect effect, such as controlling the amount of sugars flowing to meristematic regions capable of initiating floral primordia. For example, Takimoto (45) was

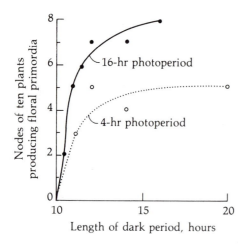

Figure 21–4. Relationship between length of dark period and initiation of floral primordia by Biloxi soybean.

Reprinted from Botanical Gazette *101:658 by K.C. Hamner by permission of The University of Chicago Press. Copyright 1940 The University of Chicago Press.*

partially successful in bringing about flowering in the dark by supplying plants with sugar solutions. In addition, the effectiveness of the photoperiod diminishes in the absence of CO_2 (49). The enhancing effect of externally supplied sugar and CO_2 certainly indicates that substrate provided by photosynthesis has some effect on the ability of the plant to produce flowers. In addition to

its indirect effect through photosynthesis, the intensity of light might be of direct importance in the synthesis of some factor or hormone necessary for floral formation.

Hamner (16) studied the quantitative effect of light duration and intensity on floral initiation by Biloxi soybean on a photoinductive cycle. He found that at light intensities below 100 footcandles, no flowers are produced. Increase in light intensity brings about an increase in the number of flowers produced (see Figure 21–6). Of the two photoperiods used in the experiment described in Figure 21–6, the longer photoperiod produces the greater number of flowers.

Photoinductive Cycles

Early workers with photoperiodism and flowering were concerned more with the number and quality of flowers obtained than the time a plant needed to be subjected to a cycle conducive to flowering in order to differentiate floral primordia. However, the number of cycles needed to induce flowering differs widely among different plant species. For example, *Xanthium pennsylvanicum* requires only one photoinductive cycle to initiate floral primordia. In contrast, *Sal-*

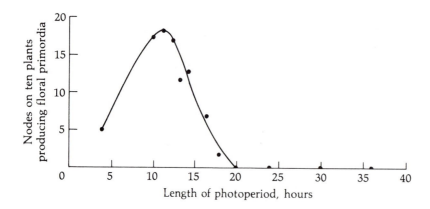

Figure 21–5. Relationship between light duration and initiation of floral primordia by Biloxi soybean. Dark period in all treatments was 16 hours.

Reprinted from Botanical Gazette *101:658 by K.C. Hamner by permission of The University of Chicago Press. Copyright 1940 The University of Chicago Press.*

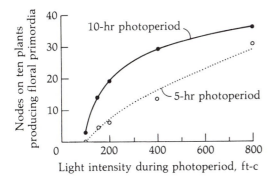

Figure 21–6. Quantitative effect of light duration and intensity on floral initiation by Biloxi soybean on a photoinductive cycle. No flowers are produced with light intensities under 100 footcandles, and more flowers are produced with longer photoperiod.

Reprinted from Botanical Gazette *101:658 by K.C. Hamner by permission of The University of Chicago Press. Copyright 1940 The University of Chicago Press.*

via occidentalis, a short-day plant, requires at least 17 photoinductive cycles to flower (49), and *Plantago lanceolata,* a long-day plant, needs to receive 25 photoinductive cycles for maximum floral response (21).

We should note that the formation of flowers by a plant is an all-or-nothing affair with respect to photoperiodism. Once a plant has received the minimum number of photoinductive cycles, it will flower even if it is returned to noninductive cycles.

Partial induction has been observed in short-day plants. The short-day plant *Impatiens balsamina,* for example, requires only three photoinductive cycles for floral bud initiation. For these buds to form flowers, however, more than eight photoinductive cycles are necessary (29).

Partial induction may also be obtained in long-day plants. The long-day plant *Plantago lanceolata* needs 25 photoinductive cycles for 100 percent inflorescence formation. If the plant is given 10 photoinductive cycles

and then subjected to a noninductive cycle, it will not flower. However, if the plant is returned to a photoinductive cycle, only 15 cycles are needed to produce 100 percent inflorescence (21). Formation of floral primordia by the aquatic plant *Lemna gibba* (duckweed) requires a minimum of 1 long day. However, at least 6 long days are required to obtain mature flowers—long days are apparently required for the early stages of flower development in this plant (11).

One implication from this discussion is that some factor involved in the flowering response is accumulated during the inductive cycle. In some plants (e.g., *Xanthium*) enough is accumulated after only one cycle to promote flowering. In other plants more than one inductive cycle is needed. In long-day plants the noninductive cycle does not appear to modify the effects of a previous exposure to an inductive cycle. The noninductive cycle in short-day plants, however, appears to be inhibitory. Schwabe (41) has shown this effect in several short-day plants by alternating inductive and noninductive cycles. The noninductive cycle inhibits the effect of the previous inductive cycle.

Perception of Photoperiodic Stimulus

A good deal of early work on photoperiodism was aimed at establishing which part of the plant receives the photoperiodic stimulus. The organs of the plant receiving the most attention were the leaves and buds.

Knott (29) demonstrated, in 1934, that in spinach, a long-day plant, the leaves are the receptors of the photoperiodic stimulus. In addition, he postulated that something is produced in the leaves in response to a photoinductive cycle and is then translocated to the apical tip, causing the initiation of floral primordia. Evidence for leaves being the organs of perception in the flowering response to photoinduction cycles is overwhelming. In many cases exposing a single leaf to photoinductive cycles while the rest of the plant is on a noninductive cycle is sufficient to cause flowering. For example, if a single leaf of *Xanthium* is exposed to short photoperiods while the rest of the plant receives long photoperiods, flowers are formed (17, 33).

Workers have shown that grafting photoinduced leaves from one plant to another plant on a noninductive cycle promotes flowering on the receptor plant (18, 35). Before grafting the photoinduced leaves, workers first defoliated the receptor plants to eliminate any influence of the noninduced leaves.

A minimum amount of leaf tissue appears necessary for flowering to occur (1, 25). The developmental stage of the leaf is also important in regard to sensitivity to photoperiodic induction. For example, partially mature *Xanthium* leaves are much less sensitive to photoperiodic induction (27).

Surprisingly enough, mature leaves also seem capable of neutralizing the flower-promoting effect of a photoperiodic stimulus. When a photoinduced leaf or branch is grafted to a plant receiving a noninductive cycle, mature leaves present on the receptor plant antagonize progress toward the flowering response. Defoliation of the receptor plant eliminates the antagonism.

Light Quality and Photoperiodism

Light has to be absorbed to be effective. Practically all of the early work on photoperiodism was concerned with the effects of white light on flowering—that is, the combined effects of all the wavelengths of the visible spectrum. However, it has become a customary practice in the investigation of photobiological reactions to find the wavelengths that are most effective, or, in other words, to develop an *action spectrum* for the process. In this manner, scientists can compare the absorption spectra of known constituents of the plant with the action spectrum of a photobiological process under investigation. If the absorption spectrum of an extracted plant constituent closely resembles the action spectrum of the process, it is a strong indication of the involvement of that constituent. The photoreceptor most likely initiates that process.

We have already seen detective work of this kind in the study of photosynthesis and auxin destruction. In photosynthesis the most effective wavelengths are found in the blue and red regions. It is in these regions that chlorophyll absorbs the most. We have also noted that the action spectrum for oat coleoptile curvature closely resembles the absorption spectrum for riboflavin. Thus riboflavin has been suspected as the photoreceptor in auxin destruction.

A group of scientists, headed by Borthwick and Hendricks (3, 5, 19), undertook this investigation into action spectra. They were interested in determining the action spectrum for the inhibitory action of light breaks during the dark period. Parker and colleagues (37) obtained the first action spectra for the control of flowering from two short-day plants, cocklebur (*Xanthium*) and Biloxi soybean. Since then, several action spectra have been measured for short- and long-day plants. All of these spectra appear essentially the same, thus suggesting a common receptor for the wavelengths of light effective in photoperiodism.

As mentioned earlier, if the long night

of a photoinductive cycle for *Xanthium* is broken by a brief flash of light (light break), the plant does not flower. An action spectrum for the effectiveness of different wavelengths of light demonstrates that the most efficient wavelengths for inhibition of flowering are found between 620 and 660 nm (orange-red) with a maximum at about 640 nm. Therefore, red light is considered to be the most efficient radiation in light-break reactions.

Far-red radiation when used alone has no effect as a light-break factor—that is, it does not break up a long night into two short nights. However, the startling discovery was made, first by Borthwick and colleagues (4) and then by Downs (14), that far-red radiation is capable of reversing the light-break effect of red light. If a brief flash of far-red radiation follows in sequence a brief flash of red light in the middle of a long night of a photoinductive cycle for short-day plants, flowering will occur. If the far-red radiation is followed further in sequence by red light, flowering will again be inhibited. In other words, the radiation used last in the sequence will determine the response of the plant (see Table 21–2).

Light Quality and Seed Germination

Let us look briefly at some early work on the effect of light on lettuce seed germination, which significantly increased our understanding of photoperiodism and the photoreceptor. Experiments by Borthwick, Hendricks, and their colleagues (6) revealed that a pigment system was involved in the germination of some seeds (lettuce, *Lactuca sativa*, variety Grand Rapids, and *Lepidium virginicum*). In their studies, these workers allowed the seeds to imbibe water in darkness, then exposed them to different wavelengths of light. They then placed the seeds back in darkness and later evaluated them for the amount of germination.

Although much of the early work on photoperiodism involved the use of white light or combined effects of all the wavelengths of the visible spectrum, Borthwick and Hendricks employed techniques to find the most effective wavelengths—that is, they made use of an action spectrum. In this manner they obtained data indicating that when the light-sensitive seeds were exposed to red light (660 nm), germination re-

Treatment	Mean Stage of Floral Development in Cocklebur	Mean Number of Flowering Nodes in Biloxi Soybean
Dark control	6.0	4.0
R	0.0	0.0
R, FR	5.6	1.6
R, FR, R	0.0	0.0
R, FR, R, FR	4.2	1.0
R, FR, R, FR, R	0.0	—
R, FR, R, FR, R, FR	2.4	0.6
R, FR, R, FR, R, FR, R	0.0	0.0
R, FR, R, FR, R, FR, R, FR	0.6	0.0

Table 21-2. Effect of daily interruptions of the dark period with several consecutive irradiations with red (R) and far-red (FR), in sequence, on flower initiation of cocklebur and soybeans.

Source: *From R. J. Downs. 1956. Photoreversibility of flower initiation.* Plant Physiol. 31:279.

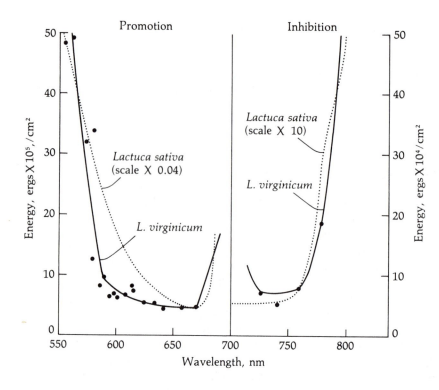

Figure 21–7. Action spectra for promotion and inhibition of germination of seeds of *Lepidium virginicum* and *Lactuca sativa* to 50 percent.

From E.H. Toole et al., eds. 1959. "Photoperiodism and Related Phenomena in Plants and Animals." American Association for the Advancement of Science 55:89. Westview Press. Copyright 1959 by the American Association for the Advancement of Science.

sulted. They also found that far-red light (730 nm), given immediately after a red-light exposure, inhibited germination (see Figure 21–7). If they again treated the seeds with red light, germination was promoted. These results indicated to Borthwick, Hendricks, and their colleagues that a reversible pigment system is operative in lettuce seeds after the seeds imbibe water. The last light treatment determines the response of the seeds (see Table 21–3).

Reversible effect. The fact that promotion and inhibition of germination by red and far-red irradiation is reversible was first discovered by Borthwick and colleagues (6). As we will discuss later, the red/far-red system active in lettuce seeds is similar and in all probability is identical to the red/far-red phytochrome system active in the flowering of

some plants, expansion of bean leaf disks, etiolation, and the unfolding of the plumular arch of bean seedlings.

Time factor. In order for good reversal of red-light promotion to occur, far-red irradiation should follow immediately. If far-red irradiation is delayed, inhibition of germination is less marked. Toole and colleagues (47) found that lettuce seeds fail to respond to far-red irradiation 12 hours after exposure to red light. At that time the processes leading to germination have, most likely, reached such an advanced state that reversal is impossible.

Phytochrome

Garner and Allard's early work led to the discovery, isolation, and much of the char-

Table 21–3. Promotion and inhibition of germination by red (R) and far-red (FR) irradiation. Seeds were irradiated at 26°C and then allowed to germinate at 20°C. Note reversible effect of one treatment on the other.

Irradiation	Germination at 20°C (percent)
R	70
R, FR	6
R, FR, R	74
R, FR, R, FR	6
R, FR, R, FR, R	76
R, FR, R, FR, R, FR	7
R, FR, R, FR, R, FR, R	81
R, FR, R, FR, R, FR, R, FR	7

Source: *Reprinted from "Action of Light on Lettuce-Seed Germination" by H.A. Borthwick, S.B. Hendricks, E.H. Toole, and V.K. Toole, Botanical Gazette 115:102 by permission of The University of Chicago Press. Copyright 1954 The University of Chicago Press.*

acterization of the pigment responsible for absorbing light involved in photoperiodic phenomena of plants. Borthwick, Hendricks, and their colleagues later termed this pigment *phytochrome*. There is now general agreement that phytochrome is a pigment involved in the perception of photoperiodic stimuli controlling flowering, lettuce seed germination, and other morphogenetic phenomena. The evidence, as indicated in absorption spectra (see Figure 21–8), shows phytochrome as existing in two forms: the phytochrome red absorbing form (Pr) and the phytochrome far-red absorbing form (Pfr). Scientists consider the Pfr form to be the physiologically active form. The two forms are photochemically interconvertible. In addition, scientists have found that the Pfr form will slowly convert to the Pr form

in darkness or will decay to an unknown inactive compound (2). The dark conversion of Pfr to the Pr form appears to be restricted to the dicotyledons (26).

Flash photolysis and low-temperature techniques have uncovered a number of short-lived forms, or intermediates, between Pr and Pfr. This discovery suggests, of course, that when one form of the pigment converts to the other form, the conversion occurs in a stepwise fashion through a number of transient intermediates. Also, there is greater quantum efficiency in the conversion of Pr to Pfr, which probably is the reason for the observed higher levels of Pfr under natural photoperiodic conditions. The Pfr form of phytochrome is unstable and undergoes a process called *decay*. The term, as used here, refers only to a loss of photoreversibility and not to actual destruction (26). Nevertheless, under these conditions the pigment cannot be detected, and there is a decrease in total phytochrome as measured by differential two-wave spectrophotometry. Thus if we subject a plant to continuous red light, the level of phytochrome in the plant decreases below a criti-

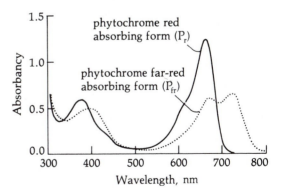

Figure 21–8. Absorption spectra of purified solution of oat phytochrome.

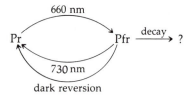

cal value and triggers the de novo synthesis of more phytochrome in the Pr form (38, 39, 40). The result is an equilibrium in the plant between the synthesis and the destruction of phytochrome.

Chemistry

The clarification of the chemical structure of phytochrome was due to the isolation efforts and purification of phytochrome from several plant sources by Borthwick,

Hendricks, and their colleagues. Phytochrome was initially isolated from cotyledons of etiolated turnip seedlings (8). Siegelman and Firer (43) were responsible for a highly purified extract that led to further purifications and analysis of the phytochrome structure.

Phytochrome is a protein with a chromophore (pigment-colored portion) prosthetic group (e.g., chromoprotein) that resembles, in basic structure, the open-chain tetrapyrrole chromophore of the algal pigment 6-phycocyanin (see Figure 21–9). Further, there is probably one chromophore for each phytochrome molecule. The chromophore is linked to the protein at ring III. Apparently, the photoconversions of the Pr and Pfr forms involve electronic changes in ring I, with either the addition or loss of a proton. Conformational (structural)

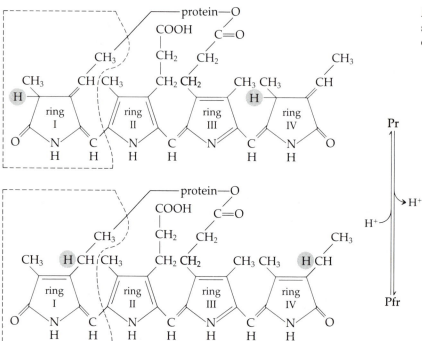

Figure 21–9. Proposed structure of phytochrome chromophore.

changes in the protein probably contribute to dark conversion and possibly decay. We do not know the exact molecular weight of the protein. However, the monomer molecular weight is believed to be approximately 120,000.

Conversions and Responses

We should reemphasize that the Pfr form of phytochrome is the active form. The Pr form is not considered to be active. Consequently, when light-sensitive lettuce seeds that are imbibing are treated with red light or white light, the Pfr form accumulates and is translated chemically into a germination response. However, if the red-light treatment is followed by far-red light, the Pfr form is converted to the inactive Pr form, and germination does not take place.

Upon exposure to white light during the day, the Pfr form of phytochrome may accumulate above a critical level and favor flowering in a long-day flowering plant, but it will not promote flowering above a critical level in a short-day flowering plant. The white light under normal environmental conditions has a net red-light effect even though far-red light wavelengths are present. The main reason for the Pfr accumulation in white light, however, is the greater efficiency (quantum efficiency) of phytochrome conversion to the Pfr form over that of light conversion from Pfr to the Pr form.

The dark period is significant in that it provides time for dark conversion of the Pfr form to the Pr form. Below a critical level of Pfr (or the ratio of Pfr to Pr), a long-day plant will remain vegetative. On the other hand, Pfr below a critical level (again a ratio of one form to the other may be important) will promote flowering in a short-day plant.

We should keep in mind that Pfr is required for flowering in both long- and short-day plants, but that some internal conditions, which are manifested somehow through a critical dark period, are responsive to different levels of Pfr.

The interruption of the dark period with red light will cause the accumulating Pr to return to the Pfr form, thus inhibiting flowering in a short-day plant. If the red light is followed by a far-red break, the red-light influence is erased.

Distribution

Evidence from numerous studies indicates that phytochrome is present in a wide variety of plants (7, 22, 42) and, indeed, may occur universally among green plants. In addition to having been isolated from such higher plants as tobacco, oat, corn, and bean, phytochrome has been isolated from the alga *Mestaenium* and the liverwort *Sphaerocarpos* (46). Not only is phytochrome widely distributed in the plant world, it is also widely distributed within the plant itself. Phytochrome has been detected in roots, stems, hypocotyls, cotyledons, coleoptiles, leaf blades, petioles, vegetative buds, developing fruits, floral receptacles, and inflorescences (22).

Scientists believe that the primary intracellular action of phytochrome is one of altering membrane permeability. Also, Borthwick and Hendricks suggested that the rapid phytochrome responses indicate that phytochrome is associated with the membranes. Phytochrome is thought to be associated with etioplasts, chloroplast membranes, and the plasmalemma. In fact, some scientists suggest that the action of Pfr is dependent on its association with and its alteration of membranes soon after Pfr is formed. Some evidence for the Pfr association with membranes exists, but its role at this level requires considerable experimentation.

Phytochrome and Endogenous Circadian Rhythms

Although in photoperiodic phenomena relative levels of phytochrome are important as an indicator of the dark period, numerous experiments indicate that phytochrome interconversions represent only part of the time-measuring mechanism in plants. One feasible theory is that phytochrome (specifically Pfr levels), indicating somehow when the dark periods occur, interacts with internal rhythmic or cyclic plant processes. These rhythmic processes, or times of cyclic cellular receptivity to Pfr, reflect the operation of a *biological clock*.

Many processes, such as the rise and fall of metabolic components, cell division, phototaxis, bioluminescence, stomatal opening and closing, and leaf movements and growth, occur with a rhythmicity that reflects an organism's adaptations to the external environment. These many endogenous rhythms are often based on a cycle of approximately 24 to 26 hours and are, therefore, called endogenous, *circadian* (about a day) *rhythms*. We can easily observe the operation of many circadian rhythms that are governed by the biological clock. For example, some plants that are taken out of their natural environmental cycles of light and temperature will continue to exhibit chemical and morphogenetic responses in a typical daily cycle, even when factors of the environment (e.g., light, temperature) are held constant. However, when plants are exposed to new fluctuating patterns of a given stimulus (light), the biological clock may be reset or conditioned with time to conform to the new pattern. This process, called *entrainment*, has obvious selective advantages in nature and indicates flexibility of the time mechanism. For instance, the phytochrome system, namely Pfr, will have an impact on the biological clock system at times when the clock is particularly sensitive or can be easily reset or modulated. As pointed out by Zeevaart (51), the effect of Pfr levels in both short-day plants and long-day plants influences the timing mechanism during the dark period.

From experimentation on *Chenopodium* and *Pharbitis* (23, 51), scientists discovered that relatively high Pfr levels are required, particularly during the photoperiod and the first part of the dark period, for a short-day plant to flower. At the last part of the dark period, Pfr must be low or absent. Conversely, long-day plants seem to require the relatively continuous presence of Pfr during the dark period and particularly higher levels during the later part of the dark period for good flowering.

We should understand from the foregoing discussion that the control of flowering in short-day and long-day plants is not due simply to the interconversion of the phytochrome pigment. Rather, both long-day and short-day plants have an endogenous rhythm of receptivity to Pfr levels, which will regulate the reception and translation of light signals from the environment into other biochemical signals, such as the synthesis of flowering hormones. We cannot cover here the vast experimental literature that deals with the complex and as yet unexplained operations of the biological clock. Therefore, the reader should consult the many excellent articles, particularly by Hillman, on this subject (13, 23, 34, 44).

Flowering Hormones and Gibberellins

When we look at the early work, we are led to believe that a flowering factor (or factors) is produced in photoinduced leaves and is transported with relative ease to the buds. In 1936 Cajlachjan, working on floral initiation, coined the term *florigen* for the unidentified hypothetical flowering hormone thought to be present in photoinduced leaves and plants.

A dramatic demonstration in support of the existence of florigen and proof of its ease of translocation comes from the earlier observations of two-branched *Xanthium* plants grafted in series (36). If the end branch of a series of plants is given a photo-inductive cycle, it will cause flowering in all plants in a chainlike reaction (Figure 21–10).

Equally dramatic were the experiments by Zeevaart (50), in which he grafted long-day plants to short-day plants and vice versa. When the long-day plant *Sedum spectabile* was grafted to the short-day plant *Kalanchoë blossfeldiana*, it flowered under short-day conditions. When the latter plant was grafted to the long-day plant, it flowered under long-day conditions. Also, experiments by Hodson and Hamner (24) demonstrated that extracts of flowering *Xanthium* could initiate flowering in *Lemna* (duckweed), but that extracts of vegetative *Xanthium* could not. In other words, these experiments demonstrated that florigen is not species-specific and that it has the same

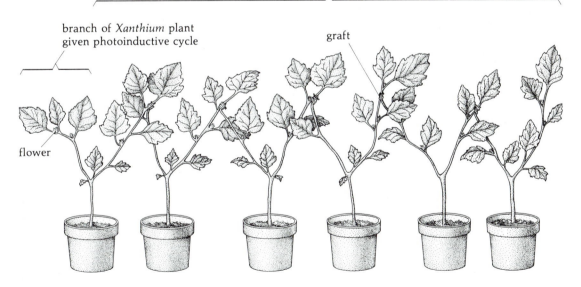

branches of *Xanthium* plants
kept on nonphotoinductive cycle

branch of *Xanthium* plant
given photoinductive cycle

graft

flower

Figure 21–10. Translocation of floral hormone. One branch of two-branched *Xanthium* plant, grafted in series to five other *Xanthium* plants, was given photoinductive cycle. Second branch of first plant and of other five plants were kept on nonphotoinductive cycle. All plants flowered.

or nearly the same properties in both long-day and short-day plants. Thus it seems reasonable to believe that florigen might someday be isolated and characterized. The economic importance of such a successful endeavor is immense.

From early and subsequent experimentation, the idea emerged that phytochrome, as the photoreactor, mediates the production of florigen in leaves, which in turn translocates to and activates the conversion of vegetative meristems to floral meristems. As a result, by the late 1950s and early 1960s, several research groups actively attempted to isolate and identify florigen. Although all attempts to date have been unsuccessful, there is an indication that florigen might be an isoprenoid or a steroidlike compound. The work leading to these suggestions, however, has not been continued. Similarly, extracts from *Lemna* (31) and *Xanthium* (32) showing florigenic activity have not yielded a definitive compound.

So far in our discussion of photoperiodism and flowering, we have ignored the role of gibberellins in flowering. As mentioned in a previous chapter, the application of gibberellin to most long-day plants will cause them to flower when they are on a noninductive cycle. However, gibberellin is not assumed to be a floral hormone, or at least it does not cause flowering directly. Two lines of evidence support this assumption. The stimulation of flowering by long-day induction and gibberellin stimulation of long-day plants appear to be different. First, in long-day induction the differentiation of floral primordia occurs simultaneously with stem elongation (12). In the gibberellin induction of flowering, the flower-bearing stem elongation, known as *bolting*, proceeds for some time before we can observe floral primordia (see 34), thus suggesting that gibberellin-stimulated growth and differentia-

tion fulfill the requirements of floral differentiation and development. Cleland and Zeevaart (12) have provided evidence that supports the idea that bolting and flowering are separate but somewhat sequential. Using Amo-1618 as an inhibitor of GA synthesis, they found that although bolting was inhibited in the photoinduced long-day plant *Silene armeria,* flowering was not. Thus the linkage of the two processes are not necessary for flowering. Also, gibberellins do not promote flowering in short-day plants on noninductive cycles.

Cajlachjan (10) made actual measurements of the gibberellin levels in leaves of both short-day and long-day plants under photoinduction cycles. The results indicate that the gibberellin content is higher under long-day conditions, regardless of the class of plant used.

Cajlachjan presented a hypothesis associating gibberellin with the floral hormone in the photoperiodic response of flowering (9). He suggested that there are two steps involved in the flowering process, the first mediated by gibberellin and the second by one or more flowering factors called *anthesins.* Together, gibberellin and anthesins constitute the true florigen. According to his ideas, long-day plants on noninductive cycles have a sufficient amount of anthesin but not enough gibberellin. This situation is reversed in short-day plants on noninductive cycles—gibberellin is high and anthesin is low. This theory would account for the promotion of flowering when gibberellins are applied to long-day plants on noninductive cycles. In addition, it accounts for the neutral effects when gibberellins are applied to short-day plants on noninductive cycles. This theory, however, is still in the realm of speculation until additional experimental evidence becomes available.

Questions

21–1. Define the following terms: photoreceptor, photoperiodism, and photoperiodic response.

21–2. Describe the major contributions to our knowledge of light-controlled processes provided by the studies of Garner and Allard.

21–3. The statement has often been made that long-day flowering plants should be more correctly called short-night flowering plants. Why?

21–4. What is the importance of the dark period on flowering? Is the photoperiod important?

21–5. Describe some of the early evidence that led to the detection and isolation of phytochrome.

21–6. Describe the action of light and darkness on lettuce seed germination. What is the role of phytochrome and which form is physiologically active? What is responsible for the decrease in the Pfr form during exposure to darkness?

21–7. Why does continuous red light lead to a decrease in the level of phytochrome in etiolated plants?

21–8. Discuss the possible roles of the tetrapyrrole chromophore and protein of the phytochrome molecule.

21–9. What is the relationship of the critical length in plants and the state of the phytochrome pigment?

21–10. Describe several morphogenetic responses that are under the control of phytochrome. Can one primary action of phytochrome influence the varied responses attributed to the pigment?

21–11. Define the following terms: circadian rhythm, entrainment, and biological clock.

21–12. Discuss the ideas that have been provided to explain the regulation of flowering by phytochrome, florigen, and gibberellic acid.

21–13. List the pigments in plants that are important to growth and survival. What are the real or suspected functions of each pigment or groups of pigments?

21–14. The Pfr form of phytochrome has a half-life of slightly less than two hours. How is this fact important to the action of phytochrome as compared with the short half-life of other pigments?

Suggested Readings

Black, M. 1969. Light controlled germination of seeds. *Symp. Soc. Exp. Biol.* 23:193–217.

Borthwick, H.A. 1972. History of phytochrome. In K. Mitrakos and W. Shropshire, Jr., eds., *Phytochrome*. New York: Academic Press.

DeGreef, J., ed. 1980. *Photoreceptors and Plant Development*. Antwerp: Antwerpen Univ. Press.

Feldman, J.F. 1982. Genetic approaches to circadian clocks. *Ann. Rev. Plant Physiol.* 33:583–608.

Holmes, M.G., and H. Smith. 1975. The function of phytochrome in plants growing in the natural environment. *Nature* 254:512–514.

Holmes, M.G., and H. Smith. 1977. The function of phytochrome in the natural environment. IV. Light quality and plant development. *Photochem. Photobiol.* 25:551–557.

Holmes, M.G., and E. Wagner. 1980. A re-evaluation of phytochrome involvement in time measurements in plants. *J. Theor. Biol.* 83:255–265.

Marmé, D. 1977. Phytochromes: membranes as possible sites of primary action. *Ann. Rev. Plant Physiol.* 28:173–198.

Moore, T.C. 1979. *Biochemistry and Physiology of Plant Hormones*. New York: Springer-Verlag.

Pratt, L.H. 1982. Phytochrome: the protein moiety. *Ann. Rev. Plant Physiol.* 33:557–582.

Schopfer, P. 1977. Phytochrome control of enzymes. *Ann. Rev. Plant Physiol.* 28:223–252.

Smith, H. 1982. Light quality, photoperception, and plant strategy. *Ann. Rev. Plant Physiol.* 33:481–518.

Chapter 22

Vernalization and Cold Tolerance

Cold-tolerant sweet Williams (*Dianthus barbatus*) flowering in field during late fall. Cold-intolerant dwarf Queen Aster (*Callastephus*) at right. *Courtesy of Chiko Haramaki, The Pennsylvania State University.*

Not all plants will flower when subjected to the correct photoperiod. In many plants, temperature has a profound influence on the initiation and development of reproductive structures. In the annual plant, growth is started in the spring, flowers are developed in the summer, and fruit and seed are produced in the fall. The influence of temperature on flowering in the annual plant is secondary to that of light, the effect of temperature being more metabolic than catalytic.

Biennials, on the other hand, present an entirely different situation. They remain vegetative the first growing season and, after prolonged exposure to the cold temperatures of winter, flower during the following season. Without exposure to a cold treatment, the majority of these plants would remain vegetative indefinitely. However, with prolonged exposure to low temperature, followed by the correct photoperiod, cold-requiring plants will flower. That a cold period is necessary was proven unequivocally when it was shown that for most biennials an "artificial" cold treatment followed by the correct photoperiod and temperature will result in flowering during the first growing season. Thus a biennial may be made to flower in the same period of time required for flowering in annuals. Chouard (4, 5, 6) defined the term used to describe this phenomenon, as the "acquisition or acceleration of the ability to flower by a chilling treatment." However, application of the concept of vernalization without the formulation of a hypothesis has been in practice for many years (44). Growers, some instinctively and some knowingly, have recognized the fact that some plants need a cold period in order to flower. In 1940 McKinney (44), in his review of the subject of vernalization, pointed out a report by Klippart in 1857 to the Ohio State Board of Agriculture that shows a remarkable application of the concepts of vernalization. This report, as quoted by McKinney, is as follows:

> To convert winter into spring wheat nothing more is necessary than that the winter wheat should be allowed to germinate slightly in the fall or winter, but kept from vegetation by a low temperature or freezing until it can be sown in the spring. This is usually done by soaking and sprouting the seed and freezing it while in this state and keeping it frozen until the season for spring sowing has arrived. Only two things seem requisite, germination and freezing. It is probable that winter wheat sown in the fall so late as only to germinate in the earth without coming up would produce a grain which would be a spring wheat if sown in April instead of September. The experiment of coverting winter wheat has met with great success. It retains many of its primitive winter wheat qualities and produces at the rate of 28 bushels per acre.

Since Klippart's report, numerous workers have pursued systematic study of the influence of temperature on flowering. In 1939 Melchers (48) coined the term *vernalin* for the hypothetical active factor that was thought to accumulate during vernalization. Later, after the gibberellins had become known as prominent phytohormones, they were implicated in the vernalization process. For reasons we shall see later, some scientists considered GA and vernalin to be one and the same.

Vernalization and Flowering

We should emphasize that vernalization per se does not induce flowering but merely prepares the plant for flowering. The effect of vernalization can be contrasted with that of photoperiod on flowering. The photoperiodic inductive cycle not only prepares the plant to flower but also initiates flowering. The classic experiments concerned with vernalization have been performed on *Hyoscyamus niger* (henbane) and *Secale cereale* (Petkus rye). Therefore, we will center our discussion around the study of these two plants.

Hyoscyamus niger (Henbane)

Often the response to cold treatments is genetically controlled. This situation exists with henbane, which produces an annual and a biennial type. The annual type flowers in one growing season, the biennial type requires a cold winter before flowering in its second growing season. Apparently, the mechanism necessary to initiate the

chemical changes required for flowering is nonoperative in the biennial henbane and may be substituted for by a cold treatment. Like the annual, the biennial henbane is a long-day plant; the vegetative state is maintained under short-day conditions, regardless of the temperature treatment it receives.

The biennial henbane exhibits a qualitative response to cold treatments—that is, unless it is exposed to low temperatures for a certain period of time, it will remain entirely vegetative. However, after the plant has reached the rosette stage and is at least 10 days old, it may be vernalized and then flower in one growing season, provided it receives the correct photoperiod. An age of 10 days and the rosette stage appear to be prerequisites for a response to cold treatment by henbane (5). Figure 22–1 shows the flowering response of a typical biennial henbane to a cold treatment. We may also ob-

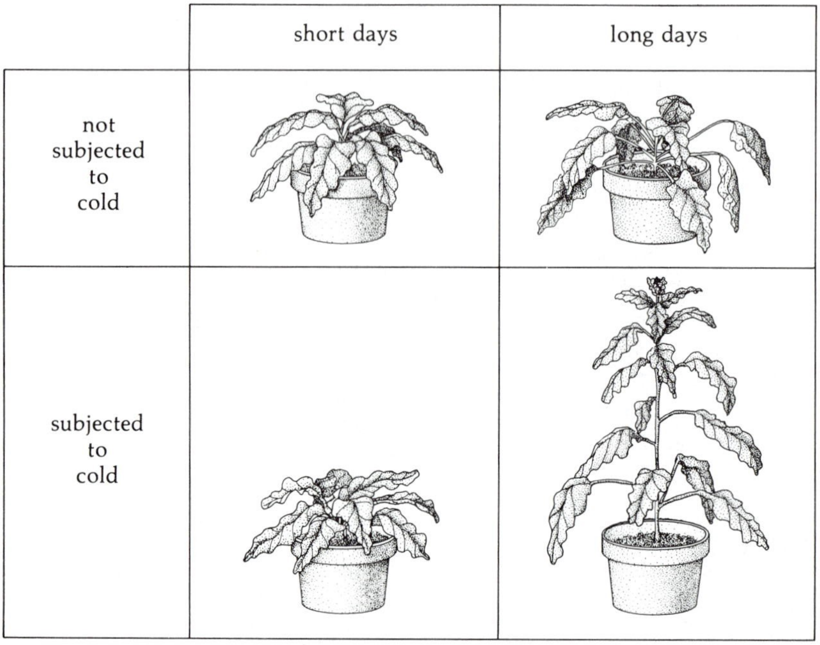

	short days	long days
not subjected to cold		
subjected to cold		

Figure 22–1. Response of henbane, long-day plant, to different temperature and photoperiod treatments.
From data of G. Melchers and A. Lang. 1948. Biol. Zentr. 67:105. Redrawn from Principles of Plant Physiology *by J. Bonner and A.W. Galston. W.H. Freeman and Company. Copyright © 1952.*

serve in Figure 22–1 the henbane's need for the correct photoperiod.

Secale cereale (Petkus Winter Rye)

As with henbane, there is also a spring and winter strain of Petkus rye. The spring strain is a typical annual rosette plant, flowering and fruiting in one growing season. The winter strain is a typical biennial rosette plant, staying vegetative the first growing season and then flowering and fruiting after a prolonged exposure to the cold temperatures of winter. The winter strain, when vernalized, resembles the spring strain in every way (51).

Although Petkus winter rye and henbane are cold-requiring plants, they differ in many ways in their response to cold treatments. Petkus winter rye may be vernalized in the seed stage (54); henbane has to be at least 10 days old and in the rosette stage. Unlike henbane, Petkus winter rye does not have an obligate requirement for vernalization. Under continuous light, unvernalized Petkus winter rye will "head" in 15 weeks. However, if it is vernalized, heading occurs in about $7\frac{1}{2}$ weeks—about the same time heading occurs in the spring valley variety under continuous light. Thus in Petkus winter rye, vernalization serves to shorten the time it takes the plant to flower and is not an absolute requirement (15). Finally, Petkus winter rye differs from henbane in that the vernalization stimulus, once received, is not transmitted across a graft union.

Research scientists develop theoretical schemes to provide a sense of direction and research approach. Purvis (51), a plant scientist who contributed greatly to our understanding of vernalization, developed one such scheme to describe flowering in cereal plants. Although the scheme was developed approximately twenty years ago, it still provides a working hypothesis and an example of the thinking that is required of research scientists when they attempt to elucidate the mechanics of natural phenomena. The scheme for flowering in cereal plants is as follows:

$$A \xrightarrow{\text{cold}} B \underset{}{\overset{\text{short-day}}{\rightleftarrows}} C \xrightarrow{\text{long-day}} D$$

$$B \xrightarrow{\text{day-neutral}} E$$

In this scheme, B is some compound that is part of a reaction system leading to flowering. This reaction system from B to D is under photoperiodic control, and possibly leads to the synthesis of a floral hormone. In spring rye, B is either present in the embryo or is produced from A at normal temperatures. However, in winter rye the production of B is retarded although not completely inhibited. It accumulates at a slow rate with the growth of the plant. Exposure to low temperatures accelerates the production of B in winter rye.

Purvis gives two reasons why she believes B is accumulated even under normal temperatures. First, flowering occurs eventually under continuous light even in the absence of a cold treatment. Second, even in those species showing an obligate requirement for vernalization (e.g., henbane), once vernalized will remain as such even though the plant is subjected to a noninductive photocycle. That is, the presence of B persists until the plant is returned to an inductive photocycle and is not diluted by the vegetative growth that takes place during the time the plant is exposed to a noninductive cycle. The persistence of B has been shown in rye by Purvis (51) and in henbane by Lang and Melchers (39). In fact, it has been suggested that B, once produced by vernalization, increases without further aid from low temperatures.

The reaction from B to C to D is under photoperiodic control. The reaction from B to E (leaf-forming substance) is day-neutral and occurs at optimal rates when the reaction from B to C is blocked or inhibited. In the scheme by Purvis, D represents the flowering hormone, and C is an intermediate capable of initiating early stages in flower initiation. In spring rye or vernalized winter rye, there is a high accumulation of B. Under continuous light, B is only slowly converted to C, which in turn is rapidly converted to D, the flowering hormone. The continual drain of C to form D keeps the reaction B to C to D going, despite the unfavorable presence of continuous light on the reaction B to C. Eventually, D reaches a critical level and flowering ensues.

Under short-day conditions, the reaction C to D is inhibited, thus forcing the back reaction, C to B to E, to occur, keeping the plant in a vegetative state. This state will persist until the inhibited reaction C to D finally produces the critical amount of D needed for floral initiation. An analysis of this scheme will show why Petkus spring rye is a long-day plant and why the vernalized winter strain resembles it.

Some of the more important aspects of the study of vernalization in Petkus rye, henbane, and related plants are site of vernalization, dependence on temperature and duration of exposure, transmission of vernalization by grafting experiments, age factor, devernalization, and substitution of gibberellin for the cold treatment. We will discuss these aspects in greater detail.

Site of Vernalization

Experiments with many different cold-requiring plants, including henbane, have strongly suggested that the site of vernalization is the growing point. This has been shown with localized low temperature treatments of different plant parts in celery (8), beets (7), and chrysanthemum (57). Melchers, as a result of grafting experiments with annual and biennial races of *Hyoscyamus niger* concluded that the stem apex is the plant part that responds initially to cold treatment (46, 47). Apparently, the stem tip is the site of perception of vernalization, and the stimulus is translocated to other parts of the plant. Schwabe (57) found that in chrysanthemum, keeping the apex warm and chilling the rest of the plant had negligible results with respect to flowering. In addition, Purvis (51, 52) showed that the dissected apices from imbibed embryos provided with sucrose and minerals can be vernalized.

That the growing tip is the only locus of perception of vernalization has been chal-

Table 22-1. Flowering percentages among regenerated leaf cuttings of *Lunaria biennis*, taken from mother plants of five ages after cold treatments during five periods.

Age of Mother Plants (weeks)	Cold Treatment (weeks)				
	0	8	12	16	20
6	0	0	0	0	3.6
8	0	0	0	0	21.4
10	0	0	0	7.1	25.0
12	0	0	12.5	40.7	40.6
14	0	0	7.5	18.4	40.0

Source: *From S.J. Wellensiek. 1964. Dividing cells as the prerequisite for vernalization.* Plant Physiol. *39:832.*

lenged by Wellensiek. He demonstrated that both isolated leaves and isolated roots of *Lunaria biennis* are capable of being vernalized (67, 68). If these isolated plants receive a cold treatment, the plants regenerating from the isolated plant parts flower. Wellensiek concluded from his experiments that dividing cells are necessary for perception of vernalization, no matter what their location in the plant. Data from more recent work by Wellensiek (69) on vernalization of isolated leaves are given in Table 22–1. We should note that the duration of cold treatment and age of the leaf are important factors in the flowering response.

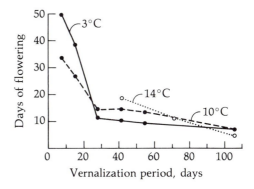

Figure 22–2. Interrelationship between temperature and time of exposure in acceleration of flowering of *Hyoscyamus niger*.
From A. Lang. 1951. Der Zuchter 21:241.

Dependence on Temperature and Duration of Exposure

Lang's work with henbane (36) illustrates the relationship between temperature and time of exposure and the influence of this relationship on the efficiency of vernalization. He exposed the cold-requiring henbane to different temperatures, ranging from 3° to 17°C for varying periods of time. The plant was then given a photoinductive cycle at 23°C until flower initiation occurred. The efficiency of the vernalization treatment was determined by the number of days to flower after treatment.

Lang found that all temperatures from 3° to 17°C are effective if the period of vernalization is 105 days. Flower initiation was observed in 8 days. However, if the vernalization period is shortened to 15 days, a separation in the effectiveness of different temperatures is observed. Under these circumstances, a temperature of 10°C during the 15-day vernalization period is the most efficient treatment, requiring 23 days to initiate flowering. If the vernalization period is

extended to 42 days, the most effective temperatures are found in the range from 3° to 6°C, thus requiring 10 days for flower initiation. Figure 22–2 shows these relationships.

Hänsel (20) studied the vernalizing effect of a wide range of temperature, including temperatures below freezing, on Petkus winter rye. He found that vernalization fails at temperatures below −4°C, but from this temperature up to 14°C, vernalization is observed. Temperatures from 1° to 7°C are equally efficient in shortening the number of days to flowering. There is a rapid fall in the rate of vernalization when temperatures are increased from 7° to 15°C. Figure 22–3 shows these relationships.

From this discussion and from Figures 22–2 and 22–3, we can clearly see that the flowering response to vernalization is dependent on the temperature used and the duration of the vernalization period. The most efficient combination of temperature and exposure time for maximum response has to be determined for each species of plant.

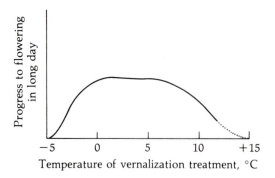

Figure 22–3. Effect of temperature on vernalization of Petkus winter rye.

From H. Hänsel. 1953. Ann. Bot. 17:417.

Grafting Experiments

In *Hyoscyamus*, Melchers (46, 47) demonstrated remarkably well the transmission of a vernalization stimulus across a graft union. If a plant part (leaf or stem) of a vernalized henbane is grafted to an unvernalized henbane, the latter part will flower. The question arises whether this is a transmission of florigen from the donor to the receptor or a transmission of a substance produced as a result of vernalization. However, florigen has been ruled out as a result of additional experiments by Melchers and Lang, which have been reviewed by Lang (37). If an unvernalized henbane plant is grafted to a Maryland Mammoth tobacco plant, the henbane plant will flower, whether or not the tobacco plant receives a photoinductive cycle. Henbane, as the receptor in this experiment, receives a stimulus from the tobacco plant, which leads to flowering. This stimulus cannot be florigen since it is transmitted from tobacco plants on noninductive cycles as well as inductive cycles. Since the tobacco plant is not a cold-requiring plant, the stimulus or substance (vernalin) produced by vernalization should

be present in the absence of cold treatments.

These experiments by Melchers and Lang do provide some evidence for the existence of vernalin. However, the examples of vernalization induction from donor to receptor are few in number. In addition, vernalin has not been extracted as yet even in crude form. Therefore, evidence for the existence of vernalin, at least in a mobile form, rests on relatively few experiments.

Age Factor

A noticeable aspect of the phenomenon of vernalization is the relationship between the age of a plant and its response to low-temperature treatments. The age at which a plant is sensitive to vernalization is quite different in different species. For example, in cereals, low-temperature treatments effectively vernalize the germinating seed and may even vernalize the embryos developing in the mother plant (38, 54). Shinohara (60) has reported the partial vernalization of the ripening seeds of garden peas, winter wheat, barley, broad beans, and Minowase radish.

In contrast to these plants, many cold-requiring plants need a certain period of growth before they become sensitive to low-temperature treatments. For example, the biennial strain of *Hyoscyamus niger* must be in the rosette stage and have completed at least 10 days of growth before sensitivity to vernalization may be observed. In fact, Sarkar (56) pointed out that maximum sensitivity is not achieved until *Hyoscyamus* has completed 30 days of growth. In other plants, sensitivity to vernalization depends on the number of leaves produced. For example, in *Oenothera* at least six to eight leaves must be present for vernalization to be effective (4), and in Brussels sprouts, 30 leaves (66).

The term *ripeness-to-flower*, first introduced by Klebs (30) in 1913 and later used to denote the time when a plant is sensitive to photoperiod, may also be used in the study of vernalization. In cold-requiring plants, the ripeness-to-flower stage is reached when the plant has fulfilled its cold requirement. The extent of vegetative growth, such as a minimal number of leaves or nodes, is used many times as a measure to determine whether or not a plant has reached the ripeness-to-flower stage.

The need for a certain amount of vegetative growth to take place suggests that the accumulation of some factor (perhaps a receptor of the vernalization stimulus) is necessary for sensitivity to be achieved. The fact that in many plants a minimal number of leaves must be present supports this concept, since the syntheses of most of the compounds found in a plant have their origin in the photosynthetic process. In plants whose seeds may be vernalized (e.g., cereals), our hypothetical substance must already be present in sufficient amounts, either donated by the mother plant or synthesized during the development of the embryo.

Study of the sensitivity to vernalization of *Arabidopsis thaliana* at different stages of growth has produced some very interesting results (20). The seed of *A. thaliana* is very sensitive to vernalization. This sensitivity decreases with the development of the seedling until a relatively low sensitivity point is reached in the second week of development. With further growth of the plant, there is a marked change in sensitivity to low-temperature treatments. The sensitivity of the plant now increases with age. Figure 22–4 shows these relationships.

We may interpret the loss of sensitivity

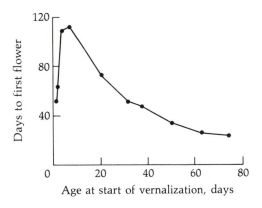

Figure 22–4. Sensitivity to vernalization of *Arabidopsis thaliana* at different stages of growth. From K. Napp-Zinn. 1960. Planta 54:409.

of *A. thaliana* in the early stages as being caused by the depletion of the stored food supply of the seed. The increase in sensitivity could be correlated with the increase in carbohydrates as a result of photosynthetic activity.

Additional evidence for the involvement of carbohydrates in the vernalization process has been provided by vernalization of Petkus winter rye embryos (54). Embryos separated from the endosperm (stored food supply) and provided sucrose and mineral nutrients will produce normal healthy plants. These embryos may also be vernalized. However, vernalization is retarded but ultimately accomplished if the embryos are denied a carbohydrate substrate (53). See Figure 22–5. As Purvis points out (53), this does not necessarily mean that only sugars accelerate the vernalization process, since the less mobile carbohydrates of the embryo (e.g., hemicellulose) may be brought into use. Although not unequivocally demonstrated as yet, a good deal of evidence supports the concept that carbohydrates are consumed in the vernalization

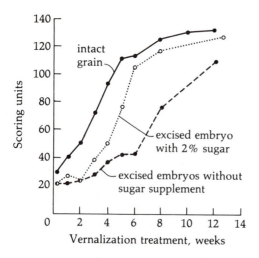

Figure 22–5. Progress of vernalization with duration of treatment.

From O.N. Purvis. 1961. The physiological analysis of vernalization. In W. Rhuland, ed., Encyclopedia of Plant Physiology *16:76. Berlin: Springer.*

process and, indeed, may be essential to that process.

Devernalization

In our discussion of photoperiodism, we saw that the promotion of flowering by red light could be reversed by far-red radiation. Just as the stimulus to flower received from a burst of red light can be reversed, so too can the stimulus to flower received from vernalization be reversed. This reversal may be accomplished with vernalized grains of Petkus winter rye by drying the grains and storing them under dry conditions for several weeks. The grains retain the vernalized condition for 6 weeks, but by 8 weeks they are almost completely devernalized (16).

The most-efficient devernalizing factor, however, is high temperature. Many instances have been recorded in which high

temperature following vernalization has erased the effect of the low-temperature treatment. In fact, even an alternation of high temperature with low temperature during a period of vernalization weakens the vernalization response.

Early reports on the reversal of vernalization in wheat claimed that the effects of vernalization could be completely erased if followed immediately by exposure to temperatures in the neighborhood of 35°C. However, Purvis and Gregory (55) found that complete reversal in winter rye could only be accomplished after a very brief vernalization period. An increase in the duration of the vernalization treatment causes an increase in the stability of the plant toward high temperature reversal (see Figure 22–6).

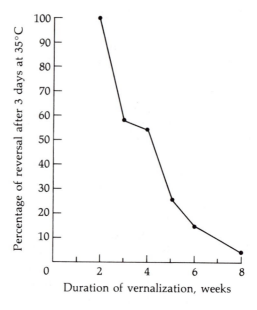

Figure 22–6. Progressive stabilization of winter rye to heat with increasing duration of vernalization treatment.

From O.N. Purvis and F.G. Gregory. 1952. Ann. Bot. 16:1.

In the biennial strain of *Hyoscyamus niger*, reversal of vernalization effects may also be accomplished. Exposure to high temperatures of about 35°C for a period of time will completely erase the vernalization effect (39). However, if the vernalized henbane is allowed a period of 3 to 4 days at 20°C, devernalization is not possible.

After the nullifying effects of high temperature, revernalization is possible in many plants. Low-temperature treatments—for example, on devernalized Petkus winter rye, sugar beet, *Arabidopsis*, henbane, and so on—will again cause the condition of vernalization.

Substitution of Gibberellin for Cold Treatment

In a previous chapter, we discussed the effect of gibberellin on bolting and flowering of rosette plants. We also mentioned that replacement of low-temperature treatments by gibberellins is only observed among rosette plants, such as henbane. However, in rosette plants, gibberellins may only promote stem elongation and not flowering. In an indirect way, through stimulation of stem growth, gibberellins may promote the elaboration of factors leading to flower formation. Among the cold-requiring caulescent plants, gibberellin has failed to replace the cold requirement for flowering.

Other Factors Modifying Vernalization Process

We might suspect that since the vernalization process is most likely dependent on a sequence of biochemical steps leading to the production of an active substance, the presence of water and oxygen are indispensable in the vernalization of seeds—water for activation of enzymes present in the seed, and oxygen for respiratory energy.

Water

Vernalization of dry seeds is impossible unless the seeds have imbibed some moisture. Purvis (54) pointed out that enough moisture must be present to initiate a small but visible degree of germination. In winter rye she found that the water imbibed must represent 50 percent of the absolute dry weight for adequate vernalization.

Oxygen

Grain kept in an atmosphere of pure nitrogen, although provided with an adequate supply of water, is unresponsive to low-temperature treatments (16). The oxygen requirement, although low, is absolute. Oxygen is also necessary for the vernalization of whole plants, such as henbane; for details, see the review by Chouard (5). Apparently, respiration is a necessary factor in the vernalization process. This conclusion has received experimental support from the study of the effect of respiratory inhibitors on vernalization. The response of winter wheat has been found to be considerably reduced by the use of these inhibitors (6).

Although the fundamental factor in the vernalization process is low temperature, it is ineffective in the absence of oxygen, water, and adequate supplies of carbohydrates for respiratory processes. Once a plant is vernalized, it may be devernalized by high temperatures and in some cases revernalized with another exposure to cold temperatures.

As with photoperiodism, we have come a long way toward understanding the process of vernalization. The physical manipulations leading to the vernalization of a plant have, for the most part, been worked out. Biochemical investigations of the process, however, have lagged behind. Understanding of the mechanism of a plant's perception of the cold temperature stimulus and the identification of the constituents involved in the sequence of reactions leading to the synthesis of the active substances are problems that need investigation. The biochemical roles of gibberellin, vernalin, and florigen need clarification.

Cold Tolerance of Plants

For some time plant scientists have realized that lower temperatures and shorter photoperiods influence metabolic changes in plants that have the genetically controlled capacity to harden, or develop *cold tolerance*. The ensuing hardening process undoubtedly insures winter survival. As Hodgson (23) suggested, "It is very reasonable to assume that natural selection would evolve a built-in, reliable, early warning of impending low temperatures, and the response to seasonal changes in photoperiod is logically such a device." Certainly the reception of photoperiodic changes and development of cold tolerance is important in many species. In alfalfa, for example, cold-tolerant cultivars are long-day flowering plants and cold-sensitive alfalfa cultivars are day-neutral plants. (24). Another important fact is that once induction has been initiated by shorter photoperiods, hardening in plants will continue at low temperatures regardless of the photoperiod (27, 59). Moreover, the expression of cold tolerance must result from the initiation of metabolic changes soon after induction by environmental factors.

Cold Tolerance, Growth, and Metabolic Components

The response of plants sensitive to the changing photoperiod is expressed as a reduction in growth rate, which is inversely proportional to the cold tolerance developed (29, 40). According to some workers, however, a reduction of top growth per se is not a requirement for hardening (9, 58). Nevertheless, the observed changes in growth rate and habit of some plants during hardening reflect metabolic modifications that are presumed to be closely related to cold-tolerance development. Siminovitch and Briggs (62) suggested that for any metabolic constituent to be directly related to cold tolerance, levels of such components should correlate with the development and loss of cold tolerance.

Lipids

Scientists have detected increased unsaturation, increased concentrations of fatty acids, and changes in the complex lipids during growth of different plants at low temperatures (11, 13, 17, 21, 35, 65, 70). The observed changes in lipids at cold temperatures suggest changes in cellular membranes. Some investigators suggest that increases in permeability, through lipid and other changes, enhance the redistribution of water between and within cells. The latter event is considered important to cold-tolerance development.

Carbohydrates

Experimenters have studied the role of carbohydrates in cold tolerance in great detail (1, 24, 27, 33, 65). Sucrose has been particularly implicated with this phenomenon (1, 24). As protectants, sugars such as sucrose, which readily form hydrogen bonds, may be important in maintaining the structural and functional integrity of proteins (glycoproteins) against freezing denaturation. They also serve as an important energy source for the numerous metabolic activities and as osmoregulators believed necessary for the development of cold tolerance.

Nucleic Acids

The nucleic acids DNA and RNA vary quantitatively with cold hardening (25, 29, 41, 58, 59), and Siminovitch and colleagues (63) postulated that the rise in RNA during hardening is a primary step in the overall mechanism of hardening. Furthermore, Li and Weiser (41) suggested that increases in nucleic acids are probably related to metabolic changes and particularly to the enzymes necessary for the synthesis of cellular components that impart resistance at freezing temperatures.

Proteins

Of all the metabolic components that reportedly increase hardening, soluble proteins appear to have the closest relationship to cold tolerance due to their dual functions as enzymes (1, 24, 28, 32, 33, 34, 40) and as possible cold protective agents (22, 64). Siminovitch and Briggs (61) provided the first indication of the relationship of soluble proteins to cold tolerance of plants. They determined that the concentration of water soluble proteins increased in the living bark of the black locust tree during fall as cold tolerance developed and then decreased in

the spring. Since then many studies have been directed to enzyme analyses of soluble proteins in a variety of hardened plants (3, 13, 25, 27, 33, 58). For example, Faw and Jung (12) separated soluble protein extracts from crown and root tissue by disk gel electrophoresis and observed more proteins in hardened than in nonhardened plants. They concluded that some of the separated proteins were correlated with cold tolerance.

Cold Tolerance and Enzymatic Activity

The enzymes that seem to play a role in cold tolerance and have been studied extensively are peroxidases, proteases, peptidases, esterases, invertases, amylases, and several dehydrogenases. See Figure 22–7 for a model outlining the major enzymatic and metabolic events possibly involved in the development of cold tolerance. This model was developed by Krasnuk, Witham, and Jung to give an overview of the concepts concerning the relationship of certain enzymes and metabolites to the cold-tolerance process.

The scheme outlines the induction process taking place after reception of photoperiodic and temperature changes in the environment. As indicated, the details of the induction process are sparse. However, according to the model, induction initially involves the activation of enzymes and protein synthesis. Protein synthesis leading to soluble enzymatic proteins is made possible with amino acids arising from the hydrolysis of insoluble proteins, one of the initial events of the induction process. We will now consider several of the pathways and enzymes shown.

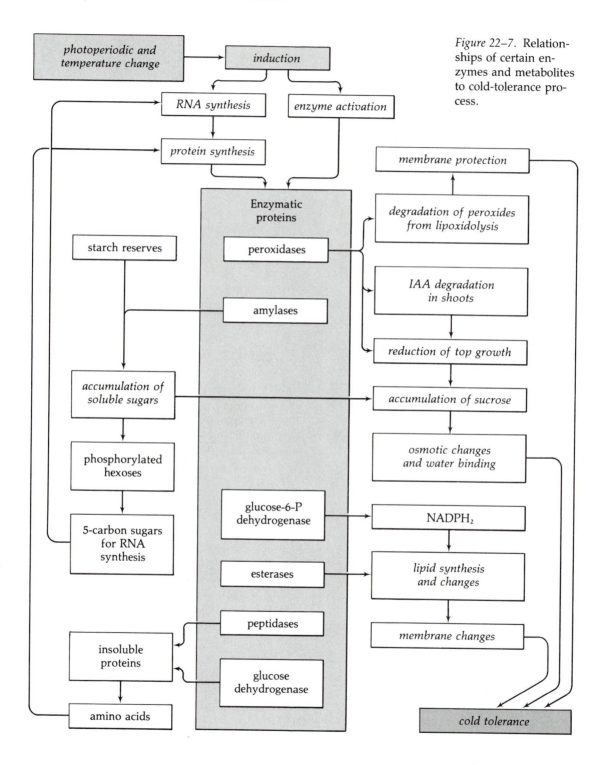

Figure 22–7. Relationships of certain enzymes and metabolites to cold-tolerance process.

Starch Hydrolysis

Starch hydrolysis is widespread in plants during growth at lower temperatures, and a precise temperature control mechanism that regulates the hydrolysis of starch to sugar seems to exist (10, 34, 45, 64). Due to the rapid rate of starch degradation during exposure, it is possible that amylases, which are inactive under summer conditions, are activated at hardening temperatures (10). The soluble sugars produced from the hydrolysis of starch might give rise to energy-releasing components or to five-carbon sugars necessary for nucleic acid synthesis. Certain dehydrogenase would have to be present and active for these conversions. One of the most important roles of sugars is their function as osmoregulators. We know that water moves into cells during hardening, probably in response to increased (more negative) water potential gradients. Undoubtedly, sugars act in this capacity and serve, additionally, as water-binding substances.

Esterases

Esterases (fatty acid hydrolyses) are thought to be important in the fatty acid and lipid transformations that occur during hardening (19). Observers have detected qualitative changes in esterases in *Salix* plants grown under hardening conditions. One esterase form, in particular, was found to be present only in cold-hardened samples (18, 19). From the model (see Figure 22–7), we observe that esterases may affect lipid and membrane changes necessary for cold tolerance.

Peroxidases

These enzymes may be especially important because experimenters have assessed changes in peroxidases in at least nine different species of plants grown under hardening conditions (14, 18, 19, 42, 43). As we indicate in Figure 22–7, these enzymes are important because of their role in the degradation of endogenous peroxides, generated during lipoxidolysis, which would otherwise adversely affect membrane integrity and permeability. Peroxidases also exhibit IAA oxidase activity, which would have effects on IAA metabolism and lignin biosynthesis.

Peroxidase activity increases with cold tolerance in alfalfa roots (14), *Sedum* (43), *Mitchella* (43), *Salix* (18), and *Dianthus* plants (42). Also, extracts of cold-resistant potatoes and wheat grown at low soil temperature should increase peroxidase activity (31). Roberts found mainly quantitative changes in anionic peroxidases of wheat grown at hardening temperatures.

In at least one study, IAA oxidase activity, due to the action of peroxidase, was found to increase tenfold in winter wheat seedlings grown at 2°C (2). If this pattern is somewhat similar in plants subjected to hardening conditions, then we can explain the reduction of top growth of such plants on the basis of IAA degradation. The diminution of growth, although not believed essential to cold tolerance by all plant scientists in this area, makes sense from the standpoints of energy, metabolite, and structural economy.

A complex interaction between photoperiod and temperature during late summer or early autumn is apparently involved in the initiation of metabolic changes necessary for cold-tolerance induction. Past investigations concerning this problem involve biochemical characterization of plants only after they have hardened. Perhaps future studies will include more dynamic considerations from induction through development of cold tolerance and will lead to an overall understanding of the details of induction and the metabolic and morphological events necessary for the establishment of cold tolerance in plants. The enzyme and metabolic changes seen in plants developing cold tolerance represent a general pattern for other kinds of stress such as drought and pollution damage. An understanding of all kinds of stress in plants is going to play a significant role in the continuing development of our agriculture.

Questions

22–1. Describe the possible mechanism of perception of temperature in a plant capable of vernalization.

22–2. What is vernalin? Is it related to gibberellic acid?

22–3. What is believed to be the primary site of vernalization in plants? Explain your answer.

22–4. Describe the role of gibberellic acid in vernalization.

22–5. Describe the relationship of growth rate and the development of cold tolerance in plants.

22–6. List the suggested major roles of nucleic acids, proteins, carbohydrates, and lipids in cold tolerance of plants.

22–7. What is the significance of increased protein synthesis during the initial stages of cold-tolerance induction in plants?

22–8. Why would the degradation of IAA be important to the development of cold tolerance in plants?

22–9. Is the development of cold tolerance in plants genetically controlled? Explain your answer.

22–10. How might you determine the conditions necessary for the induction of cold tolerance in a selected experimental plant? Outline the steps.

Suggested Readings

Berry, J., and O. Björkman. 1980. Photosynthetic response and adaptation to temperature in higher plants. *Ann. Rev. Plant Physiol.* 31:491–543.

Bixby, J.A., and G.N. Brown. 1975. Ribosomal changes during induction of cold hardiness in black locust seedlings. *Plant Physiol.* 56:617–621.

Burke, M.J., L.V. Gusta, H.A. Quamme, C.J. Weiser, and P.H. Li. 1976. Freezing injury in plants. *Ann. Rev. Plant Physiol.* 27:507–528.

Graham, D., and B.D. Patterson. 1982. Responses of plants to low, nonfreezing temperatures: proteins, metabolism and acclimation. *Ann. Rev. Plant Physiol.* 33:347–372.

Krasnuk, M., G.A. Jung, and F.H. Witham. 1975. Electrophoretic studies of the relationship of peroxidases, polyphenoloxidase and indoleacetic acid oxidase to cold tolerance of alfalfa. *Cryobiology* 12:62–80.

Laidlaw, A.S., and A.M.M. Berrie. 1977. The relative hardening of roots and shoots and the influence of day length during hardening in perennial rye grass. *Ann. Appl. Biol.* 87:443–450.

Chapter 23

Dormancy

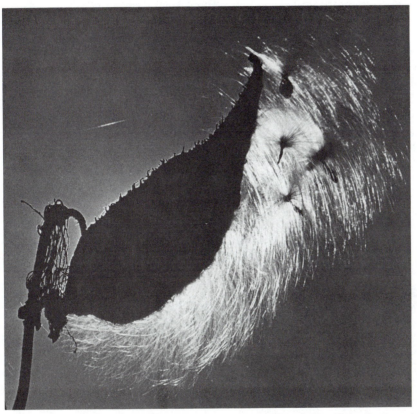

Milkweed pod releasing its seeds into cold autumn air. *Photo by Pat Little. Courtesy of* Centre Daily Times, *State College, Pennsylvania.*

We often visualize plant morphogenesis and growth as a continuous process from germination through flowering to death. The life cycles of almost all plants, however, are characterized by periods of temporarily arrested growth. They become *quiescent*— that is they continue to live, but with very low, almost nonmeasurable metabolic activity. Plants or plant parts seem to enter a state of suspended animation. An understanding of this phenomenon will be highly significant to us in agriculture and in other ways, including space travel.

Plant scientists often use the term *dormancy* to describe the arrest in growth and development of seeds (embryos), buds, and other plant parts under conditions seemingly suited for growth. Growth may also be suspended by adverse conditions in the environment. Seeds, for example, will not germinate under dry conditions, but they will readily germinate if they imbibe water. A suspension of growth may also occur because of the concentration of some growth inhibitor, or it may be caused mechanically by the mere presence of a strong, durable enclosing structure that does not allow for expansion. The presence of membranes or seed coats impermeable to water or oxygen may also keep growth in an arrested state. Finally, many seeds and buds require special conditions of light and temperature. The onset of dormancy and the deciduous habit of North Temperate zone plants serve as excellent examples of photoperiodic- and temperature-regulated growth.

Some distinction is made between a suspension of growth due to a lack of a necessary external environmental factor (e.g., water) and a suspension of growth due to internal limitations. Arrested growth because of the lack of some necessary external environmental factor is referred to by many as *quiescence*. However, as we have mentioned, many seeds and buds are unable to grow even if provided with water because of internal limitations, a situation referred to by many as dormancy (a rest stage). Since the general result, growth suspension, is the same, we include both situations under the general term *dormancy*.

In temperate zones, seasonal changes in temperature range from near 38°C in midsummer to well below freezing in midwinter. Obviously, most plants could not survive the cold temperatures of winter in the vegetative or flowering state. Thus in many plants, seed and bud dormancy begin at the onset of winter cold, thereby allowing the plant to pass through the winter with little or no damage. In the grain areas of the United States and Canada, for example, wild oat infestation is a serious problem because of the ability of the grains to survive the winter in a dormant state and then germinate the following spring. In contrast, the seeds of many other noxious weeds have only a brief dormant period, germinate in the fall, and are killed off during the severe winters that are common to the northern midwest areas.

The importance of dormancy among plants growing in arid regions is immediately apparent. For those plants, it is certainly significant if germination and growth can be managed during the relatively brief periods of rainfall in these areas. Seeds that can remain viable but dormant until sufficient water is available have a very good chance of survival. We find an even more bizarre example of the importance of dormancy in the adaptation of a plant to a dry region in the desert shrub guayule. In this plant, the chaff covering the seed contains a germination inhibitor that causes the seed to remain in a dormant state. However, with a rather strong rainfall, sufficient dilution of the inhibitor occurs to allow germination.

While talking about the beneficial role of dormancy in plants, we should also mention how seed coats impermeable to water help in the persistence of a species. Some species of *Convolvulus*, which grow in arid regions, have this type of seed coat. In order for these seeds to imbibe water and germinate, the seed coats must be mechanically broken. However, permeability to water gradually occurs over a long period of time. The advantage here is that all of the seeds will not germinate at one time, only a certain number will germinate each year. Thus it is virtually impossible for the entire species to be wiped out during the vulnerable seedling stage due to some adverse environmental condition.

Dormancy in plants is both a convenience and an inconvenience to humans. The temporary dormant period experienced by many cereal grains allows for their harvest, dry storage, and ultimate use as food. Otherwise, these grains would germinate and be useless to us. The ability, however, of certain weed seeds to lie dormant for many years in the soil has proven to be a great inconvenience. During plowing, the dormancy of many of these seeds will be broken, thereby allowing them to compete with any economic crop sown in that area. The eradication or even control of many of these weeds is almost impossible because they can never all be caught in the vulnerable seedling or vegetative state. Although some are triggered to germinate by the soil disturbances caused by plowing, there are always some that remain dormant in the soil. Therefore, each year farmers are presented with the same problem, the germination of some but not all of these weed seeds. They can only destroy those that germinate and they have almost no control over those lying dormant in the soil.

Seed Dormancy and Germination

We may define the process of germination as the sequence of steps that begins with the uptake of water and leads to the rupture of the seed coat by the radicle (embryonic root) or the shoot. Cell division, cellular enlargement, and overall increase in metabolic activity (food digestion and assimilation, for example) accompany the observable morphological steps. Even though these events commence long before the rupture of the seed coat, we usually determine the occurrence of germination visibly by noting the protrusion of the radicle. Let us consider the various factors causing dormancy and the different methods for breaking dormancy.

The absence of some external factor considered necessary for the process to occur inhibits the germination of seeds. Thus in the absence of water, the proper temperature, or the proper mixture of gases, germination is inhibited. However, many seeds may be placed in an environment considered optimum for germination and still not germinate because of some factor associated with the seeds. This factor may be a hard seed coat that is impermeable to water or gases or is physically resistant to embryo expansion, an immature embryo, a need for afterripening, a specific light requirement, a specific temperature requirement, or the presence of a substance that inhibits germination.

Hard Seed Coat

One of the most common factors associated with seed dormancy is the presence of a hard seed coat. The hard seed coat may be responsible for dormancy by preventing the absorption of water, by preventing gaseous exchange, primarily oxygen absorption, and by mechanically restricting the growth of the embryo.

Inhibition of water absorption. Many plants produce seeds with hard seed coats impermeable to water. In this respect, the Leguminosae family has by far the largest number of species (14). In addition to having hard seed coats, seeds of many members of the Leguminosae have an external waxy covering (25). Some of these seeds may be totally impermeable to water. The hardness factor in seed coats is primarily an inherited trait; however, in at least one case the hardness of a seed coat is determined by environmental conditions. Crocker (6) observed that seeds of white sweet clover are hard when they ripen during hot, dry weather, but they are soft when they ripen during rainy weather.

Hyde (21), in a study of some legume seeds, described an interesting mechanism for the control of water entering the seed. In the seeds of some legumes (e.g., *Lupinus arboreus*), water enters only through the hilum. Hyde found that the absorption of water by these seeds is controlled by hygroscopic tissue, which makes up the hilar fissure. When the relative humidity is high, this tissue swells, closes the hilar fissure, and prevents water absorption; and when the relative humidity is low, the fissure opens and allows the seed to dry out.

We can ascertain that drying out of the seed is the inevitable result under these circumstances by measuring the moisture content of scarified and unscarified seeds of white clover after they have been subjected

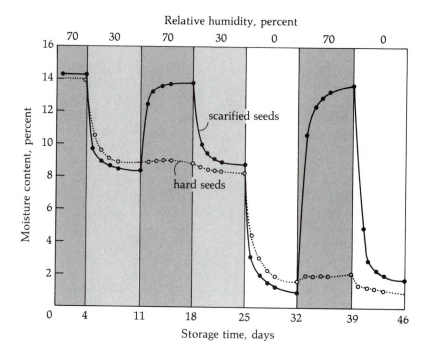

Figure 23–1. Changes in moisture content of white clover seeds transferred successively to chambers of different relative humidity.

From E.O. Hyde. 1954. Ann. Bot. *18:241.*

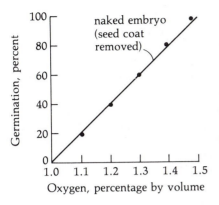

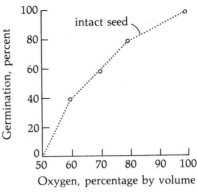

Figure 23–2. Effect of oxygen on germination of upper seed of *Xanthium*. Temperature held at 21°C. Note striking difference on oxygen requirement for germination between naked embryo and intact seed.

Data from N.C. Thornton. 1935. Contri. Boyce Thompson Inst. 10:201.

to different relative humidities. As we shall see later, scarified seeds are those in which the seed coats have been rendered permeable to water and gases. The white clover seeds contain the same type of mechanism for controlling water absorption as do the seeds of *Lupinus arboreus*. Figure 23–1 illustrates that the moisture content of the unscarified white clover seeds never rises when the seeds are transferred from a low to a high relative humidity and always falls when they are transferred from a high to a low relative humidity. The moisture content of the scarified seeds, in contrast, rises and falls relative to the humidity treatments, as we would expect of a seed that is permeable to water.

Inhibition of gas absorption. Many seeds that are permeable to water are impermeable to gases (25). We find the classical example of this type of impermeability in the cocklebur (*Xanthium*). In the burr of the cocklebur plant, there are two seeds, one borne higher up in the burr, called the *upper seed*, and one borne lower in the burr, called the *lower seed*. Crocker (6) found that the seed coats of both seeds are permeable to water. The lower seed will germinate under normal conditions of moisture and temperature,

but the upper seed will not germinate under these conditions unless the seed coat is punctured or removed. However, if the upper seed is placed under high oxygen conditions, it germinates readily. Crocker concluded that the seed coat of the upper seed limits the supply of oxygen to the embryo so that the minimum needed for germination cannot be reached. Subjection of the seed to high concentrations of oxygen overcomes this block in germination.

Later work by Shull (36, 37) and Thornton (39) demonstrated the accuracy of Crocker's observations. These two workers showed that the naked embryos of both the upper and the lower seeds have a much lower oxygen requirement than does the intact seed, and as temperature increases, the oxygen requirement decreases. For the naked embryo of the upper seed, 1.5 percent oxygen is needed at 21°C, and 0.9 percent oxygen is needed at 30°C for 100 percent germination. When the upper seed is left intact, the oxygen requirement for germination increases considerably. Pure oxygen is needed at 21°C and 80 percent oxygen is needed at 30°C to give 100 percent germination. Figure 23–2 presents some of Thornton's data. We do not as yet know whether the limiting of oxygen supply by the seed coat retards metabolic activity to the point of blocking germination or whether the high oxygen concentration has some other function that promotes germination. Wareing and Foda (53) claimed that high oxygen tensions cause the oxidation of an inhibitor present in the upper seed, thus allowing germination.

Mechanical restriction of embryo growth. Seed coats may be permeable to both oxygen and water, yet still effect a dormant state in a seed. For example, the seeds of pigweed (*Amaranthus retroflexus*) have a seed coating that is permeable to oxygen and water but that is strong enough to resist embryo expansion (26). These seeds may sometimes lie dormant but viable for many years.

Dormancy and modes of storage have maintained seeds viable for as long as 10,000 years. Dried seeds have been shown to exist in soil for thirty-one years, in dry laboratory storage for one hundred years, and in peat bogs and frozen earth considerably longer.

Scarification. Where germination is inhibited by mechanical resistance of the seed coat or impermeability of the coat to water or oxygen, dormancy may be broken by scarification. The term *scarification* refers to any method that renders the seed coat permeable to water and oxygen or breaks the seed coat so that embryo expansion is not physically retarded. The process can be accomplished in the laboratory by forms of abrasion, cutting, or chemical treatment. Scarification is roughly divided into mechanical scarification and chemical scarification.

Mechanical scarification of hard-coated seeds is effected by any treatment of the seeds that will crack or scratch the seed coats, such as shaking the seeds with some abrasive material (e.g., sand) or scratching or nicking the coat with a knife. The cracks or scratches resulting from such treatment promote germination by decreasing the resistance of the seed coat to water or oxygen absorption and to embryo expansion.

Chemical scarification is also an effective way of breaking dormancy resulting from the seed coat. Dipping seeds into strong acids, such as sulfuric acid, or into organic solvents, such as acetone or alcohol, and then rinsing the seeds with water can break this type of dormancy. Even boiling water may be a successful treatment. As in mechanical scarification, chemical scarification breaks dormancy by weakening the seed coat or by dissolving waxy material that renders the coat impervious to water.

In nature, the process is accomplished by the acid and enzymatic conditions of the digestive tracts of birds and other animals, by abrupt changes in temperature, and by the action of fungi and other microorganisms. In some areas of the world, seeds of certain species require fire for scarification. These seeds have a competitive edge immediately after the denudation of a dense vegetative area.

Immature Embryo

Failure of a seed to germinate may be a consequence of partial development of the embryo. Germination will occur only when the embryo development is complete, and embryo development may occur during or before the germination process (25). Dormancy due to immature embryos may be found in Orchidaceae and Ovobancheae, as well as some *Fraxinus* and *Ranunculus* species. Dormancy due to immature embryos can only be broken by allowing the embryo to complete development within the seed in an environment favorable to germination.

Afterripening and Stratification

A large number of plants produce seeds that do not germinate immediately but do so after a period of time under normal conditions for germination. A prerequisite to germination for this type of seed, then, is a period of *afterripening* (development of the embryo). In nature, afterripening occurs during the period between the fall of the seed to the ground in the autumn and its germination the following spring. During this time, the seeds are covered over with debris and winter snows.

Afterripening occurs for some species during dry storage. For others, moisture and low temperatures are necessary—a process called *stratification*. Natural stratification occurs when seeds shed in the fall are covered with cold soil, debris, and snow. We have learned to copy and improve on nature in this respect by devising a method of artifical stratification. In artificial stratification, layers of seeds are alternated with layers of moistened sphagnum, sand, or some other appropriate material, and stored at low temperatures. The effect of artificial stratification on the germination of *Pinus rigida* may be seen in Figure 23–3.

Because many workers refer to the period of afterripening as a dormant, or rest, period, the implication persists that nothing is occurring within the embryo during this time. However, many studies have demonstrated that considerable physiological activity may be observed during the so-called afterripening, or dormant, period (29, 31). Figure 23–4 shows the effect of afterripening time and temperature on growth of

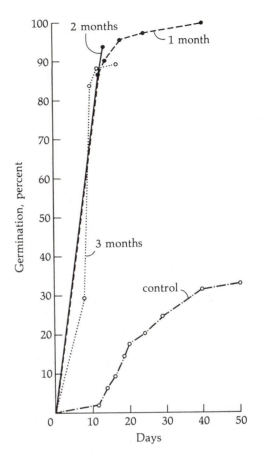

Figure 23–3. Effects of artificial stratification at 5°C for 1, 2, and 3 months on germination of *Pinus rigida* seeds.

From W. Crocker. 1948. Growth of Plants. *New York: Reinhold.*

the embryonic axis of cherry seeds. During afterripening, extensive transfer of compounds from storage cells to the embryo, sugar accumulation, and digestion of various storage lipids take place.

Stratification may effect the disappearance of inhibitors and the buildup of germination promoters such as the gibberellins and cytokinins. Certainly many changes in

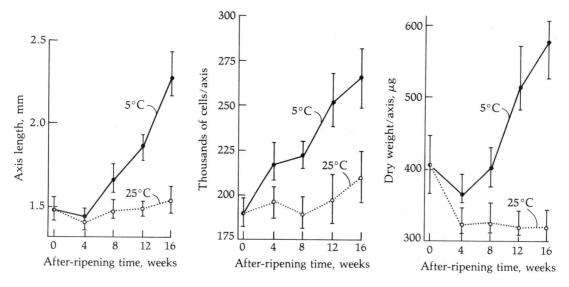

Figure 23–4. Effect of afterripening time and temperature on growth and dry weight in embryonic axis of cherry seeds.

From B.M. Pollock and H. Olney. 1959. Plant Physiol. *34:131.*

the levels and interconversions of food materials take place.

Light Requirements for Germination

With respect to germination, seeds vary considerably in their response to light. Some seeds have an absolute light requirement for germination. In other seeds, exposure to light is inhibitory to germination. And in still others, germination is associated with a photoperiodic response—that is, an alternation of light and dark periods. The determination of a seed's light requirement is made even more complex by the fact that temperature may interact with light in the germination of a seed.

As in most studies in which light is implicated as a catalytic agent, experimenters are searching for the most effective wavelengths. We have already seen that light (red and far-red) operating through phytochrome regulates germination in Grand Rapids lettuce seeds. We did not mention, however, what effect the time of water imbibition prior to light treatment and temperature changes had on Grand Rapid lettuce seed germination.

Borthwick and colleagues (4, 5) found that the response of lettuce seeds to light can be modified by the amount of time the seeds are allowed to imbibe water before exposure. Red-light promotion of germination increases with time of exposure up to 10 hours, at which point a plateau is reached. However, if the seeds are allowed to imbibe water for more than 20 hours, the germination response falls off. In contrast, the inhibitory effect of far-red irradiation has a tendency to decrease as previous imbibition time increases up to 10 hours. In keeping with this contrast, the sensitivity of lettuce seeds to far-red irradiation increases when the seeds imbibe water for more than 20 hours.

Table 23–1. Effect of temperature on photocontrol of seed germination of two varieties of lettuce seeds after exposure to red light or darkness.

Variety and Germination Temperature (°C)	Seeds Germinating under Indicated Light Condition (%)	
	Red	*Darkness*
White Boston		
10	99	95
15	99	78
20	98	57
25	1	0
Grand Rapids		
15	94	52
20	96	40
25	96	10
30	1	0

Source: *From E.H. Toole. 1959. Page 89 in R.B. Withrow, ed., Photoperiodism and Related Phenomena in Plants and Animals. Westview Press. Copyright 1959 by the American Association for the Advancement of Science.*

Effects of Temperature on Germination

Photocontrol of seed germination is, in many cases, interrelated with temperature, as evidenced by data in Table 23–1, which shows a decrease in sensitivity to light with an increase in temperature above 25°C (42).

Another, more complex, example of temperature-light interaction may be found in germination of pepper grass seeds (*Lepidium virginicum*). Maximum germination is achieved if the seeds are kept at a cool temperature before being irradiated with red light and at a relatively high temperature for a period of time after being irradiated (46). Table 23–2 shows these relationships.

In discussing the various aspects of seed germination, we have seen the importance of temperature in the prolonging or breaking of dormancy. Many seeds need a period of prechilling under moist conditions before adequate germination can take place. In natural and artificial stratification, this requirement is satisfied. After the cold requirement is satisfied, the actual germination, in most situations, takes place efficiently at about 20°C.

In some seeds, the cold requirement is modified by the age of the seed. For example, seeds of *Brassica juncea* show a definite

Temperature, First 2 Days (°C)	Temperature, 3rd–6th Day (°C)	Germination (%)	
		Irradiated Seeds	*Nonirradiated Seeds*
15	15	37	0
25	25	41	0
15	25	92	0
25	15	32	0

After E.H. Toole et al. 1955. Plant Physiol. 30:15.

Table 23–2. Light and temperature interaction in the germination of *Lepidium virginicum* seeds. Seeds were irradiated with red light in the region 5800–7000 Å.

Source: *From E.H. Toole et al. 1955. Photocontrol of* Lepidium *seed germination.* Plant Physiol. 30:15.

cold requirement immediately after harvest, and this cold requirement decreases with age (45). Immediately after harvest, 97 percent of the seeds germinate at temperatures of 10° or 15°C, 63 percent at 20°C, and only 8 percent at 25°C. However, after 3 weeks, 95 percent of the seeds germinate at 25°C. The sensitivity of seeds to high temperatures varies greatly. In some, it may persist for a long time; in others—for example, *Brassica juncea*—sensitivity is lost in 3 weeks.

In many seeds, an alternation of temperatures gives maximum germination. In some cases, such as with *Poa pratensis*, alternation of low and high temperatures repeated several times gives the best results. For example, a single alternation from 15° to 25°C in connection with light treatment of pepper grass seeds may significantly increase germination (see Table 23–3).

Low temperatures promote germination of the light-sensitive Grand Rapids lettuce seeds, and high temperatures inhibit it. Low temperatures can substitute for red-light promotion of light-sensitive lettuce seeds (22). Low-temperature promotion of lettuce seed germination is much less efficient with respect to time than is red-light promotion. We should note from Table 23–3 that far-red irradiation cannot reverse low-temperature stimulation, thus suggesting that low-temperature promotion of germination is not controlled by phytochrome (3, 22).

Chemicals and Germination

Germination Promoters

The promotion of germination by various compounds has been demonstrated a great number of times on many different seeds. Of the numerous promoters of ger-

Treatment before Final Transfer to 25°C for Germination	Germination, %	Inhibition, %
1 day at 2°C, then far-red	6	80
dark control	30	—
1.5 hours at 25°C, then far-red, then 1 day at 2°C	14	73
dark control	51	—
3 days at 2°C, then far-red	59	23
dark control	77	—
1.5 hours at 25°C, then far-red, then 3 days at 2°C	65	29
dark control	92	—
25°C control, 5 min red	92	—
25°C control, 5 min far-red	5	70
25°C control, dark control	17	—

Table 23–3. Effects of far-red light on germination of Grand Rapids lettuce seeds treated with low temperature (2°C). Irradiation with 5 min of far-red light was given at 25°C either immediately before or immediately after the cold treatment. Control seeds imbibed and were germinated at 25°C throughout, and irradiated with red or far-red light at 1.5 hours after beginning of soaking. Percent of inhibition for far-red treatment is calculated with respect to corresponding dark control.

Source: *From H. Ikuma and K.V. Thimann. 1964. Analysis of germination processes of lettuce seed by means of temperature and anaerobiosis. Plant Physiol. 39:756.*

mination known, the most popular and widely used are potassium nitrate (KNO_3),

thiourea (NH_2—$\overset{\overset{\displaystyle S}{\|}}{C}$—$NH_2$), ethylene ($C_2H_4$), gibberellins, and kinetin. Thiourea, gibberellins, and kinetin are interesting in that they have, in certain instances, substituted for light requirements in light-sensitive seeds. Whether or not this is a true substitution has been disputed. Nevertheless, these compounds have demonstrated dark promotion of germination.

Germination Inhibitors

Many compounds may inhibit germination. Any compound that is toxic, in general, to any of the essential life processes will, of course, inhibit germination and will kill the seed if it is present in a large amount. We are not concerned here with this type of inhibition, but rather with inhibition produced by natural compounds present in the seed. These compounds are often the cause of dormancy and usually act by blocking some process essential to germination. Natural germination inhibitors, however, do not reduce the viability of the seed nor do they produce any growth abnormalities in the seedling after germination.

Natural inhibitors are not confined to any particular part of the seed and may even be found in structures covering the seed (e.g., the glumes of oat grains contain an inhibitor). Germination inhibitors have been found in the pulp or juice of the fruit containing the seed, in the seed coat, in the endosperm, in the embryo, and so on. Evenari (12) studied germination inhibitors and found their presence common and widespread among plants. Some of the nat-

ural germination inhibitors that have been identified are coumarin, parascorbic acid, ammonia, phthalids, ferulic acid, and ABA. There is also probably an array of numerous inhibitors that include cyanide-releasing and ammonia-releasing substances, phenolic compounds, alkaloids, organic acids, and oils. The most potent natural inhibitor, however, is ABA.

Many of the germination inhibitors may be leached out of one plant and may inhibit the growth or seed germination of a plant from another species. Plants that exhibit this kind of competitive behavior are said to be *allelopathic*. Obviously, allelopathy has significant ecological implications and is of great interest to scientists involved with biosystematics.

Of the synthetics that retard germination, we should note that the commercial herbicides are usually germination inhibitors. For example, a herbicide such as 2,4-D will inhibit the germination of grass seedlings and should, therefore, not be used until the turf is well established.

At very low concentrations—between five and ten parts per million—ABA will completely inhibit the germination of lettuce seeds of the strains *Attraktion* and *Hohlblättriger Butter*. Sankhla and Sankhla (35) demonstrated that ABA inhibition of lettuce seed germination can be entirely counteracted by concentrations of kinetin as low as one part per million. In the same study, counteraction by GA of the ABA effect could not be demonstrated. Both kinetin and GA are well known for their stimulatory effect on lettuce seed germination.

Bud Dormancy

Before giving rise to vegetative or reproductive growth, the buds of many plant species go through a period of dormancy. A common occurrence of tree growth in temperate regions is the entrance of buds into a dormant state in late summer and their emergence from this condition the following spring to produce new leaf and flower growth. Bud dormancy of this type may usually be broken with cold temperature treatment somewhat analogous to that given to cold-requiring seed for germination. Many tree species with buds in a dormant state may be kept in that condition indefinitely if provided the artificial warmth of a greenhouse. However, if they are subjected to low temperatures (0° to 10°C) for a period of time and then returned to warm conditions, dormancy will be broken and growth will ensue.

In addition to temperature, photoperiodism and the low availability of water influence dormancy of buds. In temperate zones the most important factor regulating bud dormancy is the photoperiod.

Photoperiodism and Bud Dormancy

It is tempting to assign a universal role to low temperatures in inducing as well as in breaking dormancy. However, with respect to entering dormancy, the buds of woody species respond more to the photoperiod than to cold temperatures, as reported by Wareing (52). The shortening of the day length associated with the coming of fall and winter is an important factor in bud dormancy of woody species. Wareing (50, 51) demonstrated that bud dormancy in woody species is a photoperiodic phenomenon that is caused by short day lengths and relieved by long day lengths.

Perception of Light Stimulus

Not in every case of bud dormancy is the leaf the organ of photoperiodic perception. In some cases bud dormancy is induced and maintained after the leaves have fallen off. Wareing (51) solved the problem of how to associate photoperiodism with bud dormancy in the absence of the usual organs of photoperiodic perception. He found that the buds of leafless seedlings of beech (*Fagus sylvatica*) are capable of photoperiodic perception thereby causing a break in dormancy under long days and continuing dormant under day lengths of 12 hours or less (see Table 23–4). In addition, beech buds may break dormancy in the absence of a low-temperature treatment. The bud scales act as the photoreceptor organs. However, in many plants the leaves are the primary organ of perception and phytochrome is the primary photoreceptor chemical with respect to bud dormancy.

As in flowering, the response of buds to photoperiodism is actually controlled by the length of the dark period rather than by the length of the light period. Wareing (50) found that although buds of *Fagus sylvatica* remain dormant under short light periods, alternation of short light periods with short dark periods breaks dormancy. He also demonstrated how an interruption of a long dark period (which would normally keep buds in a dormant state) by a light break of 1 hour is sufficient to break dormancy.

Dormancy-Inducing Hormones

Evidence is strong that in bud dormancy the organs of photoperiodic perception are the leaves and buds. Since dor-

Daily Light Period (hours)	Total Number of Plants	Number of Plants Breaking Dormancy after 46 Days	Time for 50% of Plants to Show Breaking (days)
12	11	0	—
16	12	5	46
20	11	9	14
24	11	11	14

Table 23–4. Effect of length of photoperiod on breaking of dormancy in beech buds. Source: *From P.F. Wareing. 1953. Growth studies in woody species. V. Photoperiodism in dormant buds of* Fagus sylvatica. *Physiol. Plant. 6:692.*

mancy sets in after the photoperiodic stimulus has been received, a reasonable assumption would be that reception of the stimulus causes certain changes that lead to the production of a dormancy-inducing hormone. Hemberg (17, 18, 19) advanced the suggestion that bud dormancy in woody plants is controlled by growth-inhibiting substances produced in the bud. He based his suggestion on the fact that the endogenous growth inhibitor level increases with dormancy and decreases with the breaking of dormancy. Since Hemberg's work, investigators have made several studies on different woody species to correlate the level of endogenous inhibitors with the induction and the breaking of dormancy (3, 20, 23). In agreement with our earlier discussion on the role of photoperiodism in bud dormancy, investigators have shown short-day induction of bud dormancy in some species to be parallel with increased levels of inhibitors in buds and leaves (27, 30, 34).

In some plants a high correlation seems to exist between the presence of ABA and bud dormancy. Researchers have demonstrated that extracted xylem sap of plants exhibiting bud dormancy contains considerably higher levels of ABA than does the sap of plants with nondormant buds. It is only after ABA levels subside that buds will be-

gin to develop.

In their earlier work, Eagles and Wareing (11) found that the effect of ABA on inducing bud dormancy could be overcome by applications of gibberellic acid. This fact implies that endogenous gibberellins may be involved in the regulation of bud dormancy. The cold treatment required by many buds to break dormancy may be a means of bringing endogenous gibberellins to the level needed to break dormancy. The fact that higher concentrations of gibberellin-like substances have been found in bolted cold-requiring plants than in nonbolted plants (13, 28), supports the supposition that the level of endogenous gibberellins is raised as a result of low temperatures. Thus bud dormancy in some woody species may be regulated by a balance or ratio between a dormancy-inducing hormone and gibberellins.

We must emphasize again that there are some studies that show that ABA has no effect on the induction of bud dormancy. For example, direct application of ABA to shoot tips or leaves does not cause bud set. Also, in some plants (red maple, as an example), endogenous ABA does not increase with short-day treatment. Consequently, ABA may not be a general regulator of bud dormancy within the plant kingdom.

Dormancy of Potato Tuber Buds

Potato bud dormancy is a good example of bud dormancy in a nonwoody, or herbaceous, plant. The potato tuber is a modified, fleshy underground stem that contains several buds in locations commonly referred to as "eyes." If we place the newly formed potato tuber under conditions favorable to growth, sprouting will not take place. That this is not due to apical dominance, which is prevalent in potatoes, is demonstrated by the persistence of the dormant state in the individual buds when they are separated from the tuber. Dry storage at 35°C or moist storage at 20°C have been found to eliminate potato bud dormancy, low temperatures apparently have no effect (41).

Growth-Inhibiting Substances

In a series of studies on potato bud dormancy, Hemberg (15, 16, 18) demonstrated that substances extracted from potato peels of dormant tubers are capable of counteracting the effect of IAA in the *Avena* coleoptile curvature test. The group of inhibitors extracted by Hemberg were made up of acid and neutral substances. The acid inhibitors could not be detected at termination of dormancy (19). The inhibitor present in potato peels of dormant potatoes is *inhibitor-β* (3, 48), a complex of organic compounds first identified by Bennet-Clark and Kefford (1) from paper chromatograms of plant extracts. It is interesting to note that inhibitor-β has also been extracted from dormant buds of the silver maple (23).

Investigators have shown a correlation between increase and decrease of inhibitor-β with respect to onset and release of potato bud dormancy, and inhibitor-β has been ex-

tracted from the dormant buds of at least one woody species. In addition, ABA at very low dosages, almost completely inhibits the sprouting of potato buds.

Compounds Breaking Bud Dormancy

The regulation of dormancy release has academic as well as practical importance. Controlling the time of release of dormancy by artificial manipulation of the environment or by the use of active compounds may reveal some of the mechanisms involved in dormancy, thus contributing to our understanding of the overall process. The release of dormancy by some artificial means often aids the farmer in an economic way. For example, the release of potato bud dormancy of newly harvested tubers would allow some growers to produce a second crop if the length of the growing season permitted it. Some of the chemicals found useful in the release of dormancy are 2-chloroethanol, thiourea, and gibberellins.

2-Chloroethanol ($ClCH_2CH_2OH$). A thorough study of the dormancy-breaking effects of many different compounds by Denny (8, 9) brought into prominence one especially active compound: 2-chloroethanol. This compound proved highly effective in inducing sprouting in dormant potato tubers, with the added attraction of a wide safety margin between active and toxic dosages. In addition, chloroethanol proved to be very successful in breaking bud dormancy of fruit trees when it was applied in vapor form.

Several metabolic changes may take place as a result of 2-chloroethanol application. A study of its effect on dormant potato tubers showed that there is a rise in respira-

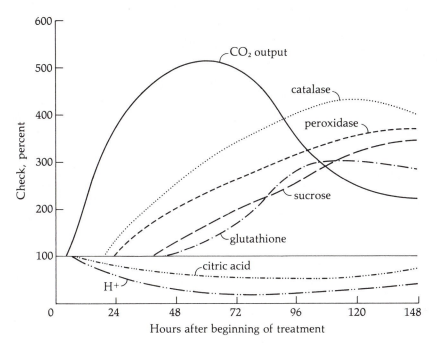

Figure 23–5. Effect of 2-chloroethanol on metabolism of potato tubers.

Data of L.P. Miller et al. 1936. Contri. Boyce Thompson Inst. 8:41. Redrawn from W. Crocker. 1948. Growth of Plants. *New York: Reinhold.*

tion, catalase and peroxidase activity, sucrose, and glutathione. There is a drop in H^+ ion and citric acid concentrations. Most likely, citric acid is being utilized as a substrate in respiration, and its decrease in content causes a decrease in the H^+ ion concentration (7). Figure 23–5 shows these relationships.

Thiourea (NH_2CSNH_2). Although not as effective as 2-chloroethanol, thiourea has proved effective in forcing sprouting in dormant potato tubers. Thiourea has an unusual effect in that it may cause the growth of several bud primordia in one eye. Observers have noted as many as eight sprouts growing from one eye (7). In contrast, 2-chloroethanol causes the growth of only one sprout in each eye. It is interesting to note that reduced oxygen pressure has an effect somewhat similar to thiourea: it causes multiple sprouting (40).

Gibberellins. A special case can be made for gibberellins acting as a bud dormancy breaker. In contrast to 2-chloroethanol and thiourea, gibberellins are a natural compound and may be a controlling factor in the overall process of bud dormancy (see Figure 23–6). In earlier discussions, we learned that gibberellins have a great influence on dormant shoots and seeds and promote the growth of both when they are applied to the plant. We would logically assume that gibberellins would also be capable of breaking bud dormancy. This influence was eventually demonstrated, with excellent success, on dormant potato tubers (32, 33) and on dormant peach buds (10). Generally, in plants that require a low-temperature period to break dormancy, gibberellins will substitute for the cold treatment and force the release of dormancy (see Figure 23–7).

Much work has been done on the effect of gibberellins on potato bud dormancy.

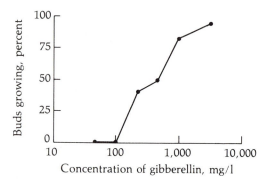

Figure 23–6. Effect of application of gibberellins on breaking of bud dormancy of Elberta peach. Buds were treated in March after 164 hours below 8°C.

Data of C.W. Donaho and D.R. Walker. 1957. Science 126:1178. Redrawn from Plant Growth and Development by A.C. Leopold. Copyright © 1964 McGraw-Hill Book Company. Used with the permission of McGraw-Hill Book Company.

Gibberellins can induce sprouting of potato tubers still on the plant (24) and in harvested tubers at any time in their dormant period. Therefore, the induction of sprouting by gibberellins may be accomplished anytime from the beginning of tuber enlargement to the end of the dormant period (38). Table 23–5 indicates the remarkable ability of gibberellins to induce sprouting when applied as a spray to potato plants 4, 2, and 1 week before harvest.

Endogenous gibberellins may play a major role in the control of dormancy. This assumption has been given strong support in several studies. For example, in 1959 gibberellin-like substances were reported present in potato tubers and higher levels were detected in sprouting tubers than in newly harvested tubers. Also, the concentration of endogenous gibberellins remains low during the dormant period, but rises thirtyfold after sprouting begins (see Figure 23–7).

Breaking of Dormancy through Gene Derepression

Evidence exists that the primary act in the breaking of potato bud dormancy may involve the derepression of repressed genes (see Chapters 17 and 19). The genome of the dormant potato bud is almost completely repressed—that is, the dormant potato bud lacks the ability to synthesize RNA in vivo, and chromatin isolated from it cannot support DNA-dependent RNA synthesis even though the potato bud is incubated with a

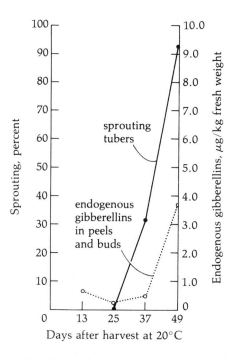

Figure 23–7. Relationship between sprouting Red Pontiac potato tubers and level of endogenous gibberellins in potato peels and buds during and after rest.

From O.E. Smith and L. Rappaport. 1961. In R.F. Gould, ed., Gibberellins. Advances in Chemistry Series. Am. Chem. Soc. 28:42.

Table 23–5. Percentage of sprouted tubers at harvest from plants that received preharvest foliar sprays of gibberellins 4, 2, and 1 week before harvest.

Gibberellin, (mg/liter)	Tubers Sprouted at Harvest (%)		
	4 Weeks	2 Weeks	1 Week
0	0.0	1.4	0.2
10	3.0	1.5	1.5
50	58.3	18.0	0.4
100	75.6	34.3	2.1
500	83.6	50.0	5.8

Source: *From L.F. Lippert, L. Rappaport, and H. Timm. 1958. Systematic induction of sprouting in white potatoes by foliar applications of gibberellin.* Plant Physiol. 33:132.

reaction mixture containing all the necessary components for this synthesis. In contrast, nondormant potato buds support an active DNA-dependent RNA synthesis.

Tuan and Bonner (47) found that ethylenechlorohydrin treatment of dormant potato buds causes a rapid synthesis of RNA in the buds (see Table 23–6). They suggested that the mechanism of action of 2-chloroethanol in the breaking of bud dormancy may be the derepression of repressed genes. The new RNA and protein

synthesis resulting from this derepression would then stimulate growth of the bud (break its dormancy).

The addition of GA to certain dormant plant tissues can stimulate the synthesis of new RNA and protein synthesis. Thus GA may also be able to derepress repressed genes. Since 2-chloroethanol mimics the action of GA on potato bud dormancy, it is entirely possible that GA, like 2-chloroethanol, breaks the dormancy of these buds by releasing repressed genes. In contrast, ABA inhibits DNA-dependent RNA synthesis and protein synthesis and counteracts the action of other phytohormones (gibberellins and cytokinins). Future studies on the interaction of these hormones with DNA should prove to be valuable to science and agriculture.

Questions

23–1. What is dormancy? Is the predisposition to dormancy in plants genetically controlled?

23–2. Discuss the significance of dormancy among plants growing in arid regions.

23–3. Describe five mechanisms of seed dormancy particularly as they involve seed germination.

Table 23–6. Effect of ethylenechlorohydrin on growth and RNA synthesis of dormant potato buds

Parameter	Days after Ethylenechloro-hydrin Treatment			
	0	2	6	10
fresh weight of buds (mg)	0.40	0.48	12.0	65.6
RNA content (μg/bud)	4.0	5.6	15.2	64.2
rate of RNA synthesis (mμg/2.5 hr/bud)	0.03	0.08	0.86	2.5
rate of RNA synthesis per unit bud DNA (mμg RNA synthesized/μg DNA/2.5 hr)	0.020	0.048	0.37	0.23

Source: *From J. Bonner. 1965.* Plant Biochemistry. *New York: Academic Press, p. 859.*

23–4. Define the following terms: scarification, afterripening, stratification, and vivipary.

23–5. Indicate some of the ways by which temperature influences seed germination.

23–6. Name some common chemical promoters of seed germination. How do you think each one functions?

23–7. Name some of the naturally occurring germination inhibitors.

23–8. Is abscisic acid a true dormancy promoter? Discuss your answer.

23–9. Budding of potatoes in storage is often an important commercial problem. What substances might we use to inhibit the bud development of potato tubers?

23–10. Discuss the evidence that suggests that bud dormancy is genetically determined.

Suggested Readings

Bewley, J.D. 1979. Physiological aspects of dessication tolerance. *Ann. Rev. Plant Physiol.* 30:195–238.

Chen, S.S.C., and J.E. Varner. 1970. Respiration and protein synthesis in dormant and afterripened seeds of *Avena fatua. Plant Physiol.* 46:108–112.

Firn, R.D., R.S. Burden, and H.F. Taylor. 1972. The detection and estimation of the growth inhibitor xanthoxin in plants. *Planta* 102:115–126.

Halloin, J.M. 1976. Inhibition of cotton seed germination with abscisic acid and its reversal. *Plant Physiol.* 57:454–455.

Pilet, P.E. 1977. Growth inhibitors in growing and geostimulated maize roots. In P.E. Pilet, ed. *Plant Growth Regulators.* New York: Springer-Verlag.

Taylorson, R.B., and S.B. Hendricks. 1977. Dormancy in seeds. *Ann. Rev. Plant Physiol.* 28:331–354.

Vegis, A. 1964. Dormancy in higher plants. *Ann. Rev. Plant Physiol.* 15:185–224.

Appendix A: Colloids

If ordinary clay soil is placed in water and agitated thoroughly, a murky liquid, which appears uniformly brown in color, results. If this mixture is allowed to stand, it clears rapidly. The larger particles settle out first, followed by the smaller particles. After a considerable length of time, however, it becomes evident that not all of the soil is going to settle out. Very minute soil particles, called *clay micelles,* remain suspended in the water. The stable, heterogeneous mixture that results is called a *colloid.* The suspended phase is called the *dispersed phase,* and the medium in which the dispersion takes place is called the *dispersion medium,* or *continuous phase.*

The term *colloid,* first suggested by Graham in 1861, is derived from the two Greek words *kolla,* which means "glue," and *eidos,* which means "like." Graham used the term to describe glue-like preparations, such as solutions of certain proteins, and liquid preparations of vegetable gums, such as gum arabic. Today the term *colloid* is used much more generally. There are many colloidal suspensions that are far from glue-like.

Colloidal suspensions are not limited to the dispersion of a solid in a liquid. The dispersion medium may be a liquid, gas, or solid. Smoke is composed of a solid dispersed in gas. Milk and mayonnaise are examples of a liquid dispersed in a liquid. Pumice stone is an example of a gas dispersed in a solid. In this discussion we are primarily interested in two general types of colloidal suspension: *sols,* colloidal systems with the property of fluidity; and *gels,* colloidal systems without the property of fluidity.

Colloidal Dimensions

Particles of colloidal dimension range in size from 1 to 200 nanometers (nm). A nanometer is one-thousandth of a micrometer (μm) or one-millionth of a millimeter. Small as this size may seem, however, it does not approach the minute size of most molecules. Colloidal particles are too small to be seen with the light microscope, but are large enough to scatter light. We can detect particles of colloidal dimension readily with the electron microscope. The size of colloidal particles is somewhere between the size of particles that form true solutions and the size of particles that are found in unstable suspensions.

Colloidal Systems

We may divide colloidal dispersions in liquids into two general classes, called *lyophilic* and *lyophobic* systems. In the lyophilic system the dispersed phase and the liquid dispersion medium are attracted to each other; in the lyophobic system the two phases actually repel each other. If the dispersion medium is water, then we use the terms *hydrophilic* (water-loving) and *hydrophobic* (water-fearing). When we add solids such as starch, gelatin, or agar to hot water, large quantities of water are taken up to form a hydrophilic colloidal sol. Sols of this type are formed rather easily, and no special method of preparation is employed. The dispersed particles of hydrophilic sol are *hydrated*—that is, water molecules are adsorbed to the surface of these particles. Water molecules nearest to the surface of the particles are adsorbed very tightly; those farther away are less tightly adsorbed.

Hydrophobic colloids are generally composed of compounds of an inorganic nature, and in most cases they are more difficult to prepare than are hydrophilic colloids. Condensation methods are frequently employed in the formation of this type of sol. In these methods smaller particles are induced to aggregate and, thereby, to form colloidal particles; they generally employ chemical reactions. For example, if we mix a concentrated solution of ferric chloride with hot water, a dark red colloidal suspension of ferric hydroxide results. The $FeCl_3$ ionizes, and hydrolysis of the ferric ion takes place to form ferric hydroxide $Fe(OH)_3$. The same reaction would take place in cold water but is considerably speeded up by the use of hot water. Aggregation of the $Fe(OH)_3$ molecules form the colloidal particles of the dispersed phase:

$$Fe^{3+} + 3Cl^- + 3H_2O \longrightarrow$$
$$Fe(OH)_3 + 3(H^+ + Cl^-)$$

In much the same manner, a colloidal suspension of arsenic sulfide is prepared by bubbling H_2S gas into a solution of arsenic oxide:

$$As_2O_3 + 3H_2S \longrightarrow As_2S_3 + 3H_2O$$

Emulsions

We may prepare an unstable emulsion by vigorously shaking two immiscible liquids together. Small droplets (dispersed phase) of one of the liquids will be dispersed throughout the other (dispersion medium). However, these small droplets have a ten-dency to coalesce and form larger and larger droplets until two distinct layers are formed and the two liquids are again separated.

An emulsion can be made stable by the addition of an *emulsifying agent*. These substances generally function either by decreasing the surface tension of the liquids, thereby reducing the tendency of the small droplets to combine, or by forming a protective layer or film around the droplets, thus making it impossible for them to combine with each other. Milk, a common emulsion, is composed of butter fat dispersed in water, with casein as an emulsifying agent.

Properties of Colloidal Suspensions

Tyndall Effect

If we pass a powerful narrow beam of light through a darkened room and view it at right angles, it is visible because some of the light is scattered in the direction of the observer. The scattering of light is due to the presence of dust particles of colloidal dimension floating in the air. This phenomenon is known as the *Tyndall effect* after its discoverer.

If we pass a narrow beam of light through a true solution, we cannot observe its path. In contrast, if we pass the light through a colloidal suspension, we can readily observe it. The particles of the dispersed phase are large enough to noticeably scatter light, but the particles of true solutions are too small to do so. We can thus use the Tyndall effect to distinguish a colloidal suspension from a true solution. Note, however, that we are not actually able to see the colloidal particles; we can only detect their presence through their ability to scatter some of the light that falls on them.

Brownian Movement

We can use the Tyndall effect with the microscope to study some of the properties of colloidal suspensions. The principle involved here is dark-field illumination. Dark-field illumination is achieved with a microscope through the use of a special condenser that focuses converging beams of light, which strike the stage at too oblique an angle to enter the objective. Because of the fact that no light enters the objective, a clean glass slide will appear completely black. However, if we allow the converging beams of light to pass through a colloidal suspension, the tiny colloidal particles will scatter some of the light that falls on them. Some of the scattered light enters the objective, thus allowing us to detect the colloidal particles as bright points of light against a black background. These bright points of light appear to move in an irregular random fashion and outline the path of the colloidal particle in suspension. The random motion is caused by the uneven bombardment of the colloidal particles by molecules of the dispersion medium. The colloidal particles are small enough to be moved by the molecules in the direction of least resistance, a direction that is changing constantly. This random motion of very small particles in suspension is called *Brownian movement*, after the botanist (Brown) who first described it. Brownian movement cannot be observed in a true solution and thus provides another way of distinguishing colloidal suspensions from true solutions.

Filtration

Although we cannot separate the dispersed phase from the dispersion medium with ordinary filter paper, we can separate colloidal particles from the dispersion medium with ultrafilters. Ultrafilters, composed of biologically inert cellulose esters (Millipore filters), have been constructed with pore sizes ranging from 10 nm to 5 μ. Since colloidal particles range in size from 1 to 200 nm, it is easy to see that we could separate the two phases of a colloidal suspension, in most cases by using these filters. However, we cannot separate the components of true solutions in this manner.

Adsorption

The tendency of molecules or ions to adhere to the surface of certain solids or liquids is known as *adsorption*. Since it is a surface phenomenon, the capacity of adsorption is dependent on the amount of surface exposed as well as on the chemical nature of the constituents involved. It is not surprising, therefore, that the adsorptive capacity of a colloidal suspension is extremely high for a given weight of colloidal particles. For example, the amount of surface area exposed by a cube measuring 1 cm on each edge is 6 cm^2. If this cube should be divided into cubes 0.1 μm on edge, the amount of surface area exposed would be 6×10^5 or 600,000 cm^2, an increase of 100,000 times over the original area. Undoubtedly, most of the important functions of colloidal systems found in the living cell are dependent on their immense adsorptive capacity.

Electrical Properties

Colloidal particles generally carry an electric charge. The charge may be either positive or negative, but in a given colloidal system it is the same for all particles. For example, all the particles of a ferric hydroxide colloidal suspension carry a positive charge; all the particles of a colloidal suspension of arsenic sulfide carry a negative charge.

The charges found on colloidal particles result from the adsorption of free ions in the dispersion medium. The preferential adsorption of positive ions by a colloidal particle will give it a positive charge; the preferential adsorption of negative ions results in a negatively charged colloidal particle. In the ferric hydroxide colloidal suspension, all of the particles have a positive charge because the ferric ion (Fe^{3+}), released in the ionization of $FeCl_3$, is preferentially adsorbed. The free chlorine ions (Cl^-) are attracted to the positive charge on the particles, and they also accumulate secondarily around the particles and form what is known as an *electrical double layer*. In the arsenic sulfide system, sulfide ions (S^{2-}) are preferentially adsorbed by the arsenic sulfide particles. Hydrogen ions, released in the ionization of H_2S, are secondarily adsorbed by the negatively charged particles.

One way of determining the electric charge on the particles of a colloidal suspension is to observe their direction of migration in an electric field. Under the influence of a direct current, all of the particles of a colloidal suspension will move in one direction. If the dispersed phase has a positive charge, the particles will collect at the cathode; and if the charge is negative, they will collect at the anode. This phenomenon was first called cataphoresis, but the term *electrophoresis* is now more widely used.

The fact that colloidal particles carry an electric charge and that all the particles of any one suspension carry the same charge is mainly responsible for the stability of colloidal suspensions. Units of like charge repel each other. If it were not for this, the colloidal particles would collide and aggregate and eventually precipitate.

Precipitation

Destruction or removal of the electric double layer will cause the dispersed particles of a colloidal suspension to collide, aggregate, and finally to precipitate. This chain of events may be accomplished by the addition of an electrolyte. For example, the addition of HCl to a colloidal suspension of arsenic sulfide will cause precipitation. Hydrogen ion concentration is increased by the addition of HCl to the extent that it causes H_2S to form ($2H^+ + S^{2-} \longrightarrow H_2S$). The removal of the negative charges on the sulfide ions causes the particles to be neutralized. The extent to which an ion will cause precipitation when added to a colloidal suspension depends on its valence. The monovalent sodium ion, for example, is less efficient than the divalent barium ion, which in turn is much less efficient than the trivalent aluminum ion.

We can observe an interesting effect of ions on colloidal suspensions at the mouth of a river. The charged ions of ocean water cause the negatively charged clay micelles of river water to lose their charge and settle out. This phenomenon leads eventually to the formation of a delta, which is often found at the mouth of a river.

Sometimes a colloid can be protected from precipitation by the presence of another colloid. Apparently, one colloid forms a protective film around the particles of the other colloid. Gelatin and gum arabic are the two colloids most widely used in this capacity. For example, the colloidal dispersion of silver halides on photographic plates is protected by the gelatin on these plates.

One of the properties of certain lyophilic sols is their ability, under certain conditions, to form an extremely viscous solidlike mass. Thus hot aqueous gelatin or agar sols will set when cooled and will form a jellylike mass called a *gel*. The conversion of a sol to a gel is called *gelation*. If a gel of agar or gelatin is again heated, it will convert back to a sol, a process known as *solation*. The addition of dilute hydrochloric acid to sodium silicate will form a *silica gel*— a colloidal dispersion of hydrated silicon dioxide.

Appendix B: Review of pH and Buffers

Acidic and basic solutions are unquestionably vital to the living system. A variety of substances that are either acids or bases are produced during the metabolic activity of the cell. Amino acids, fatty acids, and Krebs cycle intermediates are good examples of acids found in the living system. Purines and pyrimidines, organic bases that are quite common in the cell, are required for the synthesis of nucleic acids and high-energy intermediates.

Acids and bases in solution may be distinguished in several ways. Acids have a sour taste. Lemons are sour because of their high content of citric acid, and milk turns sour because of the lactic acid produced in it by bacteria. Certain natural dyes change color from blue to red when treated with an acid, and with metals such as zinc there is a release of hydrogen on contact with an acid. Finally, an acid can neutralize a base to form water and a salt.

Bases have a bitter taste and can also cause color changes in certain natural dyes. Bases in solution feel soapy to the touch. A base can neutralize an acid to form water and a salt. However, these distinguishing features do not tell us anything about the chemistry of acids and bases. We are still faced with the question of what acids and bases are.

Nature of Acids, Bases, and Salts

An *acid* is any molecule or ion that can donate a proton (H^+) to any other molecule or ion. If an acid is dissolved in water, it reacts with the water and ionization takes place. *Ionization* may be defined as a reaction between a solute and a solvent resulting in the formation of ions.

In the equation $HA \rightleftharpoons H^+ + A^-$, the acid (HA) undergoes ionization to form positive (H^+) and negative (A^-) ions. *Ions* are atoms or groups of atoms that are electrically charged. Ions that carry a positive charge are called *cations*, and ions that carry a negative charge are called *anions*. In an aqueous solution, cations will migrate toward a negative electrode (cathode), and anions will migrate toward a positive electrode (anode). The hydrogen ion (H^+) is called a *proton*.

A *base* is any molecule or ion that will accept a proton. If a base is dissolved in water, ionization takes place. In the equation $BOH \rightleftharpoons B^+ + OH^-$, the base (BOH) undergoes ionization to form positive (B^+) and negative (OH^-) ions.

Electrolytes and Nonelectrolytes

Electrolytes are substances that conduct an electric current when they are dissolved in water. The passage of an electric current through an aqueous solution of an electrolyte results in the decomposition of the electrolyte. This process is called *electrolysis*. For example, if an electric current is allowed to pass through an aqueous solution of hydrochloric acid, hydrogen gas will be released at the cathode and chlorine gas at the anode. Acids, bases, and salts are electrolytes. Their ability to conduct electricity results from the fact that electrically charged ions are formed when they are dissolved in water. Sugars and alcohols do not undergo ionization when dissolved in water and, therefore, are termed *nonelectrolytes*.

Strength of Acids or Bases

The ease with which an acid will yield a proton is a measure of its strength. Strong acids will yield protons readily, weak acids will yield protons slowly. Strong bases are compounds that accept protons readily, weak bases have only a very weak affinity for protons. Almost complete ionization takes place when either a strong acid or a strong base is dissolved in water. In contrast, only slight ionization takes place when either weak acids or weak bases are dissolved in water. Table B–1 gives a list of acids and bases of varying strength.

Amphoteric Compounds

Water can act either as a weak acid or a weak base—that is, it can either donate or accept a proton. Amino acids, components of protein molecules, are also good examples of compounds that can act as weak acids or weak bases. Compounds that act both as an acid and as a base are said to be *amphoteric*. An example of water acting as a base in the presence of HCl and as an acid in the presence of NH_3 follows:

Water as a base: $H_2O + HCl \rightleftharpoons H_3O^+ + Cl^-$

Water as an acid: $H_2O + NH_3 \rightleftharpoons NH_4^+ + OH^-$

Water in the presence of a strong acid such as HCl acts as a base and accepts a proton to form a hydronium ion (H_3O^+); water in the presence of ammonia, a base, acts as an acid and donates a proton.

Neutralization of Acids and Bases

If equivalent amounts of aqueous solutions of HCl and NaOH are mixed, acidic and ba-

Acid	Formula	Ions	Strength
hydrochloric	HCl	$H^+ + Cl^-$	strong
sulfuric	H_2SO_4	$H^+ + HSO_4^-$	strong
nitric	HNO_3	$H^+ + NO_3^-$	strong
acetic	CH_3COOH	$H^+ + CH_3COO^-$	weak
sulfurous	H_2SO_3	$H^+ + HSO_3^-$	weak

Base	Formula	Ions	Strength
sodium hydroxide	NaOH	$Na^+ + OH^-$	strong
potassium hydroxide	KOH	$K^+ + OH^-$	strong
ammonium hydroxide	NH_4OH	$NH_4^+ + OH^-$	weak

Table B–1. Strength of some common acids and bases.

Reaction	Name of Salt	Formula	
HCl + NaOH	sodium chloride	NaCl	*Table B–2.* A few common acid-base neutralizations and name and formula of the salt formed.
HCl + KOH	potassium chloride	KCl	
H_2SO_4 + 2KOH	potassium sulfate	K_2SO_4	
2HCl + $Ca(OH)_2$	calcium chloride	$CaCl_2$	
CH_3COOH + NaOH	sodium acetate	CH_3COONa	

sic properties are lost. *Neutralization* is said to have taken place. The loss of acidic and basic properties occurs because free hydrogen ions, which give a solution its acidic character, have reacted with free hydroxyl ions, which give a solution its basic character, to form water. The free sodium and chlorine do not enter into the reaction:

$$H^+ + Cl^- + Na^+ + OH^- \rightleftharpoons$$
$$H_2O + Cl^- + Na^+$$

If the water of the resulting solution is evaporated, crystals of sodium chloride salt are left. In other words, a salt is formed when acid and base solutions are mixed. For example, when a solution of NaOH is added to an acetic acid solution, the salt, sodium acetate, is formed. Table B–2 gives some acid-base neutralizations.

Normal Solutions

The dissolution of a gram equivalent weight of a substance in 1 liter of solution results in a *normal* solution of that substance. The dissolution of 2 gram equivalent weights in a liter of solution will give a 2 normal (2N) solution, and so forth. The *gram equivalent weight* of an element is the weight in grams of the element that combines with or is otherwise equivalent to 1.008 grams of hydrogen. The gram equivalent weight of a compound is the weight of the compound that will interact with one equivalent weight of an element. It is more convenient to express the concentrations of acid and base solutions in terms of normality than molarity. The gram equivalent weight of an acid or base is the quantity that will release or neutralize 1 mole of hydrogen ion. Thus, a 1M solution of HCl is also a 1N solution of the acid. When expressed in terms of normality, however, a 1M solution of H_2SO_4 is 2N. This occurs because H_2SO_4 is capable of releasing 2 moles of hydrogen ion. A 1M solution of NaOH is also 1N since the mole of hydroxide ion released in solution can neutralize 1 mole of hydrogen ion. On the other hand, a 1M solution of barium hydroxide, $Ba(OH)_2$, is 2N since in solution the 2 moles of hydroxide ion released are capable of neutralizing 2 moles of hydrogen ion. From this discussion we should realize that it takes 10 ml of a 1N solution of HCl to completely neutralize 10 ml of a 1N solution of NaOH. (What quantity of a 1N solution of H_2SO_4 would be needed to neutralize 10 ml of a 1N solution of NaOH?)

Hydrogen Ion Concentration

The acidity or basicity of a solution is determined by its hydrogen ion concentration. Conventionally, the hydrogen ion concentration of a solution is expressed as its negative logarithm or pH value.

$$pH = -\log[H^+]$$

The term *pH*, which stands for "potential of hydrogen," may be defined, therefore, as the negative logarithm of the hydrogen ion concentration.

The pH scale covers a range of values from 0 to 14. The hydrogen ion concentration in a liter of pure water is $0.0000001N$ or 10^{-7}. Since pH is equal to the negative logarithm of the hydrogen ion concentration, then

$$pH = -\log 10^{-7}$$

$$pH = \log \frac{1}{10^{-7}} = 7$$

Pure water thus has a pH of 7 and is considered neutral. Any pH values below 7 indicate acid solutions, and any pH values above 7 indicate basic solutions.

A solution with a pH of 8 has a hydrogen ion concentration ten times less than a solution with a pH of 7—that is, its hydrogen ion concentration is $0.00000001N$ or 10^{-8}. We can see that pH values differ by a factor of ten and that solutions with low pH values are strongly acidic and solutions with high pH values are strongly basic. Table B–3 provides a chart of pH values.

Buffer Solutions

A solution that contains a weak acid and its salt (e.g., acetic acid and sodium acetate) or a weak base and its salt will resist changes in hydrogen ion concentration when small amounts of a strong acid or base are added to it. These solutions are called *buffer solutions*. In biological systems the vast array of

Table B–3. pH scale.

Hydrogen Ion Concentration in Terms of Normality	Exponential Form	pH Value
1	10^{0}	0
0.1	10^{-1}	1
0.01	10^{-2}	2
0.001	10^{-3}	3
0.0001	10^{-4}	4
0.00001	10^{-5}	5
0.000001	10^{-6}	6
0.0000001	10^{-7}	7
0.00000001	10^{-8}	8
0.000000001	10^{-9}	9
0.0000000001	10^{-10}	10
0.00000000001	10^{-11}	11
0.000000000001	10^{-12}	12
0.0000000000001	10^{-13}	13
0.00000000000001	10^{-14}	14

amino acids and proteins operate as the primary buffers.

Let us use the common buffer pair, acetic acid and sodium acetate, to explain buffer action. Acetic acid is a weak acid and therefore ionizes only slightly in solution. If we add a small amount of NaOH, the hydroxyl ions released in solution are neutralized by the free hydrogen ions in the buffer solution. This phenomenon causes more of the acetic acid to ionize and thus restores the original hydrogen ion concentration. As more NaOH is added, more acetic acid ionizes until all of the acetic acid in solution is ionized. At this point, any further addition of NaOH will cause an abrupt rise in pH. If a small amount of hydrochloric acid is added to the acetic acid–sodium acetate buffer, the hydrogen ions that are released rapidly unite with free acetate ions to form undissociated acetic acid. Therefore, there is no change in hydrogen ion concentration. Let us remember, sodium acetate exists in solution as free sodium and acetate ions. As more HCl is added, more free acetate ions are converted to acetic acid, until all of the acetate ions have been converted. When this happens, any further addition of HCl will cause an abrupt drop in pH.

Buffer solutions (dissolved proteins) are abundant in living plant cells and play vital roles in their existence. Enzymes, the organic catalysts of life, generally function within narrow pH ranges. A deviation of any magnitude impairs or completely inhibits their function; living systems do not tolerate any large increases or decreases in hydrogen ion concentration. In this respect, the most fundamental action of pH on living systems is its influence on enzyme activity and reaction rates.

References

Chapter 1

1. Albersheim, P. 1958. Recent developments in the chemistry of cell walls. *Plant Physiol.* 33 (Suppl.): XIVi–XIVii.
2. Albersheim, P. 1965. The substructure of the cell wall. In J. Bonner and J.E. Varner, eds., *Plant Biochemistry.* New York: Academic Press.
3. Albersheim, P. 1975. The wall of growing plant cells. *Sci. Amer.* 232(4):80.
4. Bauer, W.D., K.W. Talmadge, K. Keegstra, and P. Albersheim. 1973. The structure of plant cell walls. II. The hemicellulose of the walls of suspension-cultured sycamore cells. *Plant Physiol.* 51:174.
5. Beer, M., and G. Setterfield. 1958. Fine structure in thickened primary walls of collenchyma cells of celery petioles. *Am. J. Bot.* 45:571.
6. Birnstiel, M., and B. Hyde. 1963. Protein synthesis by isolated pea nuclei. *J. Cell Biol.* 18:41.
7. Bishop, C., S. Bayley, and G. Setterfield. 1958. Chemical constitution of the primary cell walls of *Avena* coleoptiles. *Plant Physiol.* 33:283.
8. Bonner, J. 1965. The nucleus. In J. Bonner and J.E. Varner, eds., *Plant Biochemistry.* New York: Academic Press.
9. Brachet, J. 1957. *Biochemical Cytology.* New York: Academic Press.
10. Breidenbach, R.W., P. Castelfranco, and R.S. Criddle. 1967. Biogenesis of mitichondria in germinating peanut cotyledons. II. Changes in cytochromes and mitochondrial DNA. *Plant Physiol.* 42:1035.
11. Davson, H.A., and J.F. Danielli. 1943. The permeability of natural membranes. London: Cambridge University Press.
12. Davson, H.A., and J.F. Danielli. 1952. The permeability of natural membranes, 2nd ed. London: Cambridge University Press.
13. Gibor, A., and S. Granick. 1964. Plastids and mitochondria. *Science* 145:890.
14. Goodwin, T.W., and E.I. Mercer. 1973. *Introduction to Plant Biochemistry.* New York: Pergamon Press.
15. Green, D. 1959. Electron transport and oxidation phosphorylation. *Adv. Enzymol.* 21:73.
16. Green, D. 1959. Mitochondrial structure and function. In T. Hayashi, ed., *Subcellular Particles.* New York: Ronald Press.
17. Hodge, A., J. McLean, and F. Mercer. 1956. A possible mechanism for the morphogenesis of lamellar systems in plant cells. *J. Biophys. Biochem. Cytol.* 2:597.
18. Hoffman, H., and G. Grigg. 1958. An electron microscopic study of mitochondria formation. *Expertl. Cell Res.* 15:118.
19. Jensen, W. 1960. The composition of the developing primary wall in onion root tip cells. II. Cytochemical localization. *Am. J. Bot.* 47:287.
20. Keegstra, K., K.W. Talmadge, W.D. Baner, and P. Albersheim. 1973. The structure of plant cell walls. III. A model of the walls of suspension-cultured sycamore cells based on the interconnections of the macromolecular components. *Plant Physiol.* 51:188.
21. Kerr, T. 1951. Growth and structure of the primary wall. In F. Skoog, ed., *Plant Growth Substances.* Madison: University of Wisconsin Press.
22. Ledbetter, M.C., and K.R. Porter. 1964. Morphology of microtubules of plant cells. *Science* 144:872.
23. Mettler, I.J., and R.T. Leonard. 1979. Isolation and partial characterization of vacuoles from tobacco protoplasts. *Plant Physiol.* 64:1114.
24. Pollard, C.J., A. Stemler, and D.F. Blaydes. 1966. Ribosomal ribonucleic acids of chloroplastic and mitochondrial preparations. *Plant Physiol.* 41:1323.

25. Porter, K., and R. Machado. 1960. Studies on the endoplasmic reticulum. IV. Its form and distribution during mitosis in cells of onion root tip. *J. Biophys. Biochem. Cytol.* 7:167.

26. Preston, R. 1955. The submicroscopic structure of plant cell walls. In W. Ruhland, ed., *Encyclopedia of Plant Physiology* 1:731. Berlin: Springer.

27. Preston, R. 1955. Mechanical properties of the cell wall. In W. Ruhland, ed., *Encyclopedia of Plant Physiology* 1:745. Berlin: Springer.

28. Ramsey, J.C., and J.D. Berlin. 1976. Ultrastructural aspects of early stages in cotton fiber elongation. *Am. J. Bot.* 63:868.

29. Ray, P. 1958. Composition of cell walls of *Avena* coleoptiles. *Plant Physiol.* 33(Suppl.): XIVii.

30. Siegel, S. 1962. *The Plant Cell Wall.* New York: Macmillan.

31. Singer, S.J., and G.L. Nicolson. 1972. The fluid mosaic model of the structure of cell membranes. *Science* 175:720.

32. Suzama, Y., and W.D. Bonner, Jr. 1966. DNA from plant mitochondria. *Plant Physiol.* 41:383.

33. Wardrop, A., and D. Bland. 1959. The process of lignification in woody plants. *Proc. 4th Intl. Congr. Biochem.* New York: Pergamon Press.

34. Wareing, P.F., and I.D.J. Phillips. 1970. *The Control of Growth and Differentiation in Plants.* New York: Pergamon Press.

35. Watson, M. 1959. Further observations on the nuclear envelope of the animal cell. *J. Biophys. Biochem. Cytol.* 6:147.

36. Whaley, W., J. Kephart, and H. Mollenhauer. 1959. Developmental changes in the Golgi apparatus of maize root cells. *Am. J. Bot.* 46:743.

37. Whaley, W., H. Mollenhauer, and J. Kephart. 1959. The endoplasmic reticulum and Golgi structures in maize root cells. *J. Biophys. Biochem. Cytol.* 5:501.

38. Whaley, W., H. Mollenhauer, and J. Leech. 1960. The ultrastructure of the meristematic cell. *Am. J. Bot.* 47:401.

39. Wilder, B.M., and P. Albersheim. 1973. The structure of plant cell walls. IV. A structural comparison of the wall hemicellulose of cell suspension cultures of sycamore (*Acer pseudoplatanus*) and red kidney bean (*Phaseolus vulgaris*). *Plant Physiol.* 51:889.

40. Yatsu, L.Y., and T.J. Jacks. 1972. Spherosome membranes. *Plant Physiol.* 49:937.

41. Ziegler, D., A. Linnana, and D. Green. 1958. Studies on the electron transport system. XI. Correlation of the morphology and enzymatic properties of mitochondrial and sub-mitochondrial particles. *Biochem. Biophys. Acta* 28:524.

Chapter 2

1. Baron, W.M. 1967. *Water and Plant Life.* London: Heinemann Educational Books.

2. Dainty, J. 1963. Water relations of plant cells. In R.D. Preston, ed., *Advances in Botanical Research.* New York: Academic Press.

3. Kozlowski, T.T. 1964. *Water Metabolism in Plants.* New York: Harper & Row.

4. Kramer, P.J. 1969. *Plant and Soil Water Relationships.* New York: McGraw-Hill.

5. Schull, C.A. 1916. Measurement of the surface forces in soils. *Bot. Gaz.* 62:1.

6. Schull, C.A. 1920. Temperature and rate of moisture intake in seeds. *Bot. Gaz.* 69:361.

7. Slatyer, R.O. 1967. *Plant-Water Relationships.* New York: Academic Press.

8. Sutcliffe, J. 1968. *Plants and Water.* London: Edward Arnold.

9. Taylor, S.A. 1968. Terminology in plant and soil water relations. In T.T. Kozlowski, ed., *Water Deficits and Plant Growth.* New York: Academic Press.

10. Weatherley, P.E. 1970. Some aspects of water relations. In R.D. Preston, ed., *Advances in Botanical Research.* New York: Academic Press.

Chapter 3

1. Addoms, R.M. 1946. Entrance of water into suberized roots of trees. *Plant Physiol.* 21:109.

2. Anlel, O.M. van. 1953. The influence of salts on the exudation of tomato plants. *Acta Botan. Neerl.* 2:445.

3. Bennet-Clark, T.A., A.D. Greenwood, and J.W. Barker. 1936. Water relations of osmotic pressures of plant cells. *New Phytol.* 35:277.

4. Breazeale, E.L. 1950. Moisture absorption by plants from an atmosphere of high humidity. *Plant Physiol.* 25:413.

5. Breazeale, J.F., and W.T. McGeorge. 1953. Exudation pressure in roots of tomato plants under humid conditions. *Soil Sci.* 75:293.

6. Breazeale, E.L., W.T. McGeorge, and J. F. Breazeale. 1951. Movement of water vapor in soils. *Soil Sci.* 71:181.

7. Breazeale, E.L., W.T. McGeorge, and J. F. Breazeale. 1951. Water absorption by leaves. *Soil Sci.* 72:293.

8. Cormack, R.G.H. 1949. The development of root hairs in angiosperms. *Bot. Rev.* 15:583.

9. Crafts, A.S., and T.C. Broyer. 1938. Migration of salts and water into xylem of the roots of higher plants. *Am. J. Bot.* 25:529.

10. Dixon, H.H. 1909. Vitality and the transmission of water through the stems of plants. Notes Botany School, Trinity College, Dublin, 2:5; *Sci. Proc. Roy. Dublin Soc.* 12:21.

11. Dixon, H.H. 1910. Transpiration and the ascent of sap. *Progressus Rei Botanicae* 3:1.

12. Dixon, H.H. 1914. *Transpiration and the Ascent of Sap in Plants.* London: Macmillan.

13. Dixon, H.H. 1924. *The Transpiration Stream.* London: University of London Press.

14. Esau, K. 1958. *Plant Anatomy.* New York: Wiley.

15. Fox, D.G. 1933. Carbon dioxide narcosis. *J. Cell. Comp. Physiol.* 3:75.

16. Fritts, H.C. 1958. An analysis of radial growth of beech in a central Ohio forest during 1954–1955. *Ecology* 39:705.

17. Gessner, F. 1956. Die Wasseraufnahme durch Blätter und Samen. In W. Ruhland, ed., *Encyclopedia of Plant Physiology* 3:215. Berlin: Springer.

18. Grossenbacher, K.A. 1938. Diurnal fluctuation in root pressure. *Plant Physiol.* 13:669.

19. Grossenbacher, K.A. 1939. Autonomic cycle of rate of exudation of plants. *Am. J. Bot.* 26:107.

20. Haise, H.R., H.J. Haas, and L.R. Jensen. 1955. Soil moisture studies of some Great Plains soils. II. Field capacity as related to $\frac{1}{3}$ atmosphere percentage and "Minimum Point" as related to 15- and 26-atmosphere percentages. *Proc. Soil Sci. Soc. Am.* 10:20.

21. Kozlowski, T.T. 1964. *Water Metabolism in Plants.* New York: Harper & Row.

22. Kramer, P.J. 1937. The relation between rate of transpiration and rate of absorption of water in plants. *Am. J. Bot.* 24:10.

23. Kramer, P.J. 1956. Roots as absorbing organs. In W. Ruhland, ed., *Encyclopedia of Plant Physiology* 3:188. Berlin: Springer.

24. Kramer, P.J. 1959. Transpiration and the water economy of plants. In F.C. Steward, ed., *Plant Physiology.* New York: Academic Press.

25. Kramer, P.J. 1969. *Plant and Soil Water Relationships.* New York: McGraw-Hill.

26. Kramer, P.J., and W.T. Jackson. 1954. Causes of injury to flooded tobacco plants. *Plant Physiol.* 29:214.

27. Lott, N.A., and J.J. Darley. 1976. *A Scanning Electron Microscope Study of Green Plants.* St. Louis: Mosby.

28. McDermott, J.J. 1941. The effect of the method of cutting on the moisture content of samples from tree branches. *Am. J. Bot.* 28:506.

29. Meyer, B.S., D.B. Anderson, R.H. Bönning, D.G. Fratianne. 1973. *Introduction to Plant Physiology,* 2nd ed. New York: Van Nostrand.

30. Münch, E. 1931. *Die Stoffbenegungen in der Pflanze.* Stuttgart: Gustav Fisher Verlag.

31. Overton, J.B. 1911. Studies on the relation of the living cells to the transpiration and sap-flow in *Cyperus.* II. *Bot. Gaz.* 51:102.

32. Richards, L.A., and L.R. Weaver. 1944. Moisture retention by some irrigated soils as related to soil moisture tension. *J. Agr. Res.* 69:215.

33. Roberts, E.A., M.D. Southwick, and D.H. Palmiter. 1948. A microchemical examination of McIntosh apple leaves showing relationship of cell wall constituents to penetration of spray solutions. *Plant Physiol.* 23:557.

34. Seifriz, W. 1942. *Some Physical Properties of Protoplasm and Their Bearing on Structure: The Structure of Protoplasm.* Ames: Iowa State College Press.

35. Skoog, F., T.C. Broyer, and K.A. Grossenbacher. 1938. Effect of auxin on rates, periodicity, and osmotic relations in exudation. *Am. J. Bot.* 25:749.

36. Slatyer, R.O. 1955. Studies of the water relations of crop plants grown under natural rainfall in northern Australia. *Australian J. Agr. Research* 6:365.

37. Slatyer, R.O. 1957. The significance of the permanent wilting percentage in studies of plant and soil water relations. *Bot. Rev.* 23:585.

38. Slatyer, R.O. 1957. The influence of progressive increases in total soil moisture stress on transpiration, growth and internal water relationships of plants. *Australian J. Biol. Sci.* 10:320.

39. Stiles, W. 1924. *Permeability.* London: Wheldon & Wesley.

40. Stocking, C.R. 1956. Root pressure. In W. Ruhland, ed., *Encyclopedia of Plant Physiology* 3:583. Berlin: Springer.

41. Strasburger, E. 1891. Über den Bau und die Verrichtungen der Leitungsbahnen in den Pflanzen. *Hist. Beitr. Jena* 3:609.

42. Strasburger, E. 1893. Über das Saftsteigen. *Hist. Beitr. Jena* 5:1.

43. Thimann, K.V. 1951. Studies on the physiology of cell enlargement. *Growth Symposium* 10:5.

44. Thut, H.F. 1932. Demonstrating the lifting power of transpiration. *Am. J. Bot.* 19:358.

45. Vaadia, Y. 1960. Autonomic diurnal fluctuations in rate of exudation and root pressure of decapitated sunflower plants. *Physiol. Plant.* 13:701.

46. Wadleigh, C.H., and A.D. Ayers. 1945. Growth and biochemical composition of bean plants as conditioned by soil moisture tension and salt concentration. *Plant Physiol.* 20:106.

47. Wadleigh, C.H., H.G. Gauch, and O.C. Magistad. 1946. Growth and rubber accumulation in guayule as conditioned by soil salinity and irrigation regime. *U.S. Dept. Agr. Tech. Bull.* 925.

48. White, P.R. 1938. "Root pressure"—an unappreciated force in sap movement. *Am. J. Bot.* 25:223.

Chapter 4

1. Bailey, L.F., J.S. Rothacher, and W. H. Cummings. 1952. A critical study of the cobalt chloride method of measuring transpiration. *Plant Physiol.* 27:563.

2. Barnett, N.M., and A.W. Naylor. 1966. Amino acid and protein metabolism in Bermuda grass during water stress. *Plant Physiol.* 41:1222.

3. Baron, W.M.M. 1967. *Water and Plant Life.* London: Heinemann.

4. Black, C.C., J.F. Turner, M. Gibbs, D.W. Krogmann, and S.A. Gordon. 1962. Studies on photosynthetic processes. II. Action spectra and quantum requirement for triphosphopyridine nucleotide reduction and the formation of adenosine triphosphate by spinach chloroplasts. *J. Biol. Chem.* 237:580.

5. Brown, H.T., and F. Escombe. 1900. Static diffusion of gases and liquids in relation to the assimilation of carbon and translocation of plants. *Phil. Trans. Roy. Soc. (London)*, B, 193:223.

6. Brown, W.V., and G.A. Pratt. 1965. Stomatal inactivity in grasses. *Southwest. Natur.* 10:48.

7. Chen, D.B., B. Kessler, and S.P. Monselise. 1964. Studies on water regime and nitrogen metabolism of citrus seedlings grown under water stress. *Plant Physiol.* 39:379.

8. Cullinan, F.P. 1920. Transpiration studies

with the apple. *Proc. Am. Soc. Hort. Sci.* 17:232.

9. Cummings, W.H.A. 1941. A method of sampling the foliage of a silver maple tree. *J. Forestry* 39:382.

10. Curtis, L.C. 1943. Deleterious effects of guttated fluids on foliage. *Am. J. Bot.* 30:778.

11. Esau, K. 1965. *Plant Anatomy*, 2nd ed. New York: Wiley.

12. Fujino, M. 1959. Stomatal movement and active migration of potassium (translated). *Kagaku* 29:660.

13. Goatley, J.L., and R.W. Lewis. 1966. Composition of guttation fluid from rye, wheat, and barley seedlings. *Plant Physiol.* 41:373.

14. Griep, W. 1940. Über den Einfluss von Aussenfaktoren auf die Wirkung des Windes auf die Transpiration der Pflanzen. *Z. Bot.* 35:1.

15. Guttenberg, H. 1959. Die physiologische Anatomie der Spaltöffnungen. In W. Ruhland, ed., *Encyclopedia of Plant Physiology* 17:399. Berlin: Springer.

16. Harrison, M.A., and P.F. Saunders. 1975. The abscisic content of dormant birch buds. *Planta* 123:291.

17. Harrison, M.A., and D.C. Walton. 1975. Abscisic acid metabolism in water-stressed bean leaves. *Plant Physiol.* 56:250.

18. Hauke, R.L. 1957. The stomatal apparatus of equisetum. *Bull. Torrey Bot. Club* 84:178.

19. Heath, O.V.S. 1952. The role of starch in the light response of stomata. Part II. The light response of stomata *Allium cepal*, together with some preliminary observations on the temperature response. *New Phytol.* 51:30.

20. Heath, O.V.S. 1959. The water relations of stomatal cells and the mechanisms of stomatal movement. In F.C. Steward, ed., *Plant Physiology*. New York: Academic Press.

21. Hylmö, B. 1955. Passive components in the ion absorption of the plant. I. The zonal ion and water absorption in Brouwer's experiments. *Physiol. Plant.* 8:433.

22. Imamura, S. 1943. Investigations of the mechanisms of turgor changes in guard cells (translated). *Jap. J. Bot.* 12:251.

23. Invanoff, S.S. 1944. Guttation-salt injury on leaves of cataloupe, pepper, and onion. *Phytopathology* 34:436.

24. Kelley, V.W. 1932. The effect of pruning of excised shoots on the transpiration rate of some deciduous fruit species. *Proc. Am. Soc. Hort. Sci.* 29:71.

25. Kemble, A.R., and H.T. Macpherson. 1954. Liberation of amino acids in perennial rye grass during wilting. *Biochem J.* 58:46.

26. Kozlowski, T.T. 1955. Tree growth, action and interaction of soil and other factors. *J. Forestry* 53:508.

27. Kozlowski, T.T. 1958. Water relations and growth of trees. *J. Forestry* 56:498.

28. Kozlowski, T.T. 1964. *Water Metabolism in Plants*. New York: Harper & Row.

29. Kramer, P.J. 1957. Outer space in plants. *Science* 125:633.

30. Kramer, P.J. 1959. Transpiration and the water economy of plants. In F.C. Steward, ed., *Plant Physiology*. New York: Academic Press.

31. Levitt, J. 1974. The mechanism of stomatal movement—once more. *Protoplasma* 82:1.

32. Lloyd, F.E. 1908. The physiology of stomata. *Carnegie Inst. Wash. Publ.* 82:1.

33. Loftfield, J.V.G. 1921. The behavior of stomata. *Carnegie Inst. Wash. Publ.* 314:1.

34. Manners, D.J. 1973. Starch and inulin. In L.P. Miller, ed., *Phytochemistry*. New York: Van Nostrand Reinhold.

35. Mansfield, T.A. 1965. Responses of stomata to short duration increases in carbon dioxide concentration. *Physiol. Plant.* 18:79.

36. Martin, E.V., and F.E. Clements. 1935. Studies of the effect of artificial wind on growth and transpiration in *Helianthus annuus*. *Plant Physiol.* 10:613.

37. Maximov, N.A. 1928. *The Plant in Relation to Water*. English translation by R.H. Yapp. London: Allen & Unwin.

38. Meidner, H., and T.A. Mansfield. 1965. Stomatal responses to illumination. *Biol. Rev.* 40:483.

39. Meyer, B.S. 1956. The hydrodynamic system. In W. Ruhland, ed., *Encyclopedia of Plant Physiology* 3:596. Berlin: Springer.

40. Miller, E.C. 1938. *Plant Physiology*. New York: McGraw-Hill.

41. Möller, C.M. 1947. The effect of thinning, age, and site of foliage, increment, and loss of dry matter. *J. Forestry* 45:393.

42. Parker, J. 1949. Effects of variations in the root-leaf ratio on transpiration rate. *Plant Physiol.* 24:739.

43. Pettersson, S. 1960. Ion absorption in young sunflower plants. I. Uptake and transport mechanisms for sulphate. *Physiol. Plant.* 13:133.

44. Raschke, K. 1965. Die Stomata als Glieder eines schwengungsfahigen CO_2 Regelsystems Experimentelles Nachweis an *Zea mays*. L. Z. *Naturforsch.* 20:1261.

45. Raschke, K. 1975. Stomatal action. *Ann. Rev. Plant Physiol.* 26:309.

46. Satoo, T. 1955. The influence of wind on transpiration of some conifers. *Bull. Tokyo Univer. Forests* 50:27.

47. Satoo, T. 1962. Wind, transpiration, and tree growth. In T.T. Kozlowski, ed., *Tree Growth*. New York: Ronald Press.

48. Sayre, J.D. 1926. Physiology of the stomata of *Rumex patientia*. *Ohio J. Sci.* 26:233.

49. Scarth, G.W. 1932. Mechanism of the action of light and other factors on stomatal movement. *Plant Physiol.* 7:481.

50. Shapiro, S. 1951. Stomata on the ovules of *Zamia floridana*. *Am. J. Bot.* 38:47.

51. Small, J., M.I. Clarke, and J. Crosbie-Baird. 1942. pH phenomena in relation to stomatal opening. *Proc. Roy. Soc.* (Edinburgh) II.-V., B, 61:233.

52. Steward, F.C. 1964. *Plants at Work*. Reading, Mass.: Addison-Wesley.

53. Sutcliffe, J. 1968. *Plants and Water*. Santa Ana, Calif.: Arnold.

54. Ting, I.P., and W.E. Loomis. 1963. Diffusion through stomates. *Am. J. Bot.* 50:866.

55. Turrell, F.M. 1936. The area of the internal exposed surface of dicotyledon leaves. *Am. J. Bot.* 23:255.

56. Turrell, F.M. 1944. Correlation between internal surface and transpiration rate in mesomorphic and xeromorphic leaves grown under artificial light. *Bot. Gaz.* 105:413.

57. Wilson, C.C. 1948. The effect of some environmental factors on the movements of guard cells. *Plant Physiol.* 23:5.

58. Winneberger, J.H. 1958. Transpiration as a requirement for growth of land plants. *Physiol. Plant.* 11:56.

59. Wright, S.T.C., and R.W.P. Hiron. 1972. The accumulation of abscisic acid in plants during wilting and under other stress conditions. In D.J. Can, ed., *Plant Growth Substances*. New York: Springer-Verlag.

60. Wylie, R.B. 1948. The dominant role of the epidermis in leaves of adiatum. *Plant Physiol.* 35:465.

61. Yemm, E.W., and A.J. Willis. 1954. Stomatal movements and changes of carbohydrates in leaves of *Chrysanthemum maximum*. *New Phytologist* 53:373.

62. Yin, H.C., and Y.T. Tung. 1948. Phosphorylase in guard cells. *Science* 108:87.

63. Zelitch, I. 1961. Biochemical control of stomatal opening in leaves. *Proc. Natl. Acad. Sci., U.S.* 47:1423.

64. Zelitch, I. 1963. The control and mechanisms of stomatal movement. In I. Zelitch, ed., *Stomata and Water Relations in Plants*. New Haven: Connecticut Agr. Exp. Sta.

Chapter 5

1. Allen, M.B. 1952. The cultivation of Myxophyceae. *Archif. Mikrobiol.* 17:34.

2. Allen, M.B., and D.I. Arnon. 1955. Studies on nitrogen-fixing blue-green algae. I. Growth and nitrogen fixation by *Anabaena cylindrica* Lemm. *Plant Physiol.* 30:366.

3. Alway, F.J., A.W. Marsh, and W.J. Methley. 1937. Sufficiency of atmosphere sulfur for maximum crop yields. *Proc. Soil Sci. Soc. Am.* 2:229.

4. Amin, J.V., and H.E. Joham. 1958. A molybdenum cycle in the soil. *Soil Sci.* 85:156.

5. Arnon, D.I., and D.R. Hoagland. 1940. Crop production in artificial solutions and in soils with special reference to factors influencing yields and absorption of inorganic nutrients. *Soil Sci.* 50:463.

6. Barshad, I. 1951. Factors affecting the molybdenum content of pasture plants. I. Nature of soil molybdenum, growth of plants, and soil pH. *Soil Sci.* 71:297.

7. Beeson, K.C. 1959. Magnesium in soils—sources, availability and zonal distribution. In D.H. Horvath, ed., *Magnesium and Agriculture. Proc. West Virginia Univ. Symp.* 1–11.

8. Bertrand, G. 1905. Sur l'emploi favorable du manganèse comme engrais. *C.R. Acad. Sci. Paris* 141:1255.

9. Bingham, F.T., J.P. Martin, and J.A. Chastain. 1958. Effects of phosphorus fertilization of California soils on minor element nutrition of *Citrus*. *Soil Sci.* 86:24.

10. Bould, C. 1963. Mineral nutrition of plants in soils. In F.C. Steward, ed., *Plant Physiology*. New York: Academic Press.

11. Broyer, T.C., A.B. Carlton, C.M. Johnson, and P.R. Stout. 1954. Chlorine—a micronutrient element for higher plants. *Plant Physiol.* 29:526.

12. Camp, A.F. 1945. Zinc as a nutrient in plant growth. *Soil Sci.* 60:156.

13. Chapman, H.D. 1939. Absorption of iron from finely ground magnetite by citrus seedlings. *Soil Sci.* 49:309.

14. Cole, C.V., and M.L. Jackson. 1950. Colloidal dihydroxy dihydrogen phosphates of aluminum and iron with crystalline character established by electron and x-ray diffraction. *Physic. Colloid. Chem.* 54:128.

15. de Saussure, N.T. 1804. Recherches chimiques sur la végétation. Paris: V. Nyon.

16. Drake, M., D.H. Sieling, and G.D. Scarseth. 1941. Calcium-boron ratio as an important factor in controlling boron starvation. *J. Am. Soc. Agron.* 33:454.

17. Hanna, W.J. 1959. Magnesium as a fertilizer element. In D.J. Horvath, ed., *Magnesium and Agriculture. Proc. West Virginia Univ. Symp.* 12–19.

18. Harmer, P.M., and E.J. Benne. 1945. Sodium as a crop nutrient. *Soil Sci.* 60:137.

19. Hasler, A. 1943. Über das Verhalten des Kupfers im Boden. *Mitt. Lebensmittelunters, u. Hyg.* 34:79.

20. Hewitt, E.J. 1963. Mineral nutrition of plants in culture media. In F.C. Steward, ed., *Plant Physiology*. New York: Academic Press.

21. Hewitt, E.J., E.W. Bolle-Jones, and P. Miles. 1954. The production of copper, zinc and molybdenum deficiencies in crop plants with special reference to some effects of water supply and seed reserves. *Plant Soil* 5:205.

22. Holm-Hansen, O., G.C. Gerloff, and F. Skoog. 1954. Cobalt as an essential element for blue-green algae. *Physiol. Plant.* 7:665.

23. Kittrick, J.A., and M.L. Jackson. 1955. Common ion effect of phosphate solubility. *Soil Sci.* 79:415.

24. Leeper, G.W. 1947. The forms and reactions of manganese in the soil. *Soil Sci.* 63:79.

25. Liebig, J. 1840. *Organic Chemistry in Its Applications to Agriculture and Physiology*. L. Playfair, ed. London: Taylor and Walton.

26. Lipman, C.B. 1938. Importance of silicon, aluminum and chlorine for higher plants. *Soil Sci.* 45:189.

27. Longstaff, W.H., and E.R. Graham. 1951. Release of mineral magnesium and its effect on growth and composition of soybeans. *Soil Sci.* 71:167.

28. Lucas, R.E. 1948. Chemical and physical behavior of copper in organic soils. *Soil Sci.* 66:119.

29. Lynd, J.Q., and L.M. Turk. 1948. Overliming injury on an acid sandy soil. *J. Am. Soc. Agron.* 40:205.

30. Lyon, T.L., H.O. Buckman, and N.C. Brady. 1952. *The Nature and Properties of Soils*. New York: Macmillan.

31. Marshall, C.E. 1951. The activities of cations held by soil colloids and the chemical environment of plant roots. In E. Truog, ed., *Mineral Nutrition of Plants*. Madison: University of Wisconsin Press.

32. McAuliffe, C.D., N.S. Hall, L.A. Dean, and S.B. Hendricks. 1948. Exchange reactions between phosphates and soils: hydroxylic surfaces of soil minerals. *Proc. Soil Sci. Am.* 12:119.

33. McLean, F.T., and B.E. Gilbert. 1927. The relative aluminum tolerance of crop plants. *Soil Sci.* 24:163.

34. Menzel, R.G., and M.L. Jackson. 1950. Mechanism of sorption of hydroxy cupric ion by clays. *Proc. Soil Sci. Soc. Am.* 15:122.

35. Millar, C.E., L.M. Turk, and H.D. Foth. 1951. *Fundamentals of Soil Science.* New York: Wiley.

36. Olsen, S.R. 1953. Inorganic phosphorus in alkaline and calcareous soils. *Agronomy* 4:89.

37. Olsen, S.R. 1953. The measurement of phosphorus on the surface of soil particles and its relationship to plant available phosphorus. *Kansas Agr. Exp. Sta. Rept.* 4:59.

38. Osterhout, W.J.V. 1906. On the importance of physiologically balanced solutions for plants. I. Marine plants. *Bot. Gaz.* 42:127.

39. Osterhout, W.J.V. 1912. Plants which require sodium. *Bot. Gaz.* 54:532.

40. Piper, C.S. 1942. Investigations on copper deficiency in plants. *J. Agr. Sci.* 32:143.

41. Quastel, J.H. 1963. Microbial activities of soil as they affect plant nutrition. In F.C. Steward, ed., *Plant Physiology.* New York: Academic Press.

42. Reeve, E., and J.W. Shive. 1944. Potassium-boron and calcium-boron relationships in plant nutrition. *Soil Sci.* 57:1.

43. Rogers, L.H., and C. Wu. 1948. Zinc uptake by oats as influenced by application of lime and phosphate. *J. Am. Soc. Agron.* 40:563.

44. Sommer, A.L. 1926. Studies concerning essential nature of aluminum and silicon for plant growth. *Univ. Calif. Publ. Agr. Sci.* 5:2.

45. Steenbjerg, F. 1950. Investigations on microelements from a practical point of view. In *Trace Elements in Plant Physiology. Lotsya* 3:87.

46. Steinberg, R.A. 1938. The essentiality of gallium to growth and reproduction of *Aspergillus niger. J. Agr. Res.* 57:569.

47. Steinberg, R.A. 1941. Use of *Lemma* for nutrition studies on green plants. *J. Agr. Res.* 62:423.

48. Steinberg, R.A. 1945. Use of microorganisms to determine essentiality of minor elements. *Soil Sci.* 60:185.

49. Steinberg, R.A. 1946. Mineral requirements of *Lemma minor. Plant Physiol.* 21:42.

50. Stiles, W. 1958. Other elements. In W. Ruhland, ed., *Encyclopedia of Plant Physiology* 4:599. Berlin: Springer.

51. Stiles, W. 1961. *Trace Elements in Plants.* London: Cambridge University Press.

52. Stout, P.R., and D.I. Arnon. 1939. Experimental methods for the study of the role of copper, manganese and zinc in the nutrition of higher plants. *Am. J. Bot.* 26:144.

53. Ulrich, A., and K. Ohki. 1956. Chlorine, bromine and sodium as nutrients for sugar beet plants. *Plant Physiol.* 31:171.

54. Wiklander, L. 1958. The soil. In W. Ruhland, ed., *Encyclopedia of Plant Physiology.* 4:118. Berlin: Springer.

55. Wiklander, L., G. Hallgren, and E. Jonsson. 1950. Studies on gyttja soils. III. *Kungl. Lantbrukshogsk. Ann.* 17:425.

56. Wilson, B.D. 1926. Sulfur supplied to the soil in rainwater. *J. Am. Soc., Agron.* 18:1108.

57. Woodward, J. 1699. Some thoughts and experiments on vegetation. *Phil Trans. Roy. Soc. London* 21:382.

Chapter 6

1. Arnon, D.I., P.R. Stout, and F. Sipos. 1940. Radioactive phosphorus as an indicator of phosphorus absorption of tomato plants at various stages of development. *Am. J. Bot.* 27:791.

2. Bennet-Clark, T.A. 1956. Salt accumulation and mode of action of auxin: a preliminary hypothesis. In R.L. Wain and F. Wightman, eds., *Chemistry and Mode of Action of Plant Growth Substances.* London: Butterworth.

3. Biddulph, O. 1941. Diurnal migration of in-

jected radiophosphorus from bean leaves. *Am. J. Bot.* 28:348.

4. Biddulph, O. 1959. Translocation of inorganic solutes. In F.C. Steward, ed., *Plant Physiology.* New York: Academic Press.

5. Biddulph, O., S.F. Biddulph, R. Cory, and H. Koontz. 1958. Circulation patterns for P^{32}, S^{35}, and Ca^{45} in the bean plant. *Plant Physiol.* 33:293.

6. Biddulph, O., and R. Cory. 1957. An analysis of translocation in the phloem of the bean plant using THO, P^{32} and $C^{14}O_2$. *Plant Physiol.* 32:608.

7. Biddulph, O., and J. Markle. 1944. Translocation of radiophosphorus in the phloem of the cotton plant. *Am. J. Bot.* 31:65.

8. Biddulph, S.F. 1956. Visual indications of S^{35} and P^{32} translocation in the phloem of the cotton plant. *Am. J. Bot.* 43:143.

9. Brouwer, R. 1956. Investigations into occurrence of active and passive components in the ion uptake by *Vicia faba. Acta Bot. Néerl.* 5:287.

10. Broyer, T.C., and D.R. Hoagland. 1943. Metabolic activities of roots and their bearing on the relation of upward movements of salts and water in plants. *Am. J. Bot.* 30:261.

11. Butler, G.W. 1953. Ion uptake by young wheat plants. II. The "apparent free space" of wheat roots. *Physiol. Plant.* 5:617.

12. Clements, H.F., and C.J. Engard. 1938. Upward movement of inorganic solutes as affected by a girdle. *Plant Physiol.* 13:103.

13. Crafts, A.S. 1951. Movement of assimilates, viruses, growth regulators, and chemical indicators in plants. *Bot. Rev.* 17:203.

14. Crafts, A.S. 1961. *Translocation in Plants.* New York: Holt, Rinehart and Winston.

15. Crafts, A.S., and T.C. Broyer. 1938. Migration of salts and water into xylem of the roots of higher plants. *Am. J. Bot.* 25:529.

16. Curtis, O.F. 1935. *The Translocation of Solutes in Plants: A Critical Consideration of Evidence Bearing upon Solute Movement.* New York: McGraw-Hill.

17. Epstein, E. 1955. Passive permeation and active transport of ions in plant roots. *Plant Physiol.* 30:529.

18. Epstein, E. 1956. Mineral nutrition of plants: mechanisms of uptake and transport. *Ann. Rev. Plant Physiol.* 7:1.

19. Epstein, E., and C.E. Hagen. 1952. A kinetic study of the absorption of alkali cations by barley roots. *Plant Physiol.* 27:457.

20. Epstein, E., and J.E. Leggett. 1954. The absorption of alkaline earth cations by barley roots: kinetics and mechanism. *Am. J. Bot.* 41:788.

21. Handley, R., and R. Overstreet. 1955. Respiration and salt absorption by excised barley roots. *Plant Physiol.* 30:418.

22. Higinbotham, N. 1973. Electropotentials of plant cells. *Ann. Rev. Plant Physiol.* 24:25.

23. Higinbotham, N. 1973. The mineral absorption process in plants. *Bot. Rev.* 39:15.

24. Hoagland, D.R. 1944. Lectures on the inorganic nutrition of plants. *Chronica botanica,* Waltham, Mass.

25. Hodges, T.K. 1973. Ion absorption by plant roots. *Adv. Agron.* 25:163.

26. Honert, T.H. van den, J.J.M. Hooymans, and W.S. Volkers. 1955. Experiments on the relation between water absorption and mineral uptake by plant roots. *Acta Bot. Néerl.* 4:139.

27. Hope, A.B. 1953. Salt uptake by root tissue cytoplasm: the relation between uptake and external concentration. *Australian J. Biol. Sci.* 6:396.

28. Hope, A.B., and P.G. Stevens. 1952. Electrical potential differences in bean roots and their relation to salt uptake. *Australian J. Sci. Res.* B-1:335.

29. Hopkins, H.T. 1956. Absorption of ionic species of orthophosphate by barley roots: effects of 2,4-dinitrophenol and oxygen tension. *Plant Physiol.* 31:155.

30. Hylmö, B. 1953. Transpiration and ion absorption. *Physiol. Plant.* 6:333.

31. Hylmö, B. 1955. Passive components in the ion absorption of the plant. I. The zonal ion and water absorption in Brouwer's experiments. *Physiol. Plant.* 8:433.

32. Jenny, H. 1951. Contact phenomena be-

tween absorbents and their significance in plant nutrition. In E. Truog, ed., *Mineral Nutrition of Plants.* Madison: University of Wisconsin Press.

33. Jenny, H., and R. Overstreet. 1939. Cation interchange between plant roots and soil colloids. *Soil Sci.* 47:257.

34. Koontz, H., and O. Biddulph. 1957. Factors regulating absorption and translocation of foliar applied phosphorus. *Plant Physiol.* 32:463.

35. Kramer, P.J. 1956. Relative amounts of mineral absorption through various regions of roots. *U.S. Atomic Energy Commission Report TID–7512* 287.

36. Kylin, A., and B. Hylmö. 1957. Uptake and transport of sulfate in wheat. Active and passive components. *Physiol. Plant.* 10:467.

37. Leggett, J.E., and E. Epstein. 1956. Kinetics of sulfate absorption by barley roots. *Plant Physiol.* 31:222.

38. Levitt, J. 1957. The significance of "apparent free space" (AFS) in ion absorption. *Physiol. Plant.* 10:882.

39. Lopushinsky, W. 1964. Effect of water movement on ion movement into the xylem of tomato roots. *Plant Physiol.* 39:494.

40. Lundegårdh, H. 1950. The translocation of salts and water through wheat roots. *Physiol. Plant.* 3:103.

41. Lundegårdh, H. 1954. Anion respiration. The experimental basis of a theory of absorption, transport and exudation of electrolytes by living cells and tissues. *Symp. Soc. Exp. Biol.* 8:262.

42. Lundegårdh, H., and H. Burström. 1933. Untersuchungen über die Salzaufnahme der Pflanzen. III. Quantitative Beziehungen zwischen Atmung und Anionenaufnahme. *Biochem. Z.* 261:235.

43. Mason, T.G., and E.J. Maskell. 1931. Preliminary observations on the transport of phosphorus, potassium, and calcium. *Ann. Bot.* 45:126.

44. Mason, T.G., E.J. Maskell, and E. Phillis. 1936. Concerning the independence of solute movement in the phloem. *Ann. Bot.* 50:23.

45. Olsen, C. 1942. Water culture experiments with higher green plants in nutrient solutions having different concentrations of calcium. *C. r. Trav. Labor. Carlsberg, Sér. chim.* 24:69.

46. Overstreet, R., L. Jacobson, and R. Handley. 1952. The effect of calcium on the absorption of potassium by barley roots. *Plant Physiol.* 27:583.

47. Pfeffer, W. 1900. The mechanism of absorption and translocation. pp. 86–175 (Chapter 4). In *The Physiology of Plants,* vol. I. Translated and edited by A.J. Ewart. London: Oxford University Press.

48. Phillis, E., and T.G. Mason. 1940. The effect of ringing on the upward movement of solutes from the roots. *Ann. Bot.* 4:635.

49. Rediske, J.H., and O. Biddulph. 1953. The absorption and translocation of iron. *Plant Physiol.* 28:576.

50. Rees, W.J. 1949. The salt relations of plant tissues. IV. Some observations on the effect of the preparation of storage tissue on its subsequent absorption of manganese chloride. *Ann. Bot.* 13:29.

51. Robertson, R.N. 1958. The uptake of minerals. In W. Ruhland, ed., *Encyclopedia of Plant Physiology* 4:243 Berlin: Springer.

52. Robertson, R.N., M.J. Wilkins, and D.C. Weeks. 1951. Studies in the metabolism of plant cells. IX. The effects of 2,4-dinitrophenol on salt accumulation and salt respiration. *Australian J. Sci. Res.* B4:248.

53. Russell, R.S., and D.A. Barber. 1960. The relationship between salt uptake and the absorption of water by intact plants. *Ann. Rev. Plant Physiol.* 11:127.

54. Steward, F.C. 1935. Mineral nutrition of plants. *Ann. Rev. Biochem.* 4:519.

55. Steward, F.C., and J.F. Sutcliffe. 1959. Plants in relation to inorganic salts. In F.C. Steward, ed., *Plant Physiology.* New York: Academic Press.

56. Stout, P.R., and D.R. Hoagland. 1939. Upward and lateral movement of salt in certain plants as indicated by radioactive isotopes of potassium, sodium and phosphorus absorbed by roots. *Am. J. Bot.* 26:320.

57. Sutcliffe, J.F. 1962. *Mineral Salts Absorption in Plants.* New York: Pergamon Press.
58. Viets, F.G. 1944. Calcium and other polyvalent cations as accelerators of ion accumulation by excised barley roots. *Plant Physiol.* 19:466.

Chapter 7

1. Agarwala, S.C., and E.J. Hewitt. 1954. Molybdenum as a plant nutrient. IV. The interrelationships of molybdenum and nitrate supply in chlorophyll and ascorbic acid fractions in cauliflower plants grown in sand culture. *J. Hort. Sci.* 29:291.
2. Arnon, D.I. 1959. Chloroplasts and photosynthesis. *Brookhaven Symp. Biol.* 11:181.
3. Bandurski, R.S., L.G. Wilson, and C.L. Squires. 1956. The mechanism of "active sulfate" formation. *J. Am. Chem. Soc.* 78:6408.
4. Bennett-Clark, T.A. 1956. Salt accumulation and mode of action of auxin: a preliminary hypothesis. In R.L. Wain and F. Wightman, eds., *Chemistry and Mode of Action of Plant Growth Substances.* London: Butterworth.
5. Brown, L., and C.C. Wilson. 1952. Some effects of zinc on several species of *Gossypium* L. *Plant Physiol.* 27:812.
6. Burström, H. 1939. Über die Schwermetallkatalyze der Nitratassimilation. *Planta* 29:292.
7. Calvin, M. 1954. Chelation and catalysis. In W.D. McElroy and H.B. Glass, eds., *Mechanism of Enzyme Action.* Baltimore, Md.: Johns Hopkins University Press.
8. Davidson, F.M., and C.M. Long. 1958. The structure of the naturally occurring phosphoglycerides. 4. Action of cabbage leaf phospholipase. *Biochem. J.* 69:458.
9. Davis, D.E. 1949. Some effects of calcium deficiency on the anatomy of *Pinus taeda.* *Am. J. Bot.* 36:276.
10. Eaton, S.V. 1935. Influence of sulfur deficiency on the metabolism of the soybean. *Bot. Gaz.* 97:68.
11. Eaton, S.V. 1941. Influence of sulfur deficiency on metabolism of the sunflower. *Bot. Gaz.* 102:533.
12. Eaton, S.V. 1942. Influence of sulfur deficiency on metabolism of black mustard. *Bot. Gaz.* 104:306.
13. Eaton, S.V. 1949. Effects of phosphorus deficiency on growth and metabolism of sunflowers. *Bot. Gaz.* 110:449.
14. Eaton, S.V. 1950. Effects of phosphorus deficiency on growth and metabolism of soybean. *Bot. Gaz.* 111:426.
15. Eaton, S.V. 1951. Effects of sulfur deficiency on the growth and metabolism of the tomato. *Bot. Gaz.* 112:300.
16. Eaton, S.V. 1952. Effects of phosphorus deficiency on growth and metabolism of black mustard. *Bot. Gaz.* 113:301.
17. Eltinge, E.T. 1941. Effects of manganese deficiency upon the histology of *Lycopersicon esculentum.* *Plant Physiol.* 16:189.
18. Eversole, R.A., and E.L. Tatum. 1956. Chemical alteration of crossing over frequency in *Chlamydomonas.* *Proc. Natl. Acad. Sci., U.S.* 42:68.
19. Eyster, C., T.E. Brown, H. Tanner, and S.L. Hood, 1958. Manganese requirement with respect to growth, Hill reaction and photosynthesis. *Plant Physiol.* 33:235.
20. Florell, C. 1956. The influence of calcium on root mitochondria. *Physiol. Plant.* 9:236.
21. Florell, C. 1957. Calcium, mitochondria and anion uptake. *Physiol. Plant.* 10:781.
22. Gauch, H.G. 1957. Mineral nutrition of plants. *Ann. Rev. Plant Physiol.* 8:31.
23. Gauch, H.G., and W.M. Dugger. 1953. The role of boron in the translocation of sucrose. *Plant Physiol.* 28:457.
24. Gauch, H.G., and W.M. Dugger. 1954. *The Physiological Role of Boron in Higher Plants: A Review and Interpretation.* Tech. Bull. A–80. Agr. Exp. Sta., University of Maryland.
25. Gilbert, F.A. 1951. The place of sulfur in plant nutrition. *Bot. Rev.* 17:671.
26. Goldacre, P.L. 1961. The indole-3-acetic acid oxidase-peroxidase of peas. In R.M. Klein, ed., *Plant Growth Regulation.* Ames: Iowa State University Press.

27. Granick, S. 1950. Iron metabolism in animals and plants. *Harvey Lectures Ser.* 44:220.

28. Green, L.F., J.F. McCarthy, and C.G. King. 1939. Inhibition of respiration and photosynthesis in *Chlorella pyrenoidosa* by organic compounds that inhibit copper catalysis. *J. Biol. Chem.* 128:447.

29. Hall, J.D., R. Barr, A.H. Al-Abbas, and F.L. Crane. 1972. The ultrastructure of chloroplasts in mineral-deficient maize leaves. *Plant Physiol.* 50:404.

30. Hewitt, E.J. 1945. Marsh spot in beans. *Nature* 155:22.

31. Hewitt, E.J. 1963. The essential nutrient elements: requirements and interactions in plants. In F.C. Steward, ed., *Plant Physiology.* New York: Academic Press.

32. Hewitt, E.J., S.C. Agarwala, and E.W. Jones. 1950. Effect of molybdenum status on the ascorbic acid content of plants in sand culture. *Nature* 166:1119.

33. Hoch, F.L., and B.L. Vallee. 1958. The metabolic role of zinc. In C.A. Lamb, O.G. Bentley, and J.M. Beattie, eds., *Trace Elements.* New York: Academic Press.

34. Hyde, B.B., and R.L. Paliwal. 1958. Studies on the role of cations in the structure and behaviour of plant chromosomes. *Am. J. Bot.* 45:433.

35. Iljin, W.S. 1952. Metabolism of plants affected with lime-induced chlorosis (calciose). III. Mineral elements. *Plant Soil* 4:11.

36. Jacobson, L. 1945. Iron in the leaves and chloroplasts of some plants in relation to chlorophyll content. *Plant Physiol.* 20:233.

37. Jacobson, L., and J.J. Oertli. 1956. The relation between iron and chlorophyll contents in chlorotic sunflower leaves. *Plant Physiol.* 31:199.

38. Joham, H.E. 1957. Carbohydrate distribution as affected by calcium deficiency in cotton. *Plant Physiol.* 32:113.

39. Kalra, G.S. 1956. Responses of the tomato plant to calcium deficiency. *Bot. Gaz.* 118:18.

40. Keilin, D., and T. Mann. 1940. Carbonic anhydrase. *Biochem. J.* 34:1163.

41. Kenten, R.H. 1955. The oxidation of indole-3-acetic acid by waxpod bean root sap and peroxidase systems. *Biochem. J.* 59:110.

42. Kessler, E. 1955. On the role of manganese in the oxygen-evolving system in photosynthesis. *Arch. Biochem. Biophys.* 59:527.

43. Kessler, E., W. Arthur, and J.E. Brugger. 1957. The influence of manganese and phosphate on delayed light emission, fluorescence, photoreduction and photosynthesis in algae. *Arch. Biochem. Biophys.* 71:326.

44. Lindner, R.C., and C.P. Harley. 1944. Nutrient interrelations in lime-induced chlorosis. *Plant Physiol.* 19:420.

45. Loustalot, A.J., F.W. Burrows, S.G. Gilbert, and A. Nason. 1945. Effect of copper and zinc deficiencies on the photosynthesis activity of the foliage of young tung trees. *Plant Physiol.* 20:283.

46. Lutman, B.F. 1934. *Cell Size and Structure in Plants as Affected by Inorganic Elements.* Bull. 383. Agr. Exp. Sta. University of Vermont.

47. Lyon, C., and C.R. Garcia. 1944. Anatomical responses of tomato stems to variations in the macronutrient anion supply. *Bot. Gaz.* 105:394.

48. Lyon, C., and C.R. Garcia. 1944. Anatomical responses of tomato stems to variations in the macronutrient cation supply. *Bot. Gaz.* 105:441.

49. McElroy, W.D., and A. Nason. 1954. Mechanism of action of micronutrient elements in enzyme systems. *Ann. Rev. Plant Physiol.* 5:1.

50. Mazia, D. 1954. The particulate organization of the chromosome. *Proc. Natl. Acad. Sci., U.S.* 40:521.

51. Morton, A.G., and D.J. Watson. 1948. A physiological study of leaf growth. *Ann. Bot.* 12:281.

52. Nason, A. 1950. Effect of zinc deficiency on the synthesis of tryptophane by *Neurospora* extracts. *Science* 112:111.

53. Nason, A. 1956. Enzymatic steps in the assimilation of nitrate and nitrite in fungi and green plants. In W.D. McElroy and H.B. Glass, eds., *Inorganic Nitrogen Metabolism.*

Baltimore, M.D.: Johns Hopkins University Press.

54. Nason, A., N.O. Kaplan, and H.O. Oldewurtel. 1953. Further studies of nutritional conditions affecting enzymatic constitution in *Neurospora*. *J. Biol. Chem.* 201:435.

55. Nason, A., and W.D. McElroy. 1963. Modes of action of the essential mineral elements. In F.C. Steward, ed., *Plant Physiology*. New York: Academic Press.

56. Neish, A.C. 1939. Studies on chloroplasts. II. Their chemical composition and the distribution of certain metabolites between the chloroplasts and the remainder of the leaf. *Biochem. J.* 33:300.

57. Njoku, E. 1957. The effect of mineral nutrition and temperature on leaf shape in *Ipomoea caerulea*. *New Phytol.* 56:154.

58. Piper, C.S. 1942. Investigations on copper deficiency in plants. *J. Agr. Sci.* 32:143.

59. Possingham, J.V. 1956. The effect of mineral nutrition on the content of free amino acids and amides in tomato plants. I. A comparison of effects of deficiencies of copper, zinc, manganese, iron and molybdenum. *Australian Biol. Sci.* 9:539.

60. Price, C.A., and E.F. Carell. 1964. Control by iron of chlorophyll formation and growth in *Euglena gracilis*. *Plant Physiol.* 39:862.

61. Reed, H.S. 1946. Effects of zinc deficiency on phosphate metabolism of the tomato plant. *Am. J. Bot.* 33:778.

62. Robbins, P.W., and F. Lipmann. 1956. Identification of enzymatically active sulfate as adenosine-3'-phosphate-5'-phosphosulfate. *J. Am. Chem. Soc.* 78:2652.

63. Robbins, P.W., and F. Lipmann. 1956. The enzymatic sequence in the biosynthesis of active sulfate. *J. Am. Chem. Soc.* 78:6409.

64. Sadana, J.C., and W.D. McElroy. 1957. Nitrate reductase from *Achromobacter fischeri*. Purification and properties: functions of flavines and cytochrome. *Arch. Biochem. Biophys.* 67:16.

65. Sisler, E.C., W.M. Dugger, and H.G. Gauch. 1956. The role of boron in the trans-

location of organic compounds in plants. *Plant Physiol.* 31:11.

66. Skoog, F. 1940. Relationships between zinc and auxin in the growth of higher plants. *Am. J. Bot.* 27:939.

67. Smith, P.F., W. Reuther, and A.W. Specht. 1950. Mineral composition of chlorotic orange leaves and some observations on the relation of sample preparation technique to the interpretation of results. *Plant Physiol.* 25:496.

68. Steffensen, D. 1953. Induction of chromosome breakage at meiosis by a magnesium deficiency in *Tradescantia*. *Proc. Natl. Acad. Sci., U.S.* 39:613.

69. Steffensen, D. 1955. Breakage of chromosomes in *Tradescantia* with a calcium deficiency. *Proc. Natl. Acad. Sci., U.S.* 41:155.

70. T'so, P.O.P., J. Bonner, and J. Vinograd. 1957. Physical and chemical properties of microsomal particles from pea seedlings. *Plant Physiol. Suppl.* 32:XII.

71. Tsui, C. 1948. The role of zinc in auxin synthesis in the tomato plant. *Am. J. Bot.* 35:172.

72. Wallihan, E.F. 1955. Relation of chlorosis to concentration of iron in citrus leaves. *Am. J. Bot.* 42:101.

73. Webster, G.C. 1953. Peptide bond synthesis in higher plants. I. *Arch. Biochem. Biophys.* 47:241.

74. Webster, G.C. 1956. Effect of monovalent ions on the incorporation of amino acids into protein. *Biochem. Biophys. Acta* 20:565.

75. Webster, G.C., and J.E. Varner. 1954. Mechanism of enzymatic synthesis of gamma-glutamylcysteine. *Federation Proc.* 13:1049.

76. Weinstein, L.H., E.R. Purvis, A.N. Meiss, and R.L. Uhler. 1954. Absorption and translocation of ethylenediamine tetraacetic acid by sunflower plants. *J. Agr. Food Chem.* 2:421.

77. Wiessner, W. 1962. Inorganic micronutrients. In R.A. Lewin, ed., *Physiology and Biochemistry of Algae.* New York: Academic Press.

Chapter 8

1. Ahmed, S., and H.J. Evans. 1960. Cobalt: a micronutrient element for the growth of soybean plants under symbiotic conditions. *Soil Sci.* 90:205.
2. Ahmed, S., and H.J. Evans. 1961. The essentiality of cobalt for soybean plants grown under symbiotic conditions. *Proc. Nat. Acad. Sci., U.S.* 47:24.
3. Allen, E.K., and O.N. Allen. 1958. Biological aspects of symbiotic nitrogen fixation. In W. Ruhland, ed., *Encyclopedia of Plant Physiology* 8:48. Berlin: Springer.
4. Anfinsen, C.B. 1959. *The Molecular Basis of Evolution.* New York: Wiley.
5. Aslam, M., R.C. Huffaker, and R.L. Travis. 1973. The interaction of respiration and photosynthesis in induction of nitrate reductase activity. *Plant Physiol.* 52:137.
6. Beevers, H., L.E. Schrader, D. Flesher, and R.H. Hageman. 1965. The role of light and nitrate in the induction of nitrate reductase in radish cotyledons and maize seedlings. *Plant Physiol.* 40:691.
7. Bollard, E.G. 1959. Urease, urea and ureides in plants. *Symp. Soc. Exp. Biol.* 13:304.
8. Dalling, M.J., D.P. Hucklesby, and R.H. Hageman. 1973. A comparison of nitrite reductase enzymes from green leaves, scutella, and roots of corn (*Zea mays* L.). *Plant Physiol.* 51:481.
9. Dart, P.J. 1971. Scanning electron microscopy of plant roots. *J. Exp. Bot.* 22:163.
10. Epstein, E. 1965. Mineral metabolism. In J. Bonner and J.E. Varner, eds., *Plant Biochemistry.* New York: Academic Press.
11. Evans, H.J., and M. Kliewer. 1964. Vitamin B_{12} compounds in relation to the requirements of cobalt for higher plants and nitrogen-fixing organisms. *Ann. N.Y. Acad. Sci.* 112:735.
12. Evans, H.J., and A. Nason. 1953. Pyridine nucleotide-nitrate reductase from extracts of higher plants. *Plant Physiol.* 28:233.
13. Gest, H., J. Judis, and H.D. Peck. 1956. Reduction of molecular nitrogen and relationships with photosynthesis and hydrogen metabolism. In W.D. McElroy and B. Glass, eds., *Inorganic Nitrogen Metabolism.* Baltimore, Md.: Johns Hopkins University Press.
14. Goodwin, T.W., and E.I. Mercer. 1973. *Introduction to Plant Biochemistry.* New York: Pergamon Press.
15. Hageman, R.H., and D. Flesher. 1960. Nitrate reductase activity in corn seedlings as affected by light and nitrate content of nutrient medium. *Plant Physiol.* 35:700.
16. Harris, G.P. 1954. Amino acids as sources of nitrogen for the growth of isolated oat embryos. *New Phytol.* 55:253.
17. Hattori, A. 1958. Studies on the metabolism of urea of other nitrogenous compounds in *Chlorella ellipsoida.* II. Changes in levels of amino acids and amides during the assimilation of ammonia and urea by nitrogen-starved cells. *J. Biochem.* (Tokyo) 45:57.
18. Hewitt, E.J., and M.M.R.K. Afridi. 1959. Adaptive synthesis of nitrate reductase in higher plants. *Nature* 183:57.
19. Hinsvark, O.N., S.H. Wittwer, and H.B. Tukey. 1953. The metabolism of foliar-applied urea. I. Relative rates of $C^{14}O_2$ production by certain vegetable plants treated with labeled urea. *Plant Physiol.* 28:70.
20. Kannangara, C.G., and H.W. Woolhouse. 1967. The role of carbon dioxide, light and nitrate in the synthesis and degradation of nitrate reductase in leaves of *Perilla frutescens. New Phytol.* 66:553.
21. Kemp, J.D., D.E. Atkinson, A. Ehret, and R.A. Lazzarini. 1963. Evidence for the identity of the nicotinamide adenine dinucleotide phosphate-specific sulfite and nitrite reductase of *Escherichia coli. J. Biol. Chem.* 238:3466.
22. Medina, A., and D.J.D. Nicholas. 1957. Metallo-enzymes in the reduction of nitrite to ammonia in *Neurospora. Biochim. Biophys. Acta* 25:138.
23. Mengel, K., and E.A. Kirkby. 1978. *Principles of Plant Nutrition.* Int. Potash Inst., eds. Bern: Der Bund.
24. Murphy, M.J., L.M. Siegel, S.R. Tove, and

H. Kamin. 1974. Siroheme: a new prosthetic group participating in six-electron reduction reactions catalyzed by both sulphite and nitrite reductases. *Proc. Natl. Acad. Sci., U.S.* 71:612.

25. Nason, A., and H.J. Evans. 1954. Triphosphopyridine nucleatide-nitrate reductase in *Neurospora. J. Biol. Chem.* 202:655.

26. Nicholas, D.J.D., and A. Nason. 1954. Mechanism of action of nitrate reductase from *Neurospora. J. Biol. Chem.* 211:183.

27. Nicholas, D.J.D., and A. Nason. 1955. Role of molybdenum as a constituent of nitrate reductase from soybean leaves. *Plant Physiol.* 30:135.

28. Nightingale, G.T., L.G. Schermerhorn, and W.R. Robbins. 1928. *The Growth Status of the Tomato as Correlated with Organic Nitrogen and Carbohydrates in Roots, Stems and Leaves.* Bull. 461. N.J. Agr. Exp. Sta.

29. Paulsen, G.M., and J.E. Harper. 1968. Evidence for a role of calcium in nitrate assimilation in wheat seedlings. *Plant Physiol.* 43:775.

30. Phillips, D.A., R.M. Daniel, C.A. Appleby, and H.J. Evans. 1973. Isolation from *Rhizobium* of factors which transfer electrons to soybean nitrogenase. *Plant Physiol.* 51:136.

31. Phillips, D.A., R.L. Howard, and H.J. Evans. 1973. Studies on the genetic control of a nitrogenase component in leguminous root nodules. *Physiol. Plant.* 28:248.

32. Ritenour, G.L., K.W. Joy, J. Bunning, and R.H. Hageman. 1967. Intracellular localization of nitrate reductase, nitrite reductase, and glutamic acid dehydrogenase in green leaf tissue. *Plant Physiol.* 42:233.

33. Smillie, R.M., and B. Entach. 1971. Phytoflavin. In A. San Pietro, ed., *Methods in Enzymology,* vol. 23. New York: Academic Press.

34. Stevens, S.E., and C. Van Baalen. 1973. Characteristics of nitrate in a mutant of the blue-green alga *Agmenellum quadruplicatum. Plant Physiol.* 51:350.

35. Stiles, W. 1961. *Trace Elements in Plants,* 3rd ed. New York: Cambridge University Press.

36. Stiller, M. 1966. Hydrogenase-mediated nitrite reduction in *Chlorella. Plant Physiol.* 41:348.

37. Stiller, M., and J.K.H. Lee. 1964. Hydrogenase activity in *Chlorella. Biochim. Biophys. Acta* 93:174.

38. Street, H.E., and D.E.G. Sheat. 1958. The absorption and availability of nitrate and ammonia. In W. Ruhland, ed. *Encyclopedia of Plant Physiology* 8:150. Berlin: Springer.

39. Thimann, K.V. 1939. The physiology of nodule formation. *Trans. Third. Comm. Intern. Soc. Soil Sci.,* New Brunswick, N.J.

40. Tiedjens, V.A. 1934. Factors affecting assimilation of ammonia and nitrate nitrogen particularly in tomato and apple. *Plant Physiol.* 9:31.

41. Travis, R.L., W.R. Jordan, and R.C. Huffaker. 1970. Light and nitrate requirements for induction of nitrate reductase activity in *Hordeum vulgare. Physiol. Plant.* 23:678.

42. Travis, R.L., and J.L. Key. 1971. Correlation between polyribosome level and the ability to induce nitrate reductase in dark-grown corn seedlings. *Plant Physiol.* 48:617.

43. Verhoeven, W. 1956. Some remarks on nitrate and nitrite metabolism in microorganisms. In W.D. McElroy and B. Glass, eds., *Inorganic Nitrogen Metabolism.* Baltimore, Md.: Johns Hopkins University Press.

44. Virtanen, A.I., J. Erkama, and H. Linkola. 1947. On the relation between nitrogen fixation and leghaemoglobin content of leguminous root nodules. II. *Acta Chem. Scand.* 1:861.

45. Virtanen, A.I., and J.K. Miettinen. 1963. Biological nitrogen fixation. In F.C. Steward, ed., *Plant Physiology.* New York: Academic Press.

46. Walker, J.B. 1952. Arginosuccinic acid from *Chlorella pyrenoidosa. Proc. Natl. Acad. Sci., U.S.* 38:561.

47. Wallace, W. 1973. The distribution and characteristics of nitrate reductase and glutamate dehydrogenase in the maize seedling. *Plant Physiol.* 52:191.

48. White, P.R. 1937. Amino acids in the nutrition of excised tomato roots. *Plant Physiol.* 12:793.

49. Wilson, P.W. 1940. *The Biochemistry of Symbiotic Nitrogen Fixation.* Madison: University of Wisconsin Press.

50. Wilson, P.W. 1958. Asymbiotic nitrogen fixation. In W. Ruhland, ed., *Encyclopedia of Plant Physiology* 8:9. Berlin: Springer.

51. Wilson, P.W., and C.J. Lind. 1943. Carbon monoxide inhibition of *Azotobacter* in microrespiration experiments. *J. Bacter.* 45:219.

52. Wilson, P.W., and W.W. Umbreit. 1937. Mechanism of symbiotic nitrogen fixation. III. Hydrogen as a specific inhibitor. *Arch. Mikrobiol.* 8:440.

53. Wilson, P.W., W.W. Umbreit, and S.B. Lee. 1938 Mechanism of symbiotic nitrogen fixation. IV. Specific inhibition by hydrogen. *Biochem. J.* 32:2084.

54. Winter, H.C., and R.H. Burris. 1976. Nitrogenase. *Ann. Rev. Biochem.* 45:409.

55. Wipf, L., and D.C. Cooper. 1938. Chromosome numbers in nodules and roots of red clover, common vetch and garden peas. *Proc. Natl. Acad. Sci., U.S.* 24:87.

56. Wipf, L., and D.C. Cooper. 1940. Somatic doubling of chromosomes and nodular infection in certain *Leguminosae. Am. J. Bot.* 27:821.

Chapter 9

1. Anfinsen, C.B. 1959. *The Molecular Basis of Evolution.* New York: Wiley.

2. Berg, P., and E.J. Ofengand. 1958. An enzymatic mechanism for linking amino acids to RNA. *Proc. Natl. Acad. Sci., U.S.* 44:78.

3. Danielson, C.E. 1951. The breakdown of high molecular reserve proteins of peas during germination. *Acta Chem. Scand.* 5:551.

4. Folkes, B.F. 1959. The position of amino acids in the assimilation of nitrogen and the synthesis of proteins in plants. *S.E.B. Symposia* 13:126.

5. Folkes, B.F., and E.W. Yemm. 1958. The respiration of barley plants. X. Respiration and the metabolism of amino acids and proteins in germinating grain. *New Phytol.* 57:106.

6. Hattori, A. 1958. Studies on the metabolism of urea and other nitrogenous compounds in *Chlorella ellipsoida.* II. Changes in levels of amino acids and amides during the assimilation of ammonia and urea by nitrogen-starved cells. *J. Biochem.* (Tokyo) 45:57.

7. Hendry, L.B., and F.H. Witham. 1979. Stereochemical recognition in nucleic acid—amino acid interactions and its implications in biological coding: a model approach. *Perspect. Biol. Med.* 22:333.

8. Hendry, L.B., F.H. Witham, and O.L. Chapman. 1977. Gene regulation: the involvement of stereochemical recognition in DNA–small molecule interactions. *Perspect. Biol. Med.* 21:120.

9. Oaks, A., and H. Beevers. 1964. The requirement for organic nitrogen in *Zea mays* embryos. *Plant Physiol.* 39:37.

10. Schweet, R.S., F.C. Bovard, E. Allen, and F. Glassman. 1958. The incorporation of amino acids into ribonucleic acid. *Proc. Natl. Acad. Sci., U.S.* 44:173.

11. Synenki, R.M., C.S. Levings, III, and D.M. Shah. 1978. Physiochemical characterization of mitochondrial DNA from soybean. *Plant Physiol.* 61:460.

12. Watson, J.D., and F.H.C. Crick. 1953. Molecular structure of nucleic acids. *Nature* 171:737.

13. Wilson, D.G., K.W. King, and R.H. Burris. 1954. Transamination in plants. *J. Biol. Chem.* 208:863.

Chapter 10

1. Goodwin, T.W., and E.I. Mercer. 1973. *Introduction to Plant Biochemistry.* New York: Pergamon Press.

2. Hellerman, L., and C.C. Stock. 1938. Activation of enzymes. *J. Biol. Chem.* 125:771.

3. McGilvery, R.W., with G. Goldstein. 1979. *Biochemistry: A Functional Approach.* Philadelphia: Saunders.

4. Sumner, J.B. 1926. The isolation and crys-

tallization of the enzyme urease. *J. Biol. Chem.* 69:435.

Chapter 11

1. Akazawa, T., T. Minamikawa, and T. Murata. 1964. Enzymic mechanism of starch synthesis in ripening rice grains. *Plant Physiol.* 39:371.
2. Barker, F., H. Nasr, F. Morrice, and J. Bruce. 1950. Bacterial breakdown of structural starches in the digestive tract of ruminant and non-ruminant mammals. *J. Path.* 62:617.
3. Baum, H., and G.A. Gilbert. 1953. A simple method for the preparation of crystalline potato phosphorylase and Q-enzyme. *Nature* 17:983.
4. Bernfeld, P. 1951. Enzymes of starch degradation and synthesis. *Adv. Enzymol.* 12:379.
5. Bourne, E.J., and H. Weigel. 1954. ^{14}C-cellulose from *Acetobacter acetigenum*. *Chem. Ind.* (30 January):132.
6. Brimacombe, J.S., and M. Stacey. 1962. Cellulose, starch, and glycogen. In M. Florkin and H.S. Mason, eds., *Comparative Biochemistry*. New York: Academic Press.
7. Brummond, D.O., and A.P. Gibbons. 1964. The enzymatic synthesis of cellulose by the higher plant. *Biochem. Biophys. Res. Com.* 17:156.
8. Caputto, R., L.F. Leloir, C.E. Cardini, and A.C. Paladini. 1950. Isolation of the coenzyme of the galactose phosphate-glucose phosphate transformation. *J. Biol. Chem.* 184:333.
9. Doesburg, J.J. 1973. The pectic substances. In L.P. Miller, ed., *Phytochemistry*. New York: Van Nostrand Reinhold.
10. Edelman, J., and M.A. Hall. 1964. Effect of growth hormones on the development of invertase associated with cell walls. *Nature* 201:296.
11. Edelman, J., and T.G. Jefford. 1964. The metabolism of fructose-polymers in plants. *Biochem. J.* 93:148.
12. French, D. 1954. The raffinose family of oligosaccharides. *Adv. Carbohydrate Chem.* 9:149.
13. Gibbs, M. 1959. Metabolism of carbon compounds. *Ann. Rev. Plant Physiol.* 10:329.
14. Glaser, L. 1958. The synthesis of cellulose in cell-free extracts of *Acetobacter xylinum*. *J. Biol. Chem.* 232:627.
15. Gottschalk, A., 1958. The enzymes controlling hydrolytic phosphorolytic and transfer reactions of the oligosaccharides. In W. Ruhland, ed., *Encyclopedia of Plant Physiology* 6:87. Berlin-Springer.
16. Hanes, C.S. 1940. The reversible formation of starch from glucose-1-phosphate catalyzed by potato phosphorylase. *Proc. Roy. Soc.* (*London*) B129:174.
17. Hobson, P.N., W.J. Whelan, and S. Peat. 1951. The enzymatic synthesis and degradation of starch. XIV. R-enzyme. *J. Chem. Soc.* 1451.
18. Kaufman, P.B., N. Ghosheh, and H. Ikuma. 1968. Promotion of growth and invertase activity by gibberellic acid in developing *Avena* internodes. *Plant Physiol.* 43:29.
19. Manner, D.J. 1973. Starch and inulin. In L.P. Miller, ed., *Phytochemistry*. New York: Van Nostrand Reinhold.
20. Maruo, B., and T. Kobayaski. 1951. Enzymic scission of the branch links in amylopectin. *Nature* 167:606.
21. Mendicino, J. 1960. Sucrose phosphate synthesis in wheat germ and green leaves. *J. Biol. Chem.* 235:3347.
22. Miller, L.P. 1973. Mono- and oligosaccharides. In L.P. Miller, ed., *Phytochemistry*. New York: Van Nostrand Reinhold.
23. Murata, T., T. Minamikawa, T. Akazawa, and T. Sugiyama. 1964. Isolation of adenosine diphosphate glucose from ripening rice grains and its enzymic synthesis. *Arch. Biochem. Biophys.* 106:371.
24. Murata, T., T. Sugiyama, and T. Akazawa. 1964. Enzymic mechanism of starch synthesis in ripening rice grains. II. Adenosine diphosphate glucose pathway. *Arch. Biochem. Biophys.* 107:92.
25. Palmer, J.M. 1966. The influence of growth

regulating substances on the development of enhanced metabolic rates in thin slices of beetroot storage tissue. *Plant Physiol.* 41:1173.

26. Peat, S., W.J. Whelan, and W.R. Rees. 1953. D-Enzyme: a disproportionating enzyme in potato juice. *Nature* 172:158.

27. Ranson, S.L., and M. Thomas. 1963. Enzyme action in plant metabolism. In W.B. Turill, ed., *Vistas in Botany.* New York: Macmillan.

28. Rorem, E.S., H.G. Walker, and R.M. McCready. 1960. Biosynthesis of sucrose and sucrose-phosphate in sugar beet leaf extract. *Plant Physiol.* 35:269.

29. Scherpenberg, H. van, W. Grobner, and O. Kandler. 1965. *Beitr. Biochem. Physiol. Naturstoffen Festschr.* 387, 406.

30. Schramm, M., Z. Gromet, and S. Hestrin. 1957. Role of hexose phosphate in synthesis of cellulose by *Acetobacter xylinum. Nature.* 179:28.

31. Sellmair, J. and O. Kandler. 1970. *Z. Pflanzenphysiol.* 63:65.

32. Teng, J. and R.L. Whistler. 1973. Cellulose and chitin. In L.P. Miller, ed., *Phytochemistry.* New York: Van Nostrand Reinhold.

33. Timell, T.E. 1965. Wood and bark polysaccharides. In W.A. Coté, Jr., ed., *Cellular Ultrastructure of Woody Plants.* Syracuse, N.Y.: Syracuse University Press.

34. Walker, D.A., and W.J. Whelan. 1959. Synthesis of amylose by potato D-enzyme. *Nature* 183:46.

35. Webb, K.L., and J.W.A. Burley. 1964. Stachyose translocation in plants. *Plant Physiol.* 39:973.

36. Whelan, W.J. 1958. Starch and similar polysaccharides. In W. Ruhland, ed., *Encyclopedia of Plant Physiology* 6:154. Berlin: Springer.

37. Wolfrom, M.L., and A. Thompson. 1956. Occurrence of the $(1 \rightarrow 3)$-linkage in starches. *J. Am. Chem. Soc.* 78:4116.

38. Worth, H.G.J. 1967. The chemistry and biochemistry of pectic substances. *Chem. Rec.* 67:465.

39. Zimmermann, M.H. 1957. Translocation of organic substances in trees. I. The nature of the sugars in the sieve tube exudate of trees. *Plant Physiol.* 32:288.

40. Zimmermann, M.H. 1957. Translocation of organic substances in trees. II. On the translocation mechanism in the phloem of white ash. *Plant Physiol.* 32:399.

Chapter 12

1. Akoyunoglou, G.A., and H.W. Siegelman. 1968. Protochlorophyllide resynthesis in dark-grown bean seedlings. *Plant Physiol.* 43:66.

2. Allen, M.B. 1966. Distribution of the chlorphylls. In L.P. Vernon and G.R. Seely, eds., *The Chlorophylls.* New York: Academic Press.

3. Bamji, M.S., and N.I. Krinsky. 1965. Carotenoid de-epoxidation in algae. II. Enzymatic conversion of antheraxanthin to zeaxanthin. *J. Biol. Chem.* 240:467.

4. Bartels, P.G., K. Matsuda, A. Siegel, and T.E. Weier. 1967. Chloroplast ribosome formation: inhibition by 3-amino-1,2,4-triazole. *Plant Physiol.* 42:736.

5. Bergeron, J. 1959. The bacterial chromatophore. In *The Photochemical Apparatus—Its Structure and Function. Brookhaven Symp. Biol.* 11:118.

6. Blackman, F. 1905. Optima and limiting factors. *Ann. Bot.* 19:281.

7. Boardman, N.K. 1966. Photochlorophyll. In L.P. Vernon and G.R. Seely, eds., *The Chlorophylls.* New York: Academic Press.

8. Bogorad, L. 1965. Studies of phycobiliproteins. In D.W. Krogmann and W.H. Powers, eds., *Biochemical Dimensions of Photosynthesis.* Detroit, Mich.: Wayne State University Press.

9. Bogorad, L. 1966. The biosynthesis of chlorphylls. In L.P. Vernon and G.R. Seely, eds., *The Chlorophylls.* New York: Academic Press.

10. Bogorad, L. 1967. Chloroplast structure and

development. In A. San Pietro, F.A. Greer, and T.J. Army, eds., *Harvesting the Sun: Photosynthesis in Plant Life*. New York: Academic Press.

11. Bogorad, L., F.V. Mercer, and R. Mullens. 1963. Photosynthetic mechanisms of green plants. *Natl. Acad. Sci. Natl. Res. Council Publ.* 1145:560.

12. Calvin, M. 1955. Function of carotenoids in photosynthesis. *Nature* 176:1211.

13. Calvin, M. 1959. From microstructure of macrostructure and function in the photochemical apparatus. In *The Photochemical Apparatus—Its Structure and Function. Brookhaven Symp. Biol.* 11:160.

14. Devlin, R.M., and A.V. Barker. 1971. *Photosynthesis.* New York: Van Nostrand Reinhold.

15. Duysens, L. 1956. Energy transformations in photosynthesis. *Ann. Rev. Plant Physiol.* 7:25.

16. Gantt, E., and S.F. Conti. 1965. The ultrastructure of *Porphyridium cruentum. J. Cell Biol.* 26:365.

17. Gantt, E., and S.F. Conti. 1966. Granules associated with the chloroplast lamellae of *Porphyridium cruentum. J. Cell Biol.* 29:423.

18. Gantt, E., and S.F. Conti. 1967. Phycobiliprotein localization in algae. In *Energy conversion by the photosynthetic apparatus. Brookhaven Symp. Biol.* 19:393.

19. Gassman, M., and L. Bogorad. 1967. Control of chlorophyll production in rapidly greening bean leaves. *Plant Physiol.* 42:774.

20. Gassman, M., and L. Bogorad. 1967. Studies on the regeneration of protochlorophyllide after brief illumination of etiolated bean leaves. *Plant Physiol.* 42:781.

21. Gibson, K.D., W.G. Laver, and A. Neuberger. 1958. Initial stages in the biosynthesis of prophyrins. 2. The formation of δ-aminolevulinic acid from glycine and succinyl-coenzyme A by particles from chicken erythrocytes. *Biochem. J.* 70:71.

22. Giraud, G. 1966. In J.B. Thomas and J.C. Goedheer, eds., *Currents in Photosynthesis.* Rotterdam: Ad. Donker.

23. Glass, B. 1961. Summary. In W. McElroy and B. Glass, eds., *Light and Life.* Baltimore, Md.: Johns Hopkins University Press.

24. Goodwin, T. 1960. Chemistry, biogenesis and physiology of the carotenoids. In W. Ruhland, ed., *Encyclopedia of Plant Physiology.* 5; Part 1, 394. Berlin: Springer.

25. Granick, S. 1954. Enzymatic conversion of δ-aminolevulinic acid to porphobilinogen. *Science* 120:1105.

26. Granick, S. 1961. Magnesium protoporphyrin monoester and protoporphyrin monomethyl ester in chlorophyll biosynthesis. *J. Biol. Chem.* 236:1168.

27. Granick, S. 1961. The pigments of the biosynthetic chain of chlorophyll and their interaction with light. *Proc. 6th Int. Biochem. Congr. Biochem. Moscow* 6:176.

28. Hadziyev, D., S.L. Mehta, and S. Zalik. 1968. Studies on the ribonucleic acid from wheat leaves and chloroplasts. *Plant Physiol.* 43:229.

29. Haxo, F., and L. Blinks. 1950. Photosynthetic action spectra of marine algae. *J. Gen Physiol.* 33:389.

30. Jacobson, A.B., H. Swift, and L. Bogorad. 1963. Cytochemical studies concerning the occurrence and distribution of RNA in plastids of *Zea mays. J. Cell Biol.* 17:557.

31. Kikuchi, G., A. Kumar, P. Talmadge, and D. Shemin. 1958. The enzymatic synthesis of δ-aminolevulinic acid. *J. Biol. Chem.* 233:1214.

32. Kirk, J.T.O., R.A.E. Tilney-Bassett. 1978. *The Plastids: Their Chemistry, Structure, Growth and Inheritance.* New York: Elsevier North-Holland.

33. Klein, S., and L. Bogorad. 1964. Fine structural changes in proplastids during photodestruction of pigments. *J. Cell Biol.* 22:443.

34. Koski, V.M., and J.H.C. Smith. 1951. Chlorophyll formation in a mutant white seedling-3. *Arch. Biochem. Biophys.* 34:189.

35. Krinsky, N.I. 1966. The role of carotenoid pigments as protective agents against photosensitized oxidation in chloroplasts. In T.W. Goodwin, ed., *Biochemistry of Chloroplasts*, vol. 1. New York: Academic Press.

36. Krinsky, N.I. 1968. The protective function

of carotenoid pigments. In A.C. Giese, ed., *Photophysiology*. vol. 3. New York: Academic Press.

37. Lemberg, R. 1928. Die Chromoproteide der Rotalgen. I. Justus Liebigs. *Ann. Chem.* 461:46.

38. Loomis, W. 1960. Historical Introduction. In W. Ruhland, ed., *Encyclopedia of Plant Physiology* 5; Part 1, 85. Berlin: Springer.

39. Lundegårh, H. 1966. Action spectra and the role of carotenoids in photosynthesis. *Physiol. Plant.* 19:754.

40. Lyttleton, J.W. 1962. Isolation of ribosomes from spinach chloroplasts. *Exp. Cell Res.* 26:312.

41. Mackinney, G. 1935. Leaf carotenes. *J. Biol. Chem.* 111:75.

42. Mathis, P., and K. Sauer. 1973. Chlorophyll formation in greening bean leaves during the early stages. *Plant Physiol.* 51:115.

43. Mudrack, K. 1956. Über Grössen und Strukturänderungen der Chloroplasten in Rohrzucker und Elektrolytlosungen. *Protoplasma* (Wien) 47:461.

44. O'hEocha, C. 1962. Phycobilins. In R. Lewin, ed., *Physiology and Biochemistry of Algae*. New York: Academic Press.

45. Parenti, F., and M.M. Margulies. 1967. In vitro protein synthesis by plastids of *Phaseolus vulgaris*. I. Localization of activity in the chloroplasts of a chloroplast containing fraction from developing leaves. *Plant Physiol.* 42:1179.

46. Park, R.B. 1965. The chloroplast. In J. Bonner and J.E. Varner, eds., *Plant Biochemistry*. New York: Academic Press.

47. Possingham, J.V. 1980. Plastid replication and development in the life cycle of higher plants. *Ann. Rev. Plant Physiol.* 11:113.

48. Rebeiz, C.A., S. Larson, T.E. Weier, and P.A. Castelfranco. 1973. Chloroplast maintenance and partial differentiation in vitro. *Plant Physiol.* 51:651.

49. Ridley, S.M., and R.M. Leech. 1970. Division of chloroplasts in an artificial environment. *Nature* 227:463.

50. Sager, R. 1959. The architecture of the chloroplast in relation to its photosynthetic activities. In *The Photochemical Apparatus—Its Structure and Function. Brookhaven Symp. Biol.* 11:101.

51. Schiff, J.A., and H.T. Epstein. 1965. The continuity of the chloroplast in *Euglena*. In M. Locke, ed., *Reproduction: Molecular, Subcellular, and Cellular*. New York: Academic Press.

52. Seely, G.R. 1966. Photochemistry of chlorophylls *in vitro*. In L.P. Vernon and G.R. Seely, eds., *The Chlorophylls*. New York: Academic Press.

53. Shlyk, A.A., V.L. Kaler, L.I. Vlasenok, and V.I. Gaponenko. 1963. The final stages of biosynthesis of chlorophylls a and b in the green leaf. *Photochem. Photobiol.* 2:129.

54. Sistrom, W.R., M. Griffiths, and R.Y. Stanier. 1956. The biology of a photosynthetic bacterium which lacks carotenoids. *J. Cell. Comp. Physiol.* 48:473.

55. Strain, H.H., and W.A. Svec. 1966. Extraction, separation, estimation, and isolation of the chlorophylls. In L.P. Vernon and G.R. Seely, eds., *The Chlorophylls*. New York: Academic Press.

56. Sudyina, E.G. 1963. Chlorophyllase reaction in the last stage of biosynthesis of chlorophyll. *Photochem. Photobiol.* 2:181.

57. Sundqvist, C. 1973. The relationship between chlorophyllide accumulation, the amount of protochlorophyllide-636 and protochlorophyllide-650 in dark grown wheat leaves treated with δ-aminolevulinic acid. *Physiol. Plant.* 28:464.

58. von Wettstein, D. 1959. The formation of plastids structures. In *The Photochemical Apparatus—Its Structure and Function. Brookhaven Symp. Biol.* 11:138.

59. von Wettstein, D. 1967. Chloroplast structure and genetics. In A. San Pietro, F.A. Greer, and T.J. Army, eds., *Harvesting the Sun—Photosynthesis in Plant Life*. New York: Academic Press.

60. Weir, T., and C. Stocking. 1952. The chloroplast: structure, inheritance and enzymology. *Bot. Rev.* 18:14.

61. Wolken, J. 1961. *Euglena: An Experimental Organism for Biochemical and Biophysical Stud-*

ies. New Brunswick, N.J.: Rutgers University Press.

62. Zeldin, M.H., and J.A. Schiff. 1967. RNA metabolism during light-induced chloroplast development in euglena. *Plant Physiol.* 42:922.

63. Zscheile, F., and C. Comar. 1951. Influence of preparative procedure on the purity of chlorophyll components as shown by absorption spectra. *Bot. Gaz.* 102:463.

64. Zscheile, F., J. White, B. Beadle, and J. Roach. 1942. The preparation and absorption spectra of five pure carotenoid pigments. *Plant Physiol.* 17:331.

Chapter 13

1. Allen, M., D. Arnon, J. Capindale, F. Whatley, and L. Durham. 1955. Photosynthesis by isolated chloroplasts. III. Evidence for complete photosynthesis. *J. Am. Chem. Soc.* 77:4149.

2. Arnon, D. 1951. Extracellular photosynthetic reactions. *Nature* 167:1008.

3. Arnon, D. 1967. Photosynthetic phosphorylation: facts and concepts. In T.W. Goodwin, ed., *Biochemistry of Chloroplasts.* New York: Academic Press.

4. Arnon, D., M. Allen, and F. Whatley. 1954. Photosynthesis by isolated chloroplasts. *Nature* 174:394.

5. Arnon, D., F. Whatley, and M. Allen. 1954. Photosynthesis by isolated chloroplasts. II. Photosynthetic phosphorylation, the conversion of light into phosphate bond energy. *J. Am. Chem. Soc.* 76:6324.

6. Arnon, D., F. Whatley, and M. Allen. 1957. Triphosphopyridine nucleotide as a catalyst of photosynthetic phosphorylation. *Nature* 180:182.

7. Bachofen, R., and D.I. Arnon. 1966. Crystalline ferredoxin from the photosynthetic bacterium *Chromatium Biochim. Biophys. Acta* 120:259.

8. Barr, R., and F.L. Crane. 1967. Comparative studies on plastoquinones. III. Distribution of plastoquinones in higher plants. *Plant Physiol.* 42:1255.

9. Butler, W.L. 1966. Spectral characteristics of chlorophyll in green plants. In L.P. Vernon and G.R. Seely, eds., *The Chlorophylls.* New York: Academic Press.

10. Clayton, R.K. 1966. Physical processes involving chlorophylls *in vivo.* In L.P. Vernon and G.R. Seely, eds., *The Chlorophylls.* New York: Academic Press.

11. de Saussure, N.T. 1804. *Recherches chimiques sur la végétation.* Paris: V. Nyon.

12. Einstein, A. 1905. Über einen die Erzeugung und Verwandlung des Lichtes betreffenden heuristischen Geischtspunkt. *Ann. Physik* 17:132.

13. Emerson, R. 1958. *The Quantum Yield of Photosynthesis. Ann. Rev. Plant Physiol.* 9:1.

14. French, C.S. 1960. The chlorophylls *in vivo* and *in vitro.* In W. Ruhland, ed., *Encyclopedia of Plant Physiology* 5, part 1:252. Berlin: Springer.

15. Govindjee, R.G., and E. Rabinowitch. 1960. Two forms of chlorophyll a *in vivo* with two distinct photochemical functions. *Science* 132:355.

16. Hill, R. 1937. Oxygen evolved by isolated chloroplasts. *Nature* 139:881.

17. Homann, P.H. 1967. Studies on the manganese of the chloroplast. *Plant Physiol.* 42:997.

18. Horio, T., and A. San Pietro. 1964. Action spectrum for ferricyanide photoreduction and redox potential for chlorophyll 683. *Proc. Natl. Acad. Sci., U.S.* 51:1226.

19. Jagendorf, A.T. 1975. Mechanism of photophosphorylation. In R.G. Govindjee, ed., *Bioenergetics of Photosynthesis.* New York: Academic Press.

20. Kok, B. 1961. Partial purification and determination of oxidation reduction potential of the photosynthetic chlorophyll complex absorbing at 700 mμ. *Biochim. Biophys. Acata* 48:527.

21. Kok, B. 1967. Photosynthesis—physical aspects. In A. San Pietro, F.A. Greer, and T.J. Army, eds., *Harvesting the Sun: Photosynthesis in Plant Life.* New York: Academic Press.

22. Loomis, W. 1960. Historical introduction. In W. Ruhland, ed., *Encyclopedia of Plant Physiology* 5, part 1:85. Berlin: Springer.

23. Mitchell, P. 1961. Coupling of phosphorylation and hydrogen transfer by chemiosmotic type of mechanism. *Nature* (London) 191:144.

24. Mitchell, P. 1978. Protonmotive chemiosmotic mechanism in oxidative and photosynthetic phosphorylation. *Trends Biochem. Sci.* 3:N58.

25. Myers, J., and C.S. French. 1960. Relationship between time course, chromatic transient, and enhancement phenomena of photosynthesis. *Plant Physiol.* 35:963.

26. San Pietro, A. 1967. Electron transport in chloroplasts. In A. San Pietro, F.A. Greer, and T.J. Army, eds., *Harvesting the Sun: Photosynthesis in Plant Life.* New York: Academic Press.

27. San Pietro, A., and H.M. Lang. 1958. Photosynthetic pyridine nucleotide reductase. I. Partial purification and properties of the enzyme from spinach. *J. Biol. Chem.* 231:211.

28. Shin, M., and D.I. Arnon. 1965. Enzymic mechanisms of pyridine nucleotide reduction in chloroplasts. *J. Biol. Chem.* 240:1405.

29. Shin, M., K. Tagawa, and D.I. Arnon. 1963. Crystallization of ferredoxin-TPN reductase and its role in the photosynthetic apparatus of chloroplasts. *Biochem. Z.* 338:84.

30. Szent-Gyorgyi, A. 1941. The study of energy levels in biochemistry. *Nature* 148:157.

31. Tagawa, K., and D.I. Arnon. 1962. Ferredoxin as electron carrier in photosynthesis and in the biological production and consumption of hydrogen gas. *Nature* 195:537.

32. Van Niel, C.B. 1941. The bacterial photosyntheses and their importance for the general problem of photosynthesis. *Adv. Enzymol.* 1:263.

33. Van Niel, C.B. 1962. The present status of the comparative study of photosynthesis. *Ann. Rev. Plant Physiol.* 13:1.

34. Vernon, L.P. 1967. The photosynthetic apparatus in bacteria. In A. San Pietro, F.A. Greer, and T.J. Army, eds., *Harvesting the Sun: Photosynthesis in Plant Life.* New York: Academic Press.

Chapter 14

1. Baeyer, A. 1870. Über die Wasserentziehung und ihre Bedeutung für das Pflanzenleben und die Gährung. *Ber. Dtsch. Chem. Ges.* 3:63.

2. Barker, H. 1935. Photosynthesis in diatoms. *Arch. Mikrobiol.* 6:141.

3. Bassham, J., A. Benson, L. Kay, A. Harris, A. Wilson, and M. Calvin. 1954. The path of carbon in photosynthesis. XXI. The cyclic regeneration of carbon dioxide acceptor. *J. Am. Chem. Soc.* 76:1760.

4. Bassham, J., and M. Calvin. 1957. *The Path of Carbon in Photosynthesis.* Englewood Cliffs, N.J.: Prentice-Hall.

5. Billings, W., and R. Morris. 1951. Reflection of visible and infrared radiation from leaves of different ecological groups. *Am. J. Bot.* 38:327.

6. Bormann, F. 1956. Ecological implications of changes in photosynthetic response of *Pinus taeda* seedling during ontogeny. *Ecology* 37:70.

7. Bowes, G., and W.L. Ogren. 1972. Oxygen inhibition and other properties of soybean ribulose-1,5-diphosphate carboxylase. *J. Biol. Chem.* 247:2171.

8. Bowes, G., W.L. Ogren, and R.H. Hageman. 1975. pH dependence of the K_m (CO_2) of ribulose-1-5-diphosphate carboxylase. *Plant Physiol.* 56:630.

9. Brown, H., and F. Escombe. 1902. The influence of varying amounts of carbon dioxide in the air on the photosynthetic process of leaves and on the mode of growth of plants. *Proc. Roy. Soc.* 70B:397.

10. Calvin, M. 1956. The photosynthetic carbon cycle. *J. Am. Chem. Soc.* 78:1895.

11. Calvin, M. 1959. From microstructure to macrostructure and function in the photochemical apparatus. In *The photochemical apparatus—its structure and function. Brookhaven Symp. Biol.* 11:160.

12. Calvin, M., and A.A. Benson. 1948. The path of carbon in photosynthesis. *Science* 107:476.

13. Clayton, R.K. 1965. *Molecular Physics in Photosynthesis*. New York: Blaisdell Publishing.

14. Commoner, B. 1961. Electron spin resonance studies of photosynthetic systems. In W.D. McElroy and B. Glass, eds. *Light and Life*. Baltimore, Md.: Johns Hopkins University Press.

15. Gaffron, R. 1960. Energy storage. In F.C. Steward, ed., *Plant Physiology*. New York: Academic Press.

16. Gibbs, M., and O., Kandler. 1957. Asymmetric distribution of ^{14}C in sugars formed during photosynthesis. *Proc. Natl. Acad. Sci., U.S.* 43:446.

17. Hatch, M.D., and C.R. Slack. 1966. Photosynthesis by sugarcane leaves. A new carboxylation reaction and the pathway of sugar formation. *Biochem. J.* 101:103.

18. Hatch, M.D., C.R. Slack, and H.S. Johnson. 1967. Further studies on a new pathway of photosynthetic carbon dioxide fixation in sugarcane and its occurrence in other plant species. *Biochem. J.* 102:417.

19. Heinicke, A., and N. Childers. 1937. The daily rate of photosynthesis during the growing season of 1935, of a young apple tree of bearing age. *Cornell Univ. Agr. Expt. Sta. Mem.* 201:3.

20. Hill, R., and C. Whittingham. 1953. The induction phase of photosynthesis in *Chlorella* determined by a spectroscopic method. *New Phytol.* 52:133.

21. Kandler, O., and M. Gibbs. 1956. A symmetric distribution of C^{14} in the glucose phosphates formed during photosynthesis. *Plant Physiol.* 31:411.

22. Kandler, O., and F. Schötz. 1956. Untersuchungen über die photoxydative Farbstofzerstörung und Stoffwechselhemmung bei *Chlorella* Mutanten und panaschierten Oenotheren. *Z. Naturforsch.* 11b:708.

23. Kortschak, H.P., C.E. Hartt, and G.O. Burr. 1965. Carbon dioxide fixation in sugarcane leaves. *Plant Physiol.* 40:209.

24. Kreusler, U. 1885. Über eine Methode zur Beobachtung der Assimilation und Athmung der Pflanzen und über einige diese Vorgänge beeinflussenden Momente. *Land. Jahrb.* 14:913.

25. Kreusler, U. 1887. Beobachtungen über die Kohlensäure-Aufnahme und Ausgabe (Assimilation und Athmung) der Pflanzen. II. Mittheilung. Abhängigkeit von Entwicklungszustand-Einfluss der Temperatur. *Land Jahrb.* 16:711.

26. Laetsch, W.M. 1974. The C-4 syndrome: a structural analysis. *Ann. Rev. Plant Physiol.* 25:27.

27. Loustalot, A. 1945. Influence of soil moisture conditions on apparent photosynthesis and transpiration of pecan leaves. *J. Agr. Research* 71:519.

28. McAlister, E., and J. Myers. 1940. The time course of photosynthesis and fluorescence observed simultaneously. *Smithsonian Inst. Misc. Collection* 99, No. 6.

29. Meyer, B., and D. Anderson. 1952. *Plant Physiology*. Princeton, N.J.: Van Nostrand.

30. Mitchell, J.W. 1936. Effect of atmospheric humidity on rate of carbon fixation of plants. *Bot. Gaz.* 98:87.

31. Noddack, W., and C. Kopp. 1940. Untersuchungen über die Assimilation der Kohlensäure durch die grünen Pflanzen. IV. Assimilation und Temperatur. *Z. Physik. Chem.* 187A:79.

32. Ochoa, S. 1946. Enzymatic mechanisms of carbon dioxide assimilation. In D. Green, ed., *Currents in Biochemical Research*. New York: Interscience Publishers.

33. Ochoa, S., A. Mehler, and A. Kornberg. 1948. Biosynthesis of dicarboxylic acids by carbon dioxide fixation. I. Isolation and properties of an enzyme from pigeon liver catalyzing the reversible oxidative decarboxylation of 1-malic acid. *J. Biol. Chem.* 174:979.

34. Ochoa, S., and W. Vishniac. 1952. Carboxylation reactions and photosynthesis. *Science* 115:297.

35. Ogren, W.L., and G. Bowes. 1971. Ribulose diphosphate carboxylase regulates soybean photorespiration. *Nature New Biol.* 230:159.

36. Paechnatz, G. 1938. Zur Frage der Assimilation von Formaldehyd durch die grüne Pflanze. *Z. Bot.* 32:161.

37. Pantanelli, E. 1903. Abhängigkeit der Sauerstoffausscheidung belichteter Pflanzen von äusseren Bedingungen. *Jahrb. Wiss. Bot.* 39:167.

38. Pokrowski, G. 1925. Über die Lichtabsorption von Blättern einiger Bäume. *Biochem. Z.* 165:420.

39. Rabinowitch, E. 1945. *Photosynthesis and Related Processes,* vol. I. New York: Interscience Publishers.

40. Rabinowitch, E. 1951. *Photosynthesis and Related Processes,* vol. II, part 1. New York: Interscience Publishers.

41. Rabinowitch, E. 1956. *Photosynthesis and Related Processes,* vol. II, part 2. New York: Interscience Publishers.

42. Ruben, S., W. Hassid, and M. Kamen. 1939. Radioactive carbon in the study of photosynthesis. *J. Am. Chem. Soc.* 61:661.

43. Ruben, S., and M. Kamen. 1940. Photosynthesis with radioactive carbon. IV. Molecular weight of the intermediate products and a tentative theory of photosynthesis. *J. Am. Chem. Soc.* 62:3451.

44. Ruben, S., and M.D. Kamen. 1940. Radioactive carbon in the study of respiration in heterotrophic systems. *Proc. Natl. Acad. Sci. U.S.* 26:418.

45. San Pietro, A., and H.M. Lang. 1958. Photosynthetic pyridine nucleotide reductase. I. Partial purification and properties of the enzyme from spinach. *J. Biol. Chem.* 231:211.

46. Schneider, G., and N. Childers. 1941. Influence of soil moisture on photosynthesis, respiration, and transpiration of apple leaves. *Plant Physiol.* 16:565.

47. Seybold, A. 1932. Über die optischen Eigenschaften der Laubblätter. II. *Planta* 18:479.

48. Slack, C.R., and M.D. Hatch. 1967. Comparative studies on the activity of carboxylases and other enzymes in relation to the new pathway of photosynthetic carbon dioxide fixation in tropical grasses. *Biochem. J.* 103:660.

49. Stainer, R. 1959. Formation and function of photosynthetic pigment system in purple bacteria. *Brookhaven Symp. Biol.* 11:13.

50. Stiller, M. 1962. The path of carbon in photosynthesis. *Ann. Rev. Plant Physiol.* 13:151.

51. Talling, J. 1961. Photosynthesis under natural conditions. *Ann. Rev. Plant Physiol.* 12:133.

52. Thomas, M.D., and G.R. Hill. 1949. Photosynthesis under field conditions. In J. Franck and W.E. Loomis, eds., *Photosynthesis in Plants.* Ames: Iowa State College Press.

53. Ting, I., and W. Loomis. 1963. Diffusion through stomates. *Am. J. Bot.* 50:866.

54. Verduin, J., and W.E. Loomis. 1944. Absorption of carbon dioxide by maize. *Plant Physiol.* 19:278.

55. Vernon, L.P., and B. Ke. 1966. Photochemistry of chlorophyll in vivo. In L.P. Vernon and G.R. Seely, eds., *The Chlorophylls.* New York: Academic Press.

56. Warburg, O. 1958. Photosynthesis. *Science* 128:68.

57. Warburg, O., H. Klotzech, and G. Krippahl. 1957. Über das Verhalten einiger Aminosäuren in *Chlorella* bei Zusatz von markierter Kohlensäure. *Z. Naturf.* 126:481.

58. Wolken, J., and A. Mellon. 1957. Light and heat in the bleaching of chloroplasts in *Euglena. Biochim. Biophys. Acta* 25:267.

59. Yocum, C.F., and A. San Pietro. 1969. Ferredoxin reducing substance from spinach. *Biochem. Biophys. Res. Commun.* 36:614.

Chapter 15

1. Beer, M. 1959. Fine structure of phloem of *Cucurbita* as revealed by the electron microscope. *Proc. Int. Bot. Congr., 9th congr., Montreal, Canada* 2:26. Toronto: University of Toronto Press.

2. Biddulph, O., and R. Cory. 1957. An analysis of translocation in the phloem of the bean plant using THO, P³², and C¹⁴. *Plant Physiol.* 32:608.

3. Biddulph, O., and R. Cory. 1965. Translocation of C¹⁴ metabolites in the phloem of the bean plant. *Plant Physiol.* 40:119.

4. Biddulph, S.F. 1956. Visual indications of S³⁵ and P³² translocation in the phloem. *Am. J. Bot.* 43:143.

5. Bieleski, R.L. 1966. Sites of accumulation in excised phloem and vascular tissues. *Plant Physiol.* 41:455.

6. Booth, A., J. Moorby, C.R. Davies, H. Jones, and P.F. Wareing. 1962. Effect of indolyl-3-acetic acid on the movements of nutrients within the plant. *Nature* 194:204.

7. Bouch, G.B., and J. Cronshaw. 1965. The fine structure of differentiating sieve tube elements. *J. Cell. Biol.* 25:79.

8. Buchanan, J. 1953. The path of carbon in photosynthesis. XIX. The identification of sucrose phosphate in sugar beet leaves. *Arch. Biochem. Biophys.* 44:140.

9. Burley, J. 1961. Carbohydrate translocation in raspberry and soybean. *Plant Physiol.* 36:820.

10. Crafts, A.S. 1951. Movement of assimilates, viruses, growth regulators, and chemical indicators in plants. *Bot. Rev.* 17:203.

11. Crafts, A.S. 1961. *Translocation in Plants.* New York: Holt, Rinehart and Winston.

12. Crafts, A.S., and C.E. Crisp. 1971. *Phloem Transport in Plants.* San Francisco: Freeman.

13. Currier, H.B., and C.Y. Shih. 1968. Sieve tubes and callose in *Elodea* leaves. *Am. J. Bot.* 55:145.

14. DeStigter, H.C.M. 1961. Translocation of C¹⁴ photosynthates in the graft muskmelon *Cucurbita ficifolia. Acta Bot. Neerlandica* 10:466.

15. Duloy, M., F.V. Mercer, and N. Rathgeber. 1961. Studies in translocation. II. Submicroscopic anatomy of the phloem. *Aust. J. Biol. Sci.* 14:506.

16. Esau, K. 1947. A study of some sieve-tube inclusions. *Am. J. Bot.* 34:224.

17. Esau, K. 1950. Development and structure of the phloem tissue. II. *Bot. Rev.* 16:67.

18. Esau, K. 1960. *Anatomy of Seed Plants.* New York: Wiley.

19. Esau, K. 1965. Parenchyma cells in the conducting system (the "pumps" and "sinks"). *Plant Physiol.* 40:xxvii.

20. Evert, R.F., and L. Murmanis. 1965. Ultra structure of the secondary phloem of *Tilia americana. Am. J. Bot.* 52:95.

21. Gage, R., and S. Aronoff. 1960. Radioautography of tritiated photosynthate arising from HTO. *Plant Physiol.* 35:65.

22. Gauch, H.G., and W.M. Dugger, Jr. 1953. The role of boron in the translocation of sucrose. *Plant Physiol.* 28:457.

23. Geiger, D.R. 1966. Effect of sink region cooling on translocation of photosynthate. *Plant Physiol.* 41:1667.

24. Giaquinta, R.T., and D.R. Geiger. 1973. Mechanism of inhibition of translocation by localized chilling. *Plant Physiol.* 51:372.

25. Goren, R., and A.W. Galston. 1966. Control by phytochrome of C¹⁴-sucrose incorporation into buds of etiolated pea seedlings. *Plant Physiol.* 41:1055.

26. Goren, R., and A.W. Galston. 1967. Phytochrome controlled C¹⁴-sucrose uptake into etiolated pea buds; effects of gibberellic acid and other substances. *Plant Physiol.* 42:1087.

27. Hansen, P. 1967. C¹⁴-studies on apple trees. I. The effect of the fruit on the translocation and distribution of photosynthates. *Physiol. Plant.* 20:382.

28. Harel, S., and L. Reinhold. 1966. The effect of 2,4-dinitrophenol on translocation in the phloem. *Physiol. Plant.* 19:634.

29. Hartt, C.E. 1965. The effect of temperature upon translocation of C¹⁴ in sugarcane. *Plant Physiol.* 40:74.

30. Hartt, C.E. 1966. Translocation in colored light. *Plant Physiol.* 41:369.

31. Hartt, C.E., H.P. Kortschak, A.J. Forbes, and G.O. Burr. 1963. Translocation of C¹⁴ in sugarcane. *Plant Physiol.* 38:305.

32. Hew, C.S., C.D. Nelson, and G. Krotkov.

1967. Hormonal control of translocation of photosynthetically assimilated C^{14} in young soybean plants. *Am. J. Bot.* 54:252.

33. Hewitt, S.P., and O.F. Curtis. 1948. The effect of temperature on loss of dry matter and carbohydrate from leaves by respiration and translocation. *Am. J. Bot.* 35:746.

34. Holman, R., and W. Robbins. 1938. *Textbook of General Botany for Colleges and Universities.* New York: Wiley.

35. Joy, K.W. 1964. Translocation in sugar beet. I. Assimilation of $C^{14}O_2$ and distribution of materials from leaves. *J. Exp. Bot.* 15:485.

36. Kriedemann, P., and H. Beevers. 1967. Sugar uptake and translocation in the castor bean seedling. I. Characteristics of transfer in intact and excised seedlings. *Plant Physiol.* 42:161.

37. Kursanov, A.L. 1963. Metabolism and the transport of organic substances in the phloem. In R.D. Preston, ed., *Advances in Botanical Research.* New York: Academic Press.

38. Kursanov, A.L., and M.I. Brovchenko. 1959. *Fiziol. Rastenü.* 8:270.

39. Kursanov, A.L., M.V. Turkina, and I.M. Dubinina. 1953. Die Anwendung der Isopenmethode bei der Erforschung des Zukertransportes in der Pflanze. *C.R. Acad. Sci.* (USSR) 68:1113.

40. McNairn, R.B. 1972. Phloem translocation and heat-induced callose formation in field-grown *Gossypium hirsutum* L. *Plant Physiol.* 50:366.

41. McNairn, R.B., and H.B. Currier. 1968. Translocation blockage by sieve plate callose. *Planta* 82:369.

42. Mason, T.G., and E.J. Maskell. 1928. Studies on the transport of carbohydrates in the cotton plant. I. A study of diurnal variation in the carbohydrates of leaf, bark, and wood, and the effects of ringing. *Ann. Bot.* 42:189.

43. Mason, T.G., and E.J. Maskell. 1928. Studies on the transport of carbohydrates in the cotton plant. II. The factors determining the rate and the direction of movement of sugars. *Ann Bot.* 42:571.

44. Mason, T.G., and E. Phillis. 1937. The migration of solutes. *Bot. Rev.* 3:47.

45. Mittler, T.E. 1953. Amino acids in phloem sap and their excretion by aphids. *Nature* 172:207.

46. Mittler, T.E. 1958. Studies of the feeding and nutrition of *Tuberolachnus salignus* (Gmelin) (Homoptera, Aphidae.) II. The nitrogen and sugar composition of ingested phloem sap and excreted honeydew. *Plant Physiol.* 35:74.

47. Mothes, K., and L. Engelbrecht. 1961. Kinetin and its role in nitrogen metabolism In *Proc. Int. Bot. Congr., 9th cong., Montreal, Canada* 2:996. Toronto: University of Toronto Press.

48. Münch, E. 1930. Die *Stoffbewegungen in der Pflanze.* Stuttgart: Gustav Fisher Verlag.

49. Nelson, C.D. 1963. Effect of climate on the distribution and translocation of assimilates. In L.T. Evans, *Environmental Control of Plant Growth.* New York: Academic Press.

50. Nelson, C.D., and P.R. Gorham. 1957. Uptake and translocation of C^{14} labeled sugars applied to primary leaves of soybean seedlings. *Can. J. Bot.* 35:339.

51. Nelson, C.D., and P.R. Gorham. 1959. Translocation of C^{14}-labeled amino acids and amides in the stems of young soybean plants. *Can. J. Bot.* 37:431.

52. Peel, A.J. 1967. Demonstration of solute movement from the extracambial tissues into the xylem stream in willow. *J. Exp. Bot.* 18:600.

53. Pristupa, N.A., and A.L. Kursanov. 1957. Descending flow of assimilates and its relation to the absorbing activity of roots. *Plant Physiol.* (USSR), *Friziol. Rast.* 4:395.

54. Roeckl, B. 1949. Nachweis eines Konzentrationshubs zwischen Palisadenzellen und Siebröhren. *Planta* 36:530.

55. Seth, A.K., and P.F. Wareing. 1967. Hormone-directed transport of metabolites and its possible role in plant senescence. *J. Exp. Bot.* 18:65.

56. Shih, C.Y., and H.B. Currier. 1969. Fine structure of phloem cells in relation to

translocation in the cotton seedling. *Am. J. Bot.* 56:464.

57. Shindy, W.W., W.M. Kliewer, and R.J. Weaver. 1973. Benzyladenine-induced movement of ^{14}C-labeled photosynthate into roots of *Vitis vinifera*. *Plant Physiol.* 51:345.

58. Shiroya, M., C.D. Nelson, and G. Krotkov. 1961. Translocation of C^{14} in tobacco at different stages of development following assimilation of C^{14}O$_2$ by a single leaf. *Can. J. Bot.* 39:855.

59. Sij, J.W., and C.A. Swanson. 1973. Effect of petiole anoxia on phloem transport in squash. *Plant Physiol.* 51:368.

60. Swanson, C.A. 1959. Translocation of organic solutes. In F.C. Steward, ed., *Plant Physiology*. New York: Academic Press.

61. Swanson, C.A., and R.H. Böhning. 1951. The effect of petiole temperature on the translocation of carbohydrates from bean leaves. *Plant Physiol.* 26:557.

62. Swanson, C.A., and E.D.H. El-Shishiny. 1958. Translocation of sugars in grapes. *Plant Physiol.* 33:33.

63. Swanson, C.A., and D.R. Geiger. 1967. Time course of low temperature inhibition of sucrose translocation in sugar beets. *Plant Physiol.* 42:751.

64. Ullrich, W. 1961. Zur Sauerstofabhängigkeit des Transportes in den Siebröhren. *Planta* 57:402.

65. Vernon, L.P., and S. Aronoff. 1952. Metabolism of soybean leaves. IV. Translocation from soybean leaves. *Arch. Biochem. Biophys.* 36:383.

66. Weatherley, P.E., A.J. Peel, and G.P. Hill. 1959. The physiology of the sieve tube. Preliminary experiments using aphid mouth parts. *J. Exp. Bot.* 10:1.

67. Webb, J.A., and P.R. Gorham. 1964. Translocation of photosynthetically assimilated C^{14} in straight-necked squash. *Plant Physiol.* 39:663.

68. Willenbrink, J. 1957. Über die Hemmung des Stofftransports in den Siebröhren durch lokale Inaktivierung verschiedener Atmungenzyme. *Planta* 48:269.

69. Zimmermann, M.H. 1957. Translocation of organic substances in trees. I. The nature of the sugars in the sieve tube exudate of trees. *Plant Physiol.* 32:288.

70. Zimmermann, M.H. 1957. Translocation of organic substances in trees. II. On the translocation mechanism in the phloem of white ash. *Plant Physiol.* 32:399.

71. Zimmermann, M.H. 1958. Translocation of organic substances in the phloem of trees. In K.V. Thimann, ed., *The Physiology of Forest Trees*. New York: Ronald Press.

72. Zimmermann, M.H. 1958. Translocation of organic substances in trees. III. The removal of sugars from the sieve tubes in the white ash (*Fraxinus americana* L.). *Plant Physiol.* 33:213.

73. Zimmermann, M.H. 1960. Transport in the phloem. *Ann. Rev. Plant Physiol.* 11:167.

Chapter 16

1. Audus, L.J. 1936. Mechanical stimulation and respiration rate in cherry laurel. *New Phytol.* 34:557.

2. Audus, L.J. 1939. Mechanical stimulation and respiration in the green leaf. II. Investigation on a number of angiospermic species. *New Phytol.* 38:284.

3. Audus, L.J. 1940. Mechanical stimulation and respiration in the green leaf. III. The effect of stimulation on the rate of fermentation. *New Phytol.* 39:65.

4. Audus, L.J. 1941. Mechanical stimulation and respiration in the green leaf. Parts IV and V. *New Phytol.* 40:86.

5. Bendall, D.S., and W.D. Bonner, Jr. 1971. Cyanide-insensitive respiration in plant mitochondria. *Plant Physiol.* 47:236.

6. Breidenbach, R.W., A. Kahn, and H. Beevers. 1968. Characterization of glyoxysomes from castor bean endosperm. *Plant Physiol.* 43:705.

7. Fernandes, D.S. 1923. Aerobe und anaerobe Atmung bei Keimlingen von *Pisum sativum*. *Rec. Trav. Bot. Néerl.* 20:107.

8. Frenkel, C. 1972. Involvement of perox-

idase and indole-3-acetic acid oxidase isoenzymes from pear, tomato and blueberry fruit in ripening. *Plant Physiol.* 49:757.

9. Goodwin, T.W., and E.I. Mercer. 1972. *Introduction to Plant Biochemistry.* New York: Pergamon Press.

10. Gunsalus, I.C. 1954. Group transfer and acyl-generating functions of lipoic acid derivatives. In W.D. McElroy and B. Glass, eds., *Mechanism of Enzyme Action.* Baltimore, Md.: Johns Hopkins University Press.

11. Heath, O.V.S. 1950. Studies in stomatal behaviour. V. The role of carbon dioxide in the light response of stomata. *J. Exp. Bot.* 1:29.

12. Henry, M.F., and E.J. Nyns. 1975. Cyanide-insensitive respiration. An alternative mitochondrial pathway. *Sub-Cell Biochem.* 4:1.

13. Hopkins, E.F. 1927. Variation in sugar content in potato tubers caused by wounding and its possible relation to respiration. *Bot. Gaz.* 84:75.

14. James, W.O. 1953. *Plant Respiration.* Oxford: Clarendon Press.

15. Kidd, F. 1915. The controlling influence of carbon dioxide. III. The retarding effect of carbon dioxide on respiration. *Proc. Roy. Soc.* (London) B89:136.

16. Kornberg, H.L., and H.A. Krebs. 1957. Synthesis of cell constituents from C_2-units by a modified tricarboxylic acid cycle. *Nature* 179:988.

17. Lundegårdh, H., and H. Burström. 1933. Untersuchungen über die Salaufnahme der Pflanzen. III. Quantitative Beziehungen zwischen Atmung und Anionenaufnahme. *Biochem. Z.* 261:235.

18. Mitchell, P. 1966. Chemiosmotic coupling in oxidative and photosynthetic phosphorylation. *Biol. Rev.* 41:445.

19. Rich, P.R., and A.L. Moore. 1976. The involvement of the ubiquinone cycle in the respiratory chain of higher plants and its relation to the branchpoint of the alternative pathway. *FEBS Lett.* 65:339.

20. Solomos, T. 1977. Cyanide-resistant respiration in higher plants. *Ann. Rev. Plant Physiol.* 28:279.

21. Stiles, W. 1960. The composition of the atmosphere (oxygen content of air, water, soil, intercellular spaces, diffusion, carbon dioxide, and oxygen tensions). In W. Ruhland, ed., *Encyclopedia of Plant Physiology* 12:114. Berlin: Springer.

22. Stiles, W., and W. Leach. 1960. *Respiration in Plants.* New York: Wiley.

23. Taylor, D.L. 1942. Influence of oxygen tension on respiration, fermentation, and growth in wheat and rice. *Am. J. Bot.* 29:721.

24. Yemm, E.W. 1935. The respiration of barley plants. II. Carbohydrate concentration and carbon dioxide production in starving leaves. *Proc. Roy. Soc.* (London) B117:504.

25. Yemm, E.W. 1937. The respiration of barley plants. III. Protein catabolism in starving leaves. *Proc. Roy. Soc.* (London) B123:243.

Chapter 17

1. Audus, L.J. 1959. *Plant Growth Substances.* New York: Interscience Publishers.

2. Bayliss, W.M., and E.H. Starling. 1902. The mechanism of pancreatic secretion. *J. Physiol.* 28:325.

3. Beck, W.A. 1941. Production of solutes in growing epidermal cells. *Plant Physiol.* 16:637.

4. Beyer, A. 1928. Beiträge zum Problem der Reizleitung. *Z. Bot.* 20:321.

5. Bonner, D.M., A.J. Haagen-Smit, and F.W. Went. 1939. Leaf growth hormones. I: A bioassay and source for leaf growth factors. *Bot. Gaz.* 101:128.

6. Bonner, J. 1932. The production of growth substances by *Rhizopus suinus. Biol. Zbl.* 52:565.

7. Bonner, J. 1933. The action of the plant growth hormone. *J. Gen. Physiol.* 17:63.

8. Boysen-Jensen, P. 1910. Über die Leitung des phototripischen Reizes in Avenakeimpflanzen. *Ber. D. Bot. Ges.* 28:118.

9. Briggs, W.R., G. Morel, T.A. Steeves, I.M. Sussex, and R.H. Wetmore. 1955. Enzymatic auxin inactivation by extracts of the fern, *Osmunda cinnamomea* L. *Plant Physiol.* 30:143.

10. Burkholder, P.A., and E.S. Johnston. 1937. Inactivation of plant growth substance by light. *Smithsonian Inst. Misc. Collections* 95:20.

11. Darwin, C. 1881. *The Power of Movement in Plants.* New York: D. Appleton.

12. Devlin, R.M., and W.T. Jackson. 1961. Effect of p-chlorophenoxyisobutyric acid on rate of elongation of root hairs of *Agrostis alba.* L. *Physiol. Plant.* 14:40.

13. Dolk, H.E. 1930. Geotropic en groeistof. Dissertation, Utrecht; English transl. by F. Dolk-Hoek and K.V. Thimann, 1936. *Rec. Trav. Bot. Néerl.* 33:509.

14. DuBuy, H.G., and E. Neurenbergk. 1934. Phototropismus und Wachstum der Pflanzen. II. *Ergeb. Biol.* 10:207.

15. Fitting, H. 1909. Die Beeinflussing der Orchideenblüten durch die Bestäubung und durch andere Umstände. *Z. Bot.* 1:1.

16. Funke, H., and H. Söding. 1948. Über das Wuchsstoff-Hemmstoffsystem der Haferkoleoptile und der Kartoffelknolle. *Planta* 36:341.

17. Galston, A.W., J. Bonner, and R.S. Baker. 1953. Flavoprotein and peroxidase as components of the indoleacetic acid oxidase system of peas. *Arch. Biochem Biophys.* 49:456.

18. Galston, A.W., and L.Y. Dalberg. 1954. The adaptive formation and physiological significance of indoleacetic acid oxidase. *Am. J. Bot.* 41:373.

19. Galston, A.W., and W.S. Hillman. 1961. The degradation of auxin. In W. Ruhland, ed., *Encyclopedia of Plant Physiology* 14:647. Berlin: Springer.

20. Goldsmith, M.H. 1966. Movement of indoleacetic acid in coleoptiles of *Avena sativa* L. II. Suspension of polarity by total inhibition of the basipetal transport. *Plant Physiol.* 41:15.

21. Goldsmith, M.H.M. 1967. Movement of pulses of labeled auxin in corn coleoptiles. *Plant Physiol.* 42:258.

22. Gordon, S.A. 1956. The biogenesis of natural auxins. In R.L. Wain and F. Wightman, eds. *The Chemistry and Mode of Action of Plant Growth Substances.* London: Butterworth.

23. Gordon, S.A., and F.S. Nieva. 1949. The biosynthesis of auxin in the vegetative pineapple. I and II. *Arch. Biochem. Biophys.* 20:356.

24. Gregory, F.G., and C.R. Hancock. 1955. The rate of transport of natural auxin in woody shoots. *Ann. Bot.* N.S. 19:451.

25. Gustafson, F.G. 1941. Extraction of growth hormones from plants. *Am. J. Bot.* 28:947.

26. Haagen-Smit, A.J., W.B. Dandliker, S.H. Wittmer, and A.E. Murneek. 1946. Isolation of Indoleacetic acid from Immature Corn Kernels. *Am. J. Bot.* 33:118.

27. Haagen-Smit, A.J., and F.W. Went. 1935. A physiological analysis of the growth substance. *Proc. Kon. Nederl. Akad. Wetensch.* (Amsterdam) 38:852.

28. Haberlandt, G. 1913. Zur Physiologie der Zellteilung. *Sitzber. K. Preuss. Akad. Wiss.* 318.

29. Harrison, A. 1965. Auxanometer experiments on extension growth of *Avena* coleoptiles in different CO_2 concentrations. *Physiol. Plant.* 18:321.

30. Irvine, V.C. 1938. Studies in growth-promoting substances as related to x-radiation and photoperiodism. *Univ. Colo. Studies* 26:69.

31. Jacobs, W.P. 1961. The polar movement of auxin in the shoots of higher plants: its occurrence and physiological significance. In *Plant Growth Regulation.* Intern. Conf. Plant Growth Reg. 4th. Ames: Iowa State University Press.

32. Kögl, F., H. Erxleben, and A. Haagen-Smit. 1934. Über die Isolirung der Auxine "a" und "b" aus pflanzlichen Materialen. IX. Mitteilung. *Z. Physiol. Chem.* 225:215.

33. Kögl, F., and A. Haagen-Smith. 1931. Über die Chemie des Wuchsstoffs. *Proc. Kon.*

Akad. Nederl. Wetensch. (Amsterdam) 34:1411.

34. Kögl, F., A. Haagen-Smit, and H. Erxleben. 1934. Über ein neues Auxin (Heteroauxin) aus Harn. XI Mitteilung. *Z. Physiol. Chem.* 228:90.

35. Kögl, F., and D.G.F.R. Kostermans. 1934. Heteroauxin als Stoffwechselprodukt niederer pflanzlicher Organismen. XIII. *Z. Physiol. Chem.* 228:113.

36. Lantican, B.P., and R.M. Muir. 1967. Isolation and properties of the enzyme system forming indoleacetic acid. *Plant Physiol.* 42:1158.

37. Leopold, A.C. 1955. *Auxins and Plant Growth.* Los Angeles: University of California Press.

38. Leopold, A.C., and O.F. Hall. 1966. Mathematical model of polar auxin transport. *Plant Physiol.* 41:1476.

39. Loo, S. 1945. Cultivation of excised stem tips of asparagus *in vitro. Am. J. Bot.* 32:13.

40. Lund, E.J. 1947. *Bioelectric Fields and Growth.* Austin: University of Texas Press.

41. Moore, T.C. 1979. *Biochemistry and Physiology of Plant Hormones.* New York: Springer-Verlag.

42. Moore, T.C., and C.A. Shaner. 1967. Biosynthesis of indoleacetic acid from tryptophan-C^{14} in cell-free extracts of pea shoot tips. *Plant Physiol.* 42:1787.

43. Niedergang-Kamien, E., and A.C. Leopold. 1957. Inhibitors of polar auxin transport. *Physiol. Plant.* 10:29.

44. Paàl, A. 1919. Über phototropische Reizleitung. *Jahrb. Wiss. Bot.* 58:406.

45. Phelps, R.H., and L. Sequeira. 1967. Synthesis of indoleacetic acid via tryptamine by a cell-free system from tobacco terminal buds. *Plant Physiol.* 42:1161.

46. Pilet, P.E. 1965. Action of gibberellic acid on auxin transport. *Nature* 208:1344.

47. Pilet, P.E. 1965. Polar transport of radioactivity from C^{14}-labelled-β-indolylacetic acid in stems of *Lens culinaris. Physiol. Plant.* 18:687.

48. Popp, H.W., and H.R.C. McIlvaine. 1937. Growth substances in relation to the mechanism of the action of radiation on plants. *J. Agr. Res.* 55:931.

49. Rajagopal, R. 1967. Metabolism of indole-3-acetaldehyde. I. Distribution of indoleacetic acid and tryptophol forming activities in plants. *Physiol. Plant.* 20:982.

50. Sachs, J. 1882. Stoff und Form der Pflanzenorgane. *Arb. Bot. Inst. Wurzburg* 3:452.

51. Schrank, A.R. 1951. Electrical polarity and auxins. In F. Skoog, ed., *Plant Growth Substances.* Madison: University of Wisconsin Press.

52. Scott, T.K. 1972. Auxins and roots. *Ann. Rev. Plant Physiol.* 23:235.

53. Sherwin, J.E. 1970. A tryptophan decarboxylase from cucumber seedlings. *Plant and Cell Physiol.* 11:865.

54. Shoji, K., F.T. Addicott, and W.A. Swets. 1951. Auxin in relation to leaf blade abscission. *Plant Physiol.* 26:189.

55. Skoog, F. 1934. The effect of x-rays on growth substance and plant growth. *Science* 79:256.

56. Skoog, F. 1935. Effect of x-irradiation on auxin and plant growth. *J. Cell Comp. Physiol.* 7:227.

57. Skoog, F., and K.V. Thimann. 1940. Enzymatic liberation of auxin from plant tissues. *Science* 92:64.

58. Tang, Y.W., and J. Bonner. 1947. The enzymatic inactivation of indoleacetic acid. *Arch. Biochem. Biophys.* 13:11.

59. Thimann, K.V. 1934. Studies on the growth hormone of plants. VI. The distribution of the growth substance in plant tissues. *J. Gen. Physiol.* 18:23.

60. Thimann, K.V. 1935. In the plant growth hormone produced by *Rhizopus suinus. J. Biol. Chem.* 109:279.

61. Thimann, K.V., and F. Skoog. 1934. Inhibition of bud development and other functions of growth substance in *Vicia faba. Proc. Roy Soc.* (London) B114:317.

62. Truelsen, T.A. 1973. Indole-3-pyruvic acid as an intermediate in the conversion of tryptophan to indole-3-acetic acid. II. Distribution of tryptophan transaminase activity in plants. *Physiol. Plant.* 28:67.

63. Tukey, H.B., F.W. Went, R.M. Muir, and J. van Overbeek. 1954. Nomenclature of chemical plant regulators. *Plant Physiol.* 29:307.

64. van Overbeek, J., E.S. deVasquez, and S.A. Gordon. 1947. Free and bound auxin in the vegetative pineapple plant. *Am. J. Bot.* 34:266.

65. Went, F.W. 1926. On growth-accelerating substances in the coleoptile of *Avena sativa*. *Proc. Kon. Nederl. Akad. Wetensch.* (Amsterdam) 35:723.

66. Went, F.W. 1928. Wuchsstoff und Wachstum. *Rec. Trav. Bot. Néerl.* 25:1.

67. Went, F.W. 1934. On the pea test method for auxin, the plant growth hormone. *K. Akad. Wetenschap. Amsterdam Proc. Sect. Sci.* 37:547.

68. Went, F.W., and K.V. Thimann. 1937. *Phytohormones.* New York: Macmillan.

69. Wildman, S.G., M.G. Ferri, and J. Bonner. 1947. The enzymatic conversion of tryptophan to auxin by spinach leaves. *Arch. Biochem. Biophys.* 13:131.

70. Wildman, S.G., and R.M. Muir. 1949. Observation on the mechanism of auxin formation in plant tissues. *Plant Physiol.* 24:84.

71. Witham, F.H., and A.C. Gentile. 1961. Some characteristics and inhibitors of indole acetic acid oxidase from cultures of crown-gall. *Exp. J. Bot.* 12:188.

72. Zimmerman, P.W., and A.E. Hitchcock. 1942. Substituted phenoxy and benzoic acid growth substances and the relation of structure to physiological activity. *Contr. Boyce Thompson Inst.* 12:321.

73. Zimmerman, P.W., and A.E. Hitchcock, and F. Wilcoxon. 1936. Several esters as plant hormones. *Contr. Boyce Thompson Inst.* 8:105.

Chapter 18

1. Ables, F.B. 1967. Mechanism of action of abscission accelerators. *Physiol. Plant.* 20:442.

2. Addicott, F.T., and R.S. Lynch. 1951. Acceleration and retardation of abscission by indole-acetic acid. *Science* 114:688.

3. Addicott, F.T., and R.S. Lynch. 1955. Physiology of abscission. *Ann. Rev. Plant Physiol.* 6:211.

4. Audus, L.J. 1972. *Plant Growth Substances*, vol. 1. *Chemistry and Physiology*. London: Leonard Hill Books.

5. Beck, W.A. 1941. Production of solutes in growing epidermal cells. *Plant Physiol.* 16:637.

6. Beyer, E.M. 1973. Abscission: support for a role of ethylene modification of auxin transport. *Physiol. Plant.* 52:1.

7. Bonner, J. 1933. The action of the plant growth hormone. *J. Gen. Physiol.* 17:63.

8. Bonner, J. 1934. The relation of hydrogen ions to the growth rate of the *Avena* coleoptile. *Protoplasma* 21:406.

9. Briggs, W.R. 1964. *Phototropism in higher plants.* In A.C. Giese, ed., *Photophysiology I.* New York: Academic Press.

10. Champagnat, P. 1955. Les corrélations entre feuilles et bourgeons de la pousse herbacée du lilas. *Rev. Gen. Bot.* 62:325.

11. Chatterjee, S.K. and A.C. Leopold. 1963. Auxin structure and abscission activity. *Plant Physiol.* 38:268.

12. Chatterjee, S.K. and A.C. Leopold. 1965. Changes in abscission processes with aging. *Plant Physiol.* 40:96.

13. Cholodny, N. 1926. Beiträge zur Analyse der geotropischen Reaktion. *Jahrb. Wiss. Bot.* 65:447.

14. Cholodny, N. 1931. Zur Physiologie des pflanzlichen Wuchshormons. *Planta* 14:207.

15. Cleland, R.E., and H. Burström. 1961. Theories of the auxin action on cellular elongation. *A summary.* In W. Ruhland, ed., *Encyclopedias of Plant Physiology* 14:807. Berlin: Springer.

16. Coartney, J.S., D.J. Morré, and J.L. Key. 1967. Inhibition of RNA synthesis and auxin-induced cell wall extensibility and growth by actinomycin D. *Plant Physiol.* 42:434.

17. De Hertogh, A.A., D.C. McCune, J. Brown, and D. Antoine. 1965. The effect of antago-

nists of RNA and protein biosynthesis on IAA and 2,4-D induced growth of green pea stem sections. *Contrib. Boyce Thompson Inst.* 23:23.

18. Devlin, R.M. 1964. Effects of parachlorophenoxyisobutyric acid on abscission of debladed petioles of *Phaseolus vulgaris*. *N. Dakota Acad. Sci. Proc.* 18:75.

19. Devlin, R.M., and M.A. Hayat. 1966. Effects of indole-3-acetic acid and parachlorophenoxyisobutyric acid on abscission in petioles of debladed leaves of *Phaseolus vulgaris*. *Am. J. Bot.* 53:115.

20. Devlin, R.M., and W.T. Jackson. 1961. Effect of p-chlorophenoxyisobutyric acid on rate of elongation of root hairs of *Agrostis alba* L. *Physiol. Plant.* 14:40.

21. DuBuy, H.G., and E. Neurenbergh. 1934. Phototropismus und Wachstum der Pflanzen. II. *Ergeb. Biol.* 10:207.

22. Evans, M.L., and P.M. Ray. 1969. Timing of the auxin response in coleoptiles and its implications regarding auxin action. *J. Gen. Physiol.* 53:1.

23. Fan, D.F., and G.A. Maclachlan. 1967. Massive synthesis of ribonucleic acid and cellulose in the pea epicotyl in response to indoleacetic acid, with and without concurrent cell division. *Plant Physiol.* 42:1114.

24. Fitting, H. 1909. Die Beeinflussing der Orchideenblüten durch die Bestäubung und durch andere Umstände. *Z. Bot.* 1:1.

25. French, R.C., and H. Beevers. 1953. Respiratory and growth responses induced by growth regulators and allied compounds. *Am. J. Bot.* 40:660.

26. Galston, A.W. 1949. Indoleacetic-nicotinic acid interactions in the etiolated pea plant. *Plant Physiol.* 24:557.

27. Gordon, C.J. 1961. Morphogenetic effects of synthetic auxins. In W. Ruhland, ed., *Encyclopedia of Plant Physiology* 14:807. Berlin: Springer.

28. Gregory, F.G., and J.A. Veale. 1957. A reassessment of the problem of apical dominance. *Symp. Soc. Exp. Biol.* 11:1.

29. Gustafson, F.G. 1936. Inducement of fruit development by growth-promoting chemicals. *Proc. Natl. Acad. Sci., U.S.* 22:628.

30. Gustafson, F.G. 1939. The cause of natural parthenocarpy. *Am. J. Bot.* 26:135.

31. Haberlandt, G. 1913. Zur Physiologie der Zellteilung. *Sitzber. K. Preuss. Akad. Wiss.* 318.

32. Hardin, J.W., J.H. Cherry, D.J. Morré, and C.A. Lembi. 1972. Enhancement of RNA polymerase activity by a factor released by auxin from plasma membrane. *Proc. Natl. Acad. Sci. U.S.* 69:3146.

33. Harrison, A. 1965. Auxanometer experiments on extention growth of *Avena* coleoptiles in different CO_2 concentrations. *Physiol. Plant.* 18:321.

34. Hendry, L.B., F.H. Witham, and O.L. Chapman. 1977. Gene regulation: the involvement of stereochemical recognition in DNA–small molecule interactions. *Perspect. Biol. Med.* 21:120.

35. Iversen, T., and P. Larsen. 1973. Movement of amyloplasts in the statocytes of geotropically stimulated roots. The pre-inversion effect. *Physiol. Plant.* 28:172.

36. Jackson, W.T. 1960. Effect of indoleacetic acid on rate of elongation of root hairs on *Agrostis alba* L. *Physiol. Plant.* 13:36.

37. Juniper, B.E., and A. French. 1970. The fine structure of the cells that perceive gravity in the root tip of maize. *Planta* 95:314.

38. Juniper, B.E., S. Groves, B. Landau-Schachar and L.J. Audus. 1966. Root cap and the perception of gravity. *Nature* 209:93.

39. Key, J.L., and J.C. Shannon. 1964. Enhancement by auxin of ribonucleic acid synthesis in excised soybean hypocotyl tissue. *Plant Physiol.* 39:360.

40. Laibach, F. 1933. Wuchsstoffversuche mit lebenden Orchideen pollinien. *Ber. Dtsch. Bot. Ges.* 51:336.

41. Larsen, P. 1961. The physical phase of gravitational stimulation. In *Recent Advances in Botany*. Toronto: University of Toronto Press.

42. Larsen, P. 1965. Geotropic responses in roots as influenced by their orientation be-

fore and after stimulation. *Physiol. Plant.* 18:747.

43. Larsen, P. 1969. The optimum angle of geotropic stimulation and its relation to the starch statolith hypothesis. *Physiol. Plant.* 22:469.

44. LaRue, C.D. 1936. The effect of auxin on the abscission of petioles. *Proc. Natl. Acad. Sci., U.S.* 22:254.

45. Lund, H.A. 1956. Growth hormones in the styles and ovaries of tobacco responsible for fruit development. *Am. J. Bot.* 43:562.

46. Massart, J. 1902. Sur la pollination sans fécondation. *Bull. Jard. Bot. Brux.* 1:89.

47. Masuda, Y., E. Tanimoto, and S. Wada. 1967. Auxin-stimulated RNA synthesis in oat coleoptile cells. *Physiol. Plant.* 20:713.

48. Muir, R.M. 1942. Growth hormones as related to the setting and development of fruit in *Nicotiana tabacum. Am. J. Bot.* 29:716.

49. Muir, R.M. 1947. The relationship of growth hormones and fruit development. *Proc. Natl. Acad. Sci, U.S.* 33:303.

50. Naqvi, S.M., R.R. Dedolph, and S.A. Gordon. 1965. Auxin transport and geoelectric potential in corn coleoptile sections. *Plant Physiol.* 40:966.

51. Nooden, L. 1968. Studies on the role of RNA synthesis in auxin induction of cell enlargement. *Plant Physiol.* 43:140.

52. Rosetter, F.N., and W.P. Jacobs. 1953. Studies on abscission—the stimulating role of nearby leaves. *Am. J. Bot.* 40:276.

53. Rayle, D.L., and R. Cleland. 1970. Enhancement of wall loosening and elongation by acid solutions. *Plant Physiol.* 46:250.

54. Rayle, D.L., and R. Cleland. 1977. Control of plant cell enlargement by hydrogen ions. In A.A. Moscona and A. Morroy, eds., *Current Topics Developmental Biology,* vol. II. *Pattern Development.* New York: Academic Press.

55. Rubinstein, B., and A.C. Leopold. 1963. Analysis of the auxin control of bean leaf abscission. *Plant Physiol.* 38:262.

56. Sacher, J.A. 1967. Senescence: action of auxin and kinetin in control of RNA and protein synthesis in subcellular fractions of bean endocarp. *Plant Physiol.* 42:1334.

57. Sacher, J.A. 1967. Control of synthesis of RNA and protein in subcellular fractions of *Rhoeo discolor* leaf sections by auxin and kinetin during senescence. *Exp. Geront.* 2:261.

58. Scott, T.K. 1972. Auxins and roots. *Ann. Rev. Plant Physiol.* 23:235.

59. Scott, F.M., M.R. Schroeder, and F.M. Turrell. 1948. Development of abscission in the leaf of Valencia orange. *Bot. Gaz.* 109:381.

60. Shimoda, C., Y. Masuda, and N. Yanagishima. 1967. Nucleic acid metabolism involved in auxin-induced elongation of yeast cells. *Physiol. Plant.* 20:299.

61. Shoji, K., F.T. Addicott, and W.A. Swets. 1951. Auxin in relation to leaf blade abscission. *Plant Physiol.* 26:189.

62. Skoog, F., and K.V. Thimann. 1934. Further experiments on the inhibition of the development of lateral buds by growth hormone. *Proc. Natl. Acad. Sci., U.S.* 20:480.

63. Sonneborn, T.M. 1964. The differentiation of cells. *Proc. Natl. Acad. Sci., U.S.* 51:915.

64. Thimann, K.V. 1937. On the nature of inhibitions caused by auxin. *Am. J. Bot.* 24:407.

65. Thimann, K.V. 1956. Studies on the growth and inhibition of isolated plant parts. V. The effects of cobalt and other metals. *Am. J. Bot.* 43:241.

66. Went, F.W., and K.V. Thimann. 1937. *Phytohormones.* New York: Macmillan.

67. Wilkins, M.B., and S. Shaw. 1967. Geotropic response of coleoptiles under anaerobic conditions. *Plant Physiol.* 42:1111.

68. Witham, F.H., L.B. Hendry, and O.L. Chapman. 1978. Chirality and stereochemical recognition in DNA-phytohormone interactions: a model approach. *Origins of Life* 9:7.

69. Yasuda, S. 1934. The second report on the behaviour of the pollen tubes in the production of seedless fruits caused by interspecific pollination. *Jap. J. Genet.* 9:118.

70. Zimmerman, B.K., and W.R. Briggs. 1963.

Phototropic dosage-response curves for oat coleoptiles. *Plant Physiol.* 38:248.

71. Zimmerman, P.W., and F. Wilcoxon. 1935. Several chemical growth substances which cause initiation of roots and other responses in plants. *Contrib. Boyce Thompson Inst.* 7:209.

Chapter 19

1. Audus, L.J. 1959. *Plant Growth Substances.* New York: Interscience Publishers.

2. Bennett, P.A., and M.J. Chrispeels. 1972. De novo synthesis of ribonuclease and β-1,3-glucanase by aleurone cells of barley. *Plant Physiol.* 49:445.

3. Birch, A.J., R.W. Richards, and H. Smith. 1958. The biosynthesis of gibberellic acid. *Proc. Chem. Soc.* 192.

4. Brian, P.W., and H.G. Hemming. 1955. The effect of gibberellic acid on shoot growth of pea seedlings. *Physiol. Plant.* 8:669.

5. Brian, P.W., G.W. Elson, H.G. Hemming, and M. Radley. 1954. The plant growth-promoting properties of gibberellic acid, a metabolic product of the fungus *Gibberella fujikuroi. J. Sci. Food Agr.* 5:602.

6. Briggs, D.E. 1964. Origin and distribution of α-amylase in malt. *J. Inst. Brewing* 70:14.

7. Brown, G.N., and C.Y. Sun. 1973. Effects of abscisic acid on senescence, permeability and ribosomal patterns in mimosa hypocotyl callus tissue. *Physiol. Plant.* 28:412.

8. Chrispeels, M.J., and J.E. Varner. 1966. Inhibition of gibberellic acid–induced formation of α-amylase by abscisin II. *Nature* 212:1066.

9. Chrispeels, M.J., and J.E. Varner. 1967. Gibberellic acid–enhanced synthesis and release of α-amylase and ribonuclease by isolated barley aleurone layers. *Plant Physiol.* 42:398.

10. Chrispeels, M.J., and J.E. Varner. 1967. Hormonal control of enzyme synthesis: on the mode of action of gibberellic acid and abscisin in aleurone layers of barley. *Plant Physiol.* 42:1008.

11. Cleland, R., and N. McCombs. 1964. Gibberellic acid: action in barley endosperm does not require endogenous auxin. *Science* 150:497.

12. Crane, J.C., P.E. Primer, and R.C. Campbell. 1960. Gibberellin–induced parthenocarpy in *Prunus. Proc. Am. Soc. Hort. Sci.* 75:129.

13. Davison, R.M. 1960. Fruit-setting of apples using gibberellic acid. *Nature* 188:681.

14. Dennis, D.T., C.D. Upper, and C.A. West. 1965. An enzymic site of inhibition of gibberellin biosynthesis by AMO-1618 and other plant growth retardants. *Plant Physiol.* 40:948.

15. Dennis, D.T., and C.A. West. 1967. Biosynthesis of gibberellins. III. The conversion of (-)-kaurene to (-)-kauren-19-oic acid in endosperm of *Echinocystis macrocarpa* Greene. *J. Biol. Chem.* 242:3293.

16. Devlin, R.M., and I.E. Demoranville. 1967. Influence of gibberellic acid and gibrel on fruit set and yield in *Vaccinium macrocarpan* cv. Early Black. *Physiol. Plant.* 20:587.

17. Evins, W.H., and J.E. Varner. 1972. Hormonal control of polyribosome formation in barley aleurone layers. *Plant Physiol.* 49:348.

18. Fosket, D.E., and K.C. Short. 1973. The role of cytokinin in the regulation of growth, DNA synthesis and cell proliferation in cultured soybean tissue (*Glycine max* var. Biloxi). *Physiol. Plant.* 28:14.

19. Fries, N. 1960. The effect of adenine and kinetin on growth and differentiation of *Lupinus. Physiol. Plant.* 13:468.

20. Galston, A.W., and D.C. McCune. 1961. An analysis of gibberellin-auxin interaction and its possible metabolic basis. In R.M. Klein, ed., *Plant Growth Regulation.* Ames: Iowa State University Press.

21. Galston, A.W., and W.K. Purves. 1960. The mechanism of action of auxin. *Ann. Rev. Plant Physiol.* 11:239.

22. Harada, H., and J.P. Nitsch. 1959. Changes in endogenous growth substances during flower development. *Plant Physiol.* 34:409.

23. Harder, R., and R. Bünsow. 1956. Einfluss des Gibberellins auf die Blütenbildung bei

Kalanchoë blossfeldiana. *Naturwissenschaften* 43:544.

24. Hedden, P., J. MacMillan, and B.O. Phinney. 1978. The metabolism of the gibberellins. *Ann. Rev. Plant Physiol.* 29:149.

25. Hendry, L.B., F.H. Witham, and O.L. Chapman. 1977. Gene regulation: the involvement of stereochemical recognition in DNA–small molecule interactions. *Perspec. Biol. Med.* 21:120.

26. Hillman, W.S., and W.H. Purves. 1961. Does gibberellin act through an auxin-mediated mechanism? In R.M. Klein, ed., *Plant Growth Regulation.* Ames: Iowa State University Press.

27. Hyde, B.B., and L.G. Paleg. 1963. Ultrastructural changes in cells of isolated barley aleurone incubated with and without gibberellic acid. *Am. J. Bot.* 50:615.

28. Jacobson, J.V. 1977. Regulation of ribonucleic acid metabolism by plant hormones. *Ann. Rev. Plant Physiol.* 28:537.

29. Jacobsen, J.V., and J.E. Varner. 1967. Gibberellic acid–induced synthesis of protease by isolated aleurone layers of barley. *Plant Physiol.* 42:1596.

30. Kato, J. 1953. Studies on the physiological effect of gibberellin. I. On the differential activity between gibberellin and auxin. *Mem. Coll. Sci. Univ. Kyoto* B 29:189.

31. Kato, J. 1958. Studies on the physiological effect of gibberellin. II. On the interaction of gibberellin with auxins and growth inhibitors. *Physiol. Plant.* 11:10.

32. Kato, J. 1961. Physiological action of gibberellin with special reference to auxin. In R.M. Klein, ed., *Plant Growth Regulation.* Ames: Iowa State University Press.

33. Kende, H., and A. Lang. 1964. Gibberellin and light inhibition of stem growth in peas. *Plant Physiol.* 39:435.

34. Kende, H., H. Nunnemann, and A. Lang. 1963. Inhibition of gibberellic acid biosynthesis by AMO-1618 and CCC in *Fusarium moniliforme. Naturwissenschaften* 50:559.

35. Kessler, B. 1973. Hormonal and environmental modulation of gene expression in plant development. In J.K. Pollack and J.W. Lee, eds., *The Biochemistry of Gene Expression in Higher Organisms.* Sydney: Australia and New Zealand Book Company.

36. Kessler, B., and I. Snir. 1969. Interaction *in vitro* between gibberellin and DNA. *Biochim. Biophys. Acta.* 195:207.

37. Kohler, D., and A. Lang. 1963. Evidence for substances in higher plants interfering with response of dwarf peas to gibberellin. *Plant Physiol.* 38:555.

38. Kuraishi, S., and R.M. Muir. 1964. The relationship of gibberellin and auxin in plant growth. *Plant Cell Physiol.* 5:61.

39. Kuraishi, S., and R.M. Muir. 1964. The mechanism of gibberellic action in the dwarf pea. *Plant Cell Physiol.* 5:259.

40. Kurosawa, E. 1926. Experimental studies on the secretion of *Fusarium heterosporum* on rice plants. *Trans. Nat. Hist. Soc. Formosa* 16:213.

41. Lang, A. 1957. The effect of gibberellin upon flower formation. *Proc. Nat. Acad. Sci., U.S.* 43:709.

42. Lang, A. 1970. Gibberellins: structure and metabolism. *Ann. Rev. Plant Physiol.* 21:537.

43. Lang, A., and E. Reinhard. 1961. Gibberellins and flower formation. *Adv. Chem.* 28:71.

44. Lockhart, J.A. 1961. The hormonal mechanism of growth inhibition by visible radiation. In R.M. Klein, ed., *Plant Growth Regulation.* Ames: Iowa State University Press.

45. Lockhart, J.A. 1962. Kinetic studies of certain anti-gibberellins. *Plant Physiol.* 37:759.

46. Lockhart, J.A. 1964. Physiological studies on light-sensitive stem growth. *Planta* 62:97.

47. Luckwill, L.C. 1959. Fruit growth in relation to internal and external chemical stimuli. In D. Rudnick, ed., *Cell, Organism and Milieu, 17th Growth Symposium.* New York: Ronald Press.

48. MacLeod, A.M., and A.S. Millar. 1962. Effect of gibberellic acid on barley endosperm. *J. Inst. Brewing* 68:322.

49. Milborrow, B.V. 1974. Biosynthesis of abscisic acid by a cell-free system. *Phytochemistry* 13:131.

50. Mohr, H. 1962. Primary effects of light on growth. *Ann. Rev. Plant Physiol.* 13:465.

51. Mohr, H., and V. Appuhn. 1961. Zur Wechselwirkung von Licht and Gibberellinsaure. *Naturwissenschaften* 48:483.

52. Moore, T.C. 1979. *Biochemistry and Physiology of Plant Hormones*. New York: Springer-Verlag.

53. Nitsch, J.P. 1959. Changes in endogenous growth-regulating substances during flower initiation. *Fourth International Congress of Biochemistry*. London: Pergamon Press.

54. Ockerse, R., and A.W. Galston. 1967. Gibberellin-auxin interaction in pea stem elongation. *Plant Physiol.* 42:47.

55. Paleg, L.G. 1960. Physiological effects of gibberellic acid: I. On carbohydrate metabolism and amylase activity of barley endosperm. *Plant Physiol.* 35:293.

56. Paleg, L.G. 1960. Physiological effects of gibberellic acid: II. On starch hydrolyzing enzymes of barley endosperm. *Plant Physiol.* 35:902.

57. Paleg, L. 1964. Cellular localization of the gibberellin-induced response of barley endosperm. In J.P. Nitsch, ed., *Régulateurs naturels de la croissance végétale*. Paris: C.N.R.S.

58. Paleg, L.G. 1965. Physiological effects of gibberellins. *Ann. Rev. Plant Physiol.* 16:291.

59. Phinney, B.O., and C.A. West. 1961. Gibberellins and plant growth. In W. Ruhland, ed., *Encyclopedia of Plant Physiology* 14:1185. Berlin: Springer.

60. Purves, W.K., and W.S. Hillman. 1958. Response of pea stem sections to indoleacetic acid, gibberellic acid, and sucrose as affected by length and distance from apex. *Physiol. Plant.* 11:29.

61. Rebeiz, C.A., and J.C. Crane. 1961. Growth regulator–induced parthenocarpy in the Bing cherry. *Proc. Am. Soc. Hort. Sci.* 78:69.

62. Sachs, R.M., and A.M. Kofranek. 1963. Comparative cytohistological studies on inhibition and promotion of stem growth in *Chrysanthemum morifolium*. *Am. J. Bot.* 50:772.

63. Sawada, K. 1912. Disease of agricultural products in Japan. *Formosan Agr. Rev.* 36:10.

64. Sawada, K., and E. Kurosawa. 1924. On the prevention of the bakanae disease of rice. *Exp. Sta. Bull. Formosa* 21:1.

65. Shechter, I., and C.A. West. 1969. Biosynthesis of gibberellins. IV. Biosynthesis of cyclic diterpenes from *trans*-geranylgeranyl pyrophosphate. *J. Biol. Chem.* 244:3200.

66. Sironval, C. 1961. Gibberellins, cell division, and plant flowering. In M. Klein, ed., *Plant Growth Regulation*. Ames: Iowa State University Press.

67. Skoog, F., F.M. Strong, and C.O. Miller. 1965. Cytokinins. *Science* 148:532.

68. Snir, I., and B. Kessler. 1975. Influence of ethidium bromide on the gibberellin-induced elongation of cucumber seedlings. *Physiol. Plant.* 35:191.

69. Stodola, F.H., K.B. Roper, D.I. Fennell, H.F. Conway, V.E. Johns, C.T. Langford, and R.W. Jackson. 1955. The microbial production of gibberellins A and X. *Arch. Biochem. Biophys.* 54:240.

70. Stuart, N.W., and H.M. Cathey. 1961. Applied aspects of the gibberellins. *Ann. Rev. Plant Physiol.* 12:369.

71. Valdovinos, J.G., and L.C. Ernest. 1967. Effect of gibberellic acid and cycocel on tryptophan metabolism and auxin destruction in the sunflower seedling. *Physiol. Plant.* 20:682.

72. Varner, J.E., and D.T. Ho. 1976. Hormones. In J. Bonner and J.E. Varner, eds., *Plant Biochemistry*, 3rd ed. New York: Academic Press.

73. Varner, J.E., and G. Ram Chandra, and M.J. Chrispeels. 1965. Gibberellic acid–controlled synthesis of α-amylase in barley endosperm. *J. Cell Comp. Physiol.* 66(Suppl. 1):55.

74. West, C.A. 1973. Biosynthesis of gibberellins. In B.V. Milborrow, ed., *Biosynthesis and Its Control in Plants*. London: Academic Press.

75. Witham, F.H., and A.C. Gentile. 1961. Some characteristics and inhibitors of in-

doleacetic acid oxidase from tissue cultures of crown-gall. *J. Exp. Bot.* 12:188.

76. Witham, F.H., L.B. Hendry, and O.L. Chapman. 1978. Chirality and stereochemical recognition in DNA-phytohormone interactions: a model approach. *Origins of Life* 9:7.

77. Wittwer, S.H., and M.J. Bukovac. 1962. Exogenous plant growth substances affecting floral initiation and fruit set. *Proc. Plant Sci. Symp. Cambell Soup Company*, 65.

78. Yabuta, T. 1935. Biochemistry of the "bakanae" fungus of rice. *Agr. Hort.* (Tokyo) 10:17.

79. Yabuta, T., and T. Hayashi. 1939. Biochemical studies on "bakanae" fungus of the rice. II. Isolation of "gibberellin," the active principle which makes the rice seedlings grow slenderly. *J. Agr. Chem. Soc.* (Japan) 15:257.

Chapter 20

1. Adamson, D. 1962. Expansion and division in auxin-treated plant cells. *Can. J. Bot.* 40:719.

2. Aldwinkle, H.S., and I.W. Selman. 1967. Some effects of supplying benzyladenine to leaves and plants inoculated with viruses. *Ann. of Appl. Biol.* 60:49.

3. Armstrong, D.J., W.J. Burrows, R. Skoog, K.L. Roy, and D. Söll. 1969. Cytokinins: distribution in t-RNA species of *Escherichia coli. Proc. Natl. Acad. Sci., U.S.* 63:834.

4. Arora, N., F. Skoog, and O.N. Allen. 1959. Kinetin-induced pseudonodules on tobacco roots. *Am. J. Bot.* 46:610.

5. Banerji, D., and M.M. Laloraya. 1968. Biochemical changes accompanying kinetin-induced expansion of isolated *Cucurbita pepo* cotyledons. In S.M. Sircar, ed. *International Symposium on Plant Growth Substances*. Calcutta: University Press.

6. Bartz, J., D. Söll, W.J. Burrows, and F. Skoog. 1970. Identification of the cytokinin-active ribonucleosides in pure *Escherichia coli* t-RNA species. *Proc. Natl. Acad. Sci., U.S.* 67(3):1448.

7. Bewli, I.S., and F.H. Witham. 1976. Characterization of the kinetin-induced water uptake by detached radish cotyledons. *Bot. Gaz.* 137:58.

8. Bonner, J., and J. English. 1938. A chemical and physiological study of traumatin, a plant wound hormone. *Plant Physiol.* 13:331.

9. Bonnett, H.T., and J.G. Torrey. 1965. Chemical control of organ formation in root segments of *Convolvulus* culture *in vitro. Plant Physiol.* 40:1228.

10. Bui-Dang-Ha, D., and J.P. Nitsch. 1970. Isolation of zeatin riboside from the chickory root. *Planta* (Berlin)85:119.

11. Burg, S.P., and E.A. Burg. 1966. The interaction between auxin and ethylene and its role in plant growth. *Proc. Natl. Acad. Sci., U.S.* 55:262.

12. Burg, S.P., and E.A. Burg. 1969. Auxin-stimulated ethylene formation: its relationship to auxin-inhibited growth, root geotropism, and other plant processes. In F. Wightman and G. Setterfield, eds., *Biochemistry and Physiology of Plant Growth Substances*. Ottawa: Runge Press.

13. Burg, S.P., and C.D. Clagett. 1967. Conversion of methionine to ethylene in vegetative tissue and fruits. *Biochem. Biophys. Res. Comm.* 127:125.

14. Burrows, W.J., D.J. Armstrong, M. Kaminek, F. Skoog, R.M. Bock, S.M. Hecht, L.G. Dammann, N.J. Leonard, and J. Occolowitz. 1970. Isolation and identification of four cytokinins from wheat germ transfer ribonucleic acid. *Biochemistry* 9:1867.

15. Caplin, S.M., and F.C. Steward. 1952. Investigations on the growth and metabolism of plant cells. II. *Ann. Bot.* (London) 16:219.

16. Chadwick, A.V., and S.P. Burg. 1970. Regulation of root growth by auxin-ethylene interaction. *Plant Physiol.* 45:192.

17. Chalutz. E. 1973. Ethylene-induced pheny-

lalanine ammonia-lyase activity in carrot roots. *Plant Physiol.* 51:1033.

18. Chibnall, A.C. 1954. Protein metabolism in rooted runner-bean leaves. *New Phytol.* 53:31.

19. Conforth, J.W., B.V. Milborrow, G. Ryback, and P.F. Wareing. 1965. Identity of sycamore "dormin" with abscisin II. *Nature* (London) 205:1269.

20. Das, N.K., K. Patau, and F. Skoog. 1956. Initiation of mitosis and cell division by kinetin and indoleacetic acid in excised tobacco pith tissue. *Physiol. Plant.* 9:640.

21. Dure, L.S. 1975. Seed formation. *Ann. Rev. Plant Physiol.* 26:259.

22. Eagles, C.F., and P.F. Wareing. 1963. Experimental induction of dormancy in *Betula pubescens*. *Nature* (London) 199:874.

23. Esahi, Y., and A.C. Leopold. 1969. Cotyledon expansion as a bioassay for cytokinins. *Plant Physiol.* 44:618.

24. Fittler, F., and R.H. Hall. 1966. Selective modification of yeast seryl-t-RNA and its effect on the acceptance and binding functions. *Biochem. Biophys. Res. Comm.* 25:441.

25. Frenkel, C., I. Klein, and D.R. Dilley. 1968. Protein synthesis in relation to ripening of pome fruits. *Plant Physiol.* 43:1146.

26. Fries, N. 1960. The effect of adenine and kinetin on growth and differentiation of *Lupinus*. *Physiol. Plant.* 13:468.

27. Gefter, M.L., and R.L. Russell. 1969. Role of modifications in tyrosine t-RNA: a modified base affecting ribosome binding. *J. Mol. Biol.* 39:145.

28. Gibbons, G.S.B., and M.B. Wilkins. 1970. Growth inhibitor production by root caps in relation to geotropic responses. *Nature* 226:558.

29. Glasziou, K.T. 1957. Respiration and levels of phosphate esters during kinetin-induced cell division in tobacco pith sections. *Nature* 179:1083.

30. Glinka, Z. 1973. Abscisic acid effect on root exudation related to increased permeability to water. *Plant Physiol.* 51:217.

31. Gupta, G., R.P. Geeta, and S.C. Ma-

heshwari. 1970. Cytokinins in seeds of pumpkin. *Plant Physiol.* 45:14.

32. Haberlandt, G. 1913. Zur Physiologie der Zellteilung. *Sitzber. K. Preuss. Akad. Wiss.* 318.

33. Hall, R.H., L. Csonka, H. David, and B. McLennan. 1967. Cytokinins in the soluble RNA of plant tissues. *Science* 156:69.

34. Hall, R.H., and R.S. deRopp. 1955. Formation of 6-furfurylaminopurine from DNA breakdown products. *J. Am. Chem. Soc.* 77:6400.

35. Hall, R.H., M.J. Robbins, L. Stasiuk, and R. Thedford. 1966. Isolation of N^6-γ,γ-dimethylallyl adenosine from soluble ribonucleic acid. *J. Am. Chem. Soc.* 88:2614.

36. Hecht, S.M., N.J. Leonard, W.J. Burrows, D.J. Armstrong, F. Skoog, and J. Occolowitz. 1969. Cytokinin of wheat germ transfer RNA: 6-(4-hydroxy-3-methyl-2-butenylamino)-2-methylthio-9-B-D-ribofuranosyl purine. *Science* 166:1272.

37. Helgeson, J.P. 1968. The cytokinins. *Science* 161:974.

38. Helgeson, J.P., and N.J. Leonard. 1966. Cytokinins: identification of compounds isolated from *Corynebacterium fascians*. *Proc. Natl. Acad. Sci. U.S.* 56:60.

39. Hendry, L.B., F.H. Witham, and O.L. Chapman. 1977. Gene regulation: the involvement of stereochemical recognition in DNA–small molecule interactions. *Perspect. Biol. Med.* 21:120.

40. Huff, A.K., and C.W. Ross. 1975. Promotion of radish cotyledon enlargement and reducing sugar content by zeatin and red light. *Plant Physiol.* 56:429.

41. Hulme, A.C., M.J.C. Rhodes, T. Galliard, and L.S.C. Wooltorton. 1968. Metabolic changes in excised fruit tissue. IV. Changes occurring in discs of apple peel during the development of the respiration climacteric. *Plant Physiol.* 43:1154.

42. Jablonski, J.R., and F. Skoog. 1954. Cell enlargement and cell division in excised tobacco pith tissue. *Physiol. Plant.* 7:16.

43. Jacobsen, J.V. 1977. Regulation of ribonu-

cleic acid metabolism by plant hormones. *Ann. Rev. Plant Physiol.* 28:537.

44. Kidd, F., and C. West. 1930. Physiology of fruit. I. Changes in the respiratory activity of apples during their senescence at different temperatures. *Proc. Roy. Soc.* (London) B106:93.

45. Király, Z., and J. Szirmai. 1964. The influence of kinetin on tobacco mosaic virus production in *Nicotiana glutinosa* leaf discs. *Virology* 23:286.

46. Klämbt, D., G. Thies, and F. Skoog. 1966. Isolation of cytokinins from *Corynebacterium fascians*. *Proc. Natl. Acad. Sci., U.S.* 56:52.

47. Koshimizu, K., T. Kusaki, T. Mitsui, and S. Matsubara. 1967. Isolation of a cytokinin, (−)dihydrozeatin, from immature seeds of *Lupinus luteus*. *Tetrachdron Letters* 14:1317.

48. Krasnuk, M., F.H. Witham, and J.R. Tegley. 1971. Cytokinins extracted from pinto bean fruit. *Plant Physiol.* 48:320.

49. Letham, D.S. 1960. The separation of plant cells with ethylenediamine-tetracetic acid. *Exp. Cell Res.* 21:353.

50. Letham, D.S. 1963. Zeatin, a factor inducing cell division isolated from *Zea mays. Life Sci.* 2:569.

51. Letham, D.S. 1966. Isolation and probable identity of a third cytokinin in sweet corn extracts. *Life Sci.* 5:1999.

52. Letham, D.S. 1966. Purification and probable identity of a new cytokinin in sweet corn extracts. *Life Sci.* 5:551.

53. Letham, D.S. 1967. Chemistry and physiology of kinetin-like compounds. *Ann. Rev. Plant Physiol.* 18:349.

54. Letham, D.S. 1971. Regulators of cell division in plant tissues. XII. A cytokinins bioassay using excised radish cotyledons. *Physiol. Plant.* 25:391.

55. Letham, D.S., and C.O. Miller. 1965. Identity of kinetin-like factors from *Zea mays. Plant Cell Physiol.* 6:355.

56. Lieberman, M., L.W. Mapson, A.T. Kupnishi, and D.A. Wardale. 1966. Stimulation of ethylene production in apple tissue slices by methionine. *Plant Physiol.* 41:376.

57. Liu, W.C., and H.R. Carns. 1961. Isolation of abscisin, an abscission accelerating substance. *Science* 134:384.

58. Lyons, J.M., and H.K. Pratt. 1964. An effect of ethylene on swelling of isolated mitochondria. *Arch. Biochem. Biophys.* 104:318.

59. McGilvery, R.W. 1979. Biochemistry: a functional approach. Philadelphia: Saunders.

60. McLane, S.R., and A.E. Murneek. 1952. *The Detection of Synganin, an Indigenous Plant Hormone by Culture of Immature Corn Embryos.* Bull. 496. Agr. Exp. Sta., University of Missouri.

61. Mansfield, T.A. 1976. Delay in the response of stomata to abscisic acid in CO_2-free air. *J. Exp. Bot.* 27:559.

62. Mansfield, T.A., A.R. Wellburn, and T.J.S. Moreira. 1978. The role of abscisic acid and farnesol in the alleviation of water stress. *Philos. Trans. Roy. Soc.* (London) B284:471.

63. Matsubara, S., D.J. Armstrong, and F. Skoog. 1968. Cytokinins from t-RNA of *Corynebacterium fascians*. *Plant Physiol.* 43:451.

64. Matsubara, S., and K. Koshimizu. 1966. Factors with cytokinin activity in young *Lupinus luteus* seeds and their partial purification. *Bot. Mag.* (Tokyo) 79:389.

65. Mayak, S., and A.H. Halevy. 1972. Interrelationships of ethylene and abscisic acid in the control of rose petal senescence. *Plant Physiol.* 50:341.

66. Milborrow, B.V. 1974. Biosynthesis of abscisic acid by a cell-free system. *Phytochemistry* 13:131.

67. Milborrow, B.V. 1974. The chemistry and physiology of abscisic acid. *Ann. Rev. Plant Physiol.* 25:259.

68. Milborrow, B.V. 1978. The stability of conjugated abscisic acid during wilting. *J. Exp. Bot.* 209:1059.

69. Milborrow, B.V. 1979. Antitranspirants and regulation of abscisic acid content. *Australian J. Plant Physiol.* 6:249.

70. Miller, C.O. 1956. Similarity of some kinetin and red light effects. *Plant Physiol.* 31:318.

71. Miller, C.O. 1960. An assay for kinetin-like materials. *Plant Physiol.* (Suppl.) 35:xxvi.

72. Miller, C.O. 1961. A kinetin-like compound in maize. *Proc. Natl. Acad. Sci., U.S.* 47:170.

73. Miller, C.O. 1965. Evidence for the natural occurrence of zeatin and derivatives: compounds from maize which promote cell division. *Proc. Natl. Acad. Sci., U.S.* 54:1052.

74. Miller, C.O. 1967. Zeatin and zeatin riboside from a mycorrhizal fungus. *Science* 157:1055.

75. Miller, C.O. 1980. Cytokinin inhibition of respiration in mitochondria from six plant species. *Proc. Natl. Acad. Sci., U.S.* 77:4731.

76. Miller, C.O., F. Skoog, F.S. Okumura, M.H. von Slatza, and F.M. Strong. 1956. Isolation, structure and synthesis of kinetin, a substance promoting cell division. *J. Am. Chem. Soc.* 78:1375.

77. Miller, C.O., F. Skoog, M.H. von Saltza, and F.M. Strong. 1955. Kinetin: a cell division factor from deoxyribonucleic acid. *J. Am. Chem. Soc.* 77:1392.

78. Miller, C.O., and F.H. Witham. 1964. A kinetin-like factor from maize and other sources. *Colloq. Centre Natl. Res. Sci.* (Paris) 123:I–VI.

79. Miura, G.A., and C.O. Miller. 1969. Cytokinins from a variant strain of cultured soybean cells. *Plant Physiol.* 44:1035.

80. Moore, T.C. 1979. *Biochemistry and Physiology of Plant Hormones.* New York: Springer-Verlag.

81. Mothes, K. 1960. Über das Atern der Blätter und die Möglichkeit ihrer Wiederverjüngung. *Naturwissenschaften* 47:337–351. In Y. Oota. 1964. RNA in developing plant cells. *Ann. Rev. Plant Physiol.* 15:17.

82. Mothes, K., and L. Engelbrecht. 1961. Kinetin and its role in nitrogen metabolism. In *Recent Advances in Botany.* Toronto: University of Toronto Press.

83. Mothes, K., and L. Engelbrecht. 1961. Kinetin-induced directed transport of substances in excised leaves in the dark. *Phytochemistry* 1:58.

84. Mullins, M.G. 1967. Morphogenetic effects of roots and of some synthetic cytokinins in *Vitis vinfera* L. *J. Exp. Bot.* 18:206.

85. Nakazaki, Y. 1971. Effect of kinetin on local lesion formation on detached bean leaves inoculated with tobacco mosaic virus or its nucleic acid. *Ann. Phytopathol. Soc.* (Japan) 37:307.

86. Netien, G., and G. Beauchesne. 1952. Action d'un extrait liquide de graines de maïs immatures (lait de maïs) sur la croissance des tissues de tubercules de topinambour cultivés *in vitro. Compt. Rend.* 234:1306.

87. Netien, G. and G. Beauchesne. 1953. Différentes substances de croissance décelées dans l'extrait laiteux de graines de maïs et etudiées sur cultures *in vitro* de tissus de tubercules de topinambour. *Compt. Rend.* 237:1026.

88. Ohkuma, K., F.T. Addicott, O.E. Smith, and W.E. Thiessen. 1965. The structure of abscisin II. *Tetrahedron Lett.* 29:2529.

89. Ohkuma, K., J.L. Lyon, F.T. Addicott, and O.E. Smith. 1963. Abscisin II, an abscission-accelerating substance from young cotton fruit. *Science* 142:1592.

90. Osborne, D.J. 1959. Control of leaf senescence by auxins. *Nature* 183:1459.

91. Osborne, D.J. 1962. Effect of kinetin on protein and nucleic acid metabolism in *Xanthium* leaves during senescence. *Plant Physiol.* 37:595.

92. Osborne, D.J., and M. Hallaway. 1960. Auxin control of protein levels in detached autumn leaves. *Nature* 188:240.

93. Person, C., D.J. Samborski, and F.R. Forsyth. 1957. Effect of benzimidazole on detached wheat leaves. *Nature* 180:1294.

94. Pilet, P.E. 1972. Growth inhibitors in growing and geostimulated maize roots. In P.E. Pilet, ed., *Plant Growth Regulation.* New York: Springer-Verlag.

95. Powell, R.D., and M.M. Griffith. 1960. Some anatomical effects of kinetin and red light on disks of bean leaves. *Plant Physiol.* 35:273.

96. Raschke, K. 1975. Stomatal action. *Ann. Rev. Plant Physiol.* 26:309.

97. Raschke, K., and M. Pierce. 1973. Uptake of

sodium and chloride by guard cells of *Vicia falsa*. Plant Research "72," MSU/AEC Plant Res. Lab. Mich. State Univer. 146.

98. Reid, M., and H.K. Pratt. 1972. Effects of ethylene on potato tuber respiration. *Plant Physiol.* 49:252.

99. Richmond, A.E., and A. Lang. 1957. Effect of kinetin on protein content and survival of detached *Xanthium* leaves. *Science* 125:650.

100. Rijven, A.H.G.C., and V. Parkash. 1971. Action of kinetin on cotyledons of fenugreek. *Plant Physiol.* 47:59.

101. Robbins, M.J., R.H. Hall, R. Thedford, and L. Stasiuk. 1967. N^6-(Δ^2-isopentenyl) adenosine: a component of the transfer ribonucleic acid of yeast and mammalian tissue. Method of isolation and characterization. *Biochemistry* 6:1837.

102. Sacher, J.A. 1966. Permeability characteristics and amino acid incorporation during senescence (ripening) of banana tissue. *Plant Physiol.* 41:701.

103. Scott, T.K. 1972. Auxins and roots. *Ann. Rev. Plant Physiol.* 23:235.

104. Shaw, A., and D.V. Wilson. 1964. The synthesis of zeatin. *Proc. Chem. Soc.* 231.

105. Skoog, F., D.J. Armstrong, J.D. Cherayil, A.C. Hampel, and R.M. Bock. 1966. Cytokinin activity: localization in t-RNA preparations. *Science* 154:1354.

106. Skoog, F., and C.O. Miller. 1957. Chemical regulation of growth and organ formation in plant tissues cultured *in vivo*. *Symp. Soc. Exp. Biol.* 11:118.

107. Skoog, F., F.M. Strong, and C.O. Miller. 1965. Cytokinins. *Science* 148:532.

108. Sugiura, M., K. Umemura, and Y. Oota. 1962. The effect of kinetin on protein level of tobacco leaf disks. *Physiol. Plant.* 15:457.

109. Sveshnikova, I.N., and V.A. Kokhlova. 1969. Cytological study of the effect of 6-benzyl-aminopurine and kinetin on isolated flax cotyledons. *Soviet Plant Physiol.* 16:570.

110. Tavantzis, S.M., S.H. Smith, and F.H. Witham. 1979. The influence of kinetin on tobacco ring-spot virus infectivity and the ef-

fect of virus infection on the cytokinin activity in intact leaves of *Nicotiana glutinosa* L. *Physiol. Plant Path.* 14:227.

111. Tegley, J.R., F.H. Witham, and M. Krasnuk. 1971. Chromatographic analysis of a cytokinin from tissue cultures of crown-gall. *Plant Physiol.* 47:581.

112. Thimann, K.V. 1972. The natural plant hormones. In F.C. Steward, ed., *Plant Physiology* New York: Academic Press.

113. Torrey, J.G. 1958. Endogenous bud and root formation by isolated roots of *Convolvulus* grown *in vitro*. *Plant Physiol.* 33:258.

114. Torrey, J.G. 1962. Auxin and purine interactions in lateral root initiation in isolated pea root segments. *Physiol. Plant.* 15:177.

115. Tucker, D.J. 1977. Apical dominance in the "Rogue" tomato. *Ann. Bot.* 41:181.

116. Tucker, D.J. 1977. Hormonal regulation of lateral bud outgrowth in the tomato. *Plant Sci. Lett.* 8:105.

117. Tucker, D.J. 1978. Apical dominance in the tomato: the possible roles of auxin and abscisic acid. *Plant Sci. Lett.* 12:273.

118. Upper, C.D., J.P. Helgeson, J.D. Kemp, and C.J. Schmidt. 1970. Gas-liquid chromatographic isolation of cytokinins from natural sources. *Plant Physiol.* 45:543.

119. van Overbeek, J., M.E. Conklin, and A.F. Blakeslee. 1941. Factors in coconut milk essential for growth and development of *Datura* embryos. *Science* 94:350.

120. van Overbeek, J., R. Siu, and A.J. Haagen-Smit. 1944. Factors affecting the growth of *Datura* embryos *in vitro*. *Am. J. Bot.* 31:219.

121. Von Abrams, G.J., and H.K. Pratt. 1967. Effect of ethylene on the permeability of excised cantaloupe fruit tissue. *Plant Physiol.* 42:299.

122. Walton, D.C. 1980. Biochemistry and physiology of abscisic acid. *Ann. Rev. Plant Physiol.* 31:453.

123. Wang, C.Y., and W.M. Mellenthin. 1972. Internal ethylene levels during ripening and climacteric in Anjou pears. *Plant Physiol.* 50:311.

124. Wehnelt, B. 1927. Untersuchungen über

das Wundhormon der Pflanzen. *Jarb. Wiss. Bot.* 66:773.

125. Wickson, M., and K.V. Thimann. 1958. The antagonism of auxin and kinetin in apical dominance. *Physiol. Plant.* 11:62.

126. Witham, F.H., L.B. Hendry, and O.L. Chapman. 1978. Chirality and stereochemical recognition in DNA-phytohormone interactions: a model approach. *Origins of Life* 9:7.

127. Witham, F.H., and C.O. Miller. 1965. Biological properties of a kinetin-like substance occurring in *Zea mays. Plant Physiol.* 18:1007.

128. Yang, S.F. 1969. Biosynthesis of ethylene. In F. Wightman and G. Setterfield, eds., *Biochemistry and Physiology of Plant Growth Substances.* Ottawa: Runge Press.

129. Yang, S.F. 1980. Regulation of ethylene biosynthesis. *Hort. Sci.* 15:238.

130. Yang, S.F., H.S. Ku, and H.K. Pratt. 1966. Ethylene production from methionine by flavin mononucleotide and light. *Biochem. Biophys. Res. Comm.* 24:739.

131. Young, R.E., and J.B. Biale. 1967. Phosphorylation in avocado fruit slices in relation to the respiratory climacteric. *Plant Physiol.* 42:1357.

132. Zachau, H., D. Dutting, and H. Feldmann. 1966. Serine specific transfer ribonucleic acid. XIV. Comparison of nucleotide sequences and secondary structure models. *Cold Spr. Harb. Symp. Quant. Biol.* 31:417.

Chapter 21

1. Barber, H.N., and D.M. Paton. A gene-controlled flowering inhibitor in *Pisum. Nature* 169:592.

2. Bonner, J. 1962. *In vitro* dark conversion and other properties of phytochrome. *Plant Physiol.* (Suppl.) 37:xxvii.

3. Borthwick, H.A. 1959. Photoperiodic control of flowering. In R.B. Withrow ed., *Photoperiodism and Related Phenomena in Plants and Animals.* Washington, D.C.: American Association for the Advancement of Science.

4. Borthwick, H.A., S.B. Hendricks, and M.W. Parker. 1952. The reaction controlling floral initiation. *Proc. Natl. Acad. Sci., U.S.* 38:929.

5. Borthwick, H.A., S.B. Hendricks, and M.W. Parker. 1956. Photoperiodism. In A. Hollander ed., *Radiation Biology.* New York: McGraw-Hill.

6. Borthwick, H.A., S.B. Hendricks, M.W. Parker, E.H. Toole, and K.V. Toole. 1952. A reversible photoreaction controlling seed germination. *Proc. Natl. Acad. Sci., U.S.* 38:662.

7. Briggs, W.R., and H.W. Siegelman. 1965. Distribution of phytochrome in etiolated seedlings. *Plant Physiol.* 40:934.

8. Butler, W.L., K.H. Norris, H.W. Siegelman, and S.B. Hendricks. 1959. Detection, assay, and preliminary purification of the pigment controlling photoresponsive development of plants. *Proc. Natl. Acad. Sci. U.S.* 45:1703.

9. Cajlachjan, M.C. 1958. Hormonal factors in the flowering of plants. *Fiziol. Rast.* 5:541.

10. Cajlachjan, M.C. 1961. Effect of gibberellins and derivatives of nucleic acid metabolism on plant growth and flowering. In R.M. Klein, ed., *Plant Growth Regulation.* Ames: Iowa State University Press.

11. Cleland, C.F., and W.R. Briggs. 1967. Flowering responses of the long-day plant *Lemna gibba* G3. *Plant Physiol.* 42:1553.

12. Cleland, C.F., and J.A.D. Zeevaart. 1970. Gibberellins in relation to flowering and stem elongation in the long day plant *Silene armeria. Plant Physiol.* 46:392.

13. Cummings, B.G., and E. Wagner. 1968. Rhythmic processes in plants. *Ann. Rev. Plant Physiol.* 19:381.

14. Downs, R.J. 1956. Photoreversibility of flower initiation. *Plant Physiol.* 31:279.

15. Garner, W.W., and H.A. Allard. 1920. Effect of length of day on plant growth. *J. Agr. Res.* 18:553.

16. Hamner, K.C. 1940. Interrelation of light and darkness in photoperiodic induction. *Bot. Gaz.* 101:658.

17. Hamner, K.C., and J. Bonner. 1938. Photo-

periodism in relation to hormones as factors in floral initiation. *Bot. Gaz.* 100:388.

18. Heinze, P.H., M.W. Parker, and H.A. Borthwick. 1942. Floral initiation in Biloxi soybean as influenced by grafting. *Bot. Gaz.* 103:517.

19. Hendricks, S.B. 1958. Photoperiodism. *Agron. J.* 50:724.

20. Hendricks, S.B. 1959. The photoreaction and associated changes of plant photomorphogenesis. In R.B. Withrow, ed., *Photoperiodism and Related Phenomena in Plants and Animals.* Washington, D.C.: American Association for the Advancement of Science.

21. Hillman, W.S. 1962. *The Physiology of Flowering.* New York: Holt, Rinehart and Winston.

22. Hillman, W.S. 1967. The physiology of phytochrome. *Ann. Rev. Plant Physiol.* 18:301.

23. Hillman, W.S. 1976. Biological rhythms and physiological timing. *Ann. Rev. Plant Physiol.* 27:159.

24. Hodson, H.K, and K.C. Hamner. 1970. Floral inducing extract from *Xanthium. Science* 167:384.

25. Holdsworth, M. 1956. The concept of minimum leaf number. *J. Exp. Bot.* 7:395.

26. Kendrick, R.E., and C.J.P. Spruit. 1973. Phytochrome properties and the molecular environment. *Plant Physiol.* 52:327.

27. Khudairi, A.K., and K.C. Hamner. 1954. The relative sensitivity of *Xanthium* leaves of different ages to photoperiodic induction. *Plant Physiol.* 29:251.

28. Klebs, G. 1913. Über das Verhältnis der Aussenwelt zur Entwicklung der Pflanze. *Akad. Wiss.* (Heildelberg) B5:1.

29. Knott, J.E. 1934. Effect of localized photoperiod on spinach. *Proc. Am. Soc. Hort. Sci.* (Suppl.) 31:152.

30. Krishnamoorthy, H.N., and K.K. Nanda. 1967. Effect of intercalated long days and light interruption of dark period on flowering, extension growth and senescence of *Impatiens balsamina. Physiol. Plant.* 20:760.

31. Lincoln, R.G., A. Cunningham, B.H. Carpenter, J. Alexander, and D.L. Mayfield. 1966. Florigenic acid from fungal culture. *Plant Physiol.* 41:1079.

32. Lincoln, R.G., D.L. Mayfield, and A. Cunningham. 1961. Preparation of a floral initiating extract from *Xanthium. Science* 133:756.

33. Long. E.M. 1939. Photoperiodic induction as influenced by environmental factors. *Bot. Gaz.* 101:168.

34. Moore, T.J. 1979. *Biochemistry and Physiology of Plant Hormones.* New York: Springer-Verlag.

35. Naylor, A.W. 1953. Reactions of plants to photoperiod. In W. Loomis, ed., *Growth and Development in Plants.* Ames: University of Iowa Press.

36. Naylor, A.W. 1961. The photoperiodic control of plant behavior. In W. Ruhland, ed., *Encyclopedia of Plant Physiology.* 16:331. Berlin: Springer.

37. Parker, M.W., S.B. Hendricks, H.A. Borthwick, and N.J. Scully. 1946. Action spectrum for the photoperiodic control of floral initiation of short day plants. *Bot. Gaz.* 108:1.

38. Quail, P.H., E. Schäfer, and D. Marmé. 1972. *De novo* synthesis of phytochrome. In G.O. Schenck, ed., *Book of Abstracts.* VI. International Congress in Photobio., Biochem. 156.

39. Quail, P.H., E. Schäfer, and D. Marmé. 1973. *De novo* synthesis of phytochrome in pumpkin hooks. *Plant Physiol.* 52:124.

40. Quail, P.H., E. Schafer, and D. Marmé. 1973. Turnover of phytochrome in pumpkin cotyledons. *Plant Physiol.* 52:128.

41. Schwabe, W.W. 1959. Studies of long-day inhibition in short-day plants. *J. Exp. Bot.* 10:317.

42. Siegelman, H.W., and W.L. Butler. 1965. Properties of phytochrome. *Ann. Rev. Plant Physiol.* 16:383.

43. Siegelman, H.W., and E.M. Firer. 1964. Purification of phytochrome from oat seedlings. *Biochemistry* 3:418.

44. Salisbury, F.B., and C.W. Ross. 1978. *Plant Physiology,* 2nd ed. Belmont, Calif.: Wadsworth.

45. Takimoto, A. 1960. Effect of sucrose on

flower initiation of *Pharbitis. Plant Cell Physiol.* (Tokyo) 1:241.

46. Taylor, A.O., and B.A. Bonner. 1967. Isolation of phytochrome from the alga *Mesotaenium* and liverwort *Sphaerocarpos. Plant Physiol.* 42:762.

47. Toole, E.H., V.K. Toole, H.A. Borthwick, and S.B. Hendricks. 1955. Photocontrol of *Lepidium* seed germination. *Plant Physiol.* 30:15.

48. Tournois, J. 1912. Influence de la lumière sur la floraison du houblon japonais et du chauvre. *Comp. Rend. Acad. Sci.* (Paris) 155:297.

49. van der Veen, R., and G. Meijer. 1959. *Light and Plant Growth.* New York: Macmillan.

50. Zeevart, J.A.D. 1958. Flower formation as studied by grafting. *Med. Landbouwhogeschool Wageningen* 58:1.

51. Zeevaart, J.A.D. 1976. Physiology of flower formation. *Ann. Rev. Plant Physiol.* 27:321.

Chapter 22

1. Alden, J., and K.H. Hermann. 1971. Aspects of the cold-hardiness mechanism in plants. *Bot. Rev.* 37(1):37.

2. Bolduc, R.J., J.H. Cherry, and B.O. Blair. 1970. Increase in indoleacetic acid oxidase activity of winter wheat by cold treatment and gibberellic acid. *Plant Physiol.* 45:461.

3. Bula, R.J., D. Smith and H.J. Hodgson. 1956. Cold resistance in alfalfa at two diverse latitudes. *Agron. J.* 48:153.

4. Chouard, P. 1952. Les facteurs du milieu et les mécanismes régulateurs du développement des plantes horticoles. *Rep. Intern. Hort. Congr.* 13:17.

5. Chouard, P. 1960. Vernalization and its relations to dormancy. *Ann. Rev. Plant Physiol.* 11:191.

6. Chouard, P., and P. Poignant. 1951. Recherches préliminaires sur la vernalisation en présence d'inhibiteurs de germination et de respiration. *Compt. Rend. Acad. Sci.* (Paris) 23:103.

7. Chroboczek, E. 1934. A study of some ecological factors influencing seed-stalk development in beets (*Beta vulgaris* L.). *Mem. Cornell Agr. Expt. Sta.* 154:1.

8. Curtis, O.F., and H.T. Chang. 1930. The relative effectiveness of temperature of the crown as contrasted with that of the rest of the plant upon flowering of celery plants. *Am. J. Bot.* 17:1047.

9. Daday, H. 1964. Genetic relationship between cold hardiness and growth at low temperature in *Medicago sativa. Heredity* 19:173.

10. Dear, J. 1973. A rapid degradation of starch at hardening temperature. *Cryobiology* 10:78.

11. De La Roche, I.A., C.J. Andrews, M.K. Pomeroy, P. Weinberger, and M. Kates. 1972. Lipid changes in winter wheat seedlings (*Triticum aestivum*) at temperatures inducing cold hardiness. *Can. J. Bot.* 50(12):2401.

12. Faw, W.F., and G.A. Jung. 1972. Electrophoretic protein patterns in relation to low temperature tolerance and growth regulation of alfalfa. *Cryobiology* 9:548.

13. Gerloff, E.D., T. Richardson, and M.A. Stahmann. 1967. Changes in fatty acids of alfalfa roots during cold hardening. *Plant Physiol.* 41:1280.

14. Gerloff, E.D., M.A. Stahmann, and D. Smith. 1967. Soluble proteins in alfalfa roots as related to cold hardiness. *Plant Physiol.* 42:895.

15. Gott, M.B., F.G. Gregory, and O.N. Purvis. 1955. Studies in vernalization of cereals. XIII. Photoperidic control of stages in flowering between initiation and ear formation in vernalized and unvernalized Petkus winter rye. *Ann. Bot.* 19:87.

16. Gregory, F.G., and O.N. Purvis. 1938. Studies in the vernalization of cereals. III. The use of anaerobic conditions in the analysis of the vernalizing effect of low temperature during germination. *Ann. Bot.* 2:753.

17. Grenier, G., and C. Willemot. 1974. Lipid changes in roots of frost hardy and less hardy alfalfa varieties under hardening conditions. *Cryobiology* 11:324.

18. Hall, T.C., R.C. McLeester, B.H. McCown, and G.E. Beck. 1970. Enzyme changes during acclimation. *Cryobiology* 6:263.

19. Hall, T.C., R.C. McLeester, B.H. McCown, and G.E. Beck. 1970. Enzyme changes during deacclimation of willow stem. *Crybiology* 7:130.

20. Hänsel, H. 1953. Vernalization of winter rye by negative temperatures and the influence of vernalization upon the lamina length of the first and second leaf in winter rye, spring barley, and winter barley. *Ann. Bot.* 17:417.

21. Harris, P., and A.T. James. 1969. The effect of low temperatures on fatty acid biosynthesis in plants. *Biochem. J.* 112:325.

22. Heber, U. 1959. Beziehungen zwischen der Grösse von Chloroplasten und ihrem Gehalt und löslichen Eiweissen und Zuckern im Zusammenhang mit dem Frostresistenzproblem. *Protoplasma* 51:284.

23. Hodgson, H.J. 1965. Effect of photoperiod on development of cold resistance in alfalfa. *Crop Sci.* 4:302.

24. Jung, G.A., and K.L. Larson. 1972. Cold, drought and heat tolerance. *Agron. Monogr.* 15:185.

25. Jung, G.A., S.C. Shih, and D.C. Shelton. 1967. Influence of purines and pyrimidines on cold hardiness of plants. III. Associated changes in soluble proteins and nucleic acid content and tissue pH. *Plant Physiol.* 42:1653.

26. Jung, G.A., and D. Smith. 1960. Influence of extended storage at constant low temperature on cold resistance and carbohydrate reserves of alfalfa and medium red clover. *Plant Physiol.* 35:123.

27. Jung, G.A, and D. Smith. 1961. Trends of cold resistance and chemical changes in certain nitrogen and carbohydrate fractions. *Agron. J.* 53:359.

28. Kenefick, D.G. 1964. Cold acclimation as it relates to winter hardiness in plants. *Agri. Sci. Rev. USDA* 2:21.

29. Kenefick, D.G., and E.I. Whitehead. 1971. A search for winter hardiness. *South Dakota Farm and Home Research.* 22:36.

30. Klebs, G. 1913. Über das Verhältnis der Aussenwelt zur Entiwicklung der Pflanze. *Akad. Wiss.* (Heidelberg) B5:1.

31. Korovin, A.I., and T.A. Barskaya. 1962. Effect of soil temperature on respiration and activity of oxidative enzymes of roots in cold resistant and thermophilic plants. *Sov. Plant Physiol.* 9:331.

32. Krasnuk, M., G.A. Jung, and F.H. Witham. 1975. Electrophoretic studies of the relationship of peroxidases, polyphenol oxidase and indoleacetic acid oxidase to cold tolerance in alfalfa. *Cryobiology* 12:62.

33. Krasnuk, M., G.A. Jung, and F.H. Witham. 1976. Electrophoretic studies of several dehydrogenases in relation to the cold tolerance of alfalfa. *Cryobiology* 13:375.

34. Krasnuk, M., F.H. Witham, and G.A. Jung. 1976. Electrophoretic studies of several hydrolytic enzymes in relation to the cold tolerance of alfalfa. *Cryobiology* 13:225.

35. Kuiper, P.J.C. 1970. Lipids in alfalfa leaves in relation to cold hardiness. *Plant Physiol.* 45:684.

36. Lang, A. 1951. Untersuchungen über das Kälterbedurfnis von zweijährigen *Hyoscyamus niger. Der Zuchter.* 21:241.

37. Lang, A. 1952. Physiology of flowering. *Ann. Rev. Plant Physiol.* 3:265.

38. Lang, A. 1961. Auxins in flowering. In W. Ruhland, ed., *Encylopedia of Plant Physiology* 14:909. Berlin: Springer.

39. Lang, A., and G. Melchers. 1947. Vernalization und Devernalization bei einer zweijährigen Pflanze. *Z. Naturf.* 2b:444.

40. Levitt, J. 1969. Growth and survival of plants at extremes of temperature—a unified concept. *Symp. Soc. Exp. Biol.* 23:395.

41. Li, P.H., and C.J. Weiser. 1969. Metabolism of nucleic acids in one-year old apple twigs during cold hardening and dehardening. *Plant Cell Physiol.* 10:21.

42. McCown, B.H., T.C. Hall, and G.E. Beck. 1969. Plant leaf and stem proteins. II. Isozymes and environmental change. *Plant Physiol.* 44:210.

43. McCown, B.H., R.C. McLeester, G.E. Beck, and T.C. Hall. 1969. Environment-induced

changes in peroxidase zymograms in the stem of deciduous and evergreen plants. *Cryobiology* 5:410.

44. McKinney, H.H. 1940. Vernalization and the growth-phase concept. *Bot. Rev.* 6:25.

45. Marvin, J., and M. Morselli. 1971. Rapid low temperature hydrolysis of starch to sugars in maple stems and in maple tissue cultures. *Cryobiology* 8:339.

46. Melchers, G. 1936. Versuche zur Genetik und Entwicklungsphysiologie der Blüh-reife. *Biol. Zbl.* 56:567.

47. Melchers, G. 1937. Die Wirkung von Genen, tiefen Temperaturen und blühen-den Pfropfpartnern auf die Blühreife von *Hyoscyamus niger l. Biol. Zbl.* 57:568.

48. Melchers, G. 1939. Die Blühhormone. *Ber. Dtsch. Bot. Ges.* 57:29.

49. Melchers, G., and A. Lang. 1948. Die Physiologie der Blütenbildung. *Biol. Zentr.* 67:105.

50. Napp-Zinn, K. 1960. Vernalisation, Licht und Alter bei *Arabidopsis thaliana* (L.) Heynh. I. Licht und Dunkelheit wahrend Kalte- und Warmebehandlung. *Planta* 54:409.

51. Purvis, O.N. 1934. An analysis of the influ-ence of temperature during germination on the subsequent development of certain winter cereals and its relation to length of day. *Ann. Bot.* 48:919.

52. Purvis, O.N. 1940. Vernalization of frag-ments of embryo tissue. *Nature* 145:462.

53. Purvis, O.N. 1947. Studies in vernalization of cereals. X. The effect of depletion of car-bohydrates on the growth and vernaliza-tion response of excised embryos. *Ann. Bot.* 11:269.

54. Purvis, O.N. 1961. The physiological analy-sis of vernalization. In W. Ruhland, ed., *Encyclopedia of Plant Physiology* 16:76. Berlin: Springer.

55. Purvis, O.N., and F.G. Gregory. 1952. Studies in vernalization of cereals. XII. The reversibility by high temperature of the vernalized condition in Petkus winter rye, *Ann. Bot.* 16:1.

56. Sarkar, S. 1958. Versuche zur Physiologie der Vernalisation. *Biol. Zentralbl.* 77:1.

57. Schwabe, W.W. 1954. Factors controlling flowering in the chrysanthemum. IV. The site of vernalization and translocation of the stimulus. *J. Exp. Bot.* 5:389.

58. Shih, S.C., and G.A. Jung. 1971. Influence of purines and pyrimidines on cold hardi-ness of plants. IV. An analysis of the chem-istry of cold hardiness in alfalfa when growth is regulated by chemicals. *Cryobiology* 7:300.

59. Shih, S.C., G.A. Jung, and D.C. Shelton. 1967. Effects of temperature and photope-riod on metabolic changes in alfalfa in rela-tion to cold hardiness. *Crop Sci.* 7:385.

60. Shinohara, S. 1959. Genecological studies on the phasic development of flowering centering on the *Cruciferous* crops, espe-cially on the role of vernalization on ripen-ing seeds. Shizuoka Prefecture Agr. Expt. Sta. Tech. Bull. 6:1.

61. Siminovitch, D., and D.R. Briggs. 1949. The chemistry of the living bark of the black lo-cust tree in relation to frost hardiness. I. Seasonal variations in protein content. *Arch. Biochem.* 23:8.

62. Siminovitch, D., and D.R. Briggs. 1953. Studies on the chemistry of the living bark of the black locust tree in relation to its frost hardiness. IV. Effects of ringing on translo-cation, protein synthesis and the develop-ment of hardiness. *Plant Physiol.* 28:177.

63. Siminovitch, D., F. Gfeller, and B. Rheaume.1967. The multiple character of the biochemical mechanism of freezing re-sistance of plant cells. In E. Asahina, ed., *Cellular Injury and Resistance in Freezing Or-ganisms.* Sapporo, Japan: Institute of Low Temperature Science.

64. Siminovitch, D., C.M. Wilson, and D.R. Briggs. 1953. Studies on the chemistry of the living bark of the black locust in relation to its frost hardiness. V. Seasonal transfor-mations and variations in the carbohy-drates: starch-sucrose interconversions. *Plant Physiol.* 28:383.

65. Smith, D. 1968. Varietal chemical differences associated with the freezing resistance in forage plants. *Cryobiology* 5:148.

66. Stokes, P., and K. Verkerk. 1951. Flower formation in Brussels sprouts. *Mededel. Landbouwhogeschool Wageningen* 50:141.

67. Wellensiek, S.J. 1961. Leaf vernalization. *Nature* 192:1097.

68. Wellensiek, S.J. 1962. Dividing cells as the locus for vernalization. *Nature* 195:307.

69. Wellensiek, S.J. 1964. Dividing cells as the prerequisite for vernalization. *Plant Physiol.* 39:832.

70. Willemot, C. 1975. Stimulation of phospholipid biosynthesis during frost hardening of winter wheat. *Plant Physiol.* 55:356.

Chapter 23

1. Bennet-Clark, T.A., and N.P. Kefford. 1953. Chromatography of the growth substances in plant extracts. *Nature* 171:645.

2. Blommaert, K.L.J. 1954. Growth and inhibiting substances in relation to the rest-period of the potato tuber. *Nature* 174:970.

3. Blommaert, K.L.J. 1955. The significance of auxins and growth inhibiting substances in relation to winter dormancy of the peach. *Dept. Agr. South Africa Sci. Bull.* 368:1.

4. Borthwick, H.A., S.B. Hendricks, M.W. Parker, E.H. Toole, and V.K. Toole. 1952. A reversible photoreaction controlling seed germination. *Proc. Natl. Acad. Sci., U.S.* 38:662.

5. Borthwick, H.A., S.B. Hendricks, E.H. Toole, and V.K. Toole. 1954. Action of light on lettuce-seed germination. *Bot. Gaz.* 115:205.

6. Crocker, W. 1906. Role of seed coats in delayed germination. *Bot. Gaz.* 42:265.

7. Crocker, W. 1948. *Growth of Plants.* New York: Reinhold.

8. Denny, F.E. 1926. Hastening the sprouting of dormant potato tuber. *Am. J. Bot.* 13:118.

9. Denny, F.E. 1926. Effect of thiourea upon bud inhibition and apical dominance of potato. *Bot. Gaz.* 81:297.

10. Donaho, C.W., and D.R. Walker. 1957. Effect of gibberellic acid on breaking of the rest period in Elberta peach. *Science.* 126:1178.

11. Eagles, C.F. and P.F. Wareing. 1964. The role of growth substances in the regulation of bud dormancy. *Physiol. Plant.* 17:697.

12. Evenari, M. 1949. Germination inhibitors. *Bot. Rev.* 15:153.

13. Harada, H., and J.P. Nitsch. 1959. Changes in endogenous growth substances during flower development. *Plant Physiol.* 34:409.

14. Harrington, G.T. 1916. Agricultural value of impermeable seeds. *J. Agr. Res.* 6:761.

15. Hemberg, T. 1947. Studies of auxins and growth-inhibiting substances in the potato tuber and their significance with regard to its rest period. *Acta Hort. Berg.* 14:133.

16. Hemberg, T. 1949. The significance of growth-inhibiting substances and auxins for the rest period of the potato tuber. *Physiol. Plant.* 2:24.

17. Hemberg, T. 1949. Growth-inhibiting substances in terminal buds of *Fraxinus*. *Physiol. Plant.* 2:37.

18. Hemberg, T. 1950. The effect of glutathione on the growth-inhibiting substances in resting potato tubers. *Physiol. Plant.* 3:17.

19. Hemberg, T. 1952. The significance of the acid growth-inhibiting substances for the rest period of the potato tuber. *Physiol. Plant.* 5:115.

20. Hendershott, C.H., and L.F. Bailey. 1955. Growth inhibiting substances in dormant flower buds of peach. *Proc. Am. Soc. Hort. Sci.* 65:85.

21. Hyde, E.O. 1954. The function of the hilum in some Papilionaceae in relation to the ripening of the seed and permeability of the testa. *Ann. Bot.* 18:241.

22. Ikuma, H., and K.V. Thimann. 1964. Analysis of germination processes of lettuce seed by means of temperature and anaerobiosis. *Plant Physiol.* 39:756.

23. Lane, F.E., and L.F. Bailey. 1964. Isolation

and characterization studies on the β-inhibitor in dormant buds of the silver maple, *Acer saccharinum* L. *Physiol. Plant.* 17:91.

24. Lippert, L.F., L. Rappaport, and H. Timm. 1958. Systematic induction of sprouting in white potatoes by foliar applications of gibberellin. *Plant Physiol.* 33:132.

25. Mayer, A.M., and A. Poljakoff-Mayber. 1963. *The Germination of Seeds.* New York: MacMillan.

26. Meyer, B.S., and D.B. Anderson. 1952. *Plant Physiology.* Princeton, N.J.: Van Nostrand.

27. Nitsch, J.P. 1957. Growth responses of woody plants to photoperiodic stimuli. *Proc. Am. Soc. Hort. Sci.* 70:512.

28. Nitsch, J.P. 1959. Changes in endogenous growth regulating substances during flower initiation. *Fourth Intern. Congr. Biochem.* 6:141. London: Pergamon Press.

29. Olney, H.O., and B.M. Pollock. 1960. Studies of rest period. II. Nitrogen and phosphorus changes in embryonic organs of after-ripening cherry seed. *Plant. Physiol.* 35:970.

30. Phillips, I.D.J., and P.F. Wareing. 1958. Effect of photoperiodic conditions on the level of growth inhibitors in *Acer pseudoplatanus.* *Naturwiss.* 13:317.

31. Pollock, B.M., and H.O. Olney. 1959. Studies of the rest period. I. Growth translocation, and respiratory changes in the embryonic organs of the after-ripening cherry seed. *Plant Physiol.* 34:131.

32. Rappaport, L., L.F. Lippert, and H. Timm. 1957. Sprouting, plant growth, and tuber formation as affected by chemical treatment of white potato seed pieces. I. Breaking dormancy with gibberellic acid. *Am. Potato J.* 34:254.

33. Rappaport, L., H. Timm, and L. Lippert. 1958. Gibberellin on white potatoes. *Calif. Agr.* 12:4, 14.

34. Robinson, P.M., P.F. Wareing, and T.H. Thomas. 1963. Dormancy regulators in woody plants. Isolation of the inhibitor varying with photoperiod in *Acer pseudoplatanus.* *Nature* 199:875.

35. Sankhla, S., and D. Sankhla. 1968. Reversal of (±)-abscisin II induced inhibition of lettuce seed germination and seedling growth by kinetin. *Physiol. Plant.* 21:190.

36. Shull, C.A. 1911. The oxygen minimum and the germination of *Xanthium* seeds. *Bot. Gaz.* 52:453.

37. Shull, C.A. 1914. The role of oxygen in germination. *Bot. Gaz.* 57:64.

38. Smith, O.E., and L. Rappaport. 1961. Endogenous gibberellins in resting and sprouting potato tubers. In R.F. Gould, ed., *Gibberellins.* Am. Chem. Soc. 28:42.

39. Thornton, N.C. 1935. Factors influencing germination and development of dormancy in cocklebur seeds. *Contri. Boyce Thompson Inst.* 7:477.

40. Thornton, N.C. 1939. Carbon dioxide storage. XIII. Relationship of oxygen to carbon dioxide in breaking dormancy of potato tubers. *Contri. Boyce Thompson Inst.* 10:201.

41. Thornton, N.C. 1953. Dormancy. In W.E. Loomis, ed., *Growth and Differentiation in Plants.* Ames: Iowa State University Press.

42. Toole, E.H. 1959. Effect of light on the germination of seeds. In R.B. Withrow, ed., *Photoperiodism and Related Phenomena in Plants and Animals.* Washington, D.C.: American Association for the Advancement of Science.

43. Toole, E.H., H.A. Borthwick, S.B. Hendricks, and V.K. Toole. 1953. Physiological studies of the effects of light and temperature on seed germination. *Proc. Intern. Seed Testing Assoc.* 18(2):267.

44. Toole, E.H., S.B. Hendricks, H.A. Borthwick, and V.K. Toole. 1956. Physiology of seed germination. *Ann. Rev. Plant Physiol.* 7:299.

45. Toole, E.H., and V.K. Toole. 1939. *Proc. Intern. Seed Testing Assoc.* 11:51.

46. Toole, E.H., V.K. Toole, H.A. Borthwick, and S.B. Hendricks. 1955. Photocontrol of *Lepidium* seed germination. *Plant Physiol.* 30:15.

47. Tuan, D.Y.H., and J. Bonner. 1964. Dormancy associated with repression of genetic activity. *Plant Physiol.* 39:768.

48. Varga, M., and L. Ferenczy. 1956. Effect of "rindite" on the development of the growth substances in potato tubers. *Nature* 178:1075.

49. Walton, D.C. 1980. Biochemistry and physiology of abscisic acid. *Ann. Rev. Plant Physiol.* 31:453.

50. Wareing, P.F. 1953. Growth studies in woody species. V. Photoperiodism in dormant buds of *Fagus sylvatica. Physiol. Plant.* 6:692.

51. Wareing, P.F. 1954. Growth studies in woody species. VI. The locus of photoperiodic perception in relation to dormancy. *Physiol. Plant.* 7:261.

52. Wareing, P.F. 1956. Photoperiodism in woody plants. *Ann. Rev. Plant Physiol.* 7:191.

53. Wareing, P.F., and H.A. Foda. 1957. Growth inhibitors and dormancy in *Xanthium* seed. *Physiol. Plant.* 10:266.

Index